D1460465

Algebra
II

Algebra II

Isidore Dressler

Barnett Rich

Marilyn Davis
Contributing Editor

AMSCO SCHOOL PUBLICATIONS, INC.

315 Hudson Street / New York, N.Y. 10013

Cover design by Merrill Haber
Composition by Sierra Graphics, Inc.

Portions of this book were adapted from the following Amsco publications:

Mathematics A
  by Isidore Dressler and Edward P. Keenan

Mathematics B
  by Edward P. Keenan, Ann Xavier Gantert, and Isidore Dressler

Modern Algebra Two
  by Isidore Dressler and Barnett Rich

Please visit our Web site at: **www.amscopub.com**

When ordering this book, please specify:
R 768 P *or* **ALGEBRA II**

ISBN 978-1-56765-559-9 / NYC Item 56765-559-8

Copyright © 2003, 1973 by Amsco School Publications, Inc.

No part of this book may be reproduced in any form without written permission
from the publisher.

Printed in the United States of America

8 9 10 11 12 13 14 15   14 13 12 11 10 09

Preface

Algebra II presents a second-year course in algebra, including a thorough review of first-year algebra.

The book emphasizes an *understanding* of the structure and processes of mathematics, as well as the acquisition of the necessary manipulative skills. In short, the major objective of the text is to fuse the "why" and the "how" by developing students' understanding of the principles and procedures underlying each topic. To make mathematics teaching more effective, the book stresses basic and unifying mathematical concepts such as those of sets and function. For example, function is treated at length in a wide variety of topics such as linear functions, quadratic functions, polynomial functions, exponential functions, and logarithmic functions.

To build students' problem-solving and analytical skills, the authors have made consistent use of the laws of algebra and the application of reasoning to mathematical operations and processes. However, while presenting a challenging structural approach, the authors have used an informal treatment that avoids undue rigor.

The simplicity of language and the informal style should enable students to read and understand the text with little or no difficulty. The material is organized to promote active learning, and the exposition models pedagogically developed teaching practices.

The text, which fully covers the topics customarily contained in second-year courses in algebra, also features:

1. The presentation and use of *sets* as a unifying theme.
2. The comprehensive treatment of the *function concept* as a unifying theme.
3. The clear and precise statement and uses of the *laws of algebra* and the *properties of a field*.
4. The exhibiting of algebra as a *deductive system*.
5. The presentation of *coordinate geometry* as a graphic means of emphasizing and relating linear functions, quadratic functions, and conic sections.
6. The development of *inequalities* both algebraically and graphically.
7. The study of *logarithmic and exponential properties* and their applications.
8. The treatment of *verbal problem-solving* and the provision of a generous supply of problems necessary for its proper development.
9. The early introduction and wide use of the subject of *absolute values*.

10. The inclusion of honor units for the advanced student: complex numbers; conic sections; types of variation; sequences and series, including infinite series; binomial theorem; matrices; determinants; polynomial functions and equations of a higher degree.

Since many students who study second-year algebra have been away from algebra for a year or more, they can profit from a thorough review of the fundamental concepts and skills of first-year algebra before they begin the study of this course. Chapters I to VII provide abundant materials for such review.

An important feature of the text is the character of its organization:

1. Each chapter is divided into related and sequential learning units which, with proper application, a student can readily master.
2. The basic concepts and principles of each unit are carefully developed using precise yet simple language and symbolism.
3. New terms are clearly defined.
4. Explanations and problems suitable for teaching purposes lead to the statement of the general principles and procedures involved in a unit.
5. Model problems, whose solutions are accompanied by detailed step-by-step explanations help to show students how to apply the related principles and how to follow the necessary procedures.
6. A set of expertly selected and carefully graded exercises covering types of problems appropriate for a second course in algebra enables students to understand thoroughly the basic concepts and related procedures of each unit. The exercises enable students to test the extent of their mastery of the unit.
7. Each chapter includes exercises of more than average difficulty, as well as enrichment materials to challenge the advanced student.
8. Each chapter concludes with calculator applications, which provide detailed instructions and keystrokes for graphing and scientific calculator functions relevant to that chapter.

In summary, *Algebra II* serves either as a textbook or as a supplement to the class textbook. It makes available to both students and teachers an abundance of teaching and practice materials.

The authors are very grateful to Mr. Harry Schor, Mr. Melvin Klein, and Dr. Robert E. Dressler for their many valuable suggestions.

Isidore Dressler
Barnett Rich

Contents

III. FACTORING AND OPERATIONS ON RATIONAL EXPRESSIONS

IV. FIRST-DEGREE EQUATIONS AND INEQUALITIES IN ONE VARIABLE

V. LINEAR RELATIONS AND FUNCTIONS

VI. TRANSFORMATION GEOMETRY

XVIII. POLYNOMIAL FUNCTIONS AND POLYNOMIAL EQUATIONS OF A HIGHER DEGREE THAN SECOND DEGREE

CHAPTER ▌

The Real Number System

1. Numbers and Sets of Numbers

In the study of geometry, we deal with collections, or *sets*, of points, sets of lines, etc. In the study of arithmetic and algebra, we deal with sets of numbers, the properties of sets of numbers, and operations on numbers. Understanding sets is a basic requirement in the study of mathematics.

SETS

Originally, *number* meant one of the counting numbers: 1, 2, 3, 4, 5, ... (the three dots are read "and so on"). Today, however, the inventiveness of mathematicians has given us many other collections or sets of numbers for our study and use. In this chapter, the following sets of numbers are emphasized:

1. The set of **natural numbers,** or **counting numbers.**
2. The set of **whole numbers,** which consists of 0 and the natural numbers.
3. The set of **integers,** which consists of 0, the positive integers, and the negative integers.
4. The set of **rational numbers,** which consists of those numbers expressible as fractions whose numerator is an integer and whose denominator is a nonzero integer.
5. The set of **real numbers,** which consists of the first four sets and also the set of irrational numbers.

The branch of mathematics that studies sets is scarcely a century old. The study of sets and their properties provides a foundation for mathematics, finds application in many important fields of knowledge, and serves as a powerful aid to logical reasoning and mathematical analysis. Think of the common words we use to refer to sets, words such as *club, class, team, squad, group, regiment, family, congress, pair, trio,* and *quartet.*

Since we shall make frequent reference to collections and to objects that belong to them, we shall in the future refer to a collection as **a set**

1

and to the objects in the collection as the ***members,*** or ***elements,*** of the set.

For the most part in our study of sets, the sets which we shall consider are ***sets of numbers.***

Number and Numeral

A "number" is an abstract idea which may be represented by means of a symbol or expression called a ***numeral.*** Many different numerals may represent the same number.

For example, the numerals 5, V, ⊢⊬⊣ , and five all name the same number.

In mathematics, we perform operations on numbers, not on numerals. At times, however, to avoid awkward and cumbersome expressions, we will use the word *number* when we should really use the word *numeral.* For example, we may say, for the sake of brevity and simplicity, "write the number" instead of saying "write the numeral that represents the number." This is commonly accepted practice when the context clearly indicates whether we are referring to the number or the numeral.

Similarly, when discussing sets whose members are not numbers, we may say "write the elements of the set" rather than "write the names that represent the elements of the set."

METHODS OF DEFINING A SET

If a set is designated adequately, it is possible to determine whether a particular object does or does not belong to that set. A set is *defined* if we can identify an object as an element of that set. To designate a set and its elements, we use braces to enclose the elements; we generally use a capital letter to refer to the set.

A set may be defined in the following ways:
1. by roster or listing
2. by rule or description
3. by set-builder notation

The Roster Method

Using the ***roster method,*** a set is defined by listing its elements within braces. For example, we may indicate the set of seasons by writing $S = \{$spring, summer, fall, winter$\}$; the set of natural numbers by writing $N = \{1, 2, 3, 4, \ldots\}$; the set of whole numbers by writing $W = \{0, 1, 2, 3, 4, \ldots\}$; and the set of integers by writing $I = \{\ldots, -4, -3, -2, -1, 0, 1, 2, 3, 4, \ldots\}$.

To indicate that an element belongs to a set, we use the symbol ∈. To show that an element does not belong to a set, we use the symbol ∉. For example, if N represents the set of natural numbers, $3 \in N$ and $3\frac{1}{2} \notin N$.

Three dots are used in the roster method when it is impossible to list all the members, as in the case of an infinite set. Note this use in the case of the infinite set of natural numbers:

$$N = \{1, 2, 3, 4, 5, \ldots\}$$

Three dots are also used in the roster method when it is needless to list all the members, as in the case of a finite set with a large number of elements. For example, we may define the set of positive odd integers less than 1000 as $\{1, 3, 5, 7, \ldots, 995, 997, 999\}$. When using three dots to define a set, be sure to list enough elements of the set to make the pattern clear.

The elements of a set may be listed in any order, and each element should be listed only once. For example, the set of digits in 1,000,000 should be listed as $\{1, 0\}$ or $\{0, 1\}$.

The Rule Method

Using the *rule method,* a set may be defined by stating a rule or a condition that makes it possible to determine whether or not a given object belongs to the set. For example, if $S = \{1, 2, 3, 4, 5\}$, we may use a rule to define S as the set of the first five counting numbers, or $S = \{$the first five counting numbers$\}$. The braces are read "the set of." Using the rule method, we may define the set of rational numbers as $R = \{$numbers that may be expressed in the form of $\frac{a}{b}$ where a and b are integers and $b \neq 0\}$.

The Set-Builder Notation

A notation called the *set-builder notation* may be used to define a set. For example, if $A = \{$all natural numbers greater than 5$\}$, we may use set-builder notation to denote A as follows:

$$A = \{x \mid x > 5, x \text{ is a natural number}\}$$

or

$$A = \{x : x > 5, x \text{ is a natural number}\}$$

The following indicates the manner in which set A is read compared with the way in which it is written (note that the vertical bar, and also the colon, is read "such that"):

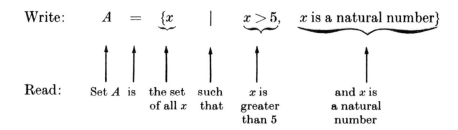

Write: A $=$ $\{x$ $|$ $x > 5,$ x is a natural number$\}$

Read: Set A is   the set   such x is   and x is
                        of all x that   greater   a natural
                                            than 5   number

The set-builder notation for $S = \{x \mid x$ is a single-digit positive odd integer$\}$ is read "S is the set of all x such that x is a single-digit positive odd integer." Using the roster method, we write $S = \{1, 3, 5, 7, 9\}$.

Exercises

In 1–4, tell whether or not the set is adequately defined. If the set is not adequately defined, state the reason why it is not.

1. $\{1, 8, 27, 64\}$ **2.** {large numbers}

3. {natural numbers less than 12}   **4.** {non-negative integers}

In 5–8, tell whether the statement is *true* or *false*.

5. $0 \in$ {whole numbers}   **6.** $-3 \notin$ {natural numbers}

7. $7 \in$ {positive odd integers}   **8.** $123 \in \{13, 23, 33\}$

In 9–13, use the roster method to define the set. Use three dots when necessary or convenient.

9. the set of days of the week that begin with the letter S

10. the set of natural numbers greater than 10 and less than 15

11. the set of odd integers greater than 100

12. the set of factors of 24 greater than 1 and less than 10

13. the set of numbers that differ from 10 by 2

In 14–19, use the rule method to define the set.

14. {Tuesday, Thursday}   **15.** $\{7, 14, 21, 28\}$

16. $\{\ldots, -7, -5, -3, -1\}$ **17.** $\{50, 51, 52, \ldots, 97, 98, 99\}$

18. $\{1, 4, 9, 16, 25, 36, 49\}$ **19.** $\{100, 200, 300, \ldots, 700, 800, 900\}$

In 20–22, use the roster method to list the elements of the set.

20. $\{x \mid x$ is a single-digit natural number$\}$

21. $\{x \mid x$ is a prime number greater than 9 and less than 25$\}$

22. $\{x \mid x^2 = 25, x$ is an integer$\}$

In 23–25, state whether the listed sets are the same set or two different sets.

23. $\{x \mid x$ is a negative integer less than $-5\}$ and $\{-4, -3, -2, -1\}$

24. $\{x \mid x$ is a natural number and a perfect square$\}$ and $\{1, 4, 9, 16, 25, \ldots\}$

25. {natural numbers less than $4\frac{1}{4}$} and $\{4, 3, 2, 1, 0\}$

2. Subsets

Consider the set $U = \{3, 5, 7\}$ and the set $B = \{3, 5\}$. Notice that every member of set B is also a member of set U. We say that set B is a **subset** of set U. We express this statement with the symbolism $B \subset U$.

In general, set A is a subset of set B ($A \subset B$) if every member of A is also a member of B.

Consider the set $U = \{3, 5, 7\}$ and the set $T = \{1, 3\}$. Observe that not every member of set T is a member of set U. Therefore, we say that set T is not a subset of set U. We express this statement with the symbolism $T \not\subset U$.

A **universal set,** or **universe,** in a discussion is a set from which elements are chosen to form subsets. A **Venn diagram** may be used to show the relationship between a subset of a set and the set itself. The Venn diagram at the right pictures the statement made earlier that the set $B = \{3, 5\}$ is a subset of the set $U = \{3, 5, 7\}$.

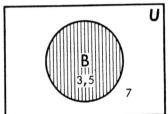

proper subset of a set is one that does not contain all the elements of that set. For example, proper subsets of set $U = \{3, 5, 7\}$ are $B = \{3, 5\}$, $C = \{3, 7\}$, $D = \{5, 7\}$, $E = \{3\}$, $F = \{5\}$, and $G = \{7\}$.

An **improper subset** of a set is one that contains all the elements of that set. For example, the set $A = \{3, 5, 7\}$ is an improper subset of the set $U = \{3, 5, 7\}$. It is obvious that every set is a subset of itself.

The **empty set,** or **null set,** is the set that has no elements. An example of the empty set is the set of all women who have been president of the United States. Either the symbol $\varnothing$ or the symbol $\{\ \}$ may be used to represent the empty set. Mathematicians have agreed:

1. There is only one empty set $\varnothing$.
2. The empty set $\varnothing$ is a subset of every set.

THE NUMBER OF SUBSETS OF A SET THAT CONTAINS *n* ELEMENTS

In our previous discussion of the subsets of a set, we saw that set $U = \{3, 5, 7\}$, which has 3 elements, has 8, or 2^3 subsets. They are the following:

$A = \{3, 5, 7\}$	$B = \{3, 5\}$	$C = \{3, 7\}$	$D = \{5, 7\}$
$E = \{3\}$	$F = \{5\}$	$G = \{7\}$	$H = \varnothing$

Rule. If a set contains n elements, the number of subsets that this set has is 2^n.

Exercises

In 1–8, tell whether the statement is *true* or *false*.
1. {cat, dog} is a subset of {dog, horse, cat}
2. {3, 4, 5, 6} is a subset of {4, 5, 6, 7}
3. {cows} $\subset$ {animals}
4. {natural numbers} $\not\subset$ {whole numbers}
5. $\varnothing \not\subset \{1\}$ **6.** $\{1\} \subset \varnothing$ **7.** $\varnothing \subset \{0\}$ **8.** $\{0\} \not\subset \varnothing$

In 9 and 10, tell whether the statement is *true* or *false*. Justify your answer using a Venn diagram.
9. If $A \subset B$ and $B \subset D$, then $A \subset D$
10. If $R \subset S$ and $T \subset S$, then $R \subset T$

11. If $U = \{r, w, b\}$, list all the subsets of U that have (*a*) one element, (*b*) two elements, (*c*) three elements, and (*d*) no elements.

In 12–15, (*a*) list all the subsets (including the empty set) of the given set and (*b*) state the number of subsets that the given set has.
12. {7}   **13.** {Tom, Harry}   **14.** {2, 4, 6}   **15.** {*a, b, c, d*}

16. Without writing the subsets of set U, determine the number of subsets (including the empty set) of U if the number of elements in U is:
     a. 1   *b.* 2   *c.* 3   *d.* 4   *e.* *n*

3. Comparing Sets

ONE-TO-ONE CORRESPONDENCE

Suppose that, in a classroom, each seat is occupied by one and only one student and each student in the class occupies one and only one seat. In this situation, each member of the set of students in the class has been paired with one and only one member of the set of seats in the room; and each member of the set of seats in the room has been paired with one and only one member of the set of students in the class. When the elements of two sets can be paired in this way, the pairing is called a ***one-to-one correspondence.*** Two such sets have the same number of elements. The number of elements that a set contains is called its ***cardinal number.***

FINITE SETS AND INFINITE SETS

A ***finite set*** is a set whose cardinal number is a natural number or zero. When the number of elements of a finite set is counted, the counting comes to an end.

For example, $A = \{2, 4, 6, 8, 10\}$ is a finite set. It has 5 elements. The cardinal number of set A is 5. Also, $B = \{$odd integers that are multiples of 2$\}$ is a finite set because B is the empty set $\emptyset$ which has zero members.

An *infinite set* is a set whose cardinal number is not a natural number or zero. When we attempt to count the number of elements of an infinite set, the counting process never comes to an end. For example, $C = \{$odd natural numbers$\}$, which may be written $\{1, 3, 5, 7, 9, \ldots\}$, is an infinite set.

EQUIVALENT SETS AND EQUAL SETS

Equivalent sets are sets whose elements can be paired so that there is a one-to-one correspondence.

For example, if $A = \{a, b, c\}$ and $B = \{1, 2, 3\}$, we can set up a one-to-one correspondence by pairing the elements of set A and set B as shown below. Then we say, "Set A is equivalent to set B," which may be written as "$A \leftrightarrow B$" or "$A \sim B$."

$$A = \{a, b, c\}$$
$$\downarrow \downarrow \downarrow$$
$$B = \{1, 2, 3\}$$

Since equivalent sets must have the same number of elements, a simple way to determine whether or not two finite sets which have few elements are equivalent is to count the number of elements in each set. Note that two equivalent sets must have the same number of elements, but need not have the same elements.

Sometimes it is possible to match two infinite sets. For example, the set of natural numbers and the set of positive even integers may be matched in a one-to-one correspondence to show that the sets are equivalent. See how this is done.

$$N = \{1, 2, 3, 4, 5, 6, 7, 8, \ldots\}$$
$$\updownarrow \updownarrow \updownarrow \updownarrow \updownarrow \updownarrow \updownarrow \updownarrow$$
$$E = \{2, 4, 6, 8, 10, 12, 14, 16, \ldots\}$$

Since the elements of set N and the elements of set E are paired in a one-to-one correspondence, then $N \leftrightarrow E$; that is, the set of natural numbers is equivalent to the set of positive even integers.

Equal sets are sets that have exactly the same elements. Recall that the order in which the elements of a set are listed does not matter. For example, if $A = \{n, o, w\}$ and $B = \{o, w, n\}$ then set A and set B are equal sets, written "$A = B$." Also, if $C = \{1, 3, 5\}$ and $D = \{5, 1, 3\}$, then $C = D$.

If two sets are equal sets, each set is an improper subset of the other.

Equivalent sets are equal sets if and only if they have the same elements. For example, if $C = \{1, 2, 3, 4, 5\}$ and $D = \{1, 2, 3, 4, 10\}$, then set C is equivalent to set D, but set C is not equal to set D, written "$C \leftrightarrow D$ but $C \neq D$."

Exercises

In 1–4, tell whether or not there is a one-to-one correspondence between the two sets.

1. $\{1, 3, 5, 7, 9\}$ and $\{0, 2, 4, 6, 8\}$
2. $\{T, A, B\}$ and $\{B, A, T\}$
3. $\{1, 2, 3, 4\}$ and $\{-1, -2, -3, -4, \ldots\}$
4. {positive odd integers} and {positive even integers}

In 5–8, state whether the set is a finite set or an infinite set. (An empty set is a finite set.)

5. the set of men 20 feet tall
6. the set of all integers less than 1
7. $\{0\}$
8. the set of rectangles having three diagonals

In 9–11, (*a*) tell whether or not the sets are equivalent sets and (*b*) tell whether or not the sets are equal sets.

9. {Jack, Harry, Ted, Martin} and {Ruth, Hilda, Rose, Sarah}
10. $\{T, E, A\}$ and $\{E, A, T\}$
11. $\{5, 6, 7, 8, 9\}$ and {natural numbers greater than 4 and less than 10}

12. Specify two finite sets that are equivalent sets but are not equal sets.
13. Specify two finite sets that are not equivalent sets.
14. Specify two infinite sets that are equivalent sets.
15. Is it possible to specify two equal sets that are not equivalent sets? Why?
16. Is it possible to specify two equivalent sets that are not equal sets? Why?

4. Using a Number Line To Graph Real Numbers and Sets of Real Numbers

GRAPHING NATURAL NUMBERS AND WHOLE NUMBERS

It is useful to associate numbers with points on a line called a *number line.* To do this, we select any two convenient points on the line. We label the point on the left "0" and call it

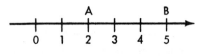

the *origin;* we label the point on the right "1." Using the distance between 0 and 1 as a unit of measure, we find points to the right of 1. We label the new points consecutively 2, 3, 4, . . . , as shown in the figure above.

On the preceding number line, which extends to the right without ending, every whole number and natural number can be associated with a point on the line. On a number line, the number associated with a point is called the **coordinate** of the point, and the point is the **graph** of the number. Hence, point A is the graph of 2; also, 5 is the coordinate of point B.

GRAPHING INTEGERS

To graph integers on a number line, we begin with a number line on which points have been associated with the set of whole numbers (natural numbers and zero). Then we use the distance between 0 and 1 as a unit of measure and find points to the *left* of 0. We label the new points consecutively -1, -2, -3, etc.

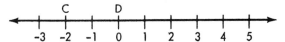

The set of integers consists of the *positive integers, zero*, and the *negative integers*. The side of the number line to the right of 0 is the positive side and contains the graphs of the positive integers. The side to the left of 0 is the negative side and contains the graphs of the negative integers. Hence, point C is the graph of -2 and point D is the graph of 0.

GRAPHING RATIONAL NUMBERS

The set of *rational numbers* consists of numbers that can be expressed in the form $\frac{a}{b}$ where a and b are integers and $b \neq 0$. (Think of a *rational* number as a *ratio* number, the ratio of a to b.) Any integer a is a rational number since it can be written in the form of $\frac{a}{1}$. For example, the integer 5, which may be written $\frac{5}{1}$, is a rational number. Other examples of rational numbers are $\frac{7}{3}$ or $2\frac{1}{3}$, $\frac{13}{10}$ or 1.3, and $\frac{-8}{+5}$ or $-1\frac{3}{5}$. Since division by 0 is impossible, $\frac{3}{0}$ is meaningless.

To graph the set of rational numbers on a number line, we begin with a number line on which points have been associated with the members of the set of integers. Then, we divide the intervals between integers into halves, thirds, quarters, etc. Now we can associate every member of the set of rational numbers with one of these points, as shown in the following figure:

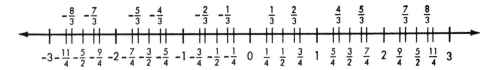

In order to graph a particular rational number whose denominator is 7, such as $\frac{18}{7}$ or $2\frac{4}{7}$, we would divide the interval between 2 and 3 into 7 equal parts.

GRAPHING REAL NUMBERS: COMPLETING THE NUMBER LINE

The graph of the set of rational numbers is a set of closely spaced points, as shown in the figure at the right. However, no 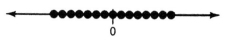 matter how close to each other two points on the number line may be, there is always an infinite number of points between them. We say that between any two rational numbers there is an infinite number of rational numbers.

For example, using two-place decimals, $\frac{1}{10}$ may be expressed as .10 and $\frac{1}{5}$ may be expressed as .20. Between .10 and .20 are the rational numbers {.11, .12, .13, ..., .18, .19}. Using three-place decimals, $\frac{1}{10}$ may be expressed as .100 and $\frac{1}{5}$ may be expressed as .200. Between .100 and .200 are the rational numbers {.101, .102, .103, ..., .198, .199}. Thus, by increasing the number of decimal places, we can obtain an infinite number of rational numbers between $\frac{1}{10}$ and $\frac{1}{5}$.

A *dense set* is any set in which there is at least one element between any two given elements of the set. In the case of the set of rational numbers, as shown above, there is not only one, but an infinite number of rational numbers between any two given rational numbers. Hence, the set of rational numbers is a dense set.

From our previous discussion, we see that every rational number can be associated with a point on the number line. It might also appear that every point on the number line can be associated with some rational number. However, this is not true. On a number line, no matter how closely packed the points which are associated with rational numbers may be, there are always points that are not associated with rational numbers. These points are associated with *irrational numbers,* numbers that are not rational numbers. Examples of irrational numbers are $\sqrt{2}, \dfrac{\sqrt{3}}{3}$, and π.

The number $\sqrt{2}$ is not a rational number. It cannot be expressed as a fraction whose numerator is an integer and whose denominator is a nonzero integer. Yet, the point that is the graph of $\sqrt{2}$ can be located on the number line by constructing a right triangle so that the length 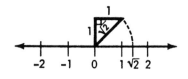 of each leg is 1 unit. According to the theorem of Pythagoras, the length of the hypotenuse is $\sqrt{2}$. As shown in the preceding figure, the length of the hypotenuse is laid off on a number line.

The set of real numbers consists of all the rational numbers and all the irrational numbers. When the set of real numbers is graphed on a number line, each point on the line can be associated with one and only one real number, and each real number can be associated with one and only one point. There are no longer any "holes" in the line. The number line for the set of real numbers is now completely filled in.

Below, we see the graphs of some members of the set of real numbers. Note the graphs of irrational numbers such as $-2\sqrt{3}$, $-\sqrt{6}$, π, and $3 + \sqrt{2}$ on this number line.

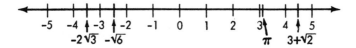

By using advanced mathematics, it can be demonstrated that there is a *one-to-one correspondence* between the set of real numbers and the set of points on a number line. This statement implies two very important ideas:

1. Each real number corresponds to a unique (one and only one) point on the number line.
2. Each point on the number line corresponds to a unique real number.

COMPARING THE GRAPHS OF SUBSETS OF THE SET OF REAL NUMBERS

Following are the graphs of sets of numbers that are subsets of the set of real numbers. Notice that the sets of natural numbers, whole numbers, integers, and rational numbers are represented by sets of points that do not fill in the entire number line. However, the graph of the set of real numbers does fill in the entire number line.

Name of Set   *Graph of Set*

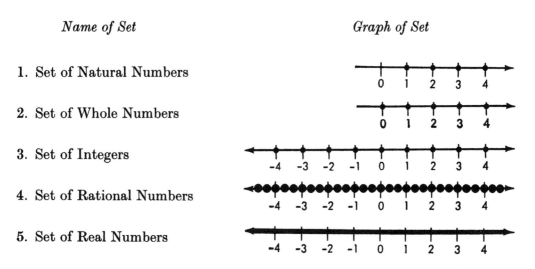

1. Set of Natural Numbers

2. Set of Whole Numbers

3. Set of Integers

4. Set of Rational Numbers

5. Set of Real Numbers

THE REAL NUMBER SYSTEM

The following diagram and the accompanying remarks will help you understand the structure of the real number system by showing you the relationships among the set of real numbers and its number subsets. Beginning at the top, note that the set of real numbers consists of the elements of two sets, the set of rational numbers and the set of irrational numbers. Continue downward and note the other relationships.

In the chart, note the following:

REAL NUMBERS

RATIONAL IRRATIONAL
NUMBERS   NUMBERS

INTEGERS   NONINTEGRAL
           RATIONALS

POSITIVE   ZERO   NEGATIVE
INTEGERS   INTEGERS

NONZERO
WHOLE NUMBERS

NATURAL
NUMBERS

1. Each *real number* is either a rational number or it is an irrational number. Thus, the real number 2 is rational, while the real number $\sqrt{2}$ is irrational.

2. Each *rational number* is either an integer or it is a nonintegral rational number. Thus, the rational number 3 (which may be expressed as $\frac{3}{1}$) is an integer, while the rational number $\frac{3}{4}$ is a nonintegral rational number.

3. Each *integer* is either a positive integer, a negative integer, or zero. Thus, the integer 3 is a positive integer, while the integer −3 is a negative integer. The integer 0 is neither positive nor negative.

4. Each *whole number* is either a natural number or zero.

5. Each *natural number* is a nonzero whole number and also corresponds to a positive integer. Thus, in counting, 3 is a natural number or a nonzero whole number. If 3 is considered as the opposite of −3, then 3 is a positive integer.

Exercises

In 1 and 2, give the coordinate of each of the labeled points on the number line.

1.
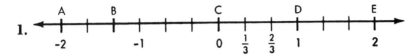

2.
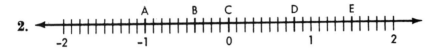

3. Draw a number line and on it locate the points whose coordinates are:

a. $-\frac{8}{5}, -\frac{2}{5}, \frac{5}{5}, \frac{7}{5}, 1\frac{4}{5}$

b. $-\frac{16}{8}, -\frac{7}{8}, \frac{5}{8}, \frac{8}{8}, \frac{15}{8}$

c. $-1.7, -.5, .3, 1.4, 1.8$

d. $-1\frac{1}{2}, -.25, \frac{3}{4}, 1\frac{1}{8}, \frac{15}{8}$

4. Express the coordinate of each labeled point on the following number line as the ratio of two integers.

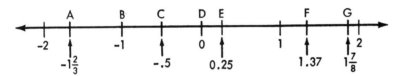

5. The distance between 0 and 1 on the following number line is a unit.

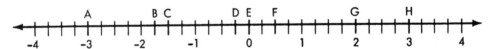

Find the coordinate of:

a. the point 3 units to the right of point E

b. each point that is 3 units from point E

c. the point $\frac{1}{4}$ of a unit to the right of point C

d. the point midway between point A and point H

In 6–9, write the rational number that is the average of the pair of given numbers.

6. 3 and 4   **7.** 2 and $2\frac{1}{2}$ **8.** $\frac{1}{2}$ and $\frac{1}{3}$ **9.** 1.2 and 1.3

In 10–13, (a) indicate whether or not the given set of numbers is dense and (b) give a reason for your answer.

10. rational numbers   **11.** rational numbers between 1 and 2

12. integers   **13.** natural numbers

5. Using a Number Line To Order Real Numbers

The statement $3 \neq 5$ is read "3 is not equal to 5." On a number line, the graphs of 3 and 5 are different points. In general, if a and b are real numbers, $a \neq b$ is read "a is not equal to b"; and on a number line, the graphs of a and b are different points.

We assume that we can arrange the set of real numbers in a definite order. For example, the statement "-2 is less than 4," symbolized by "$-2 < 4$," expresses an order relation between -2 and 4. By convention, on a number line, the graph of the smaller number, -2, is to the left of the graph of the greater number, 4. The same order relation can be expressed by the statement "4 is greater than -2," symbolized by "$4 > -2$." Observe, in the figure below, that the graph of the greater number, 4, is to the right of the graph of the smaller number, -2.

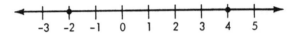

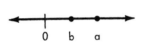

In general, if a and b are two real numbers, then $a > b$ if and only if the graph of a is to the right of the graph of b on the number line, as shown in the figure.

Note that when we use the inequality symbols $>$ and $<$ to express a true order relation between two real numbers, the narrow part of the symbol points to the numeral that represents the smaller number.

The inequalities $1 < 3$ and $3 < 5$ can be written in the compact form $1 < 3 < 5$. Similarly, $a < b$ and $b < c$ can be written $a < b < c$.

On the number line, since $1 < 3 < 5$, the graph of 1 is to the left of the graph of 3, which in turn is to the left of the graph of 5, as shown in the figure. Therefore, we say that "3 is between 1 and 5." In general, if a, b, and c are real numbers, then $a < b < c$ can be read "b is between a and c."

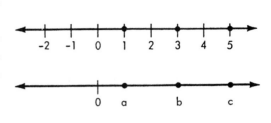

The meanings of other compact inequality symbols follow:

$a \leq b$ means $a < b$ or $a = b$. Read $a \leq b$ as "a is less than or equal to b."

$a \geq b$ means $a > b$ or $a = b$. Read $a \geq b$ as "a is greater than or equal to b."

$a < b \leq c$ means $a < b$ and $b \leq c$. Read $a < b \leq c$ as "a is less than b, and b is less than or equal to c."

Exercises

In 1–5, use a number line to determine whether the statement is *true* or *false*.

1. $3 < 5$ **2.** $3 < -2$ **3.** $-5 > -2$ **4.** $-\frac{3}{4} < \frac{1}{2}$ **5.** $-1\frac{1}{2} > -3\frac{1}{3}$

In 6–10, arrange the numbers in proper order so that they will appear from left to right on a number line.

6. $5, -2$ **7.** $-5, -8$ **8.** $3, 0, -4$ **9.** $-1, 3, -4$ **10.** $1\frac{1}{2}, -1, 2\frac{3}{4}$

In 11–13, use the symbol $<$ to write the sentence in compact form.

11. 13 is between 10 and 15. **12.** n is between -2 and 2.

13. x is between y and w, where w is greater than y.

In 14–17, order the given numbers and use the symbol $<$ to write your result in compact form.

14. $3, 5, 2$ **15.** $-6, 7, 4$ **16.** $-1, 3, -2$ **17.** $-3, -1, -6$

6. Open Sentences, Statements, Variables, and Solution Sets

Consider the sentence: It is the capital of the United States.

This sentence cannot be described as true or false until the particular city to which "It" refers is known. Such a sentence, one that is neither true nor false as it stands, is called an ***open sentence.*** Now suppose that "It" is replaced by an element of the set {London, Paris, Washington, Rome}. The resulting sentence is is called a ***statement.*** A statement is a meaningful assertion that is either true or false, but not both true and false. If "It" is replaced by Washington, the resulting statement "Washington is the capital of the United States" is true. If "It" is replaced by London, Paris, or Rome, the resulting statements are false.

A ***variable*** is a symbol that holds a place for, or represents, any one of the elements of a specified set that has more than one element. In the sentence "It is the capital of the United States," "It" is a variable. In mathematics, a letter such as x may be the variable, as in the case of $x + 5 = 8$.

The ***replacement set*** of a variable, or the ***domain*** of a variable, is the specified set whose elements may be used as replacements for the variable. In the previous discussion, {London, Paris, Washington, Rome} is the replacement set, or the domain, of the variable "It." The elements of the replacement set of a variable are called the ***values*** of the variable.

The ***truth set,*** or ***solution set,*** of an open sentence over the domain of the variable is the subset of the domain that consists of those elements for which the open sentence is true. Each element of the solution set is called a ***solution*** or a ***root*** of the open sentence and is said to ***satisfy*** the open sentence.

To solve an open sentence over a stated domain means to determine its solution set in that domain.

For example, if the domain, or replacement set, of x is $U = \{2, 3, 4, 5\}$, then the root, or solution, of $x + 5 = 8$ is 3. Hence, the solution set of $x + 5 = 8$ is $\{3\}$. In set-builder notation, we can write this result as

$$\{x \,|\, x \in U \text{ and } x + 5 = 8\} = \{3\}$$

This is read "the set of all x such that x is a member of U and $x + 5 = 8$ equals the set whose member is 3."

A change in the domain of the variable can change the solution set of an open sentence.

For example, if the replacement set of x were changed to $U = \{4, 5, 6, 7\}$, no element of this set would make the sentence $x + 5 = 8$ a true statement. As a result, the solution set of $x + 5 = 8$ would now be the empty set, $\varnothing$. To show this, we would write

$$\{x \,|\, x \in U \text{ and } x + 5 = 8\} = \varnothing$$

A **constant** is a symbol representing the one element of a set that has only a single element. Examples of a constant are 5 and π.

Exercises

In 1–8, tell whether or not the sentence is an open sentence.
1. Lyndon Johnson was a president of the United States.
2. He is 6 feet tall.
3. Henry Ford discovered radium.
4. X holds the all-time record for hitting home runs.
5. $6 + 1 = 9$ 6. $x - 1 = 4$ 7. $2 + 4 < 9$ 8. $2x > 3$

In 9–12, use roster notation to write the solution set of the sentence when the domain of the variable is $\{0, 1, 2, 3, 4, 5\}$. If the solution set is empty, write $\varnothing$ as the answer.
9. $3x + 1 = 10$ 10. $3 = 8 - 2y$ 11. $x + 2 > 5$ 12. $4 \leq 2x - 2$

In 13–16, use roster notation to write the solution set of the open sentence when the domain of the variable is the set of natural numbers. If the solution set is empty, write $\varnothing$ as the answer.
13. $x + 5 = 9$ 14. $2y + 4 = 10$ 15. $2y + 4 \geq 10$ 16. $2y + 4 < 9$

In 17–19, use roster notation to represent the set when the domain of the variable is the set of single-digit positive integers. If the given set is empty, write $\varnothing$ as the answer.
17. $\{x \,|\, 2x + 1 = 7\}$ 18. $\{w \,|\, 4w - 1 > 25\}$ 19. $\{y \,|\, 2y - 1 \leq 4\}$

In 20–23, write the set in roster notation.

20. $\{n \mid n$ is less than 5 and n is a natural number$\}$

21. $\{x \mid x > 5,\ x$ is a natural number$\}$

22. $\{y \mid 8 < y < 18,\ y$ is an odd integer$\}$

23. $\{x \mid x^2 - 9 = 0,\ x$ is a natural number$\}$

In 24–27, using set-builder notation, write the solution set of the open sentence.

24. $x + 4 = 10$ and $x \in \{5, 6, 7, 8\}$ **25.** $2y + 6 = 14$ and $y \in \{$even integers$\}$

26. $x \geq 4$ and $x \in \{$real numbers$\}$ **27.** $40 < n < 100;\ n \in \{$natural numbers$\}$

7. Graphing Sets of Numbers and Solution Sets of Open Sentences

The ***graph of a set of numbers*** is the set of points on a number line that are associated with the numbers of the set.

The ***graph of the solution set of an open sentence in one variable*** is the set of points on a number line that are associated with the members of the solution set. This graph is called the ***graph of the open sentence.***

When we make such a graph, we use:

1. A darkened circle • to represent a point on the number line that is associated with a number in the set.
2. A non-darkened circle ○ to represent a point that does not belong to the graph.
3. A darkened line ▄▄▄▄ to indicate that every point on the line is associated with a number in the set. Parts of the line that are not darkened show points that do not belong to the graph.

〰〰〰〰〰 *MODEL PROBLEMS* 〰〰〰〰〰

1. Using a number line, draw the graph of each set.

   a. $\{-2, 1, 3\}$

   Answer:

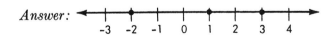

   b. $\{$real numbers between 1 and 4$\}$

   Answer:

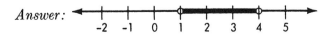

c. {real numbers greater than or equal to $-1\frac{1}{2}$}

Answer:

2. If $x \in \{0, 1, 2, 3, 4, 5\}$, (*a*) find the solution set of the open sentence $3x + 1 = 13$ and (*b*) using a number line, graph the solution set.

Solution:

a. To find the solution set, replace the variable x in the open sentence $3x + 1 = 13$ by each member of the replacement set. The number 4 is the only replacement for x that results in a true sentence, $3 \times 4 + 1 = 13$. Therefore, the solution set of $3x + 1 = 13$ is {4}.

Answer: {4}

b.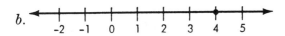

3. Graph $\{x \mid -1 < x \le 3\}$.

Solution:

1. $\{x \mid -1 < x \le 3\}$ is the set of real numbers greater than -1 and also less than or equal to 3. Hence, the solution set of the open sentence includes 3 and also all the real numbers greater than -1 and less than 3.
2. Graph the solution set. Note that the non-darkened circle at -1 indicates that -1 is not a member of the solution set; also, the darkened circle at 3 indicates that 3 is a member of the solution set.

Answer:

~~~~~~

Exercises

In 1–8, using a number line, draw the graph of the set of numbers.
1. $\{-2, 0, 3\}$ **2.** $\{-3, 1, 2\frac{1}{2}\}$ **3.** $\{-2, -1, 0, 1, \ldots 6, 7\}$ **4.** $\{0\}$
5. {whole numbers between 1 and 8}
6. {integers between 4 and 6}
7. {real numbers between 4 and 6}
8. {real numbers except -1 and 1}

In 9–12, if $R = \{-3, -1, 0, 2\}$ and $S = \{-2, -1, 0, 1, 2\}$, using a number line, graph the indicated set.
9. the set whose members are elements of set R and also of set S

10. the set of elements of set R or of set S

11. {elements of set R but not of set S}

12. {elements of set S but not of set R}

In 13–21, using a number line, draw the graph of the set of numbers.

13. $\{x \mid x + 2 = 6\}$ $\qquad$ $x \in \{1, 2, 3, 4, 5\}$

14. $\{y \mid 5 - y = 2\}$ $\qquad$ $y \in \{\text{positive integers}\}$

15. $\{y \mid y > 4\}$ $\qquad$ $y \in \{\text{positive integers less than } 10\}$

16. $\{t \mid t \leq 1\}$ $\qquad$ $t \in \{-3, -2, -1, 0, 1, 2, 3\}$

17. $\{x \mid x > -1\}$ $\qquad$ $x \in \{\text{real numbers}\}$

18. $\{x \mid x > -1\}$ $\qquad$ $x \in \{\text{integers}\}$

19. $\{x \mid x < 4\}$ $\qquad$ $x \in \{\text{real numbers}\}$

20. $\{x \mid x \leq 2\}$ $\qquad$ $x \in \{\text{negative real numbers}\}$

21. $\{x \mid x + 2 > 5\}$ $\qquad$ $x \in \{\text{real numbers}\}$

In 22–25, using $\{-3, -2, -1, 0, 1, 2, 3\}$ as the domain of the variable, (*a*) find the solution set of the open sentence and (*b*) use a number line to graph the solution set.

22. $3x - 2 = 4$ $\qquad$ **23.** $y > -2$ $\qquad$ **24.** $2t + 1 < 3$ $\qquad$ **25.** $3r - 1 \leq 5$

In 26–37, using a number line, graph the open sentence when the domain of the variable is the set of real numbers.

26. $x - 3 = 4$ $\qquad$ **27.** $3x + 1 = 7$ $\qquad$ **28.** $4x = -8$

29. $x > 1$ $\qquad$ **30.** $x < -2$ $\qquad$ **31.** $3x > 12$

32. $x \leq -2$ $\qquad$ **33.** $1 < x < 4$ $\qquad$ **34.** $-2 \leq x \leq 3$

35. $x + 2 = 2 + x$ $\qquad$ **36.** $x + 2x = 3x$ $\qquad$ **37.** $x + 1 = x + 2$

38. Using a number line, graph $x \leq 4$ when the domain of x is the set of (*a*) natural numbers, (*b*) positive odd integers, and (*c*) real numbers.

8. Postulates of the Real Number System and of a Field

BINARY OPERATIONS

In your previous study of algebra, you have learned how to perform the operations of addition, subtraction, multiplication, and division on numbers. These operations are called **binary operations.** In a binary operation, a correspondence is set up between two numbers of a given set of numbers, taken in a certain order, and another number of that set. For example, when we add the real numbers 4 and 5, we assign to those numbers the real number 9. When we multiply 4 and 5, we assign to those numbers the number 20.

In general, when any two real numbers a and b are added, the real number $a + b$ is assigned to them; when any two real numbers a and b are multiplied, the real number ab is assigned to them. In addition, the numbers a and b are called the **addends,** and $a + b$ is called the **sum;** in multiplication, the numbers a and b are called the **factors,** and ab is called the **product.**

POSTULATES FOR ADDITION AND MULTIPLICATION

In dealing with real numbers, we assume the truth of some basic statements dealing with the properties of addition and multiplication. Such statements, whose truth we assume, are called **postulates** or **axioms.**

CLOSURE POSTULATES FOR ADDITION AND MULTIPLICATION

Closure Postulate for Addition, A_1. The sum of any two real numbers is a unique (one and only one) real number.

For example, the real number 15 is the unique sum of 10 and 5.

In general, for any two real numbers a and b,

$$a + b \text{ is a unique real number}$$

We call this statement the **closure postulate for addition of real numbers.**

Closure Postulate for Multiplication, M_1. The product of any two real numbers is a unique real number.

For example, the real number 50 is the unique product of 10 and 5.

In general, for any two real numbers a and b,

$$ab \text{ is a unique real number}$$

We call this statement the **closure postulate for multiplication of real numbers.**

We also say that the set of real numbers is closed under addition and under multiplication.

COMMUTATIVE POSTULATES FOR ADDITION AND MULTIPLICATION

Commutative Postulate for Addition, A_2. The sum of any two real numbers remains the same if the numbers are interchanged.

For example, $4 + 5 = 5 + 4$.

In general, for any two real numbers a and b,

$$a + b = b + a$$

We call this statement the **commutative postulate for addition.**

Commutative Postulate for Multiplication, M_2. The product of any two real numbers remains the same if the numbers are interchanged.

For example, $4 \times 5 = 5 \times 4$.

In general, for any two real numbers a and b,

$$ab = ba$$

We call this statement the **commutative postulate for multiplication.**

We also say that, in the set of real numbers, *both addition and multiplication are commutative operations.* The word " commutative " is appropriate because it refers to "change."

ASSOCIATIVE POSTULATES FOR ADDITION AND MULTIPLICATION

Associative Postulate for Addition, A_3. The sum of any three real numbers remains the same regardless of the way in which they are grouped when their sum is found.

For example, $(3 + 4) + 5 = 3 + (4 + 5)$.

In general, for any three real numbers a, b, and c,

$$(a + b) + c = a + (b + c)$$

We call this statement the **associative postulate for addition.**

Associative Postulate for Multiplication, M_3. The product of any three real numbers remains the same regardless of the way in which they are grouped when their product is found.

For example, $(3 \cdot 4) \cdot 5 = 3 \cdot (4 \cdot 5)$.

In general, for any three real numbers a, b, and c,

$$(ab)c = a(bc)$$

We call this statement the **associative postulate for multiplication.**

We also say that, in the set of real numbers, *both addition and multiplication are associative operations.* The word "associative" is appropriate because it refers to "grouping."

IDENTITY POSTULATES FOR ADDITION AND MULTIPLICATION

Identity Postulate for Addition, A_4. The sum of 0 and any given real number is identical with that given number.

For example, $0 + 157 = 157$ and $157 + 0 = 157$.

In general, for each real number a,

$$0 + a = a \text{ and } a + 0 = a$$

The number 0 is called the **identity element for addition,** or the **additive identity.**

Identity Postulate for Multiplication, M_4. The product of 1 and any given real number is identical with that given number.

For example, $1 \cdot 157 = 157$ and $157 \cdot 1 = 157$.

In general, for each real number a,

$$1 \cdot a = a \text{ and } a \cdot 1 = a$$

The number 1 is called the **identity element for multiplication,** or the **multiplicative identity.**

INVERSE ELEMENT POSTULATES FOR ADDITION AND MULTIPLICATION

Inverse Element Postulate for Addition, A_5. For each real number a, there exists a unique real number $-a$, such that the sum $a + (-a) = 0$, the additive identity. The real number $-a$ is called the **additive inverse** of a. It is also true that a is the additive inverse of $-a$.

For example, the additive inverse of $+2$ is -2, since $(+2) + (-2) = 0$; the additive inverse of -2 is $+2$, since $(-2) + (+2) = 0$. Observe that if we are given a signed number (a number that is positive or negative), we can obtain its additive inverse by simply changing its sign.

Rule. To form the additive inverse of a signed number, change its sign.

Since $0 + 0 = 0$, we say that 0 is its own additive inverse.

Note, on the number line, the way in which a real number and its additive inverse may be paired.

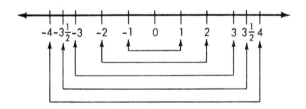

The sum of every pair of real numbers connected by the arrows is 0; that is, $(+4) + (-4) = 0$, $(+3\frac{1}{2}) + (-3\frac{1}{2}) = 0$, $(+3) + (-3) = 0$, $(-2) + (+2) = 0$, $(-1) + (+1) = 0$. Observe that each number in every pair is the *negative* of the other or the *opposite* of the other.

In general, for each real number a, there exists a unique real number $-a$ such that

$$a + (-a) = 0 \text{ and } (-a) + a = 0$$

Inverse Element Postulate for Multiplication, M_5. For each nonzero real number a, there exists a unique real number $\dfrac{1}{a}$ such that the product of a and $\dfrac{1}{a}$ is 1, the multiplicative identity. The real numbers a and $\dfrac{1}{a}$ are called ***multiplicative inverses***, or ***reciprocals***, of each other.

For example, the multiplicative inverse of 2 is $\frac{1}{2}$ since $2 \cdot \frac{1}{2} = 1$; the multiplicative inverse of $\dfrac{3}{4}$ is $\dfrac{1}{\frac{3}{4}}$, or $\dfrac{4}{3}$, since $\dfrac{3}{4} \cdot \dfrac{4}{3} = 1$.

Since $1 \cdot 1 = 1$, we say that 1 is its own multiplicative inverse.

In general, for each nonzero real number a, there exists a unique real numbe $\dfrac{1}{a}$ such that

$$a \cdot \frac{1}{a} = 1 \text{ and } \frac{1}{a} \cdot a = 1$$

The reason that 0 does not have a multiplicative inverse is that $\dfrac{1}{0}$ is not defined; $\dfrac{1}{0}$ is meaningless.

DISTRIBUTIVE POSTULATE FOR MULTIPLICATION WITH RESPECT TO ADDITION

In the set of real numbers, one property involves both multiplication and addition. For example, since $2(3 + 4) = 2(7) = 14$ and $2 \cdot 3 + 2 \cdot 4 = 6 + 8 = 14$, we know that $2(3 + 4) = 2 \cdot 3 + 2 \cdot 4$ is a true statement. This example illustrates the ***distributive postulate for multiplication with respect to addition***. The word "distributive" is appropriate because we "distribute" the multiplier 2 to each of the addends 3 and 4.

In general, we postulate that for any real numbers a, b, and c,

$$a(b + c) = ab + ac$$

By making use of the commutative property for multiplication, we can also show that

$$(b + c)a = ba + ca$$

We say that, in the set of real numbers, *multiplication is distributive with respect to addition.*

The distributive postulate may be extended to involve four or more real numbers:

$$a(b + c + d + \cdots) = ab + ac + ad + \cdots$$

FIELD POSTULATES FOR ADDITION AND MULTIPLICATION

We have considered eleven postulates of the real number system. They are properties of a mathematical system called a *field*. The following chart summarizes the eleven field postulates:

FIELD POSTULATES FOR ADDITION AND MULTIPLICATION OF REAL NUMBERS

	(*a*, *b*, and *c* represent any members of a set of real numbers.)
A_1	Closure postulate for addition $a + b$ is a unique element of the set of real numbers
A_2	Commutative postulate for addition $a + b = b + a$
A_3	Associative postulate for addition $(a + b) + c = a + (b + c)$
A_4	Additive identity postulate There exists a unique number 0 such that for each a $a + 0 = a$ and $0 + a = a$
A_5	Additive inverse postulate For each real number a, there exists a unique number $-a$ such that $a + (-a) = 0$ and $(-a) + a = 0$
M_1	Closure postulate for multiplication ab is a unique member of the set of real numbers
M_2	Commutative postulate for multiplication $ab = ba$
M_3	Associative postulate for multiplication $(ab)c = a(bc)$

M_4	Multiplicative identity postulate
	There exists a unique number 1 such that for each a
	$a \cdot 1 = a$ and $1 \cdot a = a$

M_5	Multiplicative inverse postulate
	For every real nonzero number a, there exists a unique number $\frac{1}{a}$ such that
	$a \cdot \frac{1}{a} = 1$ and $\frac{1}{a} \cdot a = 1$

	Distributive postulate for multiplication with respect to addition
	$a(b + c) = ab + ac$

SUMMARY OF THE FIELD POSTULATES AS THEY APPLY TO THE SET OF REAL NUMBERS AND ITS SUBSETS

The diagram at the right shows the development of the real number system. The system of natural numbers was extended to the system of whole numbers; the system of whole numbers was extended to the system of integers; the system of integers was extended to the system of rational numbers; the system of rational numbers was extended to the system of real numbers.

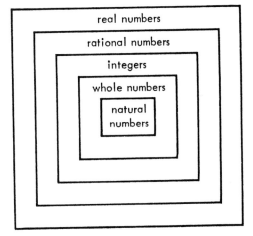

The addition and multiplication postulates of each number system are also postulates of the new number system to which it was extended. For example, the addition and multiplication postulates of the system of natural numbers are also postulates of the system of whole numbers to which the system of natural numbers was extended.

The following chart shows how the eleven field postulates for addition and multiplication apply to the set of real numbers and its subsets. A "✓" in a column naming a number system means that the set of numbers in the system satisfies the field postulate that is checked. A "No" means that the set of numbers in the system does not satisfy the postulate.

Postulate	*Natural Numbers*	*Whole Numbers*	*Integers*	*Rational Numbers*	*Real Numbers*
1. Closure under addition	✓	✓	✓	✓	✓
2. Closure under multiplication	✓	✓	✓	✓	✓
3. Commutative under addition	✓	✓	✓	✓	✓
4. Commutative under multiplication	✓	✓	✓	✓	✓
5. Associative under addition	✓	✓	✓	✓	✓
6. Associative under multiplication	✓	✓	✓	✓	✓
7. Distributive	✓	✓	✓	✓	✓
8. Multiplicative identity	✓	✓	✓	✓	✓
9. Additive identity	No	✓	✓	✓	✓
10. Additive inverse	No	No	✓	✓	✓
11. Multiplicative inverse	No	No	No	✓	✓

Note in the chart that the first eight postulates apply to each of the sets of numbers. The last three postulates do not apply to the set of natural numbers.

Since the set of whole numbers consists of 0 as well as the natural numbers, the ninth postulate, the additive identity postulate, applies to the set of whole numbers.

Since the set of integers consists of the negative integers as well as 0 and the positive integers, the tenth postulate, the additive inverse postulate, applies to the set of integers.

Since the set of rational numbers consists of numbers expressible as a fraction whose numerator is an integer and whose denominator is a nonzero integer, the eleventh postulate, the multiplicative inverse postulate, applies to the set of rational numbers. Note that the set of rational numbers and the set of real numbers are the only ones of the listed sets to which all eleven postulates apply. Hence, the sets of rational numbers and real numbers are the only fields among the listed sets.

ADDING OR MULTIPLYING THREE OR MORE NUMBERS

If we wish to add 3, 4, and 5, we define $3 + 4 + 5$ as $(3 + 4) + 5$. In general, we define $a + b + c$ to be $(a + b) + c$; that is, $a + b + c = (a + b) + c$. Likewise, we define $a + b + c + d$ to be $(a + b + c) + d$. In fact, because of the commutative and associative properties of addition, we may add the addends in a sum in any desirable groups of two numbers. For example,

$$865 + 739 + 135 + 261 = (865 + 135) + (739 + 261)$$
$$= 1000 + 1000 = 2000$$

If we wish to multiply 3, 4, and 5, we define $3 \cdot 4 \cdot 5$ as $(3 \cdot 4) \cdot 5$. In general, we define abc to be $(ab)c$; that is, $abc = (ab)c$. Likewise, we define $abcd$ to be $(abc)d$. In fact, because of the commutative and associative properties of multiplication, we can multiply the factors in a product in any convenient groups of two numbers. For example,

$$125 \cdot 25 \cdot 8 \cdot 4 = (125 \cdot 8) \cdot (25 \cdot 4) = 1000 \cdot 100 = 100{,}000$$

PROPERTIES OF EQUALITY

In our study of the real number system, we will make use of the following properties of equality, which are concerned with the use of the symbol $=$.

REFLEXIVE PROPERTY

The *reflexive property of equality* states that any number is equal to itself. For example, $2 = 2$ and $8 = 8$.

In general, for each real number a,

$$a = a$$

SYMMETRIC PROPERTY

The *symmetric property of equality* states that the sides of an equality may be interchanged. For example, if $3 + 6 = 5 + 4$, then $5 + 4 = 3 + 6$.

In general, for each real number a and each real number b,

$$\text{if } a = b, \text{ then } b = a$$

TRANSITIVE PROPERTY

The *transitive property of equality* states that if one number is equal to a second number, and the second number is equal to a third number, then the first number is equal to the third number. For example, if $7 + 3 = 6 + 4$, and $6 + 4 = 8 + 2$, then $7 + 3 = 8 + 2$.

In general, for each real number a, each real number b, and each real number c,

if $a = b$ and $b = c$, then $a = c$

The transitive property of equality is useful in the following ways:

If $a = b$ and $b = c$, the transitive property of equality makes it possible for us to replace b by c in the first equality, thus obtaining $a = c$. We call this process of replacing a number by an equal number **substitution.**

Also, the transitive property of equality allows us to state that two numbers are equal if each of them is equal to a third number. For example, if $x = y$ and $y = 5$, then $x = 5$.

Observe that we are making use of the substitution principle when we write $(5 \cdot 10) + 25 = 50 + 25$, or 75, because $5 \cdot 10 = 50$. Also, $76(85 + 15) = 76(100)$, or 7600, because $85 + 15 = 100$.

ADDITION PROPERTY

The *addition property of equality* states that if the same number is added to each of two equal numbers, the resulting numbers are equal. For example, if $5 + 2 = 6 + 1$, then $(5 + 2) + 3 = (6 + 1) + 3$.

In general, for each real number a, each real number b, and each real number c,

if $a = b$, then $a + c = b + c$

MULTIPLICATION PROPERTY

The *multiplication property of equality* states that if each of two equal numbers is multiplied by the same number, the resulting numbers are equal. For example, if $5 + 4 = 7 + 2$, then $(5 + 4) \cdot 3 = (7 + 2) \cdot 3$.

In general, for each real number a, each real number b, and each real number c,

if $a = b$, then $ac = bc$

SUMMARY OF PROPERTIES OF EQUALITY

(a, b, and c are members of the set of real numbers.)	
Reflexive property	$a = a$
Symmetric property	If $a = b$, then $b = a$.
Transitive property	If $a = b$, and $b = c$, then $a = c$.
Addition property	If $a = b$, then $a + c = b + c$.
Multiplication property	If $a = b$, then $ac = bc$.

~~~~~~~~~~~~~~~ **MODEL PROBLEM** ~~~~~~~~~~~~~~~

Name the property of the real number system illustrated in each sentence.

a. $5 + 8 = 8 + 5$ *Ans.* Commutative property for addition (A_2)

b. $(7 \times r) \times s = 7 \times (r \times s)$ *Ans.* Associative property for multiplication (M_3)

c. $(y) + (-y) = 0$ *Ans.* Additive inverse property (A_5)

d. $7 \times (a + b) = (7 \times a) + (7 \times b)$ *Ans.* Distributive property for multiplication with respect to addition

~~~~~~

Exercises

In this set of exercises, assume that no denominator is 0.

In 1–12, name the property of the real number system illustrated in the statement.

1. $6 + 4 = 4 + 6$

2. $8 \cdot (\frac{1}{8}) = 1$

3. $7 \times (9 \times 5) = (7 \times 9) \times 5$

4. $(3) + (-3) = 0$

5. $(\frac{1}{2} + \frac{1}{3}) + \frac{1}{4} = \frac{1}{2} + (\frac{1}{3} + \frac{1}{4})$

6. $\frac{2}{3}(6 + 9) = (\frac{2}{3} \times 6) + (\frac{2}{3} \times 9)$

7. $x + 3 = 3 + x$

8. $2y + (-2y) = 0$

9. $(5 + 2x) + 3x = 5 + (2x + 3x)$

10. $8 \times (\frac{1}{2} \times y) = (8 \times \frac{1}{2}) \times y$

11. $\frac{1}{4}(8x + 4y) = (\frac{1}{4} \times 8x) + (\frac{1}{4} \times 4y)$

12. $8 \times 100 = (8 \times 77) + (8 \times 23)$

In 13–18, (a) give a replacement for the question mark which makes the sentence true for all real values of the variable and (b) name the property illustrated in the sentence that is formed when the replacement is made.

13. $r \times 9 = 9 \times ?$

14. $(7 + ?) + 5 = 7 + (x + 5)$

15. $\frac{9}{4} + ? = d + \frac{9}{4}$

16. $t + ? = 0$

17. $w \times ? = w$

18. $x + (\frac{5}{9} + ?) = x + \frac{5}{9}$

In 19–27, name the property of equality that is illustrated.

19. $8 + 9 = 8 + 9$

20. If $9 + 8 = 12 + 5$, then $12 + 5 = 9 + 8$.

21. If $5 + 2 = 6 + 1$, then $(5 + 2) + 7 = (6 + 1) + 7$.

22. If $(7 + 3) = (6 + 4)$, then $(7 + 3) \times 8 = (6 + 4) \times 8$.

23. If $x = y$ and $y = 15$, then $x = 15$.

24. If $24 = 3x + 6$, then $3x + 6 = 24$.

25. If $a = b$, then $a \cdot 10 = b \cdot 10$.

26. If $a = b$, then $a + 7 = b + 7$.

27. If $40x = 200$, then $(40x) \cdot \frac{1}{40} = (200) \cdot \frac{1}{40}$.

In 28–31, find the value of the numerical expression by using the properties of the real number system to simplify the computation.

28. $125 \times 197 \times 8$ **29.** $.69 + .94 + .31$

30. $978 \times 8 + 978 \times 2$ **31.** $\frac{1}{3} \times 620 + \frac{1}{3} \times 280$

In 32–36, if r and s are elements of the given set of numbers, state whether or not each of the following must represent an element of the given set of numbers: $(a)\ r + s$ $(b)\ r - s$ $(c)\ rs$ $(d)\ r \div s,\ s \neq 0$

32. {natural numbers}   **33.** {whole numbers}   **34.** {integers}

35. {rational numbers}   **36.** {real numbers}

In 37–51, state whether or not the given set is closed under (a) addition (b) multiplication (c) subtraction (d) division.
If the answer is *no*, give an example justifying this answer.

37. {1, 3, 5}   **38.** {2, 4, 6}   **39.** {−1, 0, 1}

40. {1, 0}   **41.** {1}   **42.** {0}

43. $\{0, \frac{1}{3}, \frac{1}{9}, \frac{1}{27}, \frac{1}{81}, \ldots\}$

44. {natural numbers}

45. {even natural numbers}

46. {whole numbers}   **47.** {integers}

48. {integers divisible by 3}   **49.** {rational numbers}

50. {real numbers}   **51.** {irrational numbers}

52. List the eleven properties of both the set of rational numbers and the set of real numbers which are postulates for a field.

53. From the eleven field postulates, name those that are satisfied by each of the following sets:

    a. {the set of natural numbers}

    b. {the set of whole numbers}

    c. {the set of integers}

9. Deductive Reasoning and Proof in Algebra

In recent years, mathematicians have organized algebra into a logical structure. In this organization, as in geometry, we begin with **undefined terms,** which are used to define additional terms, and **postulates** or **axioms,** which are assumptions. By a process of logical reasoning, we can make use of the undefined terms, the defined terms, and the postulates to deduce additional number properties. Statements about numbers which have been proved are called **theorems.** Theorems, in turn, are used in proving other theorems. When we prove a theorem, we reason from the **hypothesis,** which is the given information, to the **conclusion.** This is done by using a sequence of statements, each of which is supported by the hypothesis or a definition or a postulate or a previously proved theorem.

Although we may not have realized it, the techniques which we have used in working with algebraic expressions are justified by the postulates that were assumed and the theorems that were proved.

Note. In the following model problems and exercises, the variables represent real numbers.

∿∿∿∿∿∿ *MODEL PROBLEMS* ∿∿∿∿∿∿

1. Prove: $8x + x = 9x$

Solution: *Method* 1

Statements	*Reasons*
1. $x = 1x$	1. Multiplicative identity property.
2. $8x + x = 8x + 1x$	2. Substitution principle.
3. $8x + 1x = (8 + 1)x$	3. Distributive property.
4. $8 + 1 = 9$	4. Number fact.
5. $(8 + 1)x = 9x$	5. Substitution principle.
6. $8x + x = 9x$	6. Transitive property of equality.

Method 2

Statements	*Reasons*
1. $8x + x = 8x + 1x$	1. Multiplication property of 1.
2. $8x + 1x = (8 + 1)x$	2. Distributive property.
3. $(8 + 1)x = 9x$	3. Number fact.
4. $8x + x = 9x$	4. Transitive property of equality.

Note. In the shortened version, Method 2, we have omitted steps 1 and 4 of Method 1. Also, we have not stated the substitution principle.

Method 3

Statements	*Reasons*
1. $8x + x = 8x + 1x$	1. Multiplication property of 1.
2. $\quad = (8 + 1)x$	2. Distributive property.
3. $\quad = 9x$	3. Number fact.

Note. In Method 3 we have omitted step 4 of Method 2, and have not stated the transitive property of equality. We have also omitted the rewriting of the right member of an equation as the left member in the next equation.

2. Prove: $r(x + y) = yr + xr$

Solution:

Statements	Reasons
1. $r(x + y) = rx + ry$	1. Distributive property.
2. $\qquad = ry + rx$	2. Commutative property of addition.
3. $\qquad = yr + xr$	3. Commutative property of multiplication.

3. Prove: $(5x) \cdot (7y) = 35xy$

Solution:

Statements	Reasons
1. $(5x) \cdot (7y) = [(5x) \cdot 7] \cdot y$	1. Associative property of multiplication.
2. $\qquad = [5 \cdot (x \cdot 7)] \cdot y$	2. Associative property of multiplication.
3. $\qquad = [5 \cdot (7 \cdot x)] \cdot y$	3. Commutative property of multiplication.
4. $\qquad = [(5 \cdot 7) \cdot x] \cdot y$	4. Associative property of multiplication.
5. $\qquad = (35 \cdot x) \cdot y$	5. Number fact.
6. $\qquad = 35xy$	6. Definition of multiplication.

Exercises

In 1–4, state the property that justifies the statement.

1. $x \cdot 8 = 8 \cdot x$
2. $(x + 8) + 0 = x + 8$
3. $1(x + 8) = x + 8$
4. $8 + x = x + 8$

In 5–12, prove the statement.

5. $(x + 8) + y = x + (y + 8)$
6. $x(8y) = (8x)y$
7. $(8 + x)y = y(x + 8)$
8. $y(x + 8) = xy + 8y$
9. $xy + 8y = y(x + 8)$
10. $y(8x) = (8y)x$
11. $(y + 2)(x + 8) = x(y + 2) + 8(y + 2)$
12. $(y + 2)(x + 8) = 2x + xy + 8y + 16$

In 13–18, state the property that justifies the statement.

13. $px + qx = qx + px$
14. $(px + qx) + rx = px + (qx + rx)$
15. $px + [-(px)] = 0$
16. $px \cdot \dfrac{1}{px} = 1$
17. $px + x = px + 1x$
18. $(px + qx) + 0 = px + qx$

In 19–24, prove the statement.

19. $(px)\left(\dfrac{1}{x}\right) = p$
20. $px(qx \cdot rx) = (rx \cdot qx)px$

21. $p(qx + rx) = prx + pqx$ **22.** $xp + x = x(p + 1)$

23. $p[qx + (-qx)] = 0$ **24.** $p + [qx + (-qx)] = p$

In 25–27, the statements in the proof are given. Supply the reason for each statement.

25. Prove: $r(3x + 3y + 3z) = 3(yr + xr + zr)$

1. $r(3x + 3y + 3z) = (3x + 3y + 3z)r$
2. $\quad = (3x)r + (3y)r + (3z)r$
3. $\quad = 3(xr) + 3(yr) + 3(zr)$
4. $\quad = 3(xr + yr + zr)$
5. $\quad = 3(yr + xr + zr)$

26. Prove: $p(qx + rx) = x(pr + pq)$

1. $p(qx + rx) = p(qx) + p(rx)$
2. $\quad = (pq)x + (pr)x$
3. $\quad = (pq + pr)x$
4. $\quad = x(pq + pr)$
5. $\quad = x(pr + pq)$

27. Prove: $7y + 2y + x = x + 9y$

1. $7y + 2y + x = (7y + 2y) + x$
2. $\quad = (7 + 2)y + x$
3. $\quad = 9y + x$
4. $\quad = x + 9y$

In 28–35, prove the statement.

28. $4x + 10x = 14x$ **29.** $(x + 8) + (-8) = x$

30. $5x(3) = 15x$ **31.** $(d + 7) + 2d = 3d + 7$

32. $4x(x + 5) + 10x = 4x^2 + 30x$ **33.** $(x + 4)(x + 5) = x^2 + 9x + 20$

34. $(ab)c = (ca)b$ **35.** $a(b + c + d) = ca + ba + da$

CALCULATOR APPLICATIONS

Calculators make solving mathematical problems easier and faster. Special scientific and graphing calculators combine all of the capabilities of a simple four-function calculator with many other keys and features that are very useful in algebra. However, correct use of a calculator requires understanding of the mathematics involved and good estimation skills to set up the problem and determine if the answer is reasonable.

MODEL PROBLEMS

1. Find a rational number between $\frac{1}{4}$ and $\frac{3}{5}$.

   Solution: Find the average of the two numbers.

   Enter: $\boxed{(}$ 1 $\boxed{\div}$ 4 + 3 $\boxed{\div}$ 5 $\boxed{)}$ $\boxed{\div}$ 2 $\boxed{=}$

   Display: $\boxed{0.425}$

   Note that, on a calculator, the rational number $\frac{425}{1000} = \frac{17}{40}$ is displayed in decimal form as 0.425. Some calculators will convert a decimal real number to a fraction in simplest terms. On the TI-83,

   Enter: .425 $\boxed{\text{MATH}}$ 1 $\boxed{\text{ENTER}}$

   Display: $\boxed{17/40}$

2. Write in order from least to greatest: $\frac{7}{9}, \frac{3}{4}, \frac{8}{11}$.

   Solution: One approach is to express the numbers as equivalent fractions with a common denominator and then compare numerators. With a calculator, it is easier to change the fractions to decimals by dividing each numerator by its denominator. Then compare decimal values. The answers here are from a calculator with an eight-place display.

   Enter: 7 $\boxed{\div}$ 9 $\boxed{=}$ 3 $\boxed{\div}$ 4 $\boxed{=}$ 8 $\boxed{\div}$ 11 $\boxed{=}$

   Display: $\boxed{0.7777778}$ $\boxed{0.75}$ $\boxed{0.7272727}$

   So, $\frac{8}{11} < \frac{3}{4} < \frac{7}{9}$.

3. Graph $1 \le x \le 5$ when the domain of x is the real numbers.

Solution: The expression $1 \le x \le 5$ means the set of real numbers greater than or equal to 1 *and* also less than or equal to 5. In logic, a compound sentence that combines two simple sentences using the word *and* is called a conjunction. So, the graph of $1 \le x \le 5$ is equivalent to the graph of $x \ge 1$ *and* $x \le 5$. The solution set of a conjunction of two open sentences contains only values of the variable that are true for both open sentences.

To draw this graph on a graphing calculator that has logic capabilities, such as the TI-83, we begin by setting the calculator in DOT mode for graphing and selecting a WINDOW that includes values of x from -2 to 7 and y from -4 to 4. The logic connectives are contained in a menu accessed by the TEST key.

Enter: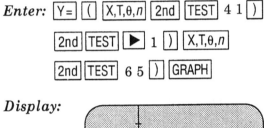

Display: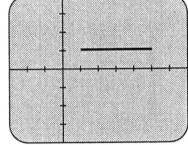

On other graphing calculators without logic capabilities, it may be possible to draw the graph as the product of the two inequalities, since this product will be 1 when both inequalities are true and 0 when one or both inequalities are false.

CHAPTER

Operations on Real Numbers, Polynomials, and Algebraic Expressions

1. Using Signed Numbers To Represent Opposite Situations

In our daily life we frequently talk about "opposite situations." For example, we may travel east or west from a starting point; we have a profit or a loss; latitude is north or south of the Equator. We see, therefore, that there is need for a number which represents both direction and size. The positive and negative real numbers, called the **signed numbers,** or **directed numbers,** meet this need.

To represent quantities that are "opposites," we use the positive (+) numbers, which are usually to the right of 0, the starting point on the real number line; and we use the negative (−) numbers, which are usually to the left of 0. Examples of positive numbers are $+4$, $+\frac{1}{2}$, $+1\frac{4}{5}$; examples of negative numbers are -2, $-\frac{2}{3}$, $-1\frac{1}{2}$. A positive number may be written without a sign. For example, $+6$ may be written as 6. Zero (0) is neither a positive number nor a negative number.

USES OF THE SYMBOL "+"

The symbol "+" may be used in two ways:

1. In $7 + 5$, the "+" indicates the operation of *addition*.
2. In $+8$, the "+" indicates that the number is *positive*.

Exercises

In 1–6, give the opposite of:
1. a rise in price
2. below sea level
3. north of the Equator
4. traveling east
5. a loss in weight
6. increasing speed

7. If $+5$ means a profit of \$5, what does -5 mean?
8. If -10 means 10 miles west, what does $+10$ mean?
9. If $+30$ means 30° north of the Equator, what does -30 mean?
10. If -2.50 means taking \$2.50 from a bank, what does $+2.50$ mean?

36

In 11–16, represent the stated expression by a signed number or by zero.

11. 100° above zero   **12.** $10\frac{1}{2}$° below zero   **13.** sea level
14. a fall of 30 feet   **15.** a height of 1000 feet   **16.** a profit of $25.75

2. The Additive Inverse of a Real Number

We have assumed that every real number a, whether it is a positive number or a negative number, has an *additive inverse* represented by $-a$, such that $a + (-a) = 0$. Observe that when a is a positive number, its additive inverse, $-a$, is a negative number. For example, the additive inverse of $+9$ is -9 because $(+9) + (-9) = 0$. When a is a negative number, its additive inverse is a positive number. For example, the additive inverse of -7 is $+7$ because $(-7) + (+7) = 0$. The additive inverse of 0 is 0 because $0 + 0 = 0$.

The additive inverse of a given number is also the *negative of the given number* as well as the *opposite of the given number*. For example,

$-(+9) = -9$ is read " the additive inverse of positive 9 is negative 9."
$-(-7) = +7$ is read " the additive inverse of negative 7 is positive 7."

The additive inverse of the additive inverse of a given real number is the given number itself. For example,

$-[-(+7)] = +7$ is read " the additive inverse of the additive inverse of $+7$ is $+7$."

In general, if a is a real number, then

$$-(-a) = a$$

━━━━━━━━━━ *MODEL PROBLEMS* ━━━━━━━━━━

In 1–3, write the additive inverse of the given number in simplest form.

1. $(3 + 6)$ *Ans.* -9 **2.** $-(-5)$ *Ans.* -5
3. $-(8 - 2)$ *Ans.* 6 *or* $+6$

4. Graph the solution set of $-a > 0$ when the domain of a is $\{-2, -1, 0, 1, 2\}$.

Solution:
1. The open sentence $-a > 0$ means, " The additive inverse of a is a positive number." Hence, the number a must be negative.
   Therefore, the solution set of $-a > 0$ is $\{-2, -1\}$.

2. Graph the solution set $\{-2, -1\}$.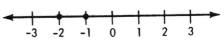

Exercises

In 1–8, write the additive inverse of the given number in simplest form.

1. $+4.5$ **2.** $-1\frac{2}{3}$ **3.** $(8+7)$ **4.** $(9-9)$

5. $-(+7)$ **6.** $-(-\frac{3}{4})$ **7.** $-[-(+3)]$ **8.** $-[-(-15)]$

In 9–16, graph the solution set of the open sentence if the domain of the variable is $\{-5, -4, -3, -2, -1, 0, 1, 2, 3, 4, 5\}$.

9. $-x > 1$ **10.** $-a < 2$ **11.** $-r \leq 0$ **12.** $-y < -2$

13. $-t \geq -1$ **14.** $-b \leq 2\frac{1}{2}$ **15.** $-3 < -x < 5$ **16.** $-3 \leq -x \leq 4$

In 17–20, graph the solution set of the open sentence if the domain of the variable is the set of real numbers.

17. $-x > 2$ **18.** $-y \leq -1$ **19.** $-6 < -m \leq 2$ **20.** $-5 \leq -x \leq 0$

In 21 and 22, graph the indicated set of numbers.

21. $\{x \mid -x < 4\}$, $x \in \{\text{real numbers}\}$

22. $\{x \mid -7 < -x \leq 5\}$, $x \in \{\text{real numbers}\}$

In 23–25, tell whether the statement is *true* or *false*.

23. If a is a positive real number, $-a$ is always a negative real number.

24. If a is a negative real number, $-a$ is always a negative real number.

25. The additive inverse of a real number is always a different real number.

3. The Absolute Value of a Real Number

In performing operations on real numbers, we will find it useful to make use of the concept of the *absolute value* of a number.

On the real number line, a pair of opposite numbers is represented by points that are equally distant from the point associated with zero. For example, the points associated with the pair of opposites $+5$ and -5 are each 5 units from 0 on the number line shown below. The real number which represents this common distance 5 is called the **absolute value** of both $+5$ and -5.

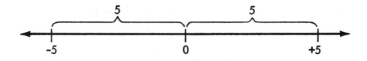

In general, the absolute value of a real number x, symbolized $|x|$, is the real number representing the non-directed distance on the number line between the zero point and the point associated with the number. For example, the absolute value of $+3$ is 3, symbolized $|+3| = 3$. The absolute value of -2 is 2, symbolized $|-2| = 2$. The absolute value of 0 is 0, symbolized $|0| = 0$.

Observe that the absolute value of a real number x, $|x|$, is a nonnegative number. Observe also that:

1. The absolute value of either a given positive number or zero is that number itself.
2. The absolute value of a given negative number is the opposite of that negative number.

In general, if x is a real number, we define $|x|$ as follows:

$$|x| = x \text{ when } x \geq 0 \qquad |x| = -x \text{ when } x < 0$$

∿∿∿∿∿∿∿ *MODEL PROBLEM* ∿∿∿∿∿∿∿

Find the value of the number expression $|15| + |-4|$.

Solution: $|15| + |-4| = 15 + 4 = 19$ *Answer:* 19

Exercises

In 1–5, (*a*) give the absolute value of the given number and (*b*) give another number—if there is such a number—that has the same absolute value as the given number.

1. $+10$ **2.** -5 **3.** 0 **4.** $6\frac{1}{2}$ **5.** -6.82

In 6–9, select the number that has the smaller absolute value.

6. $+8, +4$ **7.** $3, -20$ **8.** $-15.6, +15.06$ **9.** $-5\frac{1}{4}, 5\frac{1}{3}$

In 10–12, state whether the sentence is *true* or *false*.

10. $|-15| = -15$ **11.** $|-7| = |+7|$ **12.** $|-12| > |4|$

In 13–15, find the value of the number expression.

13. $|+7| + |-2|$ **14.** $|+8| \times |-5|$ **15.** $|+12| + |0| - |-7|$

In 16–19, state whether the sentence is *true* or *false*.

16. $|+3| + |-3| = 0$ **17.** $|5| - |-5| = 0$
18. $|+4| \times |-4| = |+16|$ **19.** $|+8| + |-8| = |+4| \times |-4|$

4. Adding Signed Numbers

ADDING SIGNED NUMBERS ON A NUMBER LINE

A sequence of directed movements on a real number line can be used to define the operation of addition for signed numbers. A positive number will be interpreted as a "movement to the right," a negative number as a "movement to the left," and 0 as "no movement at all."

〰〰〰〰〰〰〰 **MODEL PROBLEMS** 〰〰〰〰〰〰〰

1. Add $+2$ and $+4$.

Solution:

1. Start at 0 and move 2 units to the right.

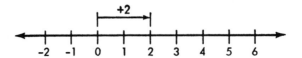

2. Move 4 more units to the right.

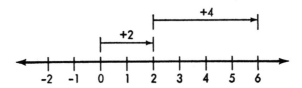

3. Read the coordinate of the point that was last reached, $+6$.

Answer: $(+2)+(+4)=+6$

2. Add $+2$ and -4.

Solution:

1. Start at 0 and move 2 units to the right.

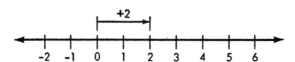

2. Then move 4 units to the left.

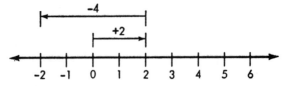

3. Read the coordinate of the point that was last reached, -2.

Answer: $(+2)+(-4)=(-2)$

Procedure. To add two signed numbers on a number line:

1. Graph the first number on the number line.
2. From this point, move to the right or to the left a number of units equal to the absolute value of the second number. If the second number is positive, move to the right (model problem 1); if the second number is negative, move to the left (model problem 2); if the second number is zero, do not move at all.
3. Read the coordinate of the point that was last reached. This coordinate represents the sum of the two numbers.

Exercises

In 1–6, use a real number line to find the sum of the real numbers.

1. $(+4) + (+3)$ 2. $(-5) + (-1)$ 3. $(-6) + (+8)$
4. $(+3) + (-8)$ 5. $(+7) + (0)$ 6. $(0) + (-5)$

In 7 and 8, use a number line to find the sum of the real numbers; state the relationship between the two results; and state the property of real numbers that is illustrated.

7. *a.* $(-3) + (-5)$ *b.* $(-5) + (-3)$
8. *a.* $[(-5) + (+6)] + (-4)$ *b.* $(-5) + [(+6) + (-4)]$

In 9 and 10, (*a*) use a number line to find the sum of the real numbers and (*b*) state the property of real numbers that is illustrated.

9. $(-5) + (0)$ 10. $(+8) + [-(+8)]$

USING RULES FOR ADDING REAL NUMBERS

Now we will define the operation of addition for real numbers, without the use of a number line, in such a way that all the properties of addition will be satisfied. Remember that we have postulated:

If a is a real number, then $a + 0 = a$ and $0 + a = a$.

ADDITION OF TWO POSITIVE NUMBERS OR TWO NEGATIVE NUMBERS

By making use of a number line, we find that $(+4) + (+5) = +9$ and $(-4) + (-5) = -9$. These examples illustrate the following rules:

Rule 1. The sum of two positive numbers is a positive number whose absolute value is found by adding the absolute values of the numbers.

In general, if a and b are both positive numbers,

$$a + b = |a| + |b|$$

Rule 2. The sum of two negative numbers is a negative number whose absolute value is found by adding the absolute values of the numbers.

In general, if a and b are both negative numbers,

$$a + b = -(|a| + |b|)$$

~~~~~~~~~ **MODEL PROBLEMS** ~~~~~~~~~

In 1–6, add.

1. $+8$	**2.** 0	**3.** -8	**4.** -9.1
$+7$	$+9$	-5	-7.5
$+15$	$+9$	-13	-16.6

5. $(-5\frac{1}{2}) + (-3\frac{1}{4}) = -8\frac{3}{4}$　　　　**6.** $(-15) + (0) = -15$

ADDITION OF A POSITIVE NUMBER AND A NEGATIVE NUMBER

By making use of a number line, we find that $(+5) + (-2) = +3$ and $(-5) + (+2) = -3$. These examples illustrate the following rule:

Rule 3. The sum of two numbers, one of which is positive or 0 and the other negative, is a number whose absolute value is found by subtracting the smaller of the absolute values of the numbers from the greater.

(1) The sum is positive if the positive number has the greater absolute value.
(2) The sum is negative if the negative number has the greater absolute value.
(3) The sum is 0 if both numbers have the same absolute value.

In general, if a is a positive number and b is a negative number:

(1) If $|a| > |b|$, then $a + b = |a| - |b|$.
(2) If $|b| > |a|$, then $a + b = -(|b| - |a|)$.
(3) If $|a| = |b|$, then $a + b = 0$.

~~~~~~~~~ **MODEL PROBLEMS** ~~~~~~~~~

In 1–6, add.

1. $+8$	**2.** -9	**3.** -7	**4.** $+3$	**5.** $-8\frac{3}{4}$	**6.** $+2.8$
-2	$+5$	$+9$	-7	$+5\frac{1}{4}$	-2.8
$+6$	-4	$+2$	-4	$-3\frac{1}{2}$	0

ADDING MORE THAN TWO NUMBERS

Since the commutative and associative properties of addition hold for signed numbers, these numbers may be rearranged in any order when they are being added.

Thus, $a+b+c=a+c+b=b+a+c=b+c+a=c+a+b=c+b+a$.

A useful method of adding signed numbers is to add all the positive numbers first, all the negative numbers second, and then add the two results. In the following example, this method is used after the terms have been rearranged in the desired order:

$$(+8)+(-3)+(+6)+(-2)=(+8)+(+6)+(-3)+(-2)$$

$$=\quad (+14)\quad +\quad (-5)\quad =9$$

In any future proof, if addends or terms are being added in a desired order, *Rearranging terms* may be used as the reason.

Exercises

In 1–36, add.

1. $+13$ $+5$	**2.** -8 -12	**3.** -14 -10	**4.** $+9$ $+15$	**5.** $+7$ 0	**6.** -19 -28
7. $+36$ -14	**8.** -50 $+33$	**9.** $+14$ -32	**10.** -25 $+25$	**11.** -15 $+35$	**12.** $+12$ -12
13. $+1.3$ $+6.4$	**14.** -6.8 $+4.7$	**15.** -8.3 $+5.9$	**16.** -3.8 -9.7	**17.** $+6.4$ -6.4	**18.** $+9.6$ -2.8
19. $+\frac{3}{8}$ $+\frac{4}{8}$	**20.** $-\frac{3}{10}$ $-\frac{2}{10}$	**21.** $+\frac{1}{4}$ $-\frac{3}{4}$	**22.** $+4\frac{1}{2}$ $-7\frac{3}{4}$	**23.** $+9\frac{2}{3}$ $-3\frac{5}{6}$	**24.** $-6\frac{3}{4}$ $-5\frac{5}{8}$

25. $(+15)+(+9)$ **26.** $(-13)+(-12)$ **27.** $(-12)+(+18)$
28. $(-16)+(+16)$ **29.** $|-15|+(-7)$ **30.** $|20|+(-|-20|)$
31. $(+15)+(-17)+(+8)$ **32.** $(-23)+(+11)+(-14)$
33. $(+8)+(+6)+(-3)+(-2)$ **34.** $(+9)+(-4)+(-6)+(+3)$
35. $|-5|+|0|+(-15)$ **36.** $|-19|+|8|+(-|-25|)$

USEFUL PROPERTIES OF ADDITION

Now we will present several theorems that state useful properties of addition of signed numbers. These theorems can be proved by deducing them from definitions that we have agreed upon, from postulates that we have assumed to be true, and from other theorems that have been proved previously.

Addition Theorem 1. If a and b are real numbers, then $(a + b) + (-b) = a$.
For example, $(7 + 3) + (-3) = 7$ and $[y + (-3)] + (+3) = y$.
The proof of this theorem is shown in exercise 8 on page 45.

Addition Theorem 2 *(Cancellation Property of Addition)*. If a, b, and c are real numbers, and if $a + c = b + c$, or $c + a = c + b$, then $a = b$.
For example, if $x + y = 5 + y$, then $x = 5$. And, if $y + (-5) = x + (-5)$, then $y = x$.
The proof of this theorem is shown in exercise 12 on page 45.

Addition Theorem 3 *(Property of the Opposite of a Sum)*. If a and b are real numbers, the opposite of their sum is equal to the sum of their opposites: $-(a + b) = (-a) + (-b)$.
For example, $-(5 + 4) = (-5) + (-4)$ and $-[5 + (-4)] = (-5) + [-(-4)]$.
The proof of this theorem is shown in exercise 13 on page 45.

MODEL PROBLEM

Prove: If c and y are real numbers, then $(-c) + (y + c) = y$.

Solution:

Statements	Reasons
1. If c and y are real numbers, then $(y + c)$ is a real number.	1. Closure property of addition.
2. If c is a real number, then $-c$ is a real number.	2. Additive inverse property of real numbers.
3. $(-c) + (y + c) = (y + c) + (-c)$	3. Commutative property of addition.
4. $(y + c) + (-c) = y$	4. Addition Theorem 1: If a and b are real numbers, then $(a + b) + (-b) = a$.

Note. This proof is a condensed or shortened version of the complete proof. To understand methods which are used to shorten proofs, see page 31.

Exercises

In 1–4, the variables represent real numbers. Give a replacement for the question mark that will make the resulting sentence true for all values of the variable.

1. $(x + 20) + (-20) = ?$ **2.** $(-y) + (a + y) = ?$
3. $(-5 + y) + 5 = ?$ **4.** $-[(-8) + (-y)] = ?$

In 5–7, give the value of the expression.
5. $-[13 + (-25)]$ **6.** $-[(+4) + (-4)]$ **7.** $-[(-7) + (-17)]$

In exercise 8, give the reason that justifies each statement in the proof.
8. Prove: If a and b are real numbers, $(a + b) + (-b) = a$.

Statements
1. If a and b are real numbers, $(a + b)$ is a real number.
2. $(a + b) + (-b) = a + [b + (-b)]$
3. $= a + 0$
4. $= a$

In 9–11, c is a real number. Prove that the statement is true.
9. $(c + 7) + (-7) = c$ **10.** $(c + 5) + (-c) = 5$ **11.** $(-c) + (5 + c) = 5$

In 12 and 13, give the reason that justifies each statement in the proof. The variables represent real numbers.
12. Prove: If a, b, and c are real numbers and $a + c = b + c$, then $a = b$.

Statements
1. If a, b, and c are real numbers, then $(a + c)$ and $(b + c)$ are real numbers.
2. $(-c)$ is a real number.
3. $a = (a + c) + (-c)$
4. $= (b + c) + (-c)$
5. $= b$

13. Prove: If a and b are real numbers, then $-(a + b) = (-a) + (-b)$.

Statements
1. If a and b are real numbers, then $(a + b)$ is a real number.
2. $(-a)$, $(-b)$, and $-(a + b)$ are real numbers.
3. $(a + b) + [(-a) + (-b)] = [(a + b) + (-a)] + (-b)$
4. $= [(-a) + (a + b)] + (-b)$
5. $= [(-a) + (a)] + [b + (-b)]$
6. $= \quad 0 \quad + \quad 0$
7. $= 0$
8. $= (a + b) + [-(a + b)]$
9. $(a + b) + [(-a) + (-b)] = (a + b) + [-(a + b)]$
10. $(-a) + (-b) = -(a + b)$

Note. Using "Rearranging terms" as a reason, statement 3 could have been written as follows, thus saving two steps:

     3. $(a + b) + [(-a) + (-b)] = [(-a) + a] + [b + (-b)]$

In 14–17, prove that the sentence is true if the variables represent real numbers. Give a reason that justifies each statement of the proof.

14. $(-a) + (x + a) = x$ **15.** $r + [y + (-r)] = y$

16. $x + y = -[(-x) + (-y)]$ **17.** $(x + y) + [(-x) + (-y)] = 0$

5. Subtracting Signed Numbers

Subtraction of real numbers is defined in terms of addition. To subtract 5 from 8, symbolized $8 - 5$, we ask, "Which number added to 5 gives 8?" That number is 3. Since $5 + 3 = 8$, we write $8 - 5 = 3$. The number 8 is the **minuend,** 5 is the **subtrahend,** and 3 is the **difference.**

In general, if a and b are real numbers, the expression $a - b$ represents a real number c (that is, $a - b = c$) such that $b + c = a$.

Thus, to find $(+3) - (-2)$, ask $(-2) + $ (what number) $= (+3)$. That number is $+5$, since $(-2) + (+5) = (+3)$. Hence, $(+3) - (-2) = +5$.

If we compare the two examples at the right, we observe that when we subtract (-2) from $(+3)$, we obtain the same result as when we add $(+2)$, the opposite or additive inverse of (-2), to $(+3)$. This example illustrates the following definition of subtraction:

If a and b are real numbers, $a - b = a + (-b)$.

Subtract:	*Add:*
$(+3)$	$(+3)$
(-2) →	$(+2)$
$(+5)$	$(+5)$

Procedure. To subtract one real number from another, add the opposite (additive inverse) of the subtrahend to the minuend.

For example, $8 - 2 = 8 + (-2) = 6$.

Since $8 - 2 = 8 + (-2)$, we may simplify the writing of $8 + (-2)$ by writing $8 - 2$. Similarly, $(+7) + (-3) + (-1)$ can be written as $7 - 3 - 1$ and $(-3) + (-4) + (-2)$ can be written as $-3 - 4 - 2$.

USES OF THE SYMBOL "−"

In the expression $9 - (-6)$, the symbol "−" is used in two different ways:

1. Between 9 and (-6), the "−" indicates the operation of *subtraction.*
2. As the sign of (-6), the "−" indicates the *negative* of 6.

Hence, when $8 + (-2)$ is written in the simplified form $8 - 2$, we may think of $8 - 2$ as $8 + (-2)$, in which case the symbol "−" indicates the *negative* of 2. Or, we may think of $8 - 2$ as $8 - (+2)$, in which case the symbol "−" indicates the *subtraction* of 2.

~~~~~~~~~~~~~~~~ **MODEL PROBLEMS** ~~~~~~~~~~~~~~~~

In 1–7, subtract the lower number from the upper number.

1. $+25$	2. $+30$	3. -13	4. -15	5. -12	6. $+9$	7. 0
$+10$	-10	$+12$	-8	-12	0	-3
$+15$	$+40$	-25	-7	0	$+9$	$+3$

Note. In each problem, the signed number is subtracted by adding its opposite to the minuend. To check the answer (difference) in subtraction, add it to the subtrahend to see whether you obtain the minuend.

~~~~~~

Exercises

In 1–12, subtract the lower number from the upper number.

1. $+35$	2. $+10$	3. $+24$	4. $+7$	5. -48	6. -32
$+25$	$+17$	-12	-7	$+13$	$+19$

7. $-14\frac{1}{2}$	8. $-5\frac{1}{4}$	9. $+5.8$	10. $+9.2$	11. -7.4	12. -8.6
$-12\frac{1}{2}$	$-5\frac{1}{4}$	$+7.6$	-5.1	-9.3	-8.6

In 13–18, perform the indicated subtraction.

13. $(+12) - (+17)$ 14. $(-13) - (-20)$ 15. $(-25) - (+3)$

16. $(+18) - (-2)$ 17. $(0) - (-4)$ 18. $(+12) - 0$

In 19–21, subtract.

19. $(+14)$ from $(+8)$ 20. (-9) from (-15) 21. (-12) from $(+5)$

In 22–25, find the value of the given expression.

22. $(+8) + (+12) - (-5)$ 23. $(-9) - (+3) + (+12)$

24. $(+12.4) - (+3.8) - (-2.5)$ 25. $42 - 9 - 20$

In 26–31, state whether or not the set of numbers is closed under subtraction. If your answer is *no*, give an example to show that the difference of two members of the set is not a unique member of the set.

26. natural numbers   27. integers   28. positive integers

29. negative integers   30. odd integers   31. even integers

32. Give an example showing that $x - y \neq y - x$ for all real numbers x and y.

33. Give an example showing that $(x - y) - z \neq x - (y - z)$ for all real values of x, y, and z.

34. Prove: If x and y are real numbers, then $x - y = -(y - x)$.

35. *a.* Prove: If x, y, and z are real numbers, then $x(y - z) = xy - xz$.

   b. What property of multiplication has been proved in part *a*?

6. Multiplying Signed Numbers

We will define multiplication of real numbers in such a way that all the properties of multiplication will be satisfied. In order to do this, we will find the following two theorems useful. Remember we have assumed the postulate that if a is a real number, then $a \times 1 = a$ and $1 \times a = a$.

Multiplication Theorem 1 (Multiplication Property of Zero). The multiplication property of zero states that the product of any given real number and 0 is 0. If a is a real number, then $a \times 0 = 0$ and $0 \times a = 0$.

For example, $(12) \times (0) = 0$ and $(0) \times (12) = 0$; also, $(-15) \times (0) = 0$ and $(0) \times (-15) = 0$.

The proof of this theorem is shown in exercise 20 on page 50.

Multiplication Theorem 2 [Multiplication Property of (-1)]. The multiplication property of (-1) states that the product of any given real number a and (-1) is the opposite of a. If a is a real number, then $a(-1) = -a$ and $(-1)(a) = -a$.

For example, $(-1)(3) = -3$; $3(-1) = -3$; and $(-1)(-3) = -(-3)$ or $+3$.

The proof of this theorem is shown in exercise 21 on page 50.

Note that since $(-1)(-1) = 1$, the reciprocal of -1 is -1; that is, $-1 = \dfrac{1}{-1}$.

Now we are ready to multiply any real numbers. For example:

$(4)(2) = 8$

$(-4)(2) = [(-1)4](2) = -1[(4)(2)] = -1(8) = -8$

$(2)(-4) = 2[(4)(-1)] = [(2)(4)](-1) = 8(-1) = -8$

$(-2)(-4) = [(-1)(2)][(-1)(4)] = [(-1)(-1)][(2)(4)] = [-(-1)](8) = (1)(8) = 8$

These four examples illustrate the following rules for multiplying real numbers:

Rule 1. The product of two positive numbers or of two negative numbers is a positive number whose absolute value is the product of the absolute values of the numbers.

In general, if a and b are both positive or both negative, then

$$ab = |a| \cdot |b|$$

Rule 2. The product of a positive number and a negative number is a negative number whose absolute value is the product of the absolute values of the numbers.

In general, if one of the two numbers a and b is positive and the other is negative, then

$$ab = -(|a| \cdot |b|)$$

Rule 3. The product of a real number and 0 is equal to 0. If a is a real number, then

$$0 \cdot a = a \cdot 0 = 0$$

The statements for the proofs of the rules for multiplying signed numbers are shown in exercises 15–17 on page 51.

~~~~~~~~~~~~~ *MODEL PROBLEMS* ~~~~~~~~~~~~~

In 1–6, multiply.

1. +10	**2.** −12	**3.** +15	**4.** −18	**5.** +7	**6.** 0
+3	−4	−3	+5	+1	−7
+30	+48	−45	−90	+7	0

~~~~~~

MULTIPLYING MORE THAN TWO NUMBERS

Since the commutative and associative properties of multiplication hold for signed numbers, these numbers may be rearranged in any order when they are being multiplied.

Thus, $abc = acb = bac = bca = cab = cba$.

For example, $(-2)(+5)(-3)(+6) = (-2)(-3)(+5)(+6)$

$$= (+6) \times (+30) = 180$$

In any future proof, if factors are being multiplied in a desired order, "Rearranging factors" may be used as the reason.

——— *KEEP IN MIND* ———

1. If, in an indicated product that has no zero factor, there is an even number of negative factors, the product is a positive number.
2. If, in an indicated product that has no zero factor, there is an odd number of negative factors, the product is a negative number.

Exercises

In 1–21, multiply.

1. +6	**2.** −8	**3.** +10	**4.** −9	**5.** +8	**6.** 0
+5	−4	−7	+5	+1	−9

7. +7 by +4   **8.** −25 by −8   **9.** +6 by −3   **10.** +15 by 0
11. (+4)(+3)   **12.** (−2)(−9)   **13.** (−7)(+4)   **14.** (+8)(−9)

15. $(+2)(+5)(+3)$ **16.** $(-1)(-6)(-9)$
17. $(-2)(+4)(-7)$ **18.** $(-5)(-5)$
19. $(-3)(-3)(-3)(-3)$ **20.** $(-5)(-1)(-2)(-10)(+4)$
21. $|+5| \cdot |-4| \cdot (6)$

In 22 and 23, state the property of real numbers that justifies the statement.
22. $(-7) \times (-4) = (-4) \times (-7)$ **23.** $[(-5) \cdot 2] \cdot 7 = (-5) \cdot (2 \cdot 7)$

24. Use the distributive property to find the result mentally.
     a. $35 \times 73 + 35 \times 27$ *b.* $97 \times (-8) + 97 \times (-2)$

ADDITIONAL PROPERTIES OF MULTIPLICATION

The following theorems can be proved using the definitions we have agreed upon and the postulates we have assumed.

Multiplication Theorem 3. If a is a real number and b is a nonzero real number, then $(ab)\dfrac{1}{b} = a$. For example, $(5 \times 7) \times \dfrac{1}{7} = 5$.

The statements for the proof of this theorem are shown in exercise 18 on page 51.

Multiplication Theorem 4 (*Cancellation Property of Multiplication*). If a and b are real numbers, and c is a nonzero real number, and if $ac = bc$ or $ca = cb$, then $a = b$. For example, if $9x = 9y$, then $x = y$.

The statements for the proof of this theorem are shown in exercise 19 on page 51.

Multiplication Theorem 5. If a and b are nonzero real numbers, the reciprocal of their product is the product of their reciprocals: $\dfrac{1}{ab} = \dfrac{1}{a} \cdot \dfrac{1}{b}$, $a \neq 0, b \neq 0$. For example, $\dfrac{1}{5 \times 3} = \dfrac{1}{5} \cdot \dfrac{1}{3}$.

The statements for the proof of this theorem are shown in exercise 22 on page 52.

Multiplication Theorem 6. If a is a nonzero real number, the reciprocal of the negative of a is the negative of the reciprocal of a: $\dfrac{1}{-a} = -\dfrac{1}{a}$, $a \neq 0$. For example, $\dfrac{1}{-9} = -\dfrac{1}{9}$.

The statements for the proof of this theorem are shown in exercise 23 on page 52.

Exercises

In 1–6, give the reciprocal of the number.

1. 7   **2.** -4 **3.** $-\frac{1}{8}$ **4.** $\frac{2}{3}$ **5.** $-\frac{3}{4}$ **6.** $-.5$

In 7–10, give the simplest replacement for the question mark that will make the resulting sentence true. The variables represent real numbers.

7. $\frac{1}{5}(5x) = ?$

8. $(-3y) \cdot (-\frac{1}{3}) = ?$

9. $(ax)\dfrac{1}{a} = ?,\ a \neq 0$

10. $\dfrac{1}{7} \cdot \dfrac{1}{x} = \dfrac{1}{?},\ x \neq 0$

In 11–14, multiply the numbers.

11. $(\frac{1}{2})(8)$ **12.** $(35)(-\frac{1}{5})$ **13.** $(\frac{1}{3})(-\frac{1}{4})$ **14.** $18(-\frac{1}{3})(+\frac{1}{6})$

In 15–23, give a reason that justifies each statement in the proof. The variables represent real numbers. Assume that all products and all additive and multiplicative inverses involved are real numbers.

15. Prove: $a(-b) = -ab$

Statements

1. $a(-b) = a[(-1)b]$
2. $ = (-1)(ab)$
3. $ = -ab$

16. Prove: $(-a)b = -ab$

Statements

1. $(-a)b = [(-1)a]b$
2. $ = (-1)ab$
3. $ = -ab$

17. Prove: $(-a)(-b) = ab$

Statements

1. $(-a)(-b) = [(-1)a][(-1)b]$
2. $ = (-1)(-1)ab$
3. $ = 1 \cdot ab$
4. $ = ab$

18. Prove: $(ab)\dfrac{1}{b} = a,\ (b \neq 0)$

Statements

1. $ab\left(\dfrac{1}{b}\right) = a\left(b \cdot \dfrac{1}{b}\right)$
2. $\phantom{ab\left(\dfrac{1}{b}\right)} = a \cdot 1$
3. $\phantom{ab\left(\dfrac{1}{b}\right)} = a$

19. Prove: If $ac = bc$, then $a = b$, $(c \neq 0)$

Statements

1. $ ac = bc$
2. $(ac) \cdot \dfrac{1}{c} = (bc) \cdot \dfrac{1}{c}$
3. $ a = b$

52   *Algebra II*

20. Prove that $a \cdot 0 = 0$
and that $0 \cdot a = 0$.

    Statements

1. $a + 0 = a$
2. $\quad\;\; = a \cdot 1$
3. $\quad\;\; = a(1 + 0)$
4. $\quad\;\; = a \cdot 1 + a \cdot 0$
5. $\quad\;\; = a + a \cdot 0$ Hence,
6. $a + 0 = a + a \cdot 0$
7. $\quad 0 = a \cdot 0$
8. $\quad a \cdot 0 = 0$
9. $\quad 0 \cdot a = 0$

21. Prove that $a(-1) = -a$
and that $(-1)a = -a$.

    Statements

1. $[a(-1) + a] = [a(-1) + a \cdot 1]$
2. $\quad\quad\quad = a[(-1) + 1]$
3. $\quad\quad\quad = a \cdot 0$
4. $\quad\quad\quad = 0$
5. $\quad\quad\quad = (-a + a)$ Hence,
6. $[a(-1) + a] = (-a + a)$
7. $\quad a(-1) = -a$
8. $\quad (-1)a = -a$

22. Prove: $\dfrac{1}{ab} = \dfrac{1}{a} \cdot \dfrac{1}{b}$, $(a \neq 0, b \neq 0)$

    Statements

1. $ab\left(\dfrac{1}{a} \cdot \dfrac{1}{b}\right) = \left(a \cdot \dfrac{1}{a}\right)\left(b \cdot \dfrac{1}{b}\right)$
2. $\quad\quad\quad = 1 \cdot 1$
3. $\quad\quad\quad = 1$
4. $\quad\quad\quad = ab\left(\dfrac{1}{ab}\right)$ Hence,
5. $ab\left(\dfrac{1}{a} \cdot \dfrac{1}{b}\right) = ab\left(\dfrac{1}{ab}\right)$
6. $\quad \dfrac{1}{a} \cdot \dfrac{1}{b} = \dfrac{1}{ab}$
7. $\quad \dfrac{1}{ab} = \dfrac{1}{a} \cdot \dfrac{1}{b}$

23. Prove: $\dfrac{1}{-a} = -\dfrac{1}{a}$, $(a \neq 0)$

    Statements

1. $\dfrac{1}{-a} = \dfrac{1}{(-1)(a)}$
2. $\quad = \dfrac{1}{-1} \cdot \dfrac{1}{a}$
3. $\quad = (-1)\dfrac{1}{a}$
4. $\quad = -\dfrac{1}{a}$

In 24 and 25, prove the theorem if a and b are nonzero real numbers.

24. If $a = b$, then $\dfrac{1}{a} = \dfrac{1}{b}$.

25. $a \cdot \dfrac{1}{ab} = \dfrac{1}{b}$

7. Dividing Signed Numbers

Division of real numbers is defined in terms of multiplication.

To divide 8 by 4, symbolized $8 \div 4$ or $\frac{8}{4}$, we ask, "Which number multiplied by 4 gives 8?" That number is 2. Since $2 \times 4 = 8$, we write $8 \div 4 = 2$. The number 8 is the **dividend,** 4 is the **divisor,** and 2 is the **quotient.**

It is impossible to divide a real number by 0. For example, to divide 8 by 0, we ask, "Which number multiplied by 0 gives 8?" There is no such number because the product of any real number and 0 is 0. Hence, $8 \div 0$ is **undefined** or meaningless.

To divide 0 by 0, we ask, "Which number multiplied by 0 gives 0?" The answer is any number. Hence, $0 \div 0$ is meaningless.

In general, for all real numbers a and b ($b \neq 0$), to divide a by b, symbolized $a \div b$ or $\frac{a}{b}$, means to find a number c such that $bc = a$. Also if $a \neq 0$, then $\frac{1}{a}$, the reciprocal of a, may represent $1 \div a$.

This definition of division leads to the following results:

$$\frac{+8}{+4} = +2 \qquad \frac{-8}{-4} = +2 \qquad \frac{-8}{+4} = -2 \qquad \frac{+8}{-4} = -2 \qquad \frac{0}{+8} = 0 \qquad \frac{0}{-8} = 0$$

These examples illustrate the following rules for dividing real numbers:

Rule 1. The quotient of two positive numbers or of two negative numbers is a positive number whose absolute value is the absolute value of the dividend divided by the absolute value of the divisor.

In general, if a and b are both positive or both negative numbers, then

$$\frac{a}{b} = \frac{|a|}{|b|}$$

Rule 2. The quotient of two numbers, one of which is a positive number and the other a negative number, is a negative number whose absolute value is the absolute value of the dividend divided by the absolute value of the divisor.

In general, if one of the numbers a and b is positive and the other is negative, then

$$\frac{a}{b} = -\left(\frac{|a|}{|b|}\right)$$

Rule 3. Zero divided by a nonzero number is zero.

In general, if a is a nonzero real number, then

$$\frac{0}{a} = 0$$

~~~~~~~~~~~~~~ **MODEL PROBLEMS** ~~~~~~~~~~~~~~

In 1–5, perform the indicated division.

1. $(+35) \div (+5) = +7$ **2.** $(-64) \div (-16) = +4$

3. $(+40) \div (-8) = -5$ **4.** $\dfrac{-36}{+4} = -9$ **5.** $\dfrac{0}{-2} = 0$

Note. To check the answer (quotient) in division, multiply the quotient by the divisor to see whether you obtain the dividend.

USING RECIPROCALS IN DIVIDING SIGNED NUMBERS

Recall that, for every nonzero real number a, $a \neq 0$, there is a unique real number called the *reciprocal* or *multiplicative inverse* of a such that $a \cdot \dfrac{1}{a} = 1$.

Using the reciprocal of a number, we can demonstrate the following relationship between multiplication and division.

For every real number a and every nonzero real number b, $b \neq 0$, a divided by b is equal to a multiplied by the reciprocal of b.

$$\frac{a}{b} = a \cdot \frac{1}{b}, \ b \neq 0$$

The statements for the proof of this sentence are shown in exercise 33 on page 55.

Rule. To divide a real number (the dividend) by a nonzero real number (the divisor), multiply the dividend by the reciprocal of the divisor.

~~~~~~~~~~~~~~ **MODEL PROBLEMS** ~~~~~~~~~~~~~~

In 1–3, perform the indicated division.

1. $\dfrac{-35}{-5} = (-35)\left(-\dfrac{1}{5}\right) = +(35)\dfrac{1}{5} = +7$ **2.** $\dfrac{0}{-6} = (0)\left(-\dfrac{1}{6}\right) = 0$

3. $(+27) \div (-\tfrac{1}{3}) = (+27)(-3) = -(27)(3) = -81$

Exercises

In 1–6, name the reciprocal of the given number.

1. 8   **2.** -7 **3.** 1   **4.** $\frac{1}{4}$ **5.** $-\frac{3}{2}$ **6.** y $(y \neq 0)$

In 7–24, perform the indicated division.

7. $\dfrac{-12}{-4}$ **8.** $\dfrac{-10}{+5}$ **9.** $\dfrac{-16}{+4}$ **10.** $\dfrac{-36}{-6}$ **11.** $\dfrac{+40}{+10}$ **12.** $\dfrac{0}{-2}$

13. $\dfrac{+8.4}{-4}$ **14.** $\dfrac{-.25}{+5}$ **15.** $\dfrac{-9.6}{-.3}$ **16.** $\dfrac{-3.6}{+1.2}$ **17.** $\dfrac{+.4}{-.8}$ **18.** $\dfrac{0}{+.7}$

19. $(+50) \div (+10)$ **20.** $(-14) \div (-7)$ **21.** $(+48) \div (-12)$

22. $(-64) \div (+16)$ **23.** $(-12) \div (+\frac{2}{3})$ **24.** $(+\frac{6}{8}) \div (-\frac{1}{4})$

25. Given the fraction $\dfrac{6}{y-x}$. Which of the following substitutions leads to an impossible operation? (1) 1 for x, -1 for y (2) 2 for x, 3 for y (3) -2 for x, -3 for y (4) 3 for x, 3 for y

In 26–29, give the multiplicative inverse of the expression and state the value of x for which the multiplicative inverse is not defined.

26. $x - 7$ **27.** $x + 5$ **28.** $3x - 1$ **29.** $2x + 1$

In 30–32, state whether or not the set of numbers is closed under division (division by zero is excluded). If your answer is *no*, give an example to show that the quotient of two members in the set is not a unique member of the set.

30. positive integers   **31.** even integers   **32.** rational numbers

33. Following are the statements in the proof of the sentence "If a and b are real numbers, and $b \neq 0$, then $\dfrac{a}{b} = a \cdot \dfrac{1}{b}$." Give a reason that justifies each statement in the proof.

Statements

1. If $\dfrac{a}{b} = c$, then $bc = a$.

2. If b is a real number, then $\dfrac{1}{b}$ is a real number.

3. $a \cdot \dfrac{1}{b} = a \cdot \dfrac{1}{b}$ 6. $= 1 \cdot c$

                                  7. $= c$

4. $= bc\left(\dfrac{1}{b}\right)$ 8. $a \cdot \dfrac{1}{b} = \dfrac{a}{b}$

5. $= \left(b \cdot \dfrac{1}{b}\right)c$ 9. $\dfrac{a}{b} = a \cdot \dfrac{1}{b}$

34. State whether the following sentences are *true* or *false*:

    a. $(18 + 6) \div 3 = (18 \div 3) + (6 \div 3)$

    b. $[(+35) - (-15)] \div (+5) = [(+35) \div (+5)] - [(-15) \div (+5)]$

35. Prove that the operation of division is distributive over addition. That is, $(x + y) \div z = x \div z + y \div z$ for all real numbers x, y, and z, $z \neq 0$.

36. Prove that the operation of division is distributive over subtraction. That is, $(x - y) \div z = x \div z - y \div z$ for all real numbers x, y, and z, $z \neq 0$.

37. Prove that the reciprocal of $\dfrac{x}{y}$ is $\dfrac{y}{x}$ for all nonzero real numbers x and y.

8. Understanding the Meanings of Important Mathematical Terms

BASE, EXPONENT, POWER

The product $2 \times 2 \times 2 \times 2 \times 2 \times 2$ may be written 2^6 to show that the same number, 2, is a factor 6 times. The value of 2^6 is 64. In $2^6 = 64$, 2 is the **base**, 6 is the **exponent** of the base, and 64 is the sixth **power** of the base.

$$\text{The Sixth Power of 2}$$
$$2^6 = 2 \times 2 \times 2 \times 2 \times 2 \times 2 = 64$$

$$\text{base} \longrightarrow \overset{\textstyle 6 \longleftarrow \text{exponent}}{2} = 64 \longleftarrow \text{power}$$

For any real number b, the powers of b are defined as follows:

first power: $b^1 = b$

second power: $b^2 = b \cdot b$, read "b-squared," or "b-second," or "b to the second."

third power: $b^3 = b \cdot b \cdot b$, read "b-cubed," or "b-cube," or "b-third," or "b to the third."

fourth power: $b^4 = b \cdot b \cdot b \cdot b$, read "$b$-fourth," or "$b$ to the fourth."

nth power: $b^n = b \cdot b \cdot b \cdots$ (n times), read "b-nth," or "b to the nth."

In general, if n is a positive integer more than 1, then b^n, the nth power of b, represents the product of n factors, each factor equal to b.

    In b^n: b is the base, n is the exponent, and b^n is the power.

For example, in 10^3, the base is 10, the exponent is 3, and 10^3 or 1000 is the power.

Caution. Distinguish between expressions such as $5y^2$ and $(5y)^2$. The expression $5y^2 = 5(y \cdot y)$. In $(5y)^2$, the entire expression is the second power of $5y$; that is, $(5y)^2 = (5y)(5y) = 25(y \cdot y) = 25y^2$.

TERM AND EXPRESSION

Understanding the meanings of *term* and *expression* and the distinction between them is important to an understanding of many of the most important ideas in mathematics. Let us begin by noting that an expression such as $6xy + \dfrac{5}{x}$, which does not contain parentheses, consists of two terms, one of which is $6xy$ and the other is $\dfrac{5}{x}$.

A *term* may be a number, a variable, a product of numbers and variables, a quotient of numbers and variables, or a combination of products and quotients.

Examples of terms are 6, x, $6xy$, $\dfrac{5}{x}$, $\dfrac{2z}{3v}$, and $.75x^2y$.

An *expression* may be one term, or the sum or difference of terms.

Examples of expressions are 6, $6xy$, $\dfrac{5}{x}$, $6xy + \dfrac{5}{x}$, and $6xy - 6$.

FACTOR OF A TERM

A *factor of a term* is any one of the numbers or variables whose product forms the term, or any product of these numbers or variables.

For example, the factors of $6xy$ are 1, 2, 3, 6, x, y, $2x$, $2y$, $3x$, $3y$, $6x$, $6y$, xy, $2xy$, $3xy$, and $6xy$. Note that the factors of 6 are $1, 2, 3,$ and 6 for the reason that, in factoring an integer such as 6, factoring is restricted to integers. Otherwise, any number except 0 could be a factor of 6; for example, $\frac{1}{2}$ and 12.

A *literal factor of a term* is the product of all the literal or variable factors. For example, the literal factor of $6xy$ is xy; the literal factor of $25c^2d$ is c^2d.

COEFFICIENT

In any product consisting of two factors, each factor is the *coefficient* of the other factor. Hence, if $6xy$ is separated into the factors 6 and xy, then 6 is the numerical coefficient of xy. In future problems, a numerical coefficient of a literal factor shall be referred to simply as the coefficient. In a term such as x^2y, the coefficient is understood to be 1, the multiplicative identity.

LIKE TERMS

Like terms or *similar terms* are terms having the identically same literal factor.

For example, $6x^2y$ and $-2x^2y$ are like terms, but $6x^2y$ and $-2xy^2$ are not like terms.

MONOMIAL

A monomial may be a number, or a variable, or the product of numbers and variables.

For example, 6, x, $6xy$, and $.75x^2y$ are monomials.

A *monomial in one variable* is of the form ax^n where n is a positive integer or zero. The degree of a monomial in one variable is n if the coefficient a is not equal to zero. The degree of a monomial that is a number is 0.

For example, the monomial $5x^2$ is a monomial in x of degree 2; the monomial $\frac{1}{4}y^5$, or $\frac{y^5}{4}$, is a monomial in y of degree 5. Since 15 may be written as $15x^0$, as will be shown in a later chapter, the monomial 15 is of degree 0. The number 0 has no degree since it may be written as $0x^2$, $0x^3$, and so on. By definition, $x = x^1$, a monomial of degree 1.

The *degree of a monomial* in more than one variable is the sum of the exponents of the variables. For example, the degree of the monomial $5x^2y^3z$ is $2 + 3 + 1$, or 6.

POLYNOMIAL

A *polynomial* may be a monomial, or the sum or difference of monomials. For example, 4, $7y + 4$, and $y^2 - 7y + 4$ are polynomials.

A *binomial* is a polynomial of two terms. For example, $7y^3 + 4$ is a binomial. Also, $\frac{x^2}{3} + \frac{x}{2}$, or $\frac{1}{3}x^2 + \frac{1}{2}x$, is a binomial.

A *trinomial* is a polynomial of three terms. For example, $x^3 - 4x + 10$ and $y^2 - 7y + 4$ are trinomials.

The *degree of a polynomial* is the degree of the monomial term of highest degree. For example, the degree of $x^2 + 5x$ is 2; the degree of $x^2 + 5xy^2$ is 3, since the degree of $5xy^2$ is 3.

Exercises

In 1–6, state the factors of the product.

1. cd **2.** $4m$ **3.** $7t$ **4.** $6mn$ **5.** $20cd$ **6.** $7xy$

In 7–12, state the base and exponent of the power.

7. c^2 **8.** e^3 **9.** x^4 **10.** 8^2 **11.** 3^5 **12.** 10^8

In 13–18, state the coefficient and the exponent.

13. $6x^2$ **14.** $2d^4$ **15.** $\frac{2}{3}t^2$ **16.** m^3 **17.** $6y^5$ **18.** $\frac{1}{8}e^3$

In 19–23, write the expression using exponents.

19. $x \cdot x \cdot x \cdot x$ **20.** $e \cdot e \cdot e \cdot e \cdot e$ **21.** $7 \cdot y \cdot y \cdot y$

22. $x \cdot x \cdot y \cdot y \cdot y$ **23.** $3 \cdot r \cdot r \cdot r \cdot s$

In 24–29, write the expression without using exponents.

24. m^3 **25.** x^4y^2 **26.** $4x^2$ **27.** $4c^2d^3$ **28.** $10x^3y^2z$ **29.** $(4x)^2$

In 30–41, find the value of the power.

30. 6^2 **31.** 2^4 **32.** 5^3 **33.** 10^5 **34.** 12^2 **35.** 1^6

36. $(\frac{1}{2})^4$ **37.** $(-2)^5$ **38.** $(-1)^4$ **39.** $(-1)^{13}$ **40.** $(-3)^4$ **41.** $(-.3)^4$

In 42–47, state the degree of the polynomial and tell whether the polynomial is a monomial, a binomial, or a trinomial.

42. $5x + 7$ **43.** $8y$ **44.** $a^2 + 4a - 5$

45. $x^2 + y^2 + z^2$ **46.** $6a - 7b$ **47.** $5x^2y^3z + 10x^2y^2z$

9. Expressing Verbal Phrases Algebraically

The verbal phrase "one more than twice a number" may be expressed algebraically as $2n + 1$, using n as a variable to represent "a number." In $2n + 1$, if the domain or replacement set of the variable n is the set of integers, then n is a *placeholder* for any integer. This means that n may be replaced by any number in the set of integers.

〜〜〜〜〜〜〜〜〜〜 *MODEL PROBLEMS* 〜〜〜〜〜〜〜〜〜〜

In 1–7, express the verbal phrase algebraically.

1. x increased by 5   *Ans.* $x + 5$

2. y decreased by 4   *Ans.* $y - 4$

3. t multiplied by 40   *Ans.* $40t$

4. d divided by r *Ans.* $d \div r$ or $\dfrac{d}{r}$

5. 10 times t, increased by u *Ans.* $10t + u$

6. 2 less than $3x$ *Ans.* $3x - 2$

7. twice the sum of x and y *Ans.* $2(x + y)$

Exercises

In 1–16, express the verbal phrase algebraically.
1. the sum of (*a*) *c* and *d*   (*b*) *x* and 2   (*c*) 3*a* and 2*b*
2. the difference between (*a*) *x* and *y*   (*b*) *m* and 1   (*c*) 6*x* and 4*y*
3. the product of (*a*) *m* and *n*   (*b*) 100 and *h*   (*c*) 5*r* and *s*
4. the quotient of (*a*) *s* and *t*   (*b*) *d* and 20   (*c*) 3*p* and 2*q*

5. *b* plus 5   6. 6 more than *x*   7. *x* more than *y*
8. *f* increased by *g*   9. *t* decreased by 7   10. 9 decreased by *w*
11. 6 less than *x*   12. *d* subtracted from 7   13. 8 times *x*
14. the product of 3*a* and *b*   15. 6 divided by *x*   16. *A* divided by *L*

In 17–22, using the letter *n* to represent the variable "a number," write the phrase as an algebraic expression.
17. the number increased by 10   18. the number decreased by 7
19. two-thirds of the number   20. 25% of the number
21. 6 more than, 4 times the number   22. 20 less than, 5 times the number

In 23–27, represent the phrase as an algebraic expression using *h* to represent the greater number and *b* to represent the smaller number.
23. the product of the two numbers
24. three times the sum of the two numbers
25. twice the greater number, increased by twice the smaller number
26. the sum of twice the greater and half the smaller
27. twice the greater, decreased by 3 times the smaller

28. A coat costs $60. Represent the cost of *c* coats.
29. A pencil costs *c* cents. Represent the cost of a dozen pencils.
30. Represent the cost of *x* oranges at *y* cents each.
31. Bill is *a* years old now. Represent his age 9 years from now.
32. Arthur is *n* years old now. Represent his age 5 years ago.
33. Margaret is 20 years old. Represent her age *y* years ago.
34. Helen is *y* years old. Her brother Jack is twice as old as Helen. Represent Jack's age *d* years ago.
35. Walter is *t* years old. His mother is 25 years older than he is. Represent his mother's age 10 years ago.
36. The length of a rectangle is *s* feet. Represent the width of the rectangle if it exceeds twice the length by 4 feet.
37. Represent the diameter of a circle if its radius is represented by *r* yards.

In 38–43, express algebraically the number of:
38. inches in *f* feet   39. yards in *f* feet   40. cents in *n* nickels
41. cents in *q* quarters   42. ounces in *p* pounds   43. days in *w* weeks

44. If n represents the smallest of three consecutive odd integers, express their sum in terms of n.

45. If N represents an even integer, which of the following represents an odd integer? (1) $3N$ (2) N^3 (3) $3N + 2$ (4) $3N + 1$

46. If n is an odd integer, which of the following represents an even integer? (1) $3n$ (2) n^3 (3) $n^2 + 1$ (4) $n + 2$

47. The units digit of a two-digit number is a and the tens digit is b. Represent the number in terms of a and b.

48. If the tens digit of a two-digit number is 5 and the units digit is represented by x, the number may be represented by (1) $5x$ (2) $50x$ (3) $50 + x$

49. A man bought n articles, each of which cost d cents. Represent the number of cents he spent.

50. A boy bought n articles, each of which cost c cents. Express, in cents, his change from a five-dollar bill.

51. If n pencils cost c cents, represent the cost of r pencils in cents.

52. If r apples cost t cents, represent the cost of m of these apples.

53. A boy was x years old 5 years ago. Represent his age 7 years from now.

54. A girl will be y years old 3 years from now. Represent her age 4 years ago.

10. Order of Operations in Evaluating Numerical Expressions

PARENTHESES MAY BE USED TO CHANGE THE ORDER OF OPERATIONS

To find the value of $(2 + 3) \times 4$, do the addition first, since it is within parentheses. Hence, the value of $(2 + 3) \times 4$ is 5×4 or 20. Similarly, to find the value of $10 - (4 + 3)$, do the addition first, since it is within parentheses. Hence, the value of $10 - (4 + 3)$ is $10 - 7$, or 3.

To find the value of a numerical expression, the order of operations that has been agreed upon by mathematicians must be observed. The correct order of operations is the order set forth in the following procedure:

Procedure. To find the value of a numerical expression:
1. Evaluate within parentheses or within other symbols of grouping.
2. Evaluate the powers and the roots.
3. Evaluate the multiplications and divisions in order from left to right.
4. Evaluate the additions and subtractions in order from left to right.

〜〜〜〜〜〜〜〜〜〜 **MODEL PROBLEMS** 〜〜〜〜〜〜〜〜〜

1. Evaluate: $8 + 4 \times 7$

Solution: Multiply before adding.

$$8 + 4 \times 7 = 8 + 28 = 36 \quad Ans.$$

2. Evaluate: $27 \div 3^2 \times 2$

Solution: Evaluate the power first.

$$27 \div 3^2 \times 2 = 27 \div 9 \times 2$$

Operate in order, from left to right, doing the division first.

$$= 3 \times 2 = 6 \quad Ans.$$

3. Evaluate: $12 - \sqrt{9} + 5$

Solution: Evaluate the root first.

$$12 - \sqrt{9} + 5 = 12 - 3 + 5$$

Operate in order, from left to right, doing the subtraction first.

$$= 9 + 5 = 14 \quad Ans.$$

4. Evaluate: $5(6 - 7)^3 + (-3)^2(-5)$

Solution: Evaluate within parentheses.

$$5(6 - 7)^3 + (-3)^2(-5) = 5(-1)^3 + (-3)^2(-5)$$

Evaluate the powers.

$$= 5(-1) + 9(-5)$$

Evaluate the multiplications; then add.

$$= (-5) + (-45) = -50 \quad Ans.$$

〜〜〜〜〜〜〜〜〜〜〜〜〜〜〜〜〜〜〜〜〜〜〜〜〜〜

Exercises

In 1–24, find the value of the numerical expression.

1. $6 + 5 \times 9$	**2.** $12 + 9 \div 3$	**3.** $30 - 15 \div 5$
4. $13 \times 3 - 2$	**5.** $81 \div 9 - 6$	**6.** $7 \times 5 - 9 \times 3$
7. $20 + 20 \div 5 + 5$	**8.** $15 + 3 \times 2 - 8$	**9.** $(3 + 2) \times 4$

10. $(18 - 12) \div 3$ **11.** $(10 - 5) + 6$ **12.** $10 - (10 \div 10)$

13. $2(8 + 6) - 4$ **14.** $2(7 + 3)(7 - 3)$ **15.** $5(3)^2 + 2$

16. $12 + (4 + 5)^2$ **17.** $12 + (-2)(-3)$ **18.** $7 - 3(4 - 6)$

19. $(-5)(+3) + (-2)(-6)$ **20.** $16 - (-3)^2$

21. $3(+5)^2 - 2(-4)^2$ **22.** $20 - 2^3(-3)^2$

23. $20(-1)^2 - 2^2(-3)^3$ **24.** $30 - \sqrt{169 - 25}$

11. Evaluating an Algebraic Expression

To evaluate an algebraic expression means to find the number that the expression represents for given values of its variables.

~~~~~~~~~~~ *MODEL PROBLEMS* ~~~~~~~~~~~

1. Evaluate $6x - 3y$ when $x = 5$ and $y = -4$.

2. Evaluate $c^2 - 2d^2$ when $c = -5$ and $d = 4$.

3. Evaluate $4x^2 + 3x - 5$ when $x = -2$.

How To Proceed	*Solution*	*Solution*	*Solution*
1. Replace the variables by their given values.	$6(5) - 3(-4)$	$(-5)^2 - 2(4^2)$	$4(-2)^2 + 3(-2) - 5$
2. Follow the correct order of operations.	$30 + 12$ 42 *Ans.*	$25 - 2(16)$ $25 - 32$ -7 *Ans.*	$4(4) + 3(-2) - 5$ $16 - 6 - 5$ $10 - 5$ 5 *Ans.*

Exercises

In 1–32, find the value of the expression when $a = 1$, $b = -2$, $c = -3$, $d = 3$, $x = 4$, $y = 5$, and $z = -1$.

1. $5a$ **2.** bx **3.** $2abz$ **4.** $\frac{1}{2}x$

5. $\frac{2}{3}b$ **6.** $\frac{1}{9}bdy$ **7.** $-d^2$ **8.** $(-d)^2$

9. y^3 **10.** $2y^2$ **11.** $-3x^2$ **12.** $4d^3$

13. $\frac{1}{2}x^2$ **14.** $\frac{2}{3}d^3$ **15.** $-\frac{1}{5}y^3$ **16.** $3y - b$

17. $b - 4z$ **18.** $5x + 2y$ **19.** $y^2 - 4y$ **20.** $a^2 - b^2$

21. $b^3 - z^3$ **22.** $2y^2 - y$ **23.** $7y - y^2$ **24.** $5z - 3z^2$

25. $x^2 + 3x + 5$ **26.** $a^2 - 5a - 3$ **27.** $-a^2 + 4a + 6$ **28.** $-b^2 + 3b - 2$
29. $15 + 5z - z^2$ **30.** $2x^2 - 5x + 15$ **31.** $4b^2 - 2b + 3$ **32.** $-2c^2 + 5c - 6$

In 33–36, find the value of the expression when $a = 12$, $b = -6$, and $c = -4$.

33. $\dfrac{a - b}{a + b}$ **34.** $\dfrac{3a + 2b}{3c}$ **35.** $\dfrac{3a - 2b}{3b + a}$ **36.** $\dfrac{a^2 + 3b^2}{c^2}$

In 37–42, find the value of the expression when $x = 8$, $y = 5$, and $z = -2$.
37. $2(x + y)$ **38.** $3(5y - 2z)$ **39.** $\frac{1}{2}x(y + 3z)$
40. $3y - (x - z)$ **41.** $2(x + z) - 5$ **42.** $4z - 5(z - x)$

12. Expressing Verbal Statements as Formulas

A *formula* is an equation that expresses one variable in terms of other variables. For example, the formula $A = bh$ expresses the variable A in terms of the variables b and h.

〰〰〰〰〰〰〰〰〰 *MODEL PROBLEM* 〰〰〰〰〰〰〰〰〰

Write a formula for the relationship: The surface, S, of a sphere is equal to the product of 4π and the square of the radius, r.

Solution: $S = 4\pi r^2$ *Ans.*

〰〰〰〰〰〰〰〰〰〰〰〰〰〰〰〰〰〰〰〰〰〰〰〰〰〰〰〰〰〰〰〰

Exercises

In 1–11, write the formula that expresses the stated relationship.
1. The selling price of an article, s, is equal to its cost, c, plus the profit, p.
2. The median of a trapezoid, M, is equal to one-half the sum of the bases, b and c.
3. The number of diagonals, d, that can be drawn from one vertex of a polygon to all the other vertices is three less than the number of sides of the polygon, n.
4. The average, M, of three numbers a, b, c is their sum divided by 3.
5. The number of degrees, d, in the central angle of a regular polygon is 360 divided by the number of sides of the polygon, n.
6. The volume of a rectangular solid, V, is equal to the product of the length, l, width, w, and height, h.
7. The surface of a cube, S, is equal to six times the square of its edge, e.
8. The volume of a sphere, V, is equal to four-thirds of the product of π and the cube of its radius, r.

9. The geometric mean, G, between two numbers, a and b, is equal to the square root of their product.
10. The Fahrenheit temperature, F, is 32° more than nine-fifths of the centigrade temperature, C.
11. The distance, s, which a body will fall from rest is one-half the product of the gravitational constant, g, and the square of the time, t.
12. Represent the cost, C, in cents, of placing an ad of n words, n being greater than 10, if the cost of the first 10 words is a cents and each additional word costs b cents.
13. Represent the cost, C, in cents, of a telephone conversation lasting 8 minutes if the charge for the first 3 minutes is x cents and the cost for each additional minute is y cents.
14. Represent the charge, C, in dollars, for renting a video for 12 days if the cost for the first 3 days is a dollars and the cost for each additional day is b dollars.

13. Evaluating the Subject of a Formula

In a formula, the variable that is expressed in terms of the remaining variables is called the **subject of the formula.** We can evaluate the subject of a formula, that is, find its value, when the values of the remaining variables in the formula are given.

◇◇◇◇◇◇◇◇◇◇ *MODEL PROBLEMS* ◇◇◇◇◇◇◇◇◇◇

1. Evaluate

$$S = \frac{n}{2}(a + l)$$

when $n = 9$, $a = -2$, $l = 22$.

2. Evaluate

$$S = \frac{n}{2}[2a + (n-1)d]$$

when $a = 5$, $n = 10$, $d = 7$.

How To Proceed	*Solution*	*Solution*
1. Replace the variables by their given values.	$S = \frac{9}{2}(-2 + 22)$	$S = \frac{10}{2}[2(5) + (10-1)7]$
2. Follow the correct order of operations.	$= \frac{9}{2}(20)$	$= 5(10 + 63)$
	$= 90$ *Ans.*	$= 5(73) = 365$ *Ans.*

Exercises

1. Using the formula $P = 2l + 2w$, find P when $l = 13$ and $w = 5$.
2. If $A = p + prt$, find A when $p = 500$, $r = .06$, and $t = 8$.
3. If $C = \frac{5}{9}(F - 32)$, find C when $F = -40$.
4. If $A = \pi r^2$, find A when $r = 20$ and $\pi = 3.14$.
5. Using the formula $V = \frac{1}{3}\pi r^2 h$, find V when $\pi = \frac{22}{7}$, $r = 3$, and $h = 2.1$.
6. If $V = \frac{4}{3}\pi r^3$, find V when $\pi = \frac{22}{7}$ and $r = 21$.
7. If $S = \dfrac{a}{1 - r}$, find S when $a = 4$ and $r = \frac{1}{2}$.
8. If $F = \dfrac{9}{5}C + 32$, find F when $C = -30°$.
9. If $v = \dfrac{MV}{M + m}$, find v when $M = 1.5$, $V = 600$, and $m = 28.5$.
10. If $C = \dfrac{nE}{R + nr}$, find C when $n = 50$, $E = 1.3$, $R = 400$, and $r = 5$.
11. If $S = \dfrac{n}{2}[2a + (n - 1)d]$, find S when $a = -4$, $n = 8$, and $d = 9$.

14. Adding Like Monomials or Like Terms

We have learned that we can transform $7x + 2x$ by using the distributive property of multiplication as follows:

$$7x + 2x = (7 + 2)x = 9x$$

Since the distributive property of multiplication guarantees that $7x + 2x = 9x$ is true for every replacement of the variable x, $7x + 2x$ and $9x$ are called *equivalent expressions.*

The previous example illustrates the following:

Procedure. To add like monomials:
1. Find the sum of their coefficients.
2. Multiply the sum by the common literal factor.

When we add like terms, we say that we *combine like terms.*
The indicated sum of two unlike terms cannot be expressed as a single term. For example, $4x + 5y$ and $2ac - bd$ cannot be simplified.

~~~~~~~~~~~~~~~~~~~ **MODEL PROBLEMS** ~~~~~~~~~~~~~~~~~~

In 1–6, add.

1. $+8a$	**2.** $-4x^2$	**3.** $+15x^2y$	**4.** $-9abc$	**5.** $-8x^a$	**6.** $-5(a+b)$
$\underline{+3a}$	$+6x^2$	$-3x^2y$	$+8abc$	$\underline{+7x^a}$	$-6(a+b)$
$+11a$	$\underline{-8x^2}$	$\underline{-x^2y}$	$\underline{+abc}$	$-x^a$	$+8(a+b)$
	$-6x^2$	$+11x^2y$	0		$\underline{-3(a+b)}$

~~~~~~

Exercises

In 1–9, add.

1. $+6a$
$\underline{+4a}$

2. $-3b$
$-b$
$\underline{-7b}$

3. $+6x^2$
$+9x^2$
$\underline{-4x^2}$

4. $-12y^2$
$+\ 8y^2$
$\underline{+\ 4y^2}$

5. $-1.5x$
$+.4x$
$\underline{-1.1x}$

6. $-8cd$
$+5cd$
$\underline{+3cd}$

7. $+6xyz$
$-4xyz$
$\underline{-7xyz}$

8. $+8z^b$
$-2z^b$
$\underline{-3z^b}$

9. $-7(m+n)$
$-9(m+n)$
$\underline{+8(m+n)}$

In 10–15, combine like terms.

10. $(+9y)+(-3y)$

11. $(-4x^2)+(-8x^2)$

12. $8r+5r$

13. $10t+(-8t)$

14. $6b^2+(-4b^2)+(-2b^2)$

15. $(9d)+(-d)$

15. Adding Polynomials

A *polynomial in x* is an expression that can be formed from the variable x and numerical coefficients using only the operations of addition, subtraction, and multiplication. All exponents of the variable must be positive integers. The variable is never in the denominator of a fraction. For example, $5x^2 - 2x + 7$ is a polynomial in x; however, $\dfrac{8}{x} + 7$ is not a polynomial because the variable x is in the denominator of a fraction.

The degree of a polynomial in one variable is the same as the greatest exponent that appears in it. For example, $3x^2 + 7x + 4$ is a polynomial of degree 2; $9x - 24$ is a polynomial of degree 1.

The polynomials $3x^2 + 7x + 4$ and $9x - 24$, or $9x + (-24)$, are written in **standard form**. The standard form of a polynomial of the second degree in x, such as $3x^2 + 7x + 4$, is $ax^2 + bx + c$, $a \neq 0$. The standard form of a polynomial of the first degree in x, such as $9x - 24$, is $ax + b$, $a \neq 0$. Note that when a polynomial in one variable is written in standard form, the term having the greatest exponent of the variable is written first and the remaining terms are written in descending powers of the variable.

The polynomial $x^2 + 2xy + 3y^2$ is a polynomial of the second degree in two variables. As it is written, $x^2 + 2xy + 3y^2$ is arranged in descending powers of x and in ascending powers of y.

A polynomial is in **simple form** if there are no like terms in it. A polynomial such as $5x^2 + 3x - x^2 - 2x + 10$ can be put in simple form by **combining like terms** or **collecting like terms**. The resulting polynomial, $4x^2 + x + 10$, is in simple form.

To add two polynomials, we use the commutative, associative, and distributive properties of real numbers to combine like terms. For example, let us add $3x + 2$ and $5x + 4$:

Statements	*Reasons*
1. $(3x + 2) + (5x + 4)$ $= (3x + 5x) + (2 + 4)$	1. Rearranging terms by means of the commutative and associative properties.
2. $= (3 + 5)x + (2 + 4)$	2. Distributive property.
3. $= 8x + 6$	3. Substitution.

This example illustrates the following:

Procedure. To add polynomials, combine like terms. For convenience, arrange the polynomials in descending or ascending powers of a particular variable so that like terms are arranged in vertical columns. Then, add each column separately.

Addition may be checked by adding again in the opposite direction.

〰〰〰〰〰〰 *MODEL PROBLEMS* 〰〰〰〰〰〰

1. Add: $(5x^2 - 6x + 3) + (4x^2 + 5x - 4) + (-3x^2 - x + 8)$

Solution:
$$5x^2 - 6x + 3$$
$$4x^2 + 5x - 4$$
$$-3x^2 - x + 8$$

Answer: $6x^2 - 2x + 7$ *Check* by adding in the opposite direction.

2. Simplify: $7b + [6b + (8 - 4b)]$

Solution: When one grouping symbol appears within another grouping symbol, first perform the operation involving the expression within the innermost grouping symbol.

$$7b + [6b + (8 - 4b)] = 7b + (6b + 8 - 4b)$$
$$= 7b + (2b + 8)$$
$$= 9b + 8 \quad Ans.$$

Exercises

In 1–4, simplify the polynomials.

1. $8x + 7y + 6x + y$ **2.** $x^2 - 5x + 3 - 2x^2 + 3x - 3$

3. $c^2 - 3cd + 8d^2 - 4c + 8cd - 8d^2$ **4.** $\frac{1}{4}y^2 + \frac{2}{3}y - 5 - \frac{5}{6}y + 1\frac{1}{2}y^2 - 4\frac{1}{2}$

In 5–7, add the polynomials.

5. $5a + 6b - 7$ **6.** $y^2 - 4y - 5$ **7.** $4r^2 - 6rs + 2s^2$
          $3a - 4b + 2$ $3y^2 - 6y + 4$ $-8r^2 \quad\quad + 7s^2$
         $-4a + 3b - 4$ $-8y^2 + 10y - 7$ $-2r^2 + 6rs$

8. Combine: $8y^2 + 6y - 3$, $-4y^2 - 7y - 2$, $-9y^2 + y + 7$

In 9–14, simplify the expression.

9. $9b + (3b - 5)$ **10.** $(-4m^2 + 7) + (8m^2)$

11. $(-7a + 4) + (4a - 8)$ **12.** $(y^2 + y - 7) + (-2x^2 - 7x + 2)$

13. $(x^2 + 8x - 3) + (-4x^2 + 5)$ **14.** $9 + [3 + (7 + y)]$

15. Find the sum of $6p - 3q + z$, $-3p - z + 9q$, and $-p + q$.

16. A boy's savings for four weeks are represented by $3x + 4y$, $9y - 2x$, $5x - 3y$, and $7x - y$. Represent his total savings.

16. Subtracting Monomials or Like Terms

We can subtract monomials in the same way that we subtracted real numbers:

$$(+9) - (-2) = (+9) + (+2) = +11$$
$$(9x) - (-2x) = (+9x) + (+2x) = +11x$$
$$(17y) - (-5x) = (17y) + (+5x) = 17y + 5x$$

Procedure. To subtract one monomial from another like monomial, add the opposite (additive inverse) of the subtrahend to the minuend.

~~~~~~~~~~~~ ***MODEL PROBLEMS*** ~~~~~~~~~~~~

In 1–5, subtract the lower number from the upper number.

1. $+7a$ **2.** $+4x^2$ **3.** $-4y^2z$ **4.** $-5abc$ **5.** $-6(a+b)$
$+4a$ $+9x^2$ $+y^2z$ $-5abc$ $-4(a+b)$
$\overline{+3a}$ $\overline{-5x^2}$ $\overline{-5y^2z}$ $\overline{0}$ $\overline{-2(a+b)}$

Exercises

In 1–12, subtract the lower number from the upper number.

1. $+10b$ **2.** $+7xy$ **3.** $+5x^2y^2$ **4.** $+7(m+n)$
$+7b$ $+xy$ $+3x^2y^2$ $+5(m+n)$

5. $-8b$ **6.** $-6rs$ **7.** $-6a^2b^2$ **8.** $-7(x-y)$
$-3b$ $-9rs$ $-8a^2b^2$ $-4(x-y)$

9. $-8m$ **10.** $-8cd$ **11.** $-7cd^2$ **12.** $+4(a+b)$
$+2m$ $+3cd$ $+7cd^2$ $+7(a+b)$

In 13–18, perform the indicated subtraction.
13. $(+12x^2)-(-4x^2)$ **14.** $(-9y^2)-(-9y^2)$ **15.** $(+7ab)-(+2ab)$
16. $(-5xy)-(-12xy)$ **17.** $(-6xy^2)-(2xy^2)$ **18.** $(+8a^2b^2)-(+8a^2b^2)$

19. Subtract $-3x$ from $-9x$.
20. From the sum of $+6w^2$ and $-8w^2$, subtract the sum of $-5w^2$ and $-w^2$.

17. Subtracting Polynomials

Procedure. To subtract one polynomial from another, add the opposite (additive inverse) of the subtrahend to the minuend.

The opposite of a given polynomial is formed by writing another polynomial whose terms are the opposites of the terms of the given polynomial. For example, the opposite of $3x^2+8x-4$ is $-3x^2-8x+4$. Thus:

$$(8x^2-6x-9)-(3x^2+8x-4)=(8x^2-6x-9)+(-3x^2-8x+4)$$
$$=5x^2-14x-5$$

For convenience, as shown at the right, we can arrange the polynomials vertically so that like terms are in the same vertical column. Then we mentally add the opposite of each term of the subtrahend to the corresponding term of the minuend.

Subtract:
$8x^2-6x-9$
$3x^2+8x-4$
$\overline{5x^2-14x-5}$

MODEL PROBLEMS

In 1 and 2, subtract the lower polynomial from the upper polynomial.

1. $7y^2 - 5y$
 $-2y^2 + 8$
 $\overline{9y^2 - 5y - 8}$

2. $-4a^2 + 3ab + 8b^2$
 $3a^2 + 9ab - 4b^2$
 $\overline{-7a^2 - 6ab + 12b^2}$

Exercises

In 1–4, write the opposite (additive inverse) of the expression.

1. $7x - 4$ **2.** $-3ab - xy$ **3.** $-a^2 + a - 5$ **4.** $x^2 - 3xy + 2y^2$

In 5–14, subtract the lower polynomial from the upper polynomial.

5. $12a + 10b$
 $5a + 6b$

6. $4x + 2y$
 $3x - y$

7. $8x - 3y$
 $-4x + 8y$

8. $4c - 7d$
 $5c - 7d$

9. $x^2 - 7x + 3$
 $4x^2 - 2x - 4$

10. $3y^2 - 2y - 1$
 $-5y^2 - 2y + 6$

11. $-6ab + 2cd$
 $3ab - 4cd$

12. $-6x^2 + 3y^2$
 $x^2 - 3y^2$

13. $3x^2 - 5xy - y^2$
 $-5x^2 + 3xy - y^2$

14. $-5x^2 - 9y^2$
 $-3x^2 + 4xy + 6y^2$

In 15–22, simplify the expression.

15. $7y - (2y - 3)$

16. $8r - (-4s - 8r)$

17. $-4d - (3c - 4d)$

18. $(3x - 7) - (8 - 9x)$

19. $(2x + 3y) - (-7x - 4y)$

20. $(5x^2 + 3x - 4) - (-2x^2 + 7x)$

21. $12a - [-5 + (6a - 9)]$

22. $3x^2 - [9x - (4x - x^2) + 8]$

23. Subtract $2x^2 + 3x - 6$ from $5x^2 - 10x - 4$.

24. From the sum of $6xy - 5yz$ and $-2xy - 3yz$, subtract $-4xy + 7yz$.

25. By how much does $a + d$ exceed $a - d$?

26. Subtract $3x^2 - y^2$ from the sum of $5xy - 2x^2 + y^2$ and $3y^2 - 4xy$.

18. Multiplying a Monomial by a Monomial

MULTIPLYING POWERS THAT HAVE THE SAME BASE

Since x^2 means $x \cdot x$ and x^4 means $x \cdot x \cdot x \cdot x$,

$$x^2 \cdot x^4 = \overbrace{x \cdot x}^{2} \cdot \overbrace{x \cdot x \cdot x \cdot x}^{4} = \overbrace{x \cdot x \cdot x \cdot x \cdot x \cdot x}^{6} = x^6$$

Observe that 6, the exponent in the product, is equal to $2 + 4$, the sum of the exponents in the factors.

In general, when x is a real number and a and b are positive integers,

$$x^a \cdot x^b = x^{a+b}$$

RULE FOR MULTIPLYING POWERS OF THE SAME BASE

When multiplying powers of the same base, find the exponent of the product by adding the exponents of the factors. The base of the power which is the product is the same as the base of the factors.

Observe that the rule for multiplying powers of the same base does not apply to $x^2 \cdot y^3$ because the powers x^2 and y^3 have different bases. The product $x^2 \cdot y^3$ represents $x \cdot x \cdot y \cdot y \cdot y$, an expression which does not have five identical factors.

To multiply a monomial by a monomial, rearrange and group the factors by making use of the commutative and associative properties of multiplication. For example,

$$(7x) \cdot (3y) = (7)(x)(3)(y) = (7 \cdot 3) \cdot (x \cdot y) = 21xy$$

$$(-2x^3)(+5x^5) = (-2)(x^3)(+5)(x^5) = [(-2) \cdot (+5)][(x^3) \cdot (x^5)] = -10x^8$$

Procedure. To multiply a monomial by a monomial:
1. **Rearrange and group the factors mentally.**
2. **Multiply the coefficients.**
3. **Multiply the factors that are powers of the same base.**
4. **Multiply the products obtained in steps 2 and 3.**

〜〜〜〜〜〜〜 *MODEL PROBLEMS* 〜〜〜〜〜〜〜

In 1–8, multiply.
1. $x^5 \cdot x^2 = x^7$ 2. $c^3 \cdot c = c^4$ 3. $a^b \cdot a^{2b} = a^{3b}$
4. $10^3 \cdot 10^2 = 10^5$ 5. $r^{2x} \cdot r^3 = r^{2x+3}$ 6. $3x^2 \cdot 4x^3 = 12x^5$
7. $(+2a^3b^2)(-3ab^4c^2) = -6a^4b^6c^2$ 8. $(-4a^{3m})(+2a^2) = -8a^{3m+2}$

Exercises

In 1–26, simplify the indicated product.
1. $c^3 \cdot c^4$ 2. $x \cdot x$ 3. $x^3 \cdot x$ 4. $2^5 \cdot 2^2$
5. $10^6 \cdot 10$ 6. $c^r \cdot c^s$ 7. $x^{2r} \cdot x^{3r}$ 8. $m^{2y} \cdot m$
9. $s^{t-1} \cdot s^{t+1}$ 10. $x^3 \cdot x^2 \cdot x^5$ 11. $2^3 \cdot 2^2 \cdot 2^4$ 12. $10^2 \cdot 10^2 \cdot 10$

13. $x^a \cdot x^{2a} \cdot x$ **14.** $6(-5x)$ **15.** $x^2(3x^3)$ **16.** $(2d^2)(4d^3)$
17. $(3x)(7x)$ **18.** $(-9y)(-2y)$ **19.** $(7e^2)(9e)$ **20.** $(-m)(+2m^3)$
21. $(3x)(4x^2)(5x)$ **22.** $(-2x)(-2x)(-2x)$ **23.** $(5x)(3y)$ **24.** $(-5m)(+3n)$
25. $(7r)(2st)$ **26.** $(-a^2)(+3a)(-5a^3)$

In 27–36, multiply.
27. $-3rs$ by $+2rs$ **28.** $-7ab^2$ by $+4a^2$ **29.** $2xy$ by $-4yz$
30. $9xz$ by $-3x^2yz$ **31.** $-2b$ by $+5a^2c^2$ **32.** $-6y^b$ by $-2y^{4b}$
33. $7a^z$ by $-4a^2$ **34.** $-3x^{2b+5}$ by $-2x^{b+1}$ **35.** $7y^{a+5}$ by $-4y^{3a}$
36. $-8b^{2d-1}$ by $+3b$

In 37–39, simplify the indicated power.
37. $(x^4)^3$ **38.** $(-3a^2)^3$ **39.** $[(x^2)^3]^4$

In 40–42, simplify the indicated product.
40. $3x(4x)^2$ **41.** $(-2x)^3(3x^2)^2$ **42.** $(x^n)^4(-5x^n)^3$

43. Select the correct answer:
   $x^m \cdot x^3$ equals   (1) x^{3m} (2) $2x^{3m}$ (3) x^{m+3} (4) $2x^{m+3}$
44. The product of 4^y and 4^y is   (1) 4^{y^2} (2) 4^{2y} (3) 16^{2y}

19. Multiplying a Polynomial by a Monomial

The distributive property of multiplication states:

$$a(b+c) = ab + ac$$

For example, $x(3x+7) = (x)(3x) + (x)(7) = 3x^2 + 7x$.
This property of multiplication justifies the following:

Procedure. To multiply a polynomial by a monomial, multiply each term of the polynomial by the monomial and add the resulting products.

~~~~~~~~~~ *MODEL PROBLEMS* ~~~~~~~~~~

In 1–4, multiply.
1. $8(2x-3) = 16x - 24$
2. $-3x(x^2 - 4x + 2) = -3x^3 + 12x^2 - 6x$
3. $-5c^2d^2(4cd^2 - 2c^3) = -20c^3d^4 + 10c^5d^2$
4. $5a^{2x}(2a^{3x} - a^x) = 10a^{5x} - 5a^{3x}$

Exercises

In 1–17, multiply.

1. $8(c+d)$ **2.** $-5(x-3y)$ **3.** $10(2x-\frac{1}{2}y)$

4. $-\frac{1}{3}(6x-3y+12z)$ **5.** $8c(2c-5)$ **6.** $-5d(d^2-3d)$

7. $xy(x-y)$ **8.** $2cd(c^2+d^2)$ **9.** $-3xy(4x^2y^2-6xy)$

10. $3x^2-5x+2$ by $3x$ **11.** x^2-7x-3 by $-5x^2$

12. $9c^2-3cd+2d^2$ by $3cd$ **13.** c^2+d^2 by $-8cd$

14. $-2ab-5ac+3bc$ by $a^2b^2c^2$ **15.** y^2+y^3 by $-y^{2c}$

16. z^b-7 by z **17.** $y^{3b}+y^{2b}-y^b$ by $-5y^2$

18. Represent the area of a rectangle whose base is $\frac{2}{3}x$ and whose height is $(9x-6)$.

19. A car travels $(3y+10)$ miles per hour. Express the distance it travels in (*a*) 4   (*b*) 10   (*c*) y (*d*) h hours.

20. Multiplying a Polynomial by a Polynomial

We can find the product $(x+5)(x+2)$ as follows:

$$(x+5)(x+2) = (x+5)(x)+(x+5)(2) \qquad \text{Distributive property.}$$
$$= x^2+5x+2x+10 \qquad \text{Distributive property.}$$
$$= x^2+7x+10 \qquad \text{Combining like terms.}$$

In general, for all real numbers, a, b, c, and d,

$$(a+b)(c+d) = (a+b)c+(a+b)d$$
$$= ac+bc+ad+bd$$

Notice that each term of the first polynomial is multiplied by each term of the second polynomial. In the model problems that follow, for convenience, we use a vertical arrangement for performing the multiplication.

Procedure. To multiply a polynomial by a polynomial:

1. Arrange the multiplicand and the multiplier according to descending or ascending powers of a common variable.

2. Using the distributive property, multiply each term of the multiplicand by each term of the multiplier.

3. Add the like terms in the partial products.

~~~~~~~~~~~~~~~ *MODEL PROBLEMS* ~~~~~~~~~~~~~~~

In 1 and 2, multiply.

1. $2x + 5$	multiplicand	**2.** $a^2 - 3ab + 2b^2$
$\underline{x - 3}$	multiplier	$\underline{a - 2b}$
$2x^2 + 5x$	partial product	$a^3 - 3a^2b + 2ab^2$
$\underline{- 6x - 15}$	partial product	$\underline{- 2a^2b + 6ab^2 - 4b^3}$
$2x^2 - x - 15$	product	$a^3 - 5a^2b + 8ab^2 - 4b^3$

~~~~~~

Exercises

In 1–12, multiply.

1. $x + 2$ by $x + 3$ **2.** $5r - 1$ by $6r + 7$ **3.** $(6 - c)(5 + c)$

4. $(4x + 3)(4x - 3)$ **5.** $(a + 2b)(a + 3b)$ **6.** $(r^2 + 5)(r^2 - 2)$

7. $(x^2 + y^2)(x^2 - y^2)$ **8.** $y^2 - 3y + 4$ by $2y - 3$

9. $2x^3 - 3x + 4$ by $2x + 1$ **10.** $m^a - 2$ by $m^a + 2$

11. $x^a + y^b$ by $x^a - y^b$ **12.** $2x^a + 1$ by $3x^a - 5$

In 13–16, perform the indicated operation and simplify the result.

13. $(3x + 4)^2$ **14.** $7a(3a + 2)^2$

15. $(x^a + y^b)^2$ **16.** $(x - y)^3$

In 17 and 18, simplify the expression.

17. $(x + 5)(x - 1) - x^2$ **18.** $(x + 3)^2 - (x - 4)^2$

19. A boy worked $(3x - 2)$ hours and earned $(4x + 1)$ dollars per hour. Represent as a polynomial the total amount he earned.

21. Using Multiplication To Simplify Algebraic Expressions Containing Symbols of Grouping

USING THE DISTRIBUTIVE PROPERTY OF MULTIPLICATION

To simplify the expression $5x + 3(2x + 7)$, (that is, to transform it to an equivalent expression that does not contain parentheses), we use the distributive property of multiplication and then combine like terms:

$$5x + 3(2x + 7) = 5x + 3(2x) + 3(7)$$
$$= 5x + 6x + 21 = 11x + 21$$

USING THE MULTIPLICATION PROPERTIES OF 1 AND (-1)

Since the multiplication property of 1 states that $1 \cdot x = x$, then $1 \cdot (3x - 7) = 3x - 7$. Hence, we can simplify $4 + (3x - 7)$ as follows:

$$4 + (3x - 7) = 4 + 1 \cdot (3x - 7) = 4 + 3x - 7 = 3x - 3$$

Since the multiplication property of (-1) states that $-x = (-1) \cdot x$, then $-(5 - 3x) = -1(5 - 3x) = -5 + 3x$. Hence, we can simplify $9x - (5 - 3x)$ as follows:

$$9x - (5 - 3x) = 9x - 1 \cdot (5 - 3x) = 9x - 5 + 3x = 12x - 5$$

~~~~~~~~~~ *MODEL PROBLEMS* ~~~~~~~~~~

In 1–4, find a polynomial in simple form that is equivalent to the given expression. [In 1 and 2, use multiplication properties of 1 and (-1).]

1. $\quad 5c + (2 - 7c)$
$\quad = 5c + 1 \cdot (2 - 7c)$
$\quad = 5c + 2 - 7c$
$\quad = -2c + 2 \ Ans.$

2. $\quad 10d - (d - 2)$
$\quad = 10d - 1 \cdot (d - 2)$
$\quad = 10d - d + 2$
$\quad = 9d + 2 \ Ans.$

3. $x^2 - (x - 3)(x - 2)$
$\quad = x^2 - 1 \cdot (x - 3)(x - 2)$
$\quad = x^2 - 1(x^2 - 5x + 6)$
$\quad = x^2 - x^2 + 5x - 6$
$\quad = 5x - 6 \ Ans.$

4. $x[x^2 - 4(x - 1)]$
$\quad = x[x^2 - 4x + 4]$
$\quad = x^3 - 4x^2 + 4x \ Ans.$

~~~~~~~~~~~~~~~~~~~~~~~~~~~~~~~~~~~~~~~

Exercises

In 1–15, find a polynomial in simple form that does not contain symbols of grouping and that is equivalent to the given expression. [In 1–5, use multiplication properties of 1 and (-1).]

1. $9 + (4x - 3)$
2. $-5x + (6 - 9x)$
3. $2x^2 + (5 - 3x - 2x^2)$
4. $15 - (10 + 3c)$
5. $6x^2 - (4x^2 - 7x - 5)$
6. $3y + 4(-8y + 7)$
7. $2x(2x - 1) + 6x$
8. $9m - 3(4 + m)$
9. $5a^2 - 3a(2a - 1)$
10. $3(a + 9) + 2(a - 6)$
11. $\frac{1}{2}(4x + 3y) - 2(6x - 4y)$
12. $m^2 - (m - 6)(m - 2)$
13. $5[20 + 3(n - 1)]$
14. $5\{2x - [4 - 5(2 - x)]\}$
15. $10y - 2\{4 - 3[y - 5(2 - 3y)]\}$

22. Dividing a Monomial by a Monomial

DIVIDING POWERS THAT HAVE THE SAME BASE

Since $x^2 \cdot x^3 = x^5$, then $x^5 \div x^3 = x^2$.

Since $x^3 \cdot x = x^4$, then $x^4 \div x = x^3$. (Remember that $x = x^1$.)

Observe that the exponent in each quotient is the difference between the exponent in the dividend and the exponent in the divisor.

In general, when x is a nonzero real number and a and b are positive integers with $a > b$,

$$x^a \div x^b = x^{a-b}$$

RULES FOR DIVIDING POWERS OF THE SAME BASE

Rule 1. In dividing powers of the same nonzero real base, find the exponent of the quotient by subtracting the exponent of the divisor from the exponent of the dividend. The base of the power which is the quotient is the same as the base of the dividend and the base of the divisor.

Since $(-7x^4)(+3x^3) = -21x^7$, then $(-21x^7) \div (+3x^3) = -7x^4$. Observe that -21 divided by $+3$ equals -7, and that x^7 divided by $x^3 = x^4$.

Rule 2. The quotient of any positive integral power of a nonzero real number divided by itself is 1.

Since $1 \cdot x = x$, then $x \div x = 1$. Also, since $1 \cdot x^3 = x^3$, then $x^3 \div x^3 = 1$.

In general, when x is a nonzero real number and a is a positive integer,

$$x^a \div x^a = 1$$

Procedure. To divide a monomial by a monomial:

1. Divide their numerical coefficients.

2. Divide the factors that are powers of the same base.

3. Multiply the quotients previously obtained.

ᴡᴡᴡ MODEL PROBLEMS ᴡᴡᴡ

In 1–9, perform the indicated division.

1. $y^8 \div y^2 = y^6$

2. $b^4 \div b = b^3$

3. $x^2 \div x^2 = 1$

4. $10^5 \div 10^3 = 10^2$

5. $x^{5a} \div x^{2a} = x^{3a}$

6. $(+8y^9) \div (+2y^2) = +4y^7$

7. $(+15x^5y^4) \div (-5x^3y^2) = -3x^2y^2$

8. $(-18x^3y^2z^2) \div (-3x^2y^2) = +6xz^2$

9. $(-24a^{3b}) \div (+6a^3) = -4a^{3b-3}$

Exercises

In 1–26, divide.

1. a^7 by a^3	**2.** b^6 by b^5	**3.** c^4 by c	**4.** x^5 by x^5
5. $(-r)^8$ by $(-r)^2$	**6.** 10^5 by 10^2	**7.** 10^4 by 10^3	**8.** 4^3 by 4
9. 2^6 by 2^6	**10.** a^{5b} by a^{3b}		**11.** x^{3y} by x^{2y}
12. x^{3a} by x	**13.** y^3 by y^a		**14.** $18c^6$ by $-9c^5$
15. $-30x^3$ by $-3x$	**16.** $-5y^4$ by $-5y^4$		**17.** $36x^3y^2$ by $6xy$
18. $+56x^3y^4b$ by $-7x^3y^2$	**19.** $13a^4b^4c$ by $13ab^3$		**20.** $-28xyz$ by $-28xyz$
21. $27r^2s^2t$ by $9r^2s^2t$	**22.** $-16y^{4b}$ by $-8y^b$		**23.** $-20a^{3c}$ by $5a^3$
24. $-10b^{c+1}$ by $5b$	**25.** $5x^{3a}$ by $-x^a$		**26.** $-9x^cy^{3c}$ by xy

27. The result of dividing x^{2m} by x is (1) x^m (2) 1^{2m} (3) x^{2m-1}

28. Divide a^{3m} by a^m.

29. Divide a^{6m} by a^{2m}.

30. $4^{2x} \div 4^x$ is (1) 1^x (2) 4^2 (3) 4^x

31. How many times $-5x^2y^3$ is $+20x^6y^9$?

32. If 10 suits cost $60y$ dollars, represent the cost of one suit.

33. If the area of a rectangle is represented by $64x^6$ and the length is represented by $16x^2$, represent the width of the rectangle.

23. Dividing a Polynomial by a Monomial

Since division is distributive over addition and subtraction,

$$(8x + 6y) \div 2 = \frac{8x}{2} + \frac{6y}{2} = 4x + 3y$$

Observe that the quotient $(4x + 3y)$ is obtained by dividing each term of $(8x + 6y)$ by 2.

In general, for all real numbers a, x, and y $(a \neq 0)$,

$$\frac{ax + ay}{a} = \frac{ax}{a} + \frac{ay}{a} = x + y \quad \text{and} \quad \frac{ax - ay}{a} = \frac{ax}{a} - \frac{ay}{a} = x - y$$

The middle steps $\dfrac{ax}{a} + \dfrac{ay}{a}$ and $\dfrac{ax}{a} - \dfrac{ay}{a}$ may be done mentally.

Procedure. To divide a polynomial by a monomial:

1. Divide each term of the polynomial by the monomial.

2. Combine the resulting quotients.

~~~~~~~~~~~~~~~~ **MODEL PROBLEMS** ~~~~~~~~~~~~~~~

In 1 and 2, perform the indicated division.
1. $(10x^5 - 25x^4 + 5x^3) \div (-5x^3) = -2x^2 + 5x - 1$
2. $(24x^3y^2 - 36xy^2z) \div (6xy^2) = 4x^2 - 6z$

~~~~~~

Exercises

In 1–12, divide.

1. $12r + 27s$ by 3
2. $18x^2 - 24y^2$ by -6
3. $c^2 + c$ by c
4. $16x^2 - 8x$ by $-4x$
5. $6r^4 + 12r^3$ by $3r^2$
6. $24c^3 - 2c^2$ by $-2c^2$
7. $16a^4 - 8a^3 - 24a^2$ by $-4a^2$
8. $6a^2b - 12ab^2$ by $2ab$
9. $a^{3x} - a^{2x}$ by a^x
10. $-9x^{2a+4} - 3x^{3a+5}$ by $-3x^{2a+1}$
11. $9x^{2a+2} - 4x^2$ by x^2
12. $-8b^{2c+2} + 4b^{2c+1}$ by $-b$

In 13–18, perform the indicated division.

13. $\dfrac{8x^2 - 10x}{2x}$
14. $\dfrac{ab^2 - 2a^2b}{ab}$
15. $\dfrac{8x^2 - 16x - 4}{4}$

16. $\dfrac{x^{2a} - 3x^a}{x^a}$
17. $\dfrac{30(a+b)^2 - 50(a+b)}{2(a+b)}$
18. $\dfrac{d^3(c-d) - d^2(c-d)}{d^2(c-d)}$

19. If $20x^2 + 10x$ represents the distance traveled by a man in $10x$ hours, represent the number of miles he travels in 1 hour.

24. Dividing a Polynomial by a Polynomial

The algebraic process of dividing a polynomial by a polynomial is very much like the arithmetic process used in the division of 1087 by 25, shown at the right. Understanding the steps used in this arithmetic process will help you understand the following steps used to divide a polynomial by a polynomial.

$$43\frac{12}{25}$$
$$25\overline{\smash{)}1087}$$
$$\underline{100}$$
$$87$$
$$\underline{75}$$
$$12$$

Procedure. To divide a polynomial by a polynomial:
1. **Arrange the terms of both the divisor and the dividend according to descending or ascending powers of one variable.**
2. **Divide the first term of the dividend by the first term of the divisor to obtain the first term of the quotient.**
3. **Multiply the complete divisor by the first term of the quotient.**
4. **Subtract this product from the dividend.**
5. **Use the remainder as the new dividend.**

6. **Repeat steps 2 to 5 until the remainder is 0 or until the degree of the remainder is less than the degree of the divisor.**
7. **If the remainder is not 0, write the division as:**

$$\frac{\text{dividend}}{\text{divisor}} = \text{quotient} + \frac{\text{remainder}}{\text{divisor}}$$

8. **To check, use the principle:**
   divisor $\times$ quotient $+$ remainder $=$ dividend

MODEL PROBLEM

Divide $t^2 + 4t - 45$ by $t + 9$. Check the answer.

Solution:

$$t + 9 \;\overline{\smash{\big)}\; t^2 + 4t - 45} \quad\quad \frac{t - 5}{}$$

$$t^2 + 9t$$
$$-5t - 45$$
$$-5t - 45$$

Divide t into t^2 to get t.
Multiply $t + 9$ by t.
Subtract.
Divide t into $-5t$ to get -5;
multiply $t + 9$ by -5.
Subtract.

Check:

$t + 9$
$t - 5$
$t^2 + 9t$
$-5t - 45$
$t^2 + 4t - 45$

Answer: Quotient is $t - 5$.

Exercises

In 1–13, divide. (Check each answer.)
1. $x^2 + 8x + 12$ by $x + 2$
2. $66 + 17x + x^2$ by $6 + x$
3. $6x^2 - 13x + 6$ by $3x - 2$
4. $2a^2 - ab - 6b^2$ by $a - 2b$
5. $4a^2 - 6 + 5a$ by $4a - 3$
6. $10y^{2c} + y^c - 3$ by $5y^c + 3$
7. $3x^3 - 19x^2 + 27x + 4$ by $x - 4$
8. $6a^3 + 27a - 19a^2 - 15$ by $3a - 5$
9. $4x^3 + 7x + 5$ by $2x + 1$
10. $x^3 + 1$ by $x + 1$
11. $y^3 + 27$ by $y + 3$
12. $a^3 - 64$ by $a - 4$
13. $8b^3 - 1$ by $2b - 1$

14. Find the dividend when the quotient is $x - 3$, the remainder is -5, and the divisor is $3x - 1$.

────── *CALCULATOR APPLICATIONS* ──────

The $\boxed{(-)}$ key is used to enter the opposite of a number. Using the $\boxed{-}$ key will result in an ERROR message.

────── *MODEL PROBLEMS* ──────

1. Show the calculator entry that would be used to find the opposite of each number.

	Entry	*Display*
a. 13	*Enter:* $\boxed{(-)}$ 1 3 $\boxed{\text{ENTER}}$ or $\boxed{=}$	$\boxed{\text{- 13}}$
b. −17	*Enter:* $\boxed{(-)}$ $\boxed{(-)}$ 17 $\boxed{\text{ENTER}}$ or $\boxed{=}$	$\boxed{17}$
c. −[−(9 − 3)]	*Enter:* $\boxed{(-)}$ $\boxed{(}$ $\boxed{(-)}$ $\boxed{(}$	
	9 $\boxed{-}$ 3 $\boxed{)}$ $\boxed{)}$ $\boxed{\text{ENTER}}$ or $\boxed{=}$	$\boxed{6}$

2. From the sum of −37 and 68, subtract −94.

   Solution:

   Enter: $\boxed{(-)}$ 37 $\boxed{+}$ 68 $\boxed{-}$ $\boxed{(-)}$ 94 $\boxed{\text{ENTER}}$ or $\boxed{=}$

   Display: $\boxed{125}$

Both scientific and graphing calculators follow the order of operations. However, with a scientific calculator, it is necessary to use the $\boxed{\times}$ key for multiplication when needed, even when that sign is not shown in the algebraic expression, such as between two variables or a number and a variable. To raise to a power, use the $\boxed{x^y}$ key on a scientific calculator or the $\boxed{\wedge}$ key on a graphing calculator, followed by the numerical value of the exponent. For a second power, the $\boxed{x^2}$ key can be used on either calculator.

3. Evaluate each expression.

   a. $6(4 + 1)^2 + (8 - 5)^3$

   b. $5 \times \dfrac{(11 - 3)^4}{16}$

   Solution:

   a. Scientific Calculator

   Enter: 6 $\boxed{\times}$ $\boxed{(}$ $\boxed{(}$ 4 $\boxed{+}$ 1 $\boxed{)}$ $\boxed{x^2}$ $\boxed{+}$
   $\boxed{(}$ 8 $\boxed{-}$ 5 $\boxed{)}$ $\boxed{x^y}$ 3 $\boxed{=}$

   Display: $\boxed{177}$

   Graphing Calculator

   Enter: 6 $\boxed{(}$ $\boxed{(}$ 4 $\boxed{+}$ 1 $\boxed{)}$ $\boxed{x^2}$ $\boxed{+}$
   $\boxed{(}$ 8 $\boxed{-}$ 5 $\boxed{)}$ $\boxed{\wedge}$ 3 $\boxed{\text{ENTER}}$

   Display: $\boxed{177}$

*b. **Enter:*** 5 ☒ ☐(11 ☐− 3 ☐) x^y 4   ***Enter:*** 5 ☐(11 ☐− 3 ☐) ∧ 4 ☐÷

   ☐÷ 16 ☐=   16 ENTER

   Display: ☐ 1280 ☐   ***Display:*** ☐ 1280 ☐

On a graphing calculator an algebraic expression can be evaluated for several different values of the variable by storing the values and entering the expression to be evaluated.

4. Evaluate $3x^2 + 4x - 5$ for $x = -7$, 3, and 6.8.

   Solution:

   Enter: (−) 7 STO▶ X,T,θ,*n* ENTER
           3 X,T,θ,*n* x^2 ☐+ 4 X,T,θ,*n* ☐− 5 ENTER

   Display: ☐ 114 ☐

   To change the value for x,

   Enter: 3 STO▶ X,T,θ,*n* ENTER

   Now press 2nd ENTRY until the original expression reappears on the screen (in this case, twice). Then press ENTER and the new computed value ☐ 34 ☐ is displayed. Repeat for 6.8 to get 160.92.

Factoring and Operations on Rational Expressions

1. Factoring Polynomials Whose Terms Have a Common Factor

Because $4 \times 7 = 28$, the integer 28 is a multiple of two integral factors, 4 and 7; also $28 \div 7 = 4$, and $28 \div 4 = 7$. This example illustrates that over the set of integers:

1. A factor of an integer is an exact divisor of the integer.
2. When the product of two integers is divided by one of its factors, the quotient is the other factor.

Factoring a number is the process of finding those numbers whose product is the given number. When we factor an integer, we deal with integral factors only. For example, although $\frac{1}{2} \times 24 = 12$, we do not say that $\frac{1}{2}$ and 24 are the factors of 12.

Because $(2x + 1)(3x - 7) = 6x^2 - 11x - 7$, we say that $(2x + 1)$ and $(3x - 7)$ are factors of the polynomial $6x^2 - 11x - 7$.

Factoring a polynomial over a specified set of numbers means to express it as a product of polynomials whose coefficients are members of that set of numbers. In future problems, if no set of numbers is specified, it is to be understood the factoring of a polynomial will be with respect to the set of integers. Exceptions will be made where there are simple nonintegral coefficients, such as in the factoring of $\frac{1}{2}x + \frac{1}{2}y$, $\sqrt{2}x - \sqrt{2}y$, and $\pi x^2 + \pi y^2$.

For polynomials, it is also true that:
1. A factor of a polynomial is an exact divisor of the polynomial.
2. When the product of two polynomials is divided by one of its factors, the quotient is the other factor.

83

For example, since 5 and $x^2 + 7$ are the factors of $5x^2 + 35$, then:

1. 5 and $x^2 + 7$ are exact divisors of $5x^2 + 35$.
2. When $5x^2 + 35$ is divided by 5, the quotient is $x^2 + 7$; when $5x^2 + 35$ is divided by $x^2 + 7$, the quotient is 5.

A positive integer is called a ***prime number*** if it is greater than 1 and it has no factors other than itself and 1. For example, 2, 3, 5, 7, and 11 are the first five prime numbers.

A ***prime polynomial*** is a polynomial other than 1 that has no factors except itself and 1 with respect to a specified set of numbers. For example, $2x + 3$ is a prime polynomial with respect to the set of integers.

Using the distributive property, $10x^2(x + 3) = 10x^3 + 30x^2$. Hence, $10x^3 + 30x^2 = 10x^2(x + 3)$. The monomial $10x^2$ is the product of 10, which is the greatest common factor of the coefficients of the polynomial $10x^3 + 30x^2$, and x^2, which is the variable factor of highest degree that is a common factor of all terms of this polynomial. Therefore, we call $10x^2$ the highest common monomial factor, or H.C.F., of the polynomial $10x^3 + 30x^2$.

The ***highest common monomial factor of a polynomial, or H.C.F.***, is the product of the greatest common factor of its coefficients and the highest power of each variable that is a common factor of each of its terms.

For example, to factor $12x^2y^2 + 18xy^3$, first we obtain the H.C.F., which is $6xy^2$. Then, we divide $12x^2y^2 + 18xy^3$ by $6xy^2$ and obtain the quotient, $2x + 3y$, which is the other factor of $12x^2y^2 + 18xy^3$. Therefore, we see that by factoring, $12x^2y^2 + 18xy^3 = 6xy^2(2x + 3y)$.

Procedure. To factor a polynomial whose terms have a common monomial factor:

1. **Find the highest monomial that is a factor of each term of the polynomial.**
2. **Divide the polynomial by the monomial factor. The quotient is the other factor.**
3. **Express the polynomial as the indicated product of the two factors.**
4. **Check by multiplying the factors to obtain the polynomial.**

〜〜〜〜〜〜〜 *MODEL PROBLEMS* 〜〜〜〜〜〜〜

In 1–5, factor.

1. $3x - 3y$ *Ans.* $3x - 3y = 3(x - y)$
2. $\frac{1}{2}na + \frac{1}{2}nl$ *Ans.* $\frac{1}{2}na + \frac{1}{2}nl = \frac{1}{2}n(a + l)$
3. $35a^2 - 7a$ *Ans.* $35a^2 - 7a = 7a(5a - 1)$
4. $25a^2b - 35ab^2$ *Ans.* $25a^2b - 35ab^2 = 5ab(5a - 7b)$
5. $x^{n+2} - 5x^n$ *Ans.* $x^{n+2} - 5x^n = x^n(x^2 - 5)$

Note. In problem 2, we are factoring over the set of rational numbers. The given polynomial is prime over the set of integers.

~~~~~~

Exercises

In 1–36, factor the polynomial.

1. $6a + 6b$
2. $7l - 7n$
3. $xc - xd$
4. $5x^2 + 5y^2$
5. $bc^2 - 2b$
6. $16a + 4b$
7. $25x^2 - 15y^2$
8. $16x + x^2$
9. $5y - 15y^3$
10. $5y^4 + 5y^2$
11. $ax + ax^3$
12. $3ab^2 - 6a^2b$
13. $21c^3d^2 - 7c^2d$
14. $12x^2y^3 - 18xy^4$
15. $40xy^2 - 32x^2y$
16. $2x^2 + 8x + 4$
17. $ay - 4aw - 12a$
18. $c^3 - c^2 + 2c$
19. $ax^3 - a^2x^2 + ax$
20. $p + prt$
21. $\pi r^2 + \pi R^2$
22. $\frac{1}{2}hb + \frac{1}{2}hc$
23. $\pi r^2 + 2\pi rh$
24. $y^{n+2} + y^2$
25. $x^{a+1} - x$
26. $x^{2a+1} - x^{2a}$
27. $z^{1+2b} + z$
28. $x^{r+5} + x^3$
29. $cdy^r - dy^r$
30. $x^{2a+2} + x^{2a+1} + x^{2a}$
31. $a^{2b+1} + 4a^{b+1} + a$
32. $x(a+b) + y(a+b)$
33. $x(2b+1) + y(2b+1)$
34. $ax + ay + bx + by$
35. $xv - xw + yv - yw$
36. $mr + nr - ms - ns$

37. Write the polynomials whose factors are:
    a. 5 and $(v + 2w)$
    b. $4xy^2$ and $(5x - 3y^2)$
    c. $3x$ and $(6x^2 - 5)$

38. Use factoring to find mentally the value of:
    a. $59 \times 37 + 41 \times 37$
    b. $64 \times 81 + 64 \times 19$

2. Multiplying the Sum and Difference of Two Terms

In the examples at the right, we have found the binomial which is the product of the sum of two terms and the difference of the same two terms. The procedure that follows will enable us to find such products mentally.

$$
\begin{array}{ll}
a + b & 3x^2 + 5y \\
\underline{a - b} & \underline{3x^2 - 5y} \\
a^2 + ab & 9x^4 + 15x^2y \\
\underline{ - ab - b^2} & \underline{ - 15x^2y - 25y^2} \\
a^2 - b^2 & 9x^4 - 25y^2
\end{array}
$$

Procedure. To multiply the sum of two terms by the difference of the same two terms:

1. **Square the first term.**
2. **From this result, subtract the square of the second term.**

┌─ *KEEP IN MIND* ─┐
$$(a + b)(a - b) = a^2 - b^2$$

∿∿∿∿∿∿ *MODEL PROBLEMS* ∿∿∿∿∿∿

In 1–3, find the product mentally.

	Problem	Think	Write
1.	$(x + 9)(x - 9)$	$= (x)^2 - (9)^2$	$= x^2 - 81$ *Ans.*
2.	$(2x + 7y)(2x - 7y)$	$= (2x)^2 - (7y)^2$	$= 4x^2 - 49y^2$ *Ans.*
3.	$(a^3 - 5b^2)(a^3 + 5b^2)$	$= (a^3)^2 - (5b^2)^2$	$= a^6 - 25b^4$ *Ans.*

Exercises

In 1–16, find the product mentally.

1. $(x + 4)(x - 4)$ **2.** $(n - 9)(n + 9)$ **3.** $(6 + b)(6 - b)$
4. $(x - y)(y + x)$ **5.** $(3x + 2)(3x - 2)$ **6.** $(8 - 3x)(8 + 3x)$
7. $(3m + 7n)(3m - 7n)$ **8.** $(2 - 5y^2)(5y^2 + 2)$ **9.** $(b + \frac{1}{3})(b - \frac{1}{3})$
10. $(.7 + d)(.7 - d)$ **11.** $(rs + 6)(rs - 6)$ **12.** $(x^2 - 8)(x^2 + 8)$
13. $(2ab + 3)(2ab - 3)$ **14.** $(m + 3)(m - 3)(m^2 + 9)$
15. $(a^x + 2)(a^x - 2)$ **16.** $(2c^{3x} - 3d^y)(2c^{3x} + 3d^y)$

In 17–20, first express the factors as the sum and difference of the same two numbers. Then multiply mentally.

17. 23×17 **18.** 49×51
19. 36×44 **20.** 88×92

3. Factoring the Difference of Two Squares

An expression of the form $a^2 - b^2$ is called a ***difference of two squares.*** Since the product of $(a + b)$ and $(a - b)$ is $a^2 - b^2$, the factors of $a^2 - b^2$ are $(a + b)$ and $(a - b)$. Therefore, $a^2 - b^2 = (a + b)(a - b)$.

~~~~~~~~~~~~~~~~~~~ *MODEL PROBLEMS* ~~~~~~~~~~~~~~~

1. Factor: $x^2 - 100$

How To Proceed	Solution
1. Express each term as the square of a monomial.	$x^2 - 100 = (x)^2 - (10)^2$
2. Apply the rule: $a^2 - b^2 = (a + b)(a - b)$	$x^2 - 100 = (x + 10)(x - 10)$ *Ans.*

In 2 and 3, factor the polynomial in the set of rational numbers.

2. Factor: $25a^2 - \frac{1}{9}b^2c^2$

Solution: $25a^2 - \frac{1}{9}b^2c^2 = (5a)^2 - (\frac{1}{3}bc)^2$
$25a^2 - \frac{1}{9}b^2c^2 = (5a + \frac{1}{3}bc)(5a - \frac{1}{3}bc)$ *Ans.*

3. Factor: $x^2 - .64y^2$

Solution: $x^2 - .64y^2 = (x)^2 - (.8y)^2$
$x^2 - .64y^2 = (x + .8y)(x - .8y)$ *Ans.*

Note. The polynomials in problems 2 and 3 are not factorable in the set of integers. They are factorable in the set of rational numbers.

In 4 and 5, factor the polynomial mentally.

Problem	Think	Write
4. $a^6 - b^4$	$= (a^3)^2 - (b^2)^2$	$= (a^3 + b^2)(a^3 - b^2)$ *Ans.*
5. $x^{2a} - y^{2b}$	$= (x^a)^2 - (y^b)^2$	$= (x^a + y^b)(x^a - y^b)$ *Ans.*

~~~~~~

Exercises

In 1–21, factor the polynomial.

1. $y^2 - 64$ **2.** $x^2 - 81$ **3.** $144 - a^2$ **4.** $4x^2 - 25y^2$

5. $49a^2 - 64b^2$ **6.** $10^2 - 81d^2$ **7.** $r^2s^2 - 144$ **8.** $a^2 - .49$

9. $36 - .49d^2$ **10.** $25x^2 - \frac{1}{9}$ **11.** $\frac{1}{4}r^2 - \frac{25}{9}s^2$ **12.** $x^2y^2 - 121a^2b^2$

13. $a^4 - c^2$ **14.** $4x^6 - 9y^4$ **15.** $x^{2c} - y^{2d}$

16. $a^{2m} - b^{4n}$ **17.** $x^{2a} - 1$ **18.** $9 - a^{2c}$

19. $\frac{1}{4}y^{2d} - .81$ **20.** $9x^{4a} - 16y^{2b}$ **21.** $(x + y)^2 - z^2$

22. Write the polynomial whose factors are:

  a. $(7x + 3y)$ and $(7x - 3y)$ *b.* $(x^m + 1)$ and $(x^m - 1)$

4. Finding the Product of Two Binomials

Let us learn how to find, mentally, the product of two binomials of the form $ax + b$ and $cx + d$.

Study the multiplication of the two binomials at the right.

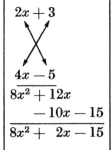

Note:
1. $8x^2$, the first term in the product, is equal to the product of $2x$ and $4x$, the first terms of the binomials.
2. -15, the last term in the product, is equal to the product of $+3$ and -5, the last terms of the binomials.
3. $+2x$, the middle term, is obtained by adding the cross-products, that is, by multiplying the first term of each binomial by the second term of the other and adding these products, $(+12x) + (-10x) = +2x$.

~~~~~~~~~~~~~ **MODEL PROBLEM** ~~~~~~~~~~~~

Multiply: $(3x - 4)(5x + 9)$

How To Proceed	*Solution*
1. Multiply the first terms.	$(3x)(5x) = \quad 15x^2$
2. Add the cross-products.	$(-20x) + (+27x) = \quad +7x$
3. Multiply the last terms.	$(-4)(+9) = -36$
4. Combine the products obtained.	*Ans.* $15x^2 + 7x - 36$

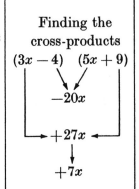

~~~~~~

Exercises

In 1–19, multiply the binomials mentally.

1. $(x + 7)(x + 2)$
2. $(a - 1)(a - 5)$
3. $(z - 2)(z - 2)$
4. $(12 - d)(4 - d)$
5. $(x + 9)(x - 2)$
6. $(5 - a)(3 + a)$
7. $(6 - m)(7 + m)$
8. $(c - 5)(3c + 1)$
9. $(5z - 7)(4z + 3)$
10. $(3x + 2)(3x + 2)$
11. $(5 - 3d)^2$
12. $(2t - 5r)(3t - r)$
13. $(7x + 5y)(7x + 5y)$
14. $(a^2 + 5)(a^2 - 2)$
15. $(2x^2 - 7)(x^2 - 9)$
16. $(x^a + 7)(x^a - 5)$
17. $(y^b - 4)(2y^b + 1)$
18. $(c^{2d} - 4)(c^{2d} - 9)$
19. $(x^{3n} + 5)(2x^{3n} - 1)$

20. Represent the area of a square whose side is represented by $(2x - 3)$.

5. Factoring Quadratic Trinomials of the Form $ax^2 + bx + c$

Because $(x+2)(x+3) = x^2 + 5x + 6$, the factors of $x^2 + 5x + 6$ are $(x+2)$ and $(x+3)$. Factoring a quadratic trinomial of the form $ax^2 + bx + c$ (a, b, and $c \neq 0$) is the reverse of multiplying binomials of the form $(ax+b)$ and $(cx+d)$. When we factor a trinomial of this form, we list the possible pairs of factors and test them one by one until we find the correct result.

For example, let us factor $2x^2 + 7x - 15$.
1. Since the product of the first terms of the binomials must be $2x^2$, one of these terms must be $2x$, the other x. We write:

$$2x^2 + 7x - 15 = (2x \quad)(x \quad)$$

2. Since the product of the last terms of the binomials must be -15, one of these last terms must be positive, the other negative. The pairs of integers whose product is -15 are $(+1)$ and (-15), (-1) and $(+15)$, $(+3)$ and (-5), (-3) and $(+5)$.
3. Because of the results obtained in steps 1 and 2, the possible pairs of factors are:

$(2x+1)(x-15)$	$(2x-1)(x+15)$
$(2x+15)(x-1)$	$(2x-15)(x+1)$
$(2x+3)(x-5)$	$(2x-3)(x+5)$
$(2x+5)(x-3)$	$(2x-5)(x+3)$

4. To discover the product in which the middle term is $+7x$, we test the middle term of each of the trinomial products. We find that only the pair $(2x-3)(x+5)$ yields a middle term of $+7x$.
5. Hence, $2x^2 + 7x - 15 = (2x-3)(x+5)$.

KEEP IN MIND

In factoring a trinomial of the form $ax^2 + bx + c$, when a is a positive integer ($a > 0$):
1. If the last term, the constant c, is positive, the last terms of the binomial factors must be either both positive or both negative.
2. If the last term, the constant c, is negative, one of the last terms in the binomial factors must be positive, the other negative.

~~~~~~~~ *MODEL PROBLEMS* ~~~~~~~~

In 1 and 2, factor.

1. $x^2 + 6x + 8$ **2.** $3x^2 - 2x - 5$

How To Proceed	*Solution*	*Solution*
	$x^2 + 6x + 8$	$3x^2 - 2x - 5$
1. Factor the first term and the last term.	$x(x)$ $(+1)(+8)$ $(+2)(+4)$ $(-1)(-8)$ $(-2)(-4)$	$(3x)x$ $(+1)(-5)$ $(-1)(+5)$
2. Write the possible pairs of binomial factors.	$(x+1)(x+8)$ $(x+2)(x+4)$ $(x-1)(x-8)$ $(x-2)(x-4)$	$(3x+1)(x-5)$ $(3x-1)(x+5)$ $(3x+5)(x-1)$ $(3x-5)(x+1)$
3. Test to find which trinomial product has the given middle term.	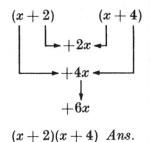 $(x+2)(x+4)$ *Ans.*	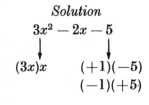 $(3x-5)(x+1)$ *Ans.*

In 3 and 4, factor.

3. $x^2 + 5xy - 6y^2 = (x + 6y)(x - y)$ *Ans.*

4. $m^4 - m^2 - 56 = (m^2 - 8)(m^2 + 7)$ *Ans.*

~~~~~~

Exercises

In 1–24, factor the trinomial.

1. $a^2 + 8a + 12$ **2.** $r^2 - 6r + 8$ **3.** $a^2 - 14a + 24$

4. $c^2 + c - 6$ **5.** $36 + 13x + x^2$ **6.** $c^2 - 16 - 6c$

7. $3x^2 - 16x + 5$ **8.** $16 - 8d + d^2$ **9.** $5x^2 - 13x + 6$

10. $12 + 7y - 10y^2$ **11.** $6y^2 + 19y - 20$ **12.** $-12 + 5x^2 - 4x$

13. $x^2 + 5xy + 4y^2$ **14.** $2a^2 - 7ab + 6b^2$ **15.** $4x^2 - 13xy + 3y^2$

16. $2x^2 - 7xy - 15y^2$ **17.** $a^4 - 7a^2 + 10$ **18.** $18 + 3y^2 - y^4$

19. $x^{2a} + 6x^a + 8$ **20.** $x^{2a} - 6x^a + 5$ **21.** $y^{2b} - 14y^b + 45$

22. $x^{2a} + 6x^a - 16$ **23.** $5x^{2a} - 11x^a - 12$ **24.** $3x^{2n} - 11x^n + 6$

6. Factoring a Polynomial Completely

To *factor a polynomial completely* means to find the *prime factors* of the polynomial with respect to a specified set of numbers. Therefore, whenever we facto a polynomial, we will continue the process of factoring until all factors other than monomial factors are prime factors with respect to the specified sets of numbers.

Procedure. To factor a polynomial completely:
1. **Factor the given polynomial. First find the H.C.F., if there is one. Then examine each factor.**
2. **Continue factoring the factors obtained in step 1 until all factors other than monomial factors are prime.**

〰〰〰〰〰〰〰〰 *MODEL PROBLEMS* 〰〰〰〰〰〰〰〰

In 1–4, factor completely.

1. Factor: $5x^3 - 45x$

Solution

$5x^3 - 45x$
$= 5x(x^2 - 9)$
$= 5x(x + 3)(x - 3)$ *Ans.*

2. Factor: $3x^2 - 6x - 24$

Solution

$3x^2 - 6x - 24$
$= 3(x^2 - 2x - 8)$
$= 3(x - 4)(x + 2)$ *Ans.*

3. Factor: $x^4 - 16$

Solution

$x^4 - 16$
$= (x^2 + 4)(x^2 - 4)$
$= (x^2 + 4)(x + 2)(x - 2)$ *Ans.*

4. Factor: $4x^{n+2} - x^n$

Solution

$4x^{n+2} - x^n$
$= x^n(4x^2 - 1)$
$= x^n(2x + 1)(2x - 1)$ *Ans.*

〰〰〰〰〰〰〰〰〰〰〰〰〰〰〰〰〰〰〰〰〰〰〰〰〰〰〰〰

Exercises

In 1–16, factor the polynomial completely.

1. $5x^2 - 5$
2. $3x^2 - 75$
3. $y - 9y^3$
4. $25x^3 - x$

5. $4 - 16x^2$
6. $xy - x^3y^3$
7. $\pi R^2 - \pi r^2$
8. $16x - x^3$

9. $ax^2 - 9a$
10. $x^3 - x$
11. $c^4 - d^4$
12. $x^3 + 7x^2 + 12x$

13. $5x^2 + 4x + x^3$
14. $x^4 + x^2 - 2x^3$
15. $x^{2a+2} - x^{2a}$
16. $ay^{2+2b} - ay^2$

17. Express as the product of three factors:

    a. $x^3 - a^2x$ *b.* $4x^{2a} - 100y^{4b}$ *c.* $by^{2x+1} - 5by^{x+1} - 6by$

7. Transforming Rational Expressions Into Equivalent Rational Expressions

Recall that a rational number, such as $\frac{5}{8}$, is expressible in the form

$$\frac{a}{b} \text{ where } a \text{ and } b \text{ are integers and } b \neq 0$$

The role that integers play in the study of rational numbers is very much like the role that polynomials play in the study of rational expressions, as can be seen from the following definition of a rational expression:

A **rational expression** is an expression that is expressible in the form

$$\frac{P}{Q} \text{ where } P \text{ and } Q \text{ are polynomials and } Q \neq 0$$

Examples of rational expressions are $\dfrac{5}{8}, \dfrac{5}{x}, \dfrac{5x}{8y}, \dfrac{x+5}{8y}, \dfrac{x^2+5x}{8-y}, \dfrac{5x^2+xy}{16y-8x}$.

Note that, since division by zero is an impossible operation, a fraction whose denominator is zero is meaningless. For example, the fraction $\dfrac{5}{x}$ has no meaning when $x = 0$. In this book, to avoid meaningless fractions, we will assume that the values of the variables are such that the denominators of rational expressions are nonzero. For example, in the case of $\dfrac{x^2+5x}{8-x}$, it is understood that $x \neq 8$; in the case of $\dfrac{x-5}{(x-3)(x+2)}$, $x \neq 3$ and $x \neq -2$.

Because of the close parallel between rational numbers and rational expressions, we shall find that the laws and properties we have used to deal with rational numbers can often be used to deal with rational expressions.

THE PRODUCT OF RATIONAL EXPRESSIONS

The example $\dfrac{3}{4} \times \dfrac{5}{7} = \dfrac{3 \times 5}{4 \times 7}$ illustrates the following rule involving the product of two rational numbers:

If a and c are real numbers, and b and d are nonzero real numbers,

$$\frac{a}{b} \cdot \frac{c}{d} = \frac{ac}{bd} \quad \text{and} \quad \frac{ac}{bd} = \frac{a}{b} \cdot \frac{c}{d}$$

The product of two rational expressions is found in the same way. For example,

$$\frac{x+5}{x-3} \cdot \frac{x+3}{x+5} = \frac{(x+5)(x+3)}{(x-3)(x+5)}$$

FUNDAMENTAL PROPERTY OF A FRACTION OR A RATIONAL EXPRESSION

Since $\dfrac{a}{b} = \dfrac{a}{b} \cdot 1$, then $\dfrac{a}{b} = \dfrac{a}{b} \cdot \dfrac{x}{x}$, or $\dfrac{a}{b} = \dfrac{ax}{bx}$.

Since $\dfrac{a}{b} = \dfrac{a}{b} \cdot 1$, then $\dfrac{a}{b} = \dfrac{a}{b} \cdot \dfrac{\frac{1}{x}}{\frac{1}{x}}$ or $\dfrac{a}{b} = \dfrac{a\left(\frac{1}{x}\right)}{b\left(\frac{1}{x}\right)}$, or $\dfrac{a}{b} = \dfrac{a \div x}{b \div x}$

Therefore, we have proved the following fundamental property of a fraction:

If the numerator and the denominator of a fraction are either multiplied or divided by the same nonzero number, the resulting fraction is equivalent to the original fraction.

Thus, $\dfrac{35}{25} = \dfrac{35 \div 5}{25 \div 5} = \dfrac{7}{5}$, and $\dfrac{\frac{1}{3}}{\frac{1}{4}} = \dfrac{\frac{1}{3} \times 12}{\frac{1}{4} \times 12} = \dfrac{4}{3}$.

Also, $\dfrac{6(x-2)}{2(x+2)} = \dfrac{6(x-2) \div 2}{2(x+2) \div 2} = \dfrac{3(x-2)}{x+2}$.

～～～～～～ *MODEL PROBLEM* ～～～～～～

Find the value of x for which $\dfrac{3x+9}{2x-8}$ is not defined.

Solution: $\dfrac{3x+9}{2x-8}$ is not defined when the denominator $2x-8$ is equal to 0.

$2x - 8 = 0$ when $x = 4$. *Ans.* 4

Exercises

In 1–3, represent the rational expression as a fraction and give the value(s) of the variable for which the fraction is not defined.

1. $9 \div 3x$ **2.** $10 \div (x - 8)$ **3.** $4y \div (3y + 1)$

In 4–8, find the value(s) of the variable for which the fraction is not defined.

4. $\dfrac{3}{x}$ **5.** $\dfrac{7}{y^2}$ **6.** $\dfrac{a^2 - 25}{2a - 1}$ **7.** $\dfrac{3m + 2}{4m + 2}$ **8.** $\dfrac{7b - 1}{b^2 - 81}$

In 9–14, replace the question mark with a number or a polynomial that will make the fractions equivalent.

9. $\dfrac{3}{5} \cdot \dfrac{7}{8} = \dfrac{3 \times 7}{5 \times ?}$ **10.** $\dfrac{c}{d} \cdot \dfrac{x}{y} = \dfrac{c \cdot ?}{d \cdot y}$

11. $\dfrac{3}{5} = \dfrac{3 \times 4}{5 \times ?}$ **12.** $\dfrac{10}{5} = \dfrac{10 \div 5}{5 \div ?}$

13. $\dfrac{4(x - 1)}{5(x - 1)} = \dfrac{4(x - 1) \div (?)}{5(x - 1) \div (x - 1)}$ **14.** $\dfrac{x + 1}{1 - x} = \dfrac{?}{x - 1}$

15. Write the fraction $\dfrac{6}{1 - x}$ as an equivalent fraction whose denominator is $x - 1$.

16. The fraction $\dfrac{b - a}{2}$ is *not* equal to (1) $-\dfrac{a - b}{2}$ (2) $\dfrac{a - b}{-2}$ (3) $\dfrac{a - b}{2}$ (4) $\dfrac{-a + b}{2}$

17. Prove: If a and c are real numbers, and b and d are nonzero real numbers, $\dfrac{ac}{bd} = \dfrac{a}{b} \cdot \dfrac{c}{d}$. $\left(Hint:\ \text{Use}\ \dfrac{ac}{bd} = (ac) \cdot \dfrac{1}{bd}.\right)$

18. Prove: If $\dfrac{a}{b}$ and $\dfrac{c}{d}$ are rational numbers, then $\dfrac{a}{b} \times \dfrac{c}{d} = \dfrac{c}{d} \times \dfrac{a}{b}$. That is, multiplication of rational numbers is commutative.

19. Prove: If $\dfrac{a}{b}, \dfrac{c}{d}$, and $\dfrac{e}{f}$ are rational numbers, then $\left(\dfrac{a}{b} \times \dfrac{c}{d}\right) \times \dfrac{e}{f} = \left(\dfrac{a}{b}\right) \times \left(\dfrac{c}{d} \times \dfrac{e}{f}\right)$. That is, multiplication of rational numbers is associative.

8. Reducing a Fraction or Rational Expression to Its Lowest Terms

A fraction is in its *lowest terms* or in its *simplest form* when its numerator and denominator have no common factor other than 1.

To reduce a fraction to its lowest terms, we make use of the division property of a fraction, $\dfrac{x}{y} = \dfrac{x \div a}{y \div a}$. For example,

$$\frac{12}{18} = \frac{12 \div 6}{18 \div 6} = \frac{2}{3} \qquad\qquad \frac{20x^3}{15x^2} = \frac{20x^3 \div 5x^2}{15x^2 \div 5x^2} = \frac{4x}{3}$$

In the following problem, observe the two ways that may be used to indicate that the numerator and the denominator of the fraction are divided by $x + 3$.

$$\frac{5(x+3)}{7(x+3)} = \frac{5(x+3) \div (x+3)}{7(x+3) \div (x+3)} = \frac{5}{7} \qquad\qquad \frac{5(x+3)}{7(x+3)} = \frac{5\overset{1}{\cancel{(x+3)}}}{7\underset{1}{\cancel{(x+3)}}} = \frac{5}{7}$$

We sometimes deal with fractions in which a factor of the numerator is the negative of a factor in the denominator. In such a case, remember that any expression divided by its negative gives the quotient -1. For example,

$$\frac{a-b}{b-a} = \frac{(a-b) \div (a-b)}{(b-a) \div (a-b)} = \frac{1}{-1} = -1 \quad \text{or} \quad \frac{a-b}{b-a} = \frac{\overset{1}{\cancel{(a-b)}}}{\underset{-1}{\cancel{(b-a)}}} = -1$$

~~~~~~~~~~~~~ **MODEL PROBLEMS** ~~~~~~~~~~~~~

In 1 and 2 reduce the fraction to its lowest terms.

1. $\dfrac{5x^2 + 10xy}{15xy + 30y^2}$ **2.** $\dfrac{2x-4}{4-x^2}$

How To Proceed

1. Factor both the numerator and the denominator completely.
2. Divide both the numerator and the denominator by their highest common factor.

Solution

$$\frac{5x^2 + 10xy}{15xy + 30y^2}$$

$$= \frac{5x(x+2y)}{15y(x+2y)}$$

$$= \frac{\overset{1}{\cancel{5x}}\overset{1}{\cancel{(x+2y)}}}{\underset{3}{\cancel{15y}}\underset{1}{\cancel{(x+2y)}}}$$

$$= \frac{x}{3y} \ \textit{Ans.}$$

Solution

$$\frac{2x-4}{4-x^2}$$

$$= \frac{2(x-2)}{(2+x)(2-x)}$$

$$= \frac{2\overset{-1}{\cancel{(x-2)}}}{(2+x)\underset{1}{\cancel{(2-x)}}}$$

$$= \frac{-2}{2+x} \ \textit{Ans.}$$

Exercises

In 1–24, reduce the fraction to its lowest terms.

1. $\dfrac{27a}{36b}$ **2.** $\dfrac{12y^2}{18y^2}$ **3.** $\dfrac{18c}{36c^2}$ **4.** $\dfrac{8xy^2}{24x^2y}$

5. $\dfrac{18rs^2}{45s^2}$ **6.** $\dfrac{3(x+3)}{5(x+3)}$ **7.** $\dfrac{20(y-2)}{15(y-2)}$ **8.** $\dfrac{6a^2(a-5)}{3a(5-a)}$

9. $\dfrac{2b(x+y)}{4b^2}$ **10.** $\dfrac{3x(x+1)}{3x}$ **11.** $\dfrac{9b-18}{4b-8}$ **12.** $\dfrac{y^2-9}{3y+9}$

13. $\dfrac{5a^2-20}{(a-2)^2}$ **14.** $\dfrac{5m}{15m^2-5m}$ **15.** $\dfrac{8a-4}{1-4a^2}$ **16.** $\dfrac{x^3-x}{2x+2}$

17. $\dfrac{3-3x}{x^2-1}$ **18.** $\dfrac{2x-2y}{3y-3x}$ **19.** $\dfrac{1-4x^2}{6x-3}$ **20.** $\dfrac{(x-5)^2}{10-2x}$

21. $\dfrac{y^2-3y}{y^2-4y+3}$ **22.** $\dfrac{r^2-25}{r^2-2r-15}$ **23.** $\dfrac{3-3y}{(y-1)^2}$ **24.** $\dfrac{6-m-m^2}{m^2-9}$

25. The fraction $\dfrac{t-1}{2-3t^2}$ is equivalent to   (1) $\dfrac{1-t}{3t^2-2}$ (2) $\dfrac{t-1}{3t^2-2}$ (3) $\dfrac{1-t}{2-3t^2}$

26. The fraction that is equivalent to $\dfrac{2x+4}{2x}$ is (1) 4   (2) $\dfrac{2x+2}{x}$ (3) $\dfrac{x+2}{x}$

27. When the fraction $\dfrac{5x+15}{5x}$ is reduced to its lowest terms, the result is

   (1) $\dfrac{5}{x}$ (2) $\dfrac{x+3}{3}$ (3) 15   (4) none of these

28. The fraction $\dfrac{8x}{8x+16}$ is equivalent to   (1) $\dfrac{1}{16}$ (2) $\dfrac{1}{17}$ (3) $\dfrac{x}{8x+2}$ (4) $\dfrac{x}{x+2}$

29. If r, s, and t are positive real numbers, no two of which are equal, then $\dfrac{r+s}{s+t}$ is (1) always   (2) sometimes   (3) never   equal to $\dfrac{r}{t}$.

In 30–35, explain why the indicated method which was used by a student in reducing the fraction is *incorrect*.

30. $\dfrac{\overset{1}{\cancel{3x}}}{\underset{1}{\cancel{3}+y}} = \dfrac{x}{1+y}$ **31.** $\dfrac{2\overset{1}{\cancel{x}}+y}{\underset{1}{\cancel{x}}} = 2+y$ **32.** $\dfrac{2a\overset{1}{\cancel{+b}}}{a\underset{1}{\cancel{+b}}} = 2$

33. $\dfrac{\overset{1}{\cancel{x+y}}+z}{\underset{1}{\cancel{x+y}}} = 1 + z$ **34.** $\dfrac{\overset{x}{\cancel{x^2}}+y^2}{\underset{1}{\cancel{x}}} = x + y^2$ **35.** $\dfrac{\overset{a}{\cancel{a^2}}+\overset{b}{\cancel{b^2}}}{\underset{1}{\cancel{a}}+\underset{1}{\cancel{b}}} = a + b$

9. Multiplying Fractions or Rational Expressions

Since $\dfrac{a}{b} \cdot \dfrac{c}{d} = \dfrac{ac}{bd}$ when a and c are real numbers and b and d are nonzero real numbers, we can state the following rule:

The product of two fractions is a fraction whose numerator is the product of the numerators of the given fractions and whose denominator is the product of the denominators of the given fractions.

Study the following two methods that may be used to multiply $\dfrac{2a^2}{3b}$ by $\dfrac{15b^2}{4a^3}$.

Method 1

$$\frac{2a^2}{3b} \cdot \frac{15b^2}{4a^3} = \frac{2a^2 \cdot 15b^2}{3b \cdot 4a^3} = \frac{30a^2b^2}{12a^3b} = \frac{30a^2b^2 \div 6a^2b}{12a^3b \div 6a^2b} = \frac{5b}{2a}$$

Method 2

$$\frac{2a^2}{3b} \cdot \frac{15b^2}{4a^3} = \frac{\overset{1}{\cancel{2a^2}}}{\underset{1}{\cancel{3b}}} \cdot \frac{\overset{5b}{\cancel{15b^2}}}{\underset{2a}{\cancel{4a^3}}} = \frac{5b}{2a}$$

Observe that method 2 requires less computation than method 1 because the reduced form of the product was obtained by dividing the numerator and the denominator by a common factor before the product was found.

⌇⌇⌇⌇⌇⌇⌇⌇⌇ *MODEL PROBLEM* ⌇⌇⌇⌇⌇⌇⌇⌇⌇

Multiply $\dfrac{4x+8}{4x^2-25}$ by $\dfrac{6x+15}{2x^2+4x}$ and express the product in reduced form.

How To Proceed	*Solution*
1. Factor the numerators and denominators of the fractions.	$\dfrac{4x+8}{4x^2-25}\cdot\dfrac{6x+15}{2x^2+4x}$ $=\dfrac{4(x+2)}{(2x+5)(2x-5)}\cdot\dfrac{3(2x+5)}{2x(x+2)}$
2. Divide both the numerator and denominator by all common factors.	$=\dfrac{\overset{2}{\cancel{4}}\overset{1}{\cancel{(x+2)}}}{\underset{1}{\cancel{(2x+5)}}(2x-5)}\cdot\dfrac{3\overset{1}{\cancel{(2x+5)}}}{\underset{1}{\cancel{2}}x\underset{1}{\cancel{(x+2)}}}$
3. Multiply the remaining factors of the numerator; then multiply the remaining factors of the denominator.	$=\dfrac{2\cdot3}{x(2x-5)}=\dfrac{6}{x(2x-5)}$ *Ans.*

Exercises

In 1–15, perform the multiplication and express the product in its simplest form.

1. $\dfrac{36}{30}\cdot\dfrac{12}{8}$

2. $\dfrac{5}{d}\cdot d^2$

3. $\dfrac{3x}{7y}\cdot\dfrac{7a}{3b}$

4. $\dfrac{30a^2}{18b}\cdot\dfrac{6b}{5a}$

5. $\dfrac{24x^3y^2}{7z}\cdot\dfrac{21z^2}{12xy}$

6. $\dfrac{12x}{(5y)^2}\cdot\dfrac{15y^2}{36x^2}$

7. $\dfrac{5x}{x^2-9}\cdot\dfrac{6x+18}{15x^3}$

8. $\dfrac{3a-6}{6a+12}\cdot\dfrac{3a^2-12}{a^2-5a+6}$

9. $\dfrac{4x-6}{4x+8}\cdot\dfrac{6x+12}{4x^2-9}$

10. $\dfrac{b^2+b-2}{b^2-7b}\cdot\dfrac{b^2-13b+42}{b+2}$

11. $\dfrac{4x+8y}{6x+18y}\cdot\dfrac{5x+15y}{x^2-4y^2}$

12. $\dfrac{a^2-7ab+12b^2}{a^2-4b^2}\cdot\dfrac{2a+4b}{a-3b}$

13. $\dfrac{4-2x}{6x+30}\cdot\dfrac{x^2-25}{x^2-7x+10}$

14. $\dfrac{4y^3}{8-2y-y^2}\cdot\dfrac{3y+12}{6y}$

15. $\dfrac{b-a}{a+b}\cdot\dfrac{a^2-b^2}{a^2-2ab+b^2}$

16. Multiply $\dfrac{x^2-y^2}{2xy}$ by $\dfrac{x^2}{x-y}$. Express the result in simplest form.

17. Express the product $\dfrac{6}{3x-1}\cdot(1-3x)$ in simplest form.

10. Dividing Fractions or Rational Expressions

Since we have defined $a \div b$ as $a \times \dfrac{1}{b}$, a quotient may be expressed as the product of the dividend and the multiplicative inverse, or reciprocal, of the divisor. Thus, $9 \div 5 = 9 \times \dfrac{1}{5} = \dfrac{9}{5}$ and $\dfrac{4}{5} \div \dfrac{3}{7} = \dfrac{4}{5} \times \dfrac{1}{\frac{3}{7}} = \dfrac{4}{5} \times \dfrac{7}{3} = \dfrac{28}{15}$.

In general, if a is a real number and b, c, and d are nonzero real numbers,

$$\frac{a}{b} \div \frac{c}{d} = \frac{a}{b} \cdot \frac{d}{c} = \frac{ad}{bc}$$

The quotient of two rational expressions can be found in exactly the same way.

~~~~~~~~~~~ *MODEL PROBLEMS* ~~~~~~~~~~~

In 1 and 2, perform the division and represent the quotient in its simplest form.

1. $\dfrac{8x^3}{27y^2} \div \dfrac{20x^4}{9y^3}$ **2.** $\dfrac{5x+15}{x^2-9} \div \dfrac{10x^2+10x}{4x-12}$

How To Proceed	*Solution*	*Solution*
1. Find the reciprocal of the divisor.	The reciprocal of $\dfrac{20x^4}{9y^3}$ is $\dfrac{9y^3}{20x^4}$.	The reciprocal of $\dfrac{10x^2+10x}{4x-12}$ is $\dfrac{4x-12}{10x^2+10x}$.
2. Multiply the dividend by the reciprocal of the divisor.	$\dfrac{8x^3}{27y^2} \div \dfrac{20x^4}{9y^3}$ $= \dfrac{8x^3}{27y^2} \cdot \dfrac{9y^3}{20x^4}$	$\dfrac{5x+15}{x^2-9} \div \dfrac{10x^2+10x}{4x-12}$ $= \dfrac{5x+15}{x^2-9} \cdot \dfrac{4x-12}{10x^2+10x}$

$$= \dfrac{\overset{2}{\cancel{8x^3}}}{\underset{3}{\cancel{27y^2}}} \cdot \dfrac{\overset{y}{\cancel{9y^3}}}{\underset{5x}{\cancel{20x^4}}}$$

$$= \dfrac{2y}{15x} \ Ans.$$

$$= \dfrac{\overset{1}{\cancel{5}}\overset{1}{(\cancel{x+3})}}{\underset{1}{(\cancel{x+3})}(\cancel{x-3})} \cdot \dfrac{\overset{2}{\cancel{4}}\overset{1}{(\cancel{x-3})}}{\underset{\underset{1}{\cancel{2}}}{\cancel{10}x(x+1)}}$$

$$= \dfrac{2}{x(x+1)} \ Ans.$$

~~~~~~

Exercises

In 1–19, perform the division and represent the quotient in its simplest form.

1. $\dfrac{12}{35} \div \dfrac{4}{7}$ **2.** $15 \div \tfrac{1}{3}$ **3.** $\dfrac{6}{y^2} \div \dfrac{6}{y}$ **4.** $\dfrac{21x}{20y} \div \dfrac{3x}{5y}$

5. $\dfrac{7xy^2}{10rs} \div \dfrac{14y^2}{5r^2s^2}$ **6.** $\dfrac{ab^2}{a^2b} \div \dfrac{a}{b^3}$ **7.** $8xy \div \dfrac{24x}{y}$ **8.** $\dfrac{a-5}{a} \div \dfrac{a^2-25}{a^2}$

9. $\dfrac{4-b^2}{b^2} \div \dfrac{2+b}{b}$ **10.** $\dfrac{12+6a}{15-3a} \div \dfrac{4-a^2}{25-a^2}$ **11.** $\dfrac{x^2-9y^2}{8x+4y} \div \dfrac{3x-9y}{12x+6y}$

12. $\dfrac{y^2-49}{(y+7)^2} \div \dfrac{3y-21}{2y+14}$ **13.** $\dfrac{25a^2}{(a-b)^2} \div \dfrac{5a}{a-b}$ **14.** $\dfrac{-3}{2-x} \div \dfrac{9}{x-2}$

15. $\dfrac{x^2-1}{x^2+1} \div \dfrac{1-x}{3}$ **16.** $\dfrac{r^2-3rs-10s^2}{r^2-25s^2} \div \dfrac{10r+20s}{2r+10s}$

17. $\dfrac{x-7}{3x^2-8x+4} \div \dfrac{3x-21}{6x^2-24}$ **18.** $\dfrac{3x^2+2y}{x-y} \cdot \dfrac{x^2-y^2}{9x^4-4y^2} \div \dfrac{2x+2y}{12x^2-8y}$

19. $\dfrac{x^2-4}{2x+14} \cdot \dfrac{x^2+10x+21}{6x-12} \div \dfrac{x^2+5x+6}{12x}$

20. If the area of a rectangle is represented by $\dfrac{x+y}{x^2-xy}$ and the length is repre-

sented by $\dfrac{1}{x^2-y^2}$, what expression represents the width?

11. Adding or Subtracting Fractions or Rational Expressions That Have the Same Denominator

When we add two fractions that have the same denominator, we make use of the definition of division and the distributive property of multiplication over addition. For example,

$$\frac{3}{11} + \frac{4}{11} = 3 \cdot \left(\frac{1}{11}\right) + 4 \cdot \left(\frac{1}{11}\right) = (3+4) \cdot \frac{1}{11} = \frac{3+4}{11} = \frac{7}{11}$$

In general, if a and b are real numbers and x is a nonzero real number,

$$\frac{a}{x} + \frac{b}{x} = \frac{a+b}{x}$$

Rational expressions that have the same divisor are added in the same way.

~~~~~~~~~~~~~~~~ **MODEL PROBLEMS** ~~~~~~~~~~~~~~~~

In 1 and 2, add or subtract as indicated. Reduce answers to lowest terms.

1. $\dfrac{5a}{8} - \dfrac{2a}{8} + \dfrac{3a}{8}$

2. $\dfrac{5}{2x - 2} - \dfrac{2x - 1}{2 - 2x}$

How To Proceed	*Solution*	*Solution*
	$\dfrac{5a}{8} - \dfrac{2a}{8} + \dfrac{3a}{8}$	$\dfrac{5}{2x - 2} - \dfrac{2x - 1}{2 - 2x}$
1. Write a fraction whose numerator is the sum (or difference) of the numerators and whose denominator is the denominator of the given fractions.	$= \dfrac{5a - 2a + 3a}{8}$ $= \dfrac{6a}{8}$	$= \dfrac{5}{2x - 2} - \dfrac{-2x + 1}{2x - 2}$ $= \dfrac{5 - (-2x + 1)}{2x - 2}$ $= \dfrac{5 + 2x - 1}{2x - 2}$ $= \dfrac{2x + 4}{2x - 2}$
2. Reduce the resulting fraction to its lowest terms.	$= \dfrac{\overset{3}{\cancel{6}}a}{\underset{4}{\cancel{8}}}$ $= \dfrac{3a}{4}$ *Ans.*	$= \dfrac{\overset{1}{\cancel{2}}(x + 2)}{\underset{1}{\cancel{2}}(x - 1)}$ $= \dfrac{x + 2}{x - 1}$ *Ans.*

~~~~~~

Exercises

In 1–15, add or subtract (combine) the fractions as indicated. Reduce answers to lowest terms.

1. $\dfrac{3}{4} + \dfrac{5}{4} + \dfrac{1}{4}$

2. $\dfrac{7}{12} - \dfrac{1}{12} - \dfrac{2}{12}$

3. $\dfrac{5x}{4} + \dfrac{x}{4} + \dfrac{3x}{4}$

4. $\dfrac{11r}{6} - \dfrac{r}{6} - \dfrac{7r}{6}$

5. $\dfrac{9}{8x} + \dfrac{3}{8x} - \dfrac{2}{8x}$

6. $\dfrac{a}{5y} + \dfrac{b}{5y} + \dfrac{c}{5y}$

7. $\dfrac{x}{x + 1} + \dfrac{1}{x + 1}$

8. $\dfrac{a}{a^2 - b^2} - \dfrac{b}{a^2 - b^2}$

9. $\dfrac{3}{x - 4} + \dfrac{1}{4 - x}$

10. $\dfrac{x}{x - 3} + \dfrac{3}{3 - x}$

11. $\dfrac{4x}{2x - 5} + \dfrac{10}{5 - 2x}$

12. $\dfrac{5a}{a^2 - b^2} - \dfrac{5b}{b^2 - a^2}$

13. $\dfrac{3r+6}{2} + \dfrac{2r+1}{2}$ **14.** $\dfrac{12x-15}{12x} - \dfrac{9x-6}{12x}$ **15.** $\dfrac{6r-5}{r^2-1} - \dfrac{5r-6}{r^2-1}$

12. Adding or Subtracting Fractions or Rational Expressions That Have Different Denominators

If we wish to add (or subtract) fractions that have different denominators, we first transform them into equivalent fractions that have the same denominator by using the fundamental multiplication property of a fraction. Then we add (or subtract) the resulting equivalent fractions.

For example, to add $\frac{7}{24}$ and $\frac{11}{36}$, we first transform them to equivalent fractions which have a common denominator. Any integer which has both 24 and 36 as factors could become a common denominator. To simplify our work, we use the lowest common denominator (L.C.D.), which can be found in the following manner:

1. Express each denominator as a product of prime factors.

$$24 = 2 \cdot 2 \cdot 2 \cdot 3 = 2^3 \cdot 3 \qquad 36 = 2 \cdot 2 \cdot 3 \cdot 3 = 2^2 \cdot 3^2$$

2. Write the product of the highest power of each of the different prime factors of the denominators.

$$\text{L.C.D.} = 2^3 \cdot 3^2 = 8 \cdot 9 = 72$$

We now can add $\frac{7}{24}$ and $\frac{11}{36}$ as follows:

$$\frac{7}{24} + \frac{11}{36} = \frac{7 \cdot 3}{24 \cdot 3} + \frac{11 \cdot 2}{36 \cdot 2} = \frac{21}{72} + \frac{22}{72} = \frac{21+22}{72} = \frac{43}{72}$$

Rational expressions written as fractions are added in the same manner.

Procedure. To add (or subtract) fractions that have different denominators:
1. Factor each denominator in order to find the lowest common denominator, **L.C.D.**
2. Transform each fraction to an equivalent fraction by multiplying its numerator and its denominator by the quotient that is obtained when the L.C.D. is divided by the denominator of the fraction.
3. Write a fraction whose numerator is the sum (or difference) of the numerators of the new fractions and whose denominator is the L.C.D.
4. Reduce the resulting fraction to its lowest terms.

See how these steps are used in the following two model problems:

~~~~~~~~~~~~~~~~ *MODEL PROBLEMS* ~~~~~~~~~~~~~~

In 1 and 2, combine (add or subtract) the fractions.

1. $\dfrac{3x+4}{4} + \dfrac{x-3}{6}$

2. $\dfrac{3}{x^2-4} - \dfrac{4}{2-x}$

Solution:

Step 1

$$4 = 2 \cdot 2 = 2^2; \; 6 = 2 \cdot 3$$

$$\text{L.C.D.} = 2^2 \cdot 3 = 12$$

$$12 \div 4 = 3$$
$$12 \div 6 = 2$$

$$\dfrac{3}{x^2-4} - \dfrac{4}{2-x}$$

$$= \dfrac{3}{x^2-4} - \dfrac{-4}{x-2}$$

$$x^2 - 4 = (x-2)(x+2)$$
$$x - 2 = 1 \cdot (x-2)$$
$$\text{L.C.D.} = (x-2)(x+2)$$
$$(x-2)(x+2) \div (x-2) = (x+2)$$

Step 2

$$\dfrac{3x+4}{4} + \dfrac{x-3}{6}$$

$$= \dfrac{3(3x+4)}{3 \cdot 4} + \dfrac{2(x-3)}{2 \cdot 6}$$

$$= \dfrac{9x+12}{12} + \dfrac{2x-6}{12}$$

$$\dfrac{3}{x^2-4} - \dfrac{-4}{x-2}$$

$$= \dfrac{3}{(x-2)(x+2)} - \dfrac{-4}{(x-2)}$$

$$= \dfrac{3}{(x-2)(x+2)} - \dfrac{-4(x+2)}{(x-2)(x+2)}$$

Step 3

$$= \dfrac{(9x+12)+(2x-6)}{12}$$

$$= \dfrac{9x+12+2x-6}{12}$$

$$= \dfrac{3-(-4x-8)}{(x-2)(x+2)}$$

$$= \dfrac{3+4x+8}{(x-2)(x+2)}$$

Step 4

$$= \dfrac{11x+6}{12} \; Ans.$$

$$= \dfrac{4x+11}{(x-2)(x+2)} \; \text{or} \; \dfrac{4x+11}{x^2-4} \; Ans.$$

~~~~~~

Exercises

In 1–4, find the L.C.D. for two fractions whose denominators are:

1. $12x^2y^3$; $18xy^2$ **2.** x; $x+5$ **3.** $5x+15$; $4x+12$ **4.** y^2-4; $3y+6$

In 5–8, transform the given fractions into equivalent fractions that have the L.C.D. as their denominators.

5. $\dfrac{7y}{12}$; $\dfrac{3y}{90}$ **6.** $\dfrac{5}{4x^2}$; $\dfrac{7}{xy^2}$ **7.** $\dfrac{5}{d}$; $\dfrac{d-2}{d+2}$ **8.** $\dfrac{2y+1}{y^2-9}$; $\dfrac{-7}{y+3}$

In 9–36, combine the fractions.

9. $\frac{1}{2}+\frac{3}{4}+\frac{7}{8}$ **10.** $\frac{3}{4}-\frac{2}{3}+\frac{5}{6}$ **11.** $\frac{1}{2}-\frac{3}{4}+\frac{4}{5}$

12. $\dfrac{9x}{4}+\dfrac{x}{2}+\dfrac{3x}{8}$ **13.** $\dfrac{5a}{6}-\dfrac{3a}{4}+\dfrac{a}{12}$ **14.** $\dfrac{7}{4y}-\dfrac{2}{6y}-\dfrac{1}{9y}$

15. $\dfrac{5}{x}-\dfrac{6}{y}$ **16.** $\dfrac{n}{s}-\dfrac{m}{t}$ **17.** $\dfrac{3}{y^2}-\dfrac{2}{y}$

18. $\dfrac{1}{5x^2}+\dfrac{3}{10xy}-\dfrac{4}{x^2y^2}$ **19.** $\dfrac{y-2}{3}+\dfrac{y+1}{6}$ **20.** $\dfrac{2a+1}{5a}-\dfrac{4a-2}{4a}$

21. $\dfrac{x-2}{4x}+\dfrac{2x-3}{3x}$ **22.** $\dfrac{r+x}{r^2x}-\dfrac{r-2x}{rx^2}$ **23.** $\dfrac{3}{x-1}-\dfrac{1}{x}$

24. $\dfrac{4}{x-3}-\dfrac{3}{x}$ **25.** $\dfrac{4}{x+1}+\dfrac{2}{x-1}$ **26.** $\dfrac{1}{a+b}-\dfrac{1}{a-b}$

27. $\dfrac{5}{x^2-4}+\dfrac{3}{x-2}$ **28.** $\dfrac{m}{a^2-ab}-\dfrac{m}{b^2-ab}$ **29.** $\dfrac{3}{x^2-25}+\dfrac{4}{5-x}$

30. $\dfrac{2}{y^2-4}+\dfrac{1}{2y-y^2}$ **31.** $\dfrac{1+x}{1-x}+\dfrac{1-x}{1+x}$ **32.** $\dfrac{a+b}{a-b}+\dfrac{a-b}{a+b}$

33. $\dfrac{3}{x^2-3x+2}-\dfrac{2}{x^2-1}$ **34.** $\dfrac{x}{x^2-3x-10}-\dfrac{2}{x^2-6x+5}$

35. $\dfrac{4}{x^2-x-6}+\dfrac{5}{4-x^2}$ **36.** $\dfrac{5}{x+2}+\dfrac{5}{2-x}-\dfrac{6}{x^2-4}$

13. Multiplying and Dividing Mixed Expressions

The mixed number $2\frac{1}{4}$, which means the sum of the integer 2 and the fraction $\frac{1}{4}$, $2+\frac{1}{4}$, can be expressed as the fraction $\frac{9}{4}$.

A *mixed expression* is the indicated sum or difference of a polynomial and a rational expression written as a fraction. For example, $x+\dfrac{2}{x}$ is a mixed expression. A mixed expression can be transformed into an equivalent fraction. For example,

$$x+\frac{2}{x}=\frac{x}{1}+\frac{2}{x}=\frac{x\cdot x}{x\cdot 1}+\frac{2}{x}=\frac{x^2}{x}+\frac{2}{x}=\frac{x^2+2}{x}$$

~~~~~~~~~~~~ *MODEL PROBLEM* ~~~~~~~~~~~~

Simplify: $\left(2 + \dfrac{a}{b}\right) \div \left(4 - \dfrac{a^2}{b^2}\right)$

How To Proceed	*Solution*
	$\left(2 + \dfrac{a}{b}\right) \div \left(4 - \dfrac{a^2}{b^2}\right)$
1. Combine the terms in each mixed expression into a single fraction.	$= \left(\dfrac{2 \cdot b}{1 \cdot b} + \dfrac{a}{b}\right) \div \left(\dfrac{4 \cdot b^2}{1 \cdot b^2} - \dfrac{a^2}{b^2}\right)$
	$= \dfrac{2b + a}{b} \div \dfrac{4b^2 - a^2}{b^2}$
2. Perform the indicated division.	$= \dfrac{2b + a}{b} \cdot \dfrac{b^2}{4b^2 - a^2}$
	$= \dfrac{\overset{1}{(2b+a)}}{\underset{1}{\cancel{b}}} \cdot \dfrac{\overset{b}{\cancel{b^2}}}{\underset{1}{(2b+a)(2b-a)}}$
	$= \dfrac{b}{2b - a} \quad Ans.$

~~~~~~

Exercises

In 1–12, perform the indicated operations on the mixed expressions. Express the result in its simplest form.

1. $\left(\dfrac{x}{y} + 2\right)\left(\dfrac{x}{y} - 2\right)$

2. $\left(1 + \dfrac{a}{b}\right)\left(\dfrac{b^2}{b^2 - a^2}\right)$

3. $\left(1 - \dfrac{a}{a + b}\right)\left(\dfrac{a^2}{b^2} - 1\right)$

4. $\left(1 - \dfrac{x}{y}\right) \div \left(y - \dfrac{x^2}{y}\right)$

5. $\left(5 + \dfrac{5}{3x}\right) \div \left(\dfrac{1}{9x} - x\right)$

6. $\left(1 - \dfrac{5}{x^2 - 4}\right) \div (x + 3)$

7. $\left(r - 4 + \dfrac{3}{r}\right)\left(\dfrac{r^2}{r - 3}\right)$

8. $\left(x + 5 - \dfrac{14}{x}\right) \div \left(x - 4 + \dfrac{4}{x}\right)$

9. $\left(\dfrac{1}{y} - \dfrac{1}{x}\right) \div \left(1 - \dfrac{x}{y}\right)$

10. $\left(\dfrac{r^2}{s^2} - \dfrac{a^2}{b^2}\right) \div \left(\dfrac{a}{b} - \dfrac{r}{s}\right)$

11. $\left(1 - \dfrac{2}{x^2 + 1}\right) \div (x - 1)$

12. $\left(x - 1 - \dfrac{x^2 - 1}{x}\right) \times \left(1 + \dfrac{1}{x - 1}\right)$

14. Simplifying Complex Fractions

A *complex fraction* is a fraction that contains one or more fractions in either its numerator or denominator or in both its numerator and denominator. Thus, each of the following is a complex fraction:

$$\frac{\frac{1}{2}}{\frac{1}{3}} \qquad \frac{\frac{1}{y}}{2y} \qquad \frac{x}{1+\frac{1}{x}} \qquad \frac{1+\frac{4}{x}}{1-\frac{16}{x^2}}$$

A complex fraction may be simplified using either of the two methods shown in the following model problems.

~~~~~~ *MODEL PROBLEMS* ~~~~~~

METHOD 1

Simplify:
1. $\dfrac{\dfrac{a}{b}+1}{\dfrac{a}{b}-1}$
2. $\dfrac{x+\frac{1}{2}}{x^2-\frac{1}{4}}$

How To Proceed	Solution	Solution
1. Find the L.C.D. of all fractions that appear in the complex fraction.	L.C.D. for $\dfrac{a}{b}$ and $\dfrac{a}{b}$ is b.	L.C.D. for $\frac{1}{2}$ and $\frac{1}{4}$ is 4.
2. Multiply the numerator and the denominator of the complex fraction by this L.C.D.	$\dfrac{\dfrac{a}{b}+1}{\dfrac{a}{b}-1}$ $=\dfrac{b\left(\dfrac{a}{b}+1\right)}{b\left(\dfrac{a}{b}-1\right)}$	$\dfrac{x+\frac{1}{2}}{x^2-\frac{1}{4}}$ $=\dfrac{4(x+\frac{1}{2})}{4(x^2-\frac{1}{4})}=\dfrac{4x+2}{4x^2-1}$ $=\dfrac{\overset{1}{2(\cancel{2x+1})}}{\underset{1}{(\cancel{2x+1})(2x-1)}}$
3. Simplify and reduce the resulting simple fraction.	$=\dfrac{a+b}{a-b}$ *Ans.*	$=\dfrac{2}{2x-1}$ *Ans.*

METHOD 2

Simplify:

1. $\dfrac{\dfrac{a}{b}+1}{\dfrac{a}{b}-1}$

2. $\dfrac{x+\frac{1}{2}}{x^2-\frac{1}{4}}$

How To Proceed	*Solution*	*Solution*
	$\dfrac{\dfrac{a}{b}+1}{\dfrac{a}{b}-1}$	$\dfrac{x+\frac{1}{2}}{x^2-\frac{1}{4}}$
1. Combine the terms in the numerator into a single fraction, and do the same in the denominator.	$=\dfrac{\dfrac{a}{b}+\dfrac{b}{b}}{\dfrac{a}{b}-\dfrac{b}{b}}$ $=\dfrac{\dfrac{a+b}{b}}{\dfrac{a-b}{b}}$	$=\dfrac{\dfrac{2x}{2}+\dfrac{1}{2}}{\dfrac{4x^2}{4}-\dfrac{1}{4}}$ $=\dfrac{\dfrac{2x+1}{2}}{\dfrac{4x^2-1}{4}}$
2. Divide the numerator of the resulting fraction by its denominator.	$=\dfrac{(a+b)}{b}\div\dfrac{(a-b)}{b}$ $=\dfrac{(a+b)}{\cancel{b}}\cdot\dfrac{\overset{1}{\cancel{b}}}{(a-b)}$ $=\dfrac{a+b}{a-b}$ *Ans.*	$=\dfrac{2x+1}{2}\div\dfrac{4x^2-1}{4}$ $=\dfrac{\overset{1}{\cancel{(2x+1)}}}{\cancel{2}}\cdot\dfrac{\overset{2}{\cancel{4}}}{\cancel{(2x+1)}(2x-1)}$ $=\dfrac{2}{2x-1}$ *Ans.*

Exercises

In 1–20, simplify the complex fraction.

1. $\dfrac{4}{1-\frac{1}{2}}$ **2.** $\dfrac{3-\frac{1}{2}}{10}$ **3.** $\dfrac{2+\frac{1}{4}}{1+\frac{1}{2}}$ **4.** $\dfrac{5-\frac{1}{5}}{3-\frac{3}{10}}$

5. $\dfrac{\dfrac{b^2}{y}}{\dfrac{b}{y^2}}$

6. $\dfrac{\dfrac{x}{x+y}}{\dfrac{y}{x+y}}$

7. $\dfrac{\dfrac{r^2-4s^2}{s^2}}{\dfrac{r+2s}{s}}$

8. $\dfrac{x+\frac{1}{4}}{x^2-\frac{1}{16}}$

9. $\dfrac{3-\dfrac{1}{x}}{3+\dfrac{1}{x}}$

10. $\dfrac{1-\dfrac{3}{y}}{1-\dfrac{9}{y^2}}$

11. $\dfrac{\dfrac{1}{x}-\dfrac{1}{y}}{\dfrac{1}{x}+\dfrac{1}{y}}$

12. $\dfrac{\dfrac{x}{y}+\dfrac{y}{x}}{\dfrac{1}{xy}}$

13. $\dfrac{\dfrac{1}{a^2}+\dfrac{1}{b^2}}{\dfrac{2}{ab}}$

14. $\dfrac{\dfrac{y}{y+3}}{1-\dfrac{y}{y+3}}$

15. $\dfrac{\frac{1}{9}-x^2}{x-\frac{1}{3}}$

16. $\dfrac{5+\dfrac{5}{2x}}{\dfrac{1}{4x}-x}$

17. $\dfrac{1-\dfrac{6}{x}+\dfrac{9}{x^2}}{\dfrac{3}{x}-1}$

18. $\dfrac{\dfrac{r}{s}-\dfrac{s}{r}}{\dfrac{s}{r}-1}$

19. $\dfrac{1-\dfrac{5}{x}+\dfrac{6}{x^2}}{1-\dfrac{6}{x}+\dfrac{8}{x^2}}$

20. $\dfrac{\dfrac{2}{a+b}-\dfrac{2}{a-b}}{\dfrac{4}{a^2-b^2}}$

In 21–24, write in simplest form the reciprocal of the given expression.

21. $\left(\dfrac{1}{x}+5\right)$

22. $\dfrac{c+\dfrac{c}{b}}{b-\dfrac{1}{b}}$

23. $\dfrac{\dfrac{1}{a}-b}{\dfrac{1}{b}-a}$

24. $\dfrac{\dfrac{a}{b}-\dfrac{b}{a}}{\dfrac{a}{b}-1}$

━━━━━ *CALCULATOR APPLICATIONS* ━━━━━

Calculators can be used to check the results of factoring and operations on rational expressions.

━━━━━ *MODEL PROBLEMS* ━━━━━

1. A student found the following product:

$$\frac{x^2 - 25}{x^2 - 5x} \cdot \frac{x}{3x + 15}$$

$$= \frac{\overset{1}{\cancel{(x-5)}}\overset{1}{\cancel{(x+5)}}}{\underset{1}{\cancel{x(x-5)}}} \cdot \frac{\overset{1}{\cancel{x}}}{\underset{1}{3\cancel{(x+5)}}} = \frac{1}{3} \quad Answer$$

If the answer is correct, this means that for any value of x, the value of the original expression will be $\frac{1}{3}$. Check by evaluating the expression for one or more values of x. For example, to find the value when $x = 7$,

Enter: $\boxed{(}\ 7\ \boxed{x^2}\ \boxed{-}\ 25\ \boxed{)}\ \boxed{\div}\ \boxed{(}\ \boxed{(}\ 7\ \boxed{x^2}\ \boxed{-}\ 5\ \boxed{\times}\ 7\ \boxed{)}$
$\boxed{\times}\ 7\ \boxed{\div}\ \boxed{(}\ 3\ \boxed{\times}\ 7\ \boxed{+}\ 15\ \boxed{)}\ \boxed{=}$

Display: $\boxed{\text{0.3333333}}$

The display shows the decimal form of $\frac{1}{3}$. On a graphing calculator, the key strokes after the final parenthesis could be

$\boxed{\text{ENTER}}\qquad \boxed{\text{MATH}}\ 1\ \boxed{\text{ENTER}}$

Display: $\boxed{\text{1/3}}$

A graphing calculator can be used to check if the factored form of a polynomial is correct. Graph the polynomial and then the factored form. If the two graphs correspond exactly, then the factoring is probably correct. However, because parts of the graph may not be shown, or the appearance of the graphs on the screen may be misleading (only seeming to coincide), it is advisable to check by multiplying, too.

2. Decide if each polynomial is factored correctly. If the factorization is incorrect, write the correct form.

   a. $10x^2 + 11x - 6;\ (2x + 2)(5x - 3)$

   b. $6x^2 - 5x - 4;\ (2x + 1)(3x - 4)$

Solution:

a. Adjust the $\boxed{\text{WINDOW}}$ to include y-values from -50 to 100 with $\text{Yscl} = 5$.

Enter: $\boxed{\text{Y=}}$ 10 $\boxed{\text{X,T,θ,}n}$ $\boxed{x^2}$ $\boxed{+}$ 11 $\boxed{\text{X,T,θ,}n}$ $\boxed{-}$ 6 $\boxed{\text{ENTER}}$
$\boxed{(}$ 2 $\boxed{\text{X,T,θ,}n}$ $\boxed{+}$ 2 $\boxed{)}$ $\boxed{(}$ 5 $\boxed{\text{X,T,θ,}n}$ $\boxed{-}$ 3 $\boxed{)}$ $\boxed{\text{GRAPH}}$

Display:

The two graphs do not correspond exactly, so the factored form is incorrect. The correct factoring is

$$(2x + 3)(5x - 2)$$

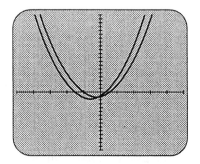

Correct this entry for Y_2 and then press $\boxed{\text{GRAPH}}$. The two graphs appear identical.

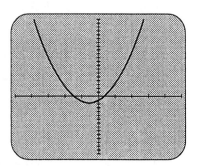

b. *Enter:* $\boxed{\text{Y=}}$ 6 $\boxed{\text{X,T,θ,}n}$ $\boxed{x^2}$ $\boxed{-}$ 5 $\boxed{\text{X,T,θ,}n}$ $\boxed{-}$ 4 $\boxed{\text{ENTER}}$
$\boxed{(}$ 2 $\boxed{\text{X,T,θ,}n}$ $\boxed{+}$ 1 $\boxed{)}$ $\boxed{(}$ 3 $\boxed{\text{X,T,θ,}n}$ $\boxed{-}$ 4 $\boxed{)}$ $\boxed{\text{GRAPH}}$

Display:

The two graphs appear identical. Multiplying confirms the factored form is correct.

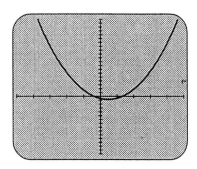

CHAPTER IV

First-Degree Equations and Inequalities in One Variable

1. Understanding the Meaning of Solving an Equation

An equation is a sentence which states that two expressions have the same value. We use the symbol $=$, read "equals" or "is equal to," to indicate that two expressions have the same value. For example $5 + 4 = 7 + 2$ is an equation in which $5 + 4$ is called the **left side** or **left member** and $7 + 2$ is called the **right side** or **right member**.

An equation may be a true sentence, such as $5 + 4 = 9$, or a false sentence, such as $5 - 4 = 2$. As we know, an equation may be an *open sentence*, such as $x + 4 = 9$. If x is replaced by a number, a new sentence is formed which is true or false depending on the selected number.

Consider the equation $x + 4 = 9$ when the *replacement set* or *domain* of x is the set of real numbers. Only when x is replaced by 5 does $x + 4 = 9$ become a true sentence: $5 + 4 = 9$. The number 5, which *satisfies* the equation $x + 4 = 9$, is called a *root* or a *solution* of the equation. The set consisting of all the solutions of an equation is called its *solution set* or *truth set*. Thus, the solution set of $x + 4 = 9$ is $\{5\}$.

The solution set of an equation is a subset of the domain of the variable. Hence, the solution set of an equation depends on the domain of the variable. When the domain of x is {real numbers}, the solution set of $x + 4 = 9$ is $\{5\}$. However, when the domain of x is {even integers}, the solution set of $x + 4 = 9$ is the empty set, $\emptyset$, because there is no even integer that can replace x in $x + 4 = 9$ and make the resulting statement true.

To **solve an equation** means to find its solution set.

An **identity** is an equation that becomes a true statement for every value of the variable over the domain of the variable. For example, $9 + x = x + 9$ is an identity if the domain of x is the set of real numbers.

A **conditional equation** or, more simply, an **equation** is an equation that becomes a true statement for some but not for every value of the variable over the domain of the variable. For example, if the domain of x is {real numbers}, $x + 4 = 9$ is a conditional equation because it is true only when $x = 5$.

There are some equations which are not satisfied by any element of the domain. For example, $x = x + 1$ is never true if the domain of x is the set of real numbers. The solution set of such an equation is the empty set, $\varnothing$.

Exercises

In 1–3, tell whether the number in parentheses is a root of the given equation.
1. $2x + 5 = 13$ (4)   **2.** $20 = 4x - 8$ (3)   **3.** $\frac{1}{2}x = 36$ (18)

In 4–6, using the domain {1, 2, 3, 4, 5, 6}, find the solution set of the equation. If the equation has no roots, indicate the solution set as the empty set, $\varnothing$.
4. $4x + 5 = 21$ **5.** $6 - 3x = 0$ **6.** $\frac{1}{2}x + 8 = 20$

7. Find the solution set of $2x + 4 = 6$ when the domain of x is (*a*) {positive integers} and (*b*) {negative integers}.
8. What might be the domain of x in order that the solution set of $x + 5 = 4$ be the empty set?
9. What is the solution set of $2x + 5 = 7 + 2x$ when the domain of x is {real numbers}?
10. What is the solution set of $4x + 2 = 2(2x + 1)$ when the domain of x is {natural numbers}?

In 11–16, using the domain {0, 1, 2, 3, 4, 5}, tell whether the equation is a conditional equation or an identity.
11. $2x + 1 = 11$ **12.** $2x + 1 = 1 + 2x$ **13.** $r - 2 - 3 = r - 5$
14. $7b = b \times 7$ **15.** $7b = 56 - 7b$ **16.** $3 \times 2 \times t = t \times 2 \times 3$

2. Solving Simple First-Degree Equations

In this chapter, we will study only *first-degree equations* in one variable that are expressible in the form $ax + b = c$, $a \neq 0$, such as $2x + 3 = 5$. If the domain of the variable is not stated, we will assume it to be {real numbers}.

Equivalent equations are equations that have the same solution set. For example, $x + 4 = 9$ and $x = 5$ are equivalent equations because the solution set of each equation is {5}.

When we solve the equation $x + 4 = 9$, we obtain the equivalent equation $x = 5$. In general, when we solve an equation, we transform it into a simpler equivalent equation of the form $x = n$ whose solution set, $\{n\}$, is also the solution set of the original given equation. In order to transform an equation of the form $ax + b = c$ into an equivalent equation of the form $x = n$ whose solution set is $\{n\}$, we will make use of certain postulates.

POSTULATES INVOLVING EQUIVALENT EQUATIONS

We will assume five postulates, which correspond to the addition, subtraction, multiplication, and division properties of equality.

Postulate 1. Substituting an equivalent expression for an expression in an equation results in an equation that is equivalent to the given equation.

Thus, $2x + 3x = 10$ and $5x = 10$ are equivalent equations.

Postulate 2. If the same number or expression is added to both members of an equation, the resulting equation is equivalent to the original equation.

Thus, if $x - 4 = 5$, then $(x - 4) + 4 = 5 + 4$, and $x = 5 + 4$. Hence, $x - 4 = 5$ has been transformed into $x = 5 + 4$ by Postulate 2. In this way, *the operation of subtraction is undone by addition*, its inverse operation. Therefore, the solution set of $x - 4 = 5$ is $\{9\}$.

Postulate 3. If the same number or expression is subtracted from both members of an equation, the resulting equation is equivalent to the original equation.

Thus, if $x + 8 = 14$, then $(x + 8) - 8 = 14 - 8$, and $x = 14 - 8$. Hence, $x + 8 = 14$ has been transformed into $x = 14 - 8$ by Postulate 3. In this way, *the operation of addition is undone by subtraction*, its inverse operation. Therefore, the solution set of $x + 8 = 14$ is $\{6\}$.

Postulate 4. If both members of an equation are multiplied by the same non-zero number or expression, the resulting equation is equivalent to the original equation.

Thus, if $\dfrac{x}{2} = 4$, then $2\left(\dfrac{x}{2}\right) = 2(4)$ and $x = 2(4)$. Hence, $\dfrac{x}{2} = 4$ has been transformed into $x = 2(4)$ by Postulate 4. In this way, *the operation of division has been undone by multiplication*, its inverse operation. Therefore, the solution set of $\dfrac{x}{2} = 4$, or $\tfrac{1}{2}x = 4$, is $\{8\}$.

Postulate 5. If both members of an equation are divided by the same nonzero number or expression, the resulting equation is equivalent to the original equation.

Thus, if $3x = 15$, then $\dfrac{3x}{3} = \dfrac{15}{3}$ and $x = \dfrac{15}{3}$. Hence, $3x = 15$ has been trans-

formed into $x = \dfrac{15}{3}$ by Postulate 5. In this way, *the operation of multiplication has been undone by division*, its inverse operation. Therefore, the solution set of $3x = 15$ is $\{5\}$.

In the examples in which Postulates 2–5 were applied, notice that the inverse operation of the operation associated with the variable was used to transform the given simple equation into an equivalent equation of the form $x = a$. In this respect, the four fundamental operations are paired as follows:

Addition and subtraction are inverse operations of each other.

That is, addition undoes subtraction, and subtraction undoes addition.

Multiplication and division are inverse operations of each other.

That is, multiplication undoes division, and division undoes multiplication.

Exercises

In 1–28, solve for the variable and check.

1. $y + 5 = 6$
2. $x - 3 = 9$
3. $36 = y + 10$
4. $x - 5 = -8$
5. $5 + r = -7$
6. $-3 = s - 7$
7. $3 = c + 12$
8. $.6 + m = .9$
9. $x + 2\frac{1}{2} = 6$
10. $x - 1\frac{1}{5} = -8$
11. $-\frac{3}{4} = x - 1\frac{1}{4}$
12. $4\frac{1}{2} = x + 6\frac{3}{4}$
13. $5a = 40$
14. $12x = -48$
15. $-63 = 9t$
16. $-x = -3$
17. $.7b = 84$
18. $.01x = 25$
19. $.002y = .4$
20. $1.4 = .5t$
21. $\frac{1}{4}x = 8$
22. $\frac{3}{5}b = 21$
23. $\frac{5}{8}x = -20$
24. $-\frac{3}{4}x = 36$
25. $-\frac{3}{2}y = -1\frac{1}{2}$
26. $\frac{2}{3}x = -1.8$
27. $\dfrac{y}{3} = -12$
28. $\dfrac{2x}{3} = \dfrac{4}{9}$

In 29–34, determine the elements of the set if $x \in \{\text{real numbers}\}$.

29. $\{x \mid x - 4 = 7\}$
30. $\{x \mid x + 15 = -9\}$
31. $\{x \mid 15 = x + 15\}$
32. $\{x \mid 4x = 28\}$
33. $\{x \mid \frac{1}{5}x = -5\}$
34. $\left\{x \mid \dfrac{x}{3} = -6\right\}$

3. Solving Equations by Using Several Operations

Procedure. To solve an equation by performing several operations upon a variable, use inverse operations to obtain an equivalent equation of the form $x = a$.

~~~~~~~~~~~~~~ *MODEL PROBLEMS* ~~~~~~~~~~~~~~

1. Solve and check: $\frac{2}{3}x + 5 = 11$

How To Proceed	*Solution*
	$\frac{2}{3}x + 5 = 11$
1. To undo addition, subtract the same number from each side. (*Note.* S_5 means "subtract 5 from each side.")	S_5: $(\frac{2}{3}x + 5) - 5 = (11) - 5$ $\frac{2}{3}x = 6$
2. To undo multiplication, divide each side by the same number. (*Note.* $D_{\frac{2}{3}}$ means "divide each side by $\frac{2}{3}$.")	$D_{\frac{2}{3}}$: $\frac{2}{3}x \div \frac{2}{3} = 6 \div \frac{2}{3}$ $x = 6 \times \frac{3}{2}$ $x = 9$

$$\text{Check}$$
$$\frac{2}{3}x + 5 = 11$$
$$\text{Let } x = 9: \ \tfrac{2}{3}(9) + 5 \overset{?}{=} 11$$
$$6 + 5 \overset{?}{=} 11$$
$$11 = 11 \text{ (true)}$$

Answer: $x = 9$, or solution set is $\{9\}$.

In model problems, if the check is not included, it is left to the student to perform the check.

2. Solve and check: $5x - 4 = 3x + 10$

How To Proceed	*Solution*
	$5x - 4 = 3x + 10$
1. To undo addition, subtract the same number from each side.	S_{3x}: $(5x - 4) - 3x = (3x + 10) - 3x$ $2x - 4 = 10$
2. To undo subtraction, add the same number to each side. (*Note.* A_4 means "add 4 to each side.")	A_4: $(2x - 4) + 4 = (10) + 4$ $2x = 14$
3. To undo multiplication, divide each side by the same number.	D_2: $\dfrac{2x}{2} = \dfrac{14}{2}$ $x = 7$

Answer: $x = 7$, or solution set is $\{7\}$.

~~~~~~

Exercises

In 1–30, solve the equation and check.

1. $11y + 12 = 39$ **2.** $5s - s = 36$ **3.** $2m - \frac{1}{2} = 3\frac{1}{2}$

4. $.8z - 5 = .6$ **5.** $90 = 4x - 10$ **6.** $3y - 8 = 25$

7. $6x + x = 49$ **8.** $8b - b - 2 = 18$ **9.** $38 = 9x + x + 3$

10. $\dfrac{a}{4} - 3 = 5$ **11.** $\frac{1}{2}c + \frac{1}{4}c = 6$ **12.** $3n + 4n - 5 = 30$

13. $\dfrac{3x}{5} + 4 = 34$ **14.** $\dfrac{m}{5} + 3\frac{1}{2} = 6\frac{1}{2}$ **15.** $5x - 3x + 5 = 21$

16. $.4c - 4 = 3.6$ **17.** $\dfrac{b}{4} - \dfrac{5}{2} = 4\frac{1}{2}$ **18.** $21 = \dfrac{4x}{9} - 7$

19. $\dfrac{x}{10} + \dfrac{3x}{5} = 14$ **20** $\dfrac{3y}{2} = 16\frac{1}{2}$ **21.** $5b + 3 - b = 22$

22. $17 = \frac{1}{3}x + 5$ **23.** $\dfrac{3a}{4} - 3 = 9$ **24.** $7x - 3 = 5x + 7$

25. $5x = 48 - x$ **26.** $3y + 12 = 7y$ **27.** $4y + 5 = y + 35$

28. $2m = 5m - 30$ **29.** $8c - 21 = c$ **30.** $3b - 3 = 37 - 2b$

In 31–33, determine the elements of the set if $x \in$ {real numbers}.

31. $\{x \mid 3x + 5 = -7\}$ **32.** $\{x \mid 5x = 3x + 4\}$ **33.** $\{x \mid 5x - 8 = 7x - 4\}$

4. Solving Equations by Using Transposition

By using the addition postulate to solve $5x - 3 = 10$, we obtain an equivalent equation $5x = 10 + 3$. If we compare the two equations, we notice that -3 has disappeared from the left member of the original equation, and its opposite, $+3$, now appears in the right member of the equivalent equation.

By using the subtraction postulate to solve $4y = 20 + 2y$, we obtain an equivalent equation $4y - 2y = 20$. If we compare the two equations, we notice that $+2y$ has disappeared from the right member of the original equation and its opposite, $-2y$, now appears in the left member of the equivalent equation.

These examples illustrate that we can make a term disappear from one member of an equation by making the opposite of that term appear in the other member of the equation. In this way, we obtain an equation that is equivalent to the original equation. This process is called ***transposition***. It must be kept in mind that transposition is not a mathematical operation. Transposition is merely a short-cut device for transforming an equation into an equivalent equation by using the postulates involving the inverse operations of addition and subtraction (postulates 2 and 3).

~~~~~~~~~~~~~ *MODEL PROBLEM* ~~~~~~~~~~~~~

Solve the equation $9x - 4 = 12 + 5x$.

How To Proceed	*Solution*
	$9x - 4 = 12 + 5x$
1. Transpose.	$9x - 5x = 12 + 4$
2. Collect like terms.	$4x = 16$
3. Solve the resulting equation.	$D_4: \dfrac{4x}{4} = \dfrac{16}{4}$
	$x = 4$

Check: The check, which is performed by substituting 4 for x in the given equation, is left to the student.

Answer: $x = 4$, or the solution set is $\{4\}$.

~~~~~~

Exercises

In 1–26, solve the equation and check.

1. $5x + 3 = 23$
2. $6x - 5 = 31$
3. $4c = 5 - 6c$
4. $7d = d - 12$
5. $-3x - 5 = 10$
6. $3x + 2\frac{1}{2} = 5\frac{1}{2}$
7. $19 = 4x - 6.5$
8. $2x + \frac{1}{4} = \frac{3}{4}$
9. $1.4x = .49 - .7x$
10. $.9x = .6x - 3$
11. $15 = 20 + 15n$
12. $5n = 7n - 12$
13. $9x - 5 = 7x + 3$
14. $11y - 7 = 6y - 22$
15. $14x + 3 = 8 - x$
16. $5y - 1 = 15y + 4$
17. $5 - 4c = c + 20$
18. $7x - 4 + 3x = 26$
19. $25 + 7x - 1 = 4x$
20. $1.4x - 7 = 5 - .6x$
21. $1.2x - 3.6 + .3x = 2.4$
22. $3.7y = 4.5y - 10 - 5.8y$
23. $2\frac{1}{2}x - 5 + 1\frac{1}{4}x = 10 - \frac{3}{4}x$
24. $2x - 1\frac{2}{3} = 8\frac{1}{3} + 7x$
25. $10x - 5 + 12 = 7x + 21 - 4x$
26. $8x - 9 + x - 17 = 3x + 4 + 8x - 12$

5. Solving Equations That Contain Parentheses

To solve an equation that contains parentheses, such as $2(x + 3) - 5 = 17 - (4x - 2)$, transform it into an equivalent equation that does not have parentheses. Note how this is done in the following model problem:

~~~~~~~~~ **MODEL PROBLEM** ~~~~~~~~~

Solve for x: $2(x+3) - 5 = 17 - (4x - 2)$

How To Proceed	*Solution*
	$2(x+3) - 5 = 17 - (4x - 2)$
1. Transform the equation by performing the indicated operations.	$2x + 6 - 5 = 17 - 4x + 2$
2. Collect like terms.	$2x + 1 = 19 - 4x$
3. Solve the resulting equation.	$6x = 18$
	$\text{D}_6: \dfrac{6x}{6} = \dfrac{18}{3}$
	$x = 3$

Answer: $x = 3$, or solution set is $\{3\}$.

~~~~~~~~~~~~~~~~~~~~~~~~~~~~~~

Exercises

In 1–18, solve the equation and check.

1. $3(x - 5) = 6$
2. $3(2x + 1) = -15$
3. $4(3 - x) = 2x$
4. $7(a + 2) = 5(a + 4)$
5. $5x + (x - 2) = 1$
6. $6 = 5 - (3x - 4)$
7. $4(x - 3) + 3x = 16$
8. $6x - 3(x - 4) = 6$
9. $13 + 2(3x + 1) = 9x + 12$
10. $10 - y = 3 - 6(y - 2)$
11. $17 - 8(2 + 3x) = 5$
12. $4x = 11 - 3(4x + 5)$
13. $3y + (2y - 5) = 13 - 2(y + 2)$
14. $2(y - 3) - 17 = 13 - 3(y + 2)$
15. $3(x - 2) - 2 = 5(x + 3) - 7(x - 1)$
16. $(x + 3)(x + 2) = x(x + 7)$
17. $(4 - c)(6 + c) = 40 - (c^2 - 4c - 18)$
18. $(x - 1)^2 - 3x(x - 2) = 19 - 2x(x + 1)$

In 19–22, determine the element of the set if the domain of the variable is {real numbers}.

19. $\{x \mid 3x + (2x - 1) = 4\}$
20. $\{y \mid 8y = 15 - (2y + 35)\}$
21. $\{c \mid 5c - 2(c - 5) = 17\}$
22. $\{z \mid 6z = 10 + 4(2z - 6)\}$

6. Graphing the Solution Set of a First-Degree Equation

~~~~~~~~~~~~~~ **MODEL PROBLEMS** ~~~~~~~~~~~~~~

In 1 and 2, using the set of real numbers as the domain, (*a*) find the solution set of the equation and (*b*) graph the solution set. The checks are left to the student.

1. $4x - 1 = 5$

Solution:

a. $4x - 1 = 5$

$4x = 5 + 1$

$4x = 6$

$x = 1.5$

Answer: Solution set is {1.5}.

b.

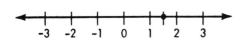

2. $\dfrac{x}{2} + 1 = 3$

Solution:

a. $\dfrac{x}{2} + 1 = 3$

$\dfrac{x}{2} = 3 - 1$

$\dfrac{x}{2} = 2$

$x = 4$

Answer: Solution set is {4}.

b.

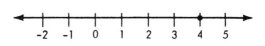

Exercises

In 1–18, find and graph the solution set. Use the set of real numbers as the domain.

1. $x - 3 = 2$

2. $x + 12 = 8$

3. $y - 1\frac{1}{2} = 2\frac{1}{2}$

4. $-3m = 6$

5. $\frac{1}{2}x = -2$

6. $3y + 4 = 13$

7. $2x + 18 = 8$

8. $5x - 2 = -22$

9. $\frac{1}{5}y - 3 = -4$

10. $8y = 90 - 2y$

11. $5a - 14 = 3a$

12. $y = 9y - 72$

13. $6y - 7 = 4y + 3$

14. $11x + 7 = 4x - 21$

15. $2(10 - 3c) = 22$

16. $3(x - 5) = 2(2x + 1)$

17. $5x - 2(x - 5) = 10$

18. $15y - 4(3y + 2) = -14$

In 19–24, graph the element of the set if $x \in$ {real numbers}.

19. $\{x \mid 3x + 2 = 20\}$

20. $\{x \mid 5x + 2x = -21\}$

21. $\{x \mid 5x = 7x + 8\}$

22. $\{x \mid 7x - (9x + 1) = -5\}$

23. $\{x \mid 4x + 2(x - 1) = 7\}$

24. $\{x \mid 2(4x + 3) = 3(3x + 3)\}$

7. Solving Equations Containing Fractions

The following are examples of equations that contain fractions:

$$\tfrac{1}{2}x + 20 = \tfrac{3}{4}x \qquad \frac{x}{2} + 20 = \frac{3x}{4} \qquad \frac{6}{x-1} = 3$$

To solve such equations, we transform the equation that contains fractions into a simpler equivalent equation that does not contain fractions. We can do this by multiplying both members of the equation by the least common denominator, L.C.D., of its denominators. This process is called *clearing the equation of fractions*.

EQUATIONS CONTAINING FRACTIONS WITH NUMERICAL DENOMINATORS

~~~~~~~~~ *MODEL PROBLEM* ~~~~~~~~~

Solve: $\dfrac{4x+1}{7} - \dfrac{2x-1}{6} = \dfrac{3}{2}$

How To Proceed	*Solution*
	$\dfrac{4x+1}{7} - \dfrac{2x-1}{6} = \dfrac{3}{2}$
1. Find the L.C.D.	L.C.D. $= 42$
2. Clear the equation of fractions by multiplying both of its members by the L.C.D.	$M_{42}: \; 42\left(\dfrac{4x+1}{7} - \dfrac{2x-1}{6}\right) = 42\left(\dfrac{3}{2}\right)$
3. Solve the resulting equation.	$\overset{6}{\cancel{42}}\left(\dfrac{4x+1}{\cancel{7}}\right) - \overset{7}{\cancel{42}}\left(\dfrac{2x-1}{\cancel{6}}\right) = \overset{21}{\cancel{42}}\left(\dfrac{3}{\cancel{2}}\right)$
	$6(4x+1) - 7(2x-1) = 63$
	$24x + 6 - 14x + 7 = 63$
	$10x + 13 = 63$
	$10x = 50$
	$x = 5$

Check: Substituting 5 for x in the given equation results in the true statement $3 - 1\tfrac{1}{2} = \tfrac{3}{2}$, or $1\tfrac{1}{2} = 1\tfrac{1}{2}$.

Answer: $x = 5$, or solution set is $\{5\}$.

Exercises

In 1–22, solve and check.

1. $\dfrac{5x}{7} = 35$

2. $\dfrac{3a-1}{4} = 2$

3. $\dfrac{1-2x}{5} = 1$

4. $\dfrac{y}{4} + \dfrac{y}{3} = \dfrac{7}{12}$

5. $\dfrac{a}{3} - \dfrac{a}{4} = \dfrac{2}{3}$

6. $\dfrac{17}{4} + \dfrac{3a}{8} = \dfrac{4a}{5}$

7. $\dfrac{x}{2} + \dfrac{x}{3} + \dfrac{x}{4} = 26$

8. $\tfrac{5}{8}x - \tfrac{1}{3}x = \tfrac{5}{6}x - 13$

9. $y + 1 - \tfrac{3}{4}y = \tfrac{1}{5}y$

10. $\dfrac{x}{6} + \dfrac{3x}{8} = \dfrac{5x}{8} - 5\tfrac{1}{4}$

11. $\dfrac{y}{4} - \dfrac{1}{2} = \dfrac{5}{12} - \dfrac{2y}{3}$

12. $\dfrac{4y}{15} + 2\dfrac{5}{12} = 3 - \dfrac{y}{5}$

13. $\dfrac{3y+1}{4} = \dfrac{13-y}{2}$

14. $\dfrac{4y+6}{5} = \dfrac{8y+5}{3}$

15. $\dfrac{a+3}{3} + \dfrac{a-3}{6} = 5$

16. $\dfrac{4}{3} + \dfrac{x-3}{4} = \dfrac{3x-1}{6}$

17. $\dfrac{x+2}{4} - \dfrac{x-3}{3} = \dfrac{1}{2}$

18. $\dfrac{3x+1}{4} = 2 - \dfrac{3-2x}{6}$

19. $\dfrac{x+4}{4} - \dfrac{3x-9}{7} = \dfrac{1}{2}$

20. $\dfrac{7x+5}{8} - \dfrac{3x+15}{10} = 2$

21. $\tfrac{2}{5}(4x-1) - \tfrac{3}{5}(x+1) = 7$

22. $\tfrac{3}{2}(2x+1) - \tfrac{1}{3}(4x-1) = -3\tfrac{1}{6}$

SOLVING FRACTIONAL EQUATIONS

A *fractional equation* is an equation that contains at least one fraction with the variable in the denominator. To solve such an equation, we use the same methods as those used in solving equations that contain fractions with numerical denominators. (See the model problems on the following page.)

EXTRANEOUS ROOTS OF A FRACTIONAL EQUATION

Sometimes, when we multiply both members of an equation by a polynomial that is the L.C.D. of the denominators of the equation, the resulting equation is not equivalent to the given equation. Consider the solution of the following equation:

$$\frac{m}{m-2} - \frac{m+4}{3(m-2)} = \frac{1}{3}$$

Multiply by the L.C.D., $3(m-2)$.

$$3(m-2)\left[\frac{m}{m-2} - \frac{m+4}{3(m-2)}\right] = 3(m-2)(\tfrac{1}{3})$$
$$3m - (m+4) = m - 2$$
$$3m - m - 4 = m - 2$$
$$m = 2$$

If we now check the solution by letting $m = 2$ in the given equation, on the preceding pages, we obtain:

$$\frac{2}{2-2} + \frac{2+4}{3(2-2)} \overset{?}{=} \frac{1}{3} \quad \text{or} \quad \frac{2}{0} - \frac{6}{0} \overset{?}{=} \frac{1}{3}$$

Since division by 0 is not defined, $\frac{2}{0}$ and $\frac{6}{0}$ are meaningless, Therefore, $m = 2$ is not a root of the given equation. However, $m = 2$ is a root of the derived equation $3m - (m+4) = m - 2$. Such a number, which is a root of the derived equation but is not a root of the original given equation, is called an **extraneous root**.

The equation $\dfrac{m}{m-2} - \dfrac{m+4}{3(m-2)} = \dfrac{1}{3}$ has no roots. Therefore, its solution set is the empty set, $\varnothing$

We see that when we transform a fractional equation by multiplying both of its members by a polynomial multiplier which represents zero, the resulting derived equation and the original equation may not be equivalent equations.

~~~~~~~~~~~~~~~ **MODEL PROBLEMS** ~~~~~~~~~~~~~~~

1. *Solve:* $\dfrac{x+5}{2x} = \dfrac{15}{x} - 2$

Solution:

1. $\dfrac{x+5}{2x} = \dfrac{15}{x} - 2$

2. L.C.D. $= 2x$

3. $2x\left(\dfrac{x+5}{2x}\right) = 2x\left(\dfrac{15}{x} - 2\right)$

4. $\overset{1}{2x}\left(\dfrac{x+5}{\underset{1}{2x}}\right) = \overset{2}{2x}\left(\dfrac{15}{\underset{1}{x}}\right) - 2x(2)$

5. $1(x+5) = 2(15) - 4x$

6. $x + 5 = 30 - 4x$

7. $x + 4x = 30 - 5$

8. $\quad 5x = 25$

9. $\quad\quad x = 5$

Check: $\dfrac{x+5}{2x} = \dfrac{15}{x} - 2$

Let $x = 5$: $\dfrac{5+5}{2(5)} \overset{?}{=} \dfrac{15}{5} - 2$

$\dfrac{10}{10} \overset{?}{=} 3 - 2$

$1 = 1$ (true)

Answer: $x = 5$, or solution set is $\{5\}$.

2. *Solve:* $\dfrac{7c-4}{c^2-c} = \dfrac{5}{c-1}$

Solution:

1. $\dfrac{7c-4}{c^2-c} = \dfrac{5}{c-1}$

2. L.C.D. $= c(c-1)$

3. $\overset{1}{c(c-1)}\left(\dfrac{7c-4}{c(c-1)}\right) = c(c-1)\left(\dfrac{5}{(c-1)}\right)$

4. $1(7c-4) = c(5)$

5. $7c - 4 = 5c$

6. $2c = 4$

7. $c = 2$

Check: $\dfrac{7c-4}{c^2-c}=\dfrac{5}{c-1}$

Let $c=2$: $\dfrac{7(2)-4}{(2)^2-2}\overset{?}{=}\dfrac{5}{2-1}$

$\dfrac{14-4}{4-2}\overset{?}{=}\dfrac{5}{1}$

$5=5$ (true)   *Answer:* $c=2$, or solution set is $\{2\}$.

KEEP IN MIND

When solving a fractional equation, check each solution of the derived equation in the original equation to see whether or not it belongs to the solution set of the original equation.

Exercises

In 1–27, solve and check.

1. $\dfrac{18}{x}-4=2$

2. $\dfrac{30}{x}-\dfrac{18}{2x}=7$

3. $\dfrac{3x-1}{4x}=\dfrac{2x+3}{3x}$

4. $\dfrac{2+x}{6x}=\dfrac{3}{5x}+\dfrac{1}{30}$

5. $\dfrac{9}{2b+1}=3$

6. $\dfrac{6}{3a-1}=\dfrac{3}{4}$

7. $\dfrac{3}{5-3t}=\dfrac{1}{2}$

8. $\dfrac{x}{x+2}=\dfrac{2}{3}$

9. $\dfrac{4x}{7+5x}=\dfrac{1}{3}$

10. $\dfrac{3}{x}=\dfrac{2}{5-x}$

11. $\dfrac{4}{3c}=\dfrac{3}{2c-1}$

12. $\dfrac{x+3}{x+1}=\dfrac{x+2}{x+4}$

13. $\dfrac{d-2}{d+1}=\dfrac{d+1}{d-2}$

14. $\dfrac{3x+2}{2x-3}=\dfrac{3x-2}{2x-5}$

15. $\dfrac{3r}{3r-1}=\dfrac{r+3}{r+2}$

16. $\dfrac{4}{x-1}=\dfrac{5}{x^2-x}$

17. $\dfrac{x^2+6x}{x^2-4}=\dfrac{x+3}{x-2}$

18. $\dfrac{7}{x-1}=5+\dfrac{7}{x-1}$

19. $\dfrac{4m}{m-2}-\dfrac{13}{3m-6}=\dfrac{1}{3}$

20, $\dfrac{1}{t-5}+\dfrac{1}{t+5}=\dfrac{8}{t^2-25}$

21. $\dfrac{w+1}{w-3}-\dfrac{2w-1}{w^2-9}=1$

22. $\dfrac{6}{x^2-1}=\dfrac{-3}{x+1}+\dfrac{5}{x-1}$

23. $\dfrac{8}{x^2-1}=\dfrac{7}{x-1}-\dfrac{4}{x+1}$

24. $\dfrac{3}{2b+4}-\dfrac{4}{b-2}=\dfrac{3}{2b^2-8}$

25. $\dfrac{t+1}{2t+6}-\dfrac{t-2}{2t-6}=\dfrac{9}{t^2-9}$

26. $\dfrac{20}{y^2-4}=\dfrac{1}{y-2}-\dfrac{5}{y+2}$

27. $\dfrac{30}{9-y^2}=\dfrac{5}{3+y}+\dfrac{2}{3-y}$

8. Solving Verbal Problems Using One Variable and a First-Degree Equation

Now we are ready to solve verbal problems algebraically.

To solve a verbal problem by using a first-degree equation involving one variable, first read the problem carefully until you understand what is given in the problem and what is to be found. Then use the following:

Procedure. To solve a verbal problem:
1. **Represent the unknowns in terms of a variable.**
2. **Translate the relationships stated in the problem into an equation.**
3. **Solve the equation to find its root.**
4. **Check the answer by testing it in the word statement of the original problem to see that it satisfies all the required conditions.**

NUMBER PROBLEMS

~~~~~~~~~~~~~~~ *MODEL PROBLEM* ~~~~~~~~~~~~~~~

Five times a number, decreased by 18, equals three times the number, increased by 6. Find the number.

How To Proceed	*Solution*
1. *Represent* the unknown by a variable.	Let $x =$ the number.
2. *Translate* the relationship involving the unknown into an equation.	Five times the number, decreased by 18 $\quad$ equals $\quad$ three times the number, increased by six.
	$5x - 18 \qquad = \qquad 3x + 6$
3. *Solve* the equation.	$5x - 18 = 3x + 6$
	$2x = 24$
	$x = 12$
4. *Check* in the original problem.	Does 12 check in the original problem?
	$5(12) - 18 \stackrel{?}{=} 3(12) + 6$
	$42 = 42$ (true)

Answer: The number is 12.

~~~~~~

Exercises

1. The larger of two numbers is twice the smaller. If the sum of the numbers is 120, find the numbers.
2. The larger of two numbers is 1 less than three times the smaller. The difference between the two numbers is 9. Find the numbers.
3. The larger of two numbers exceeds twice the smaller by 6. If the sum of the numbers is 30, find the numbers.
4. If 10 times a number is increased by 4, the result is 12 more than 9 times the number. Find the number.
5. If 14 is added to a certain number and the sum is multiplied by 2, the result is equal to 8 times the number, decreased by 14. Find the number.
6. The larger of two numbers is 1 more than 3 times the smaller. If 8 times the smaller is decreased by 2 times the larger, the result is 10. Find the numbers.
7. The larger of two numbers is 1 more than twice the smaller. Three times the larger exceeds 5 times the smaller by 10. Find the numbers.
8. Separate 144 into two parts such that one part will be 12 less than twice the other.

CONSECUTIVE INTEGER PROBLEMS

PREPARING TO SOLVE CONSECUTIVE INTEGER PROBLEMS

An *integer* is a member of set I where $I = \{\ldots, -3, -2, -1, 0, 1, 2, 3, \ldots\}$.

An *even integer* is an integer which is twice another integer. For example, 4 and -12 are even integers.

An *odd integer* is an integer which is not an even integer. For example, 3 and -9 are odd integers.

Consecutive integers are integers which differ by 1. Each of the following is a set of three consecutive integers arranged in increasing order from left to right:

$$\{2, 3, 4\} \qquad \{-7, -6, -5\}$$

In general, a set of three consecutive integers may be represented by:

$$\{n, n + 1, n + 2\} \text{ if } n \in \{\text{integers}\}$$

Consecutive even integers are even integers which differ by 2. Each of the following is a set of three consecutive even integers arranged in increasing order from left to right:

$$\{4, 6, 8\} \qquad \{-6, -4, -2\}$$

In general, a set of three consecutive even integers may be represented by:

$$\{n, n + 2, n + 4\} \text{ if } n \in \{\text{even integers}\}$$

Consecutive odd integers are odd integers which differ by 2. Each of the following is a set of three consecutive odd integers arranged in increasing order from left to right:

$$\{1, 3, 5\} \qquad \{-9, -7, -5\}$$

In general, a set of three consecutive odd integers may be represented by:

$$\{n, n+2, n+4\} \text{ if } n \in \{\text{odd integers}\}$$

KEEP IN MIND

1. Consecutive integers differ by 1.
2. Consecutive even integers and also consecutive odd integers differ by 2.

SOLVING CONSECUTIVE INTEGER PROBLEMS

∾∾∾∾∾ MODEL PROBLEM ∾∾∾∾∾

Find 3 consecutive odd integers such that 5 times the first, decreased by the second exceeds twice the third by 16.

Solution:

Let n = the first consecutive odd integer.

Then $n + 2$ = the second consecutive odd integer.

And $n + 4$ = the third consecutive odd integer.

Five times the first, decreased by the second	is	16 more than twice the third.

$$5n - (n+2) = 2(n+4) + 16$$
$$5n - n - 2 = 2n + 8 + 16$$
$$4n - 2 = 2n + 24$$
$$2n = 26$$
$$n = 13$$
$$n + 2 = 15, \; n + 4 = 17$$

Check in the original problem: The integers 13, 15, 17 are consecutive odd integers. They satisfy the condition in the given problem, 5(13), or 65, decreased by 15, that is, 50, does exceed 2(17), or 34, by 16.

Answer: The odd integers are 13, 15, and 17.

Exercises

1. Find three consecutive integers whose sum is (*a*) 99 and (*b*) −57.
2. Find three consecutive even integers whose sum is (*a*) 48 and (*b*) −60.
3. Find four consecutive odd integers whose sum is (*a*) 112 and (*b*) −136.
4. Find three consecutive even integers such that the sum of the smallest and twice the second is 20 more than the third.
5. Find three consecutive integers such that the first increased by twice the second exceeds the third by 24.
6. Find three consecutive even integers such that twice the sum of the second and the third exceeds 3 times the first by 34.
7. Find three consecutive odd integers such that the product of the second and third exceeds the square of the first by 50.
8. Prove that the sum of 3 consecutive even integers cannot be 40.
9. Prove that the sum of 3 consecutive odd integers cannot be 40.
10. Prove that the sum of three consecutive integers is equal to 3 times the middle integer.

COIN OR STAMP PROBLEMS

PREPARING TO SOLVE COIN OR STAMP PROBLEMS

The total value, *T*, of a number of coins or stamps of the same kind equals the number of the coins or stamps, *N*, multiplied by the value of each coin or stamp, *V*.

This relationship may be expressed by the formula $T = NV$.

Thus, 6 nickels have a total value of 6(5) = 30 cents.

Twelve 37-cent stamps have a total value of 12(37) = 444 cents or \$4.44.

Note. In $T = NV$, *T* and *V* must be expressed in the same unit of money; that is, if *V* is in dollars, then *T* must be in dollars.

In solving problems which deal with groups of coins or stamps of different denominations—for example, nickels, dimes, and quarters—it is often helpful to represent the values of the different groups of coins or stamps in terms of the same unit of money. In the following examples, the unit of money is cents:

The value of 7 nickels in cents is 7(5), or 35 cents.

The value of *d* dimes in cents is *d*(10), or 10*d* cents.

The value of $(3q + 1)$ quarters in cents in $(3q + 1)25$, or $(75q + 25)$ cents.

The relationship involving the number of coins, *N*, each having the same value, *V*, and the total value of all the coins, *T*, may be expressed as $NV = T$.

```
┌──────────────── KEEP IN MIND ────────────────┐
│                                               │
│   Number of coins × Value of each = Total value of all │
│     in the group       coin in cents    the coins in cents │
│        (N)                (V)                (T)          │
│                                               │
│            NV = T      V = T/N      N = T/V   │
│                                               │
└───────────────────────────────────────────────┘
```

SOLVING COIN OR STAMP PROBLEMS

~~~~~~~~~~~~~~~~~~~~ *MODEL PROBLEM* ~~~~~~~~~~~~~~~~~~~~

In a child's bank, there is a collection of nickels, dimes, and quarters which amounts to $4.15. There are 4 times as many dimes as nickels, and there are 3 less quarters than nickels. How many coins of each kind are there?

Solution:

Let $n =$ the number of nickels.

	Number of coins ×	($\cent$) Value of each coin	($\cent$) = Total value
Nickel	n	5	$5n$
Dime	$4n$	10	$10(4n)$
Quarter	$n - 3$	25	$25(n - 3)$

The total value of all the coins is 415 cents.

$$5n + 10(4n) + 25(n - 3) = 415$$
$$5n + 40n + 25n - 75 = 415$$
$$70n - 75 = 415$$
$$70n = 490$$
$$n = 7$$
$$4n = 28$$
$$n - 3 = 4$$

Answer: There are 7 nickels, 28 dimes, and 4 quarters.

~~~~~~

Exercises

1. Will has $2.05 in dimes and quarters. He has 4 more quarters than dimes. Find the number he has of each kind of coin.
2. Harry has $2.30 in nickels and dimes. The number of dimes is 7 less than the number of nickels. Find the number he has of each kind of coin.
3. Tilda deposited in her savings account $4.50 in nickels, quarters, and dimes. The number of dimes exceeded the number of nickels by 5, and the number of quarters was 16 less than the number of nickels. Find the number of each kind of coin she deposited.
4. A class contributed $8.40 in dimes and quarters to the Red Cross. In all, there were 45 coins. How many coins of each kind were there?
5. A purse contains $4.70 in nickels and quarters. There are 30 coins in all. How many coins of each kind are there?
6. A postal clerk sold 25 stamps for $7.71. Some were 23-cent stamps and some were 37-cent stamps. How many of each kind of stamp did he sell?
7. In Yvette's bank, there is $2.60 in pennies, nickels, and dimes. In all, there are 45 coins. If there are twice as many nickels as pennies, how many coins of each kind are there in the bank?
8. Rose counted her money and found that her 25 coins which were nickels, dimes, and quarters were worth $3.20. The number of dimes exceeded the number of nickels by 4. How many coins of each kind did she have?
9. Prove that it is impossible to have $4.50 in dimes and quarters with the number of quarters being twice the number of dimes.

AGE PROBLEMS

PREPARING TO SOLVE AGE PROBLEMS

If Paul is 15 years old now, 2 years from now he will be $15 + 2$, or 17 years old; 2 years ago, Paul was $15 - 2$, or 13 years old.

```
┌──────────── KEEP IN MIND ────────────┐
│  1. To represent a person's age a number of years hence, │
│     add that number of years to his present age.          │
│  2. To represent a person's age a number of years ago,    │
│     subtract that number of years from his present age.    │
└───────────────────────────────────────┘
```

~~~~~~~~~~ *MODEL PROBLEMS* ~~~~~~~~~~

Represent the age of a person in years:
a. 5 years hence if his present age is x years — *Ans.* $x + 5$
b. 5 years ago if his present age is x years — *Ans.* $x - 5$
c. in y years if his present age is 25 years — *Ans.* $25 + y$
d. y years ago if his present age is 25 years — *Ans.* $25 - y$
e. y years ago if his present age is n years — *Ans.* $n - y$

SOLVING AGE PROBLEMS

~~~~~~~~~~ *MODEL PROBLEM* ~~~~~~~~~~

Ray is 20 years older than Bill. Five years ago, Ray was 5 times as old as Bill was then. Find the present age of both Ray and Bill.

Solution:

Let $x =$ Bill's present age in years.
Then $x + 20 =$ Ray's present age in years.
And $x - 5 =$ Bill's age 5 years ago.
And $x + 15 =$ Ray's age 5 years ago.

Ray was 5 times as old as Bill was 5 years ago.

$x + 15 = 5(x - 5)$	*Check* in the original problem:
$x + 15 = 5x - 25$	Is Ray 20 years older than Bill?
$40 = 4x$	$30 \overset{?}{=} 10 + 20,\ 30 = 30$ (true)
$10 = x$	Was Ray 5 times as old as Bill 5 years ago?
$x + 20 = 30$	$30 - 5 \overset{?}{=} 5(10 - 5),\ 25 = 25$ (true)

Answer: Bill's age is 10 years. Ray's age is 30 years.

Exercises

1. Walter is 3 times as old as Martin. Ten years from now, Walter will be twice as old as Martin will be then. How old is each now?
2. Vanessa is 3 times as old as Lillian. Ten years from now, Vanessa's age will exceed twice Lillian's age at that time by 1 year. Find the present age of both Lillian and Vanessa.
3. Sarah's age exceeds Dana's age by 16 years. Four years ago, Sarah was twice as old as Dana was then. Find the present age of each.
4. The sum of Finn's age and Ellen's age is 40 years. Finn's age 10 years from

now will be 1 year less than 4 times Ellen's age 6 years ago. Find their present ages.

5. Delroy is 30 years old and Edward is 15 years old. In how many years will Delroy be $1\frac{1}{2}$ times as old as Edward will be then?

RATIO PROBLEMS

PREPARING TO SOLVE RATIO PROBLEMS

Two numbers whose ratio is $5:2$ may be represented by $5x$ and $2x$ because the ratio of $5x$ to $2x$ may be written $5x:2x$, or $\dfrac{5x}{2x}$, which, simplified, becomes $\frac{5}{2}$, or $5:2$.

In general, if a, b, and x are numbers ($b \neq 0$ and $x \neq 0$), ax and bx represent two numbers whose ratio is $a:b$.

Likewise, three numbers whose continued ratio is $3:2:1$ may be represented by $3x$, $2x$, and $1x$. This is so because the ratio $3x:2x = 3:2$, and the ratio $2x:1x = 2:1$. The separate ratios $3:2$ and $2:1$ may be written as the continued ratio $3:2:1$.

SOLVING RATIO PROBLEMS

~~~~~~~~~~~~~ *MODEL PROBLEM* ~~~~~~~~~~~~~

An angle whose degree measure is 120 is divided into three angles in the ratio $2:3:7$. Find the degree measure of each angle.

Solution:
    Let $2x =$ the degree measure of the first angle.
    Then $3x =$ the degree measure of the second angle.
    And $7x =$ the degree measure of the third angle.

The sum of the degree measures of the three angles is 120.

$2x + 3x + 7x = 120$	*Check* in the original problem:
$12x = 120$	Is the sum of the degree measures of the three
$x = 10$	angles 120?
	$20 + 30 + 70 = 120$ (true)
$2x = 20,\ 3x = 30,\ 7x = 70$	Is the ratio of the degree measures of the three
	angles $2:3:7$?
	$20:30:70 = 2:3:7$ (true)

Answer: The degree measures of the three angles are 20, 30, 70.

~~~~~~

Exercises

1. Two numbers have the ratio 5 : 3. Their sum is 88. Find the numbers.
2. Find two positive numbers whose ratio is 4 : 1 and whose difference is 36.
3. A line segment 64 inches in length is divided into two parts which are in the ratio 1 : 7. Find the length of each part.
4. The ratio of Paul's age to Luis's age is 2 : 3. The sum of their ages is 35 years. Find the age of each one.
5. Divide $3300 into three parts whose ratio is 1 : 3 : 7.
6. An angle, whose degree measure is 180, is divided into three angles. The degree measures of the first two angles are in the ratio 2 : 3. The degree measure of the third angle is equal to the sum of the degree measures of the first two angles. Find the degree measure of each angle.
7. The ratio of the amount of money Carl has to the amount of money Donald has is 7 : 3. If Carl gives Donald $20, the two will then have equal amounts. Find the original amount that each one has.
8. Two positive numbers have the ratio 2 : 3. The larger number is 30 more than one-half of the smaller number. Find the numbers.
9. Three positive numbers are in the ratio 7 : 3 : 2. The sum of the smallest number and the largest number exceeds twice the remaining number by 30. Find the three numbers.
10. Sean's age and Helen's age are in the ratio of 3 : 5. Two years ago, Helen was twice as old as Sean was then. Find their present ages.
11. The ratio of Zia's age to Joe's age is 4 : 1. Twenty years from now, Zia will be twice as old as Joe will be then. Find their present ages.
12. Four years ago, the ratio of Grace's age to Marta's age was 5 : 2. At present, Grace's age exceeds twice Marta's age by 2 years. How old were Grace and Marta 4 years ago?

UNIFORM MOTION PROBLEMS

PREPARING TO SOLVE UNIFORM MOTION PROBLEMS

If a car traveled at the rate of 40 miles per hour, in 3 hours it traveled 3(40), or 120 miles. In this case, the three related quantities are:

1. The *distance* traveled, 120 miles (mi.).
2. The rate of speed, or *rate*, 40 miles per hour (mph).
3. The *time* traveled, 3 hours (hr.).

The relation involving the distance, D, the rate, R, and the time, T, may be expressed in the following ways:

$$D = RT \qquad T = \frac{D}{R} \qquad R = \frac{D}{T}$$

In our work, *rate* will represent either of the following:

1. The **uniform rate of speed**, which represents a rate of speed that does not change throughout a trip.

2. The **average rate of speed**, which represents the total distance traveled divided by the total time traveled. Thus, a car which traveled 100 miles in 2 hours was traveling at an average rate of $100 \div 2$, or 50 miles per hour.

Note that the rate, time, and distance must be expressed in corresponding units. For example, if the rate is measured in miles per hour, the time must be measured in hours and the distance must be measured in miles.

KEEP IN MIND

$$D = RT \qquad T = \frac{D}{R} \qquad R = \frac{D}{T}$$

SOLVING UNIFORM MOTION PROBLEMS

In solving a uniform motion problem, we can use a chart to organize the facts in the problem and to make the relationships more evident. It is also helpful to draw a diagram to show the relationship of the distances which are involved in the problem.

〰〰〰〰〰〰〰〰 *MODEL PROBLEMS* 〰〰〰〰〰〰〰〰

1. Two trains whose rates differ by 6 miles per hour start at the same time from stations which are 273 miles apart. They meet in $3\frac{1}{2}$ hours. Find the rate of each train.

Solution: Let $r =$ the rate of the slow train.
    Then $r + 6 =$ the rate of the fast train.
First fill in the time and rate of each train.
Then represent the distance for each train.

	(mph) Rate	×	(hr.) Time	=	(mi.) Distance
Slow train	r		3.5		$3.5r$
Fast train	$r+6$		3.5		$3.5(r+6)$

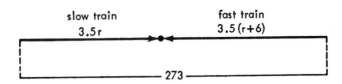

The total distance between the stations was 273 miles.

$$3.5r + 3.5(r+6) = 273$$
$$3.5r + 3.5r + 21 = 273$$
$$7r = 252$$
$$r = 36$$
$$r + 6 = 42$$

Check in the original problem:

The distance the slow train traveled was 3.5(36) or 126 miles.

The distance the fast train traveled was 3.5(42) or 147 miles.

Was the total distance traveled $\overline{273}$ miles? (true)

Answer: The rate of the slow train is 36 mph. The rate of the fast train is 42 mph.

2. How far can a man drive out into the country at an average rate of 50 mph and return over the same road at the average rate of 40 mph if he travels a total of 9 hours.

Solution: Let $h =$ the number of hours he spent traveling out.

Then $9 - h =$ the number of hours he spent traveling back.

First fill in the rate and time for each trip.

Then represent the distance for each trip.

	(mph) Rate	×	(hr.) Time	=	(mi.) Distance
Trip out	50		h		$50h$
Trip back	40		$9-h$		$40(9-h)$

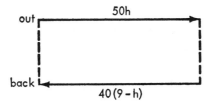

The distance out is the same as the distance back.

$$50h = 40(9 - h)$$
$$50h = 360 - 40h$$
$$50h + 40h = 360$$
$$90h = 360$$
$$h = 4$$

The distance out is $50h = 200$

Check in the original problem:
The distance traveled out is 50(4), or 200 miles.
The distance traveled back is 40(5), or 200 miles.
Are the distances the same? $200 = 200$ (true)
Is the total time 9 hours? $4 + 5 = 9$ (true)

Answer: He can travel 200 miles out into the country.

~~~~~~

Exercises

1. Two automobiles start at the same time and travel in opposite directions. The first averages 28 miles per hour and the second averages 35 miles per hour. In how many hours are they 189 miles apart?
2. Two planes start at 9 A.M. at two airports which are 2100 miles apart and fly toward each other at average rates of 150 mph and 200 mph. At what time will they pass each other?
3. A passenger train and a freight train start from the same point at the same time and travel in opposite directions. The passenger train traveled 3 times as fast as the freight train. In 5 hours, they were 360 miles apart. Find the rate of each train.
4. Marie and Anne start from the same point at the same time and travel in the same direction at rates which are in the ratio of 3 : 2. In $2\frac{1}{2}$ hours, they are 30 miles apart. Find the rate of each girl.
5. Arthur and Jim are 292 miles apart. At 10 A.M., they start toward each other; and at 1 P.M., they are 40 miles apart. If Arthur's average rate exceeds Jim's average rate by 8 mph., find the rate of each.

6. Two cars start from the same place at the same time and travel in opposite directions. At the end of 5 hours, they are 340 miles apart. If the average rate of the slower car exceeds $\frac{1}{2}$ of the average rate of the faster car by 11 mph, find the rate of each car.

7. Ms. Atkins spent 7 hours in driving from her home to Albany, a distance of 238 miles. Before noon, she averaged 32 miles per hour. After noon, she averaged 39 miles per hour. Find the number of hours she traveled at each rate of speed.

8. James and Thomas are in two cities which are 186 miles apart and travel toward each other. James' average rate was 32 mph and Thomas' average rate was 36 mph. If James started at 9:00 A.M. and Thomas started at 9:30 A.M., at what time did they meet?

9. Two trains, starting at the same time from stations 396 miles apart, meet in $4\frac{1}{2}$ hours. How far has the faster train traveled when it meets the slower one, if the difference in their rates is 8 miles per hour?

10. Portia leaves a place 3 hours before Harry. Both travel on the same road and in the same direction. If Portia travels 20 miles per hour and Harry travels 25 miles per hour, in how many hours will Harry overtake Portia?

11. Mr. Jackson leaves on a trip planning to travel 24 miles per hour. How fast must his son James travel to overtake him in 5 hours if he starts $2\frac{1}{2}$ hours after his father?

12. A speedboat traveling at 30 miles per hour traveled the length of a lake in 20 minutes less time than when traveling at 24 miles per hour. Find the length of the lake.

13. A man hiked a distance of 9 miles into the country. He rode back in a car at the rate of 27 miles per hour. If the entire trip took 3 hours and 20 minutes, find his rate while walking.

14. Mr. Fox drove from his home to his factory at the rate of 25 miles per hour and returned by a different route at the rate of 30 miles per hour. The route by which he returned was 5 miles longer than the route by which he went. The return trip took 10 minutes less than the trip out. Find the distance Mr. Fox traveled each way.

15. Tanya walked from A to B at the rate of $3\frac{1}{3}$ miles per hour and returned by a different route at the rate of 4 miles per hour. The route by which she returned was one mile shorter than the route by which she went and the return trip took 45 minutes less time. Find the distance between A and B by the shorter route.

SOLVING UNIFORM MOTION PROBLEMS INVOLVING FRACTIONS

〰〰〰〰〰〰〰 *MODEL PROBLEM* 〰〰〰〰〰〰〰

A man traveled 150 miles at a certain average rate. By increasing his average rate by 20 mph, he traveled 250 miles in the same time that he spent on the 150-mile trip. Find his average rate on the first trip.

Solution:

Let $r =$ the average rate on the first trip in miles per hour.
Then $r + 20 =$ the average rate on the second trip in miles per hour.

	(mi.) Distance	$\div$ (mph) Rate	= (hr.) Time
First Trip	150	r	$\dfrac{150}{r}$
Second Trip	250	$r + 20$	$\dfrac{250}{r + 20}$

$$\text{Time} = \frac{\text{Distance}}{\text{Rate}}$$

The time for the first trip was the same as the time for the second trip.

$$\frac{150}{r} = \frac{250}{r + 20}$$

Multiply both members of the equation by $r(r + 20)$.

$$r(r + 20)\left(\frac{150}{r}\right) = r(r + 20)\left(\frac{250}{r + 20}\right)$$

$$150(r + 20) = 250r$$
$$150r + 3000 = 250r$$
$$3000 = 100r$$
$$30 = r$$
$$r + 20 = 50$$

Answer: Average rate on the first trip was 30 mph.

Exercises

1. On a trip, a motorist drove 150 miles in the morning and 50 miles in the afternoon. His average rate in the morning was twice his average rate in the afternoon. He spent 5 hours in driving. Find his average rate on each part of the trip.

2. Ms. Bliss traveled 150 miles at a certain average rate. By increasing her rate 15 mph, she traveled 225 miles in the same time that she spent on the 150-mile trip. Find her average rate on the first trip.

3. How far can Mr. Rabin drive into the country at the average rate of 30 mph and return over the same road at the rate of 40 mph if he travels a total of 7 hours.

4. Carla rode away from home at the rate of 32 mph. She walked back on the same road at the rate of 4 mph. The round trip required $2\frac{1}{4}$ hours. How far did Carla walk?

5. A pilot plans to make a flight lasting 2 hours and 30 minutes. How far can he fly from his base at the rate of 300 mph and return over the same route at the rate of 200 mph?

6. Mr. Sawyer drove his car from his home to New York at the rate of 45 mph and returned over the same road at the rate of 40 mph. If his time returning exceeded his time going by 30 minutes, find his time going and his time returning.

7. The rate of a passenger train exceeds the rate of a freight train by 20 mph. It takes the passenger train $\frac{1}{2}$ as much time to travel 160 miles as it does the freight train. Find the rate of each train.

8. Stanley drove 240 miles at a certain rate of speed. If he had traveled 15 mph faster, he would have been able to travel 30 miles further in $\frac{3}{4}$ of the time that he spent on his trip. What was Stanley's rate on the trip?

9. Each of two cars, whose average rates are in the ratio of 4 : 5, travels a distance of 160 miles. If the fast car travels $\frac{1}{2}$ hour less than the slow car, find the average rate of each car.

STREAM AND WIND MOTION PROBLEMS
(MOTION PROBLEMS INVOLVING CURRENTS)

PREPARING TO SOLVE STREAM AND WIND MOTION PROBLEMS

Let us suppose that the motors of a boat supply enough power for the boat to travel at the rate of 25 mph in a body of still water, where there is no current. When the boat travels in a body of water where there is a current, the boat will move faster than 25 mph when it is traveling downstream with the current. It will move slower than 25 mph when it is traveling upstream against the current. For example, if the boat is traveling in a river that flows at the rate of 3 mph, the rate of the boat traveling downstream with the current will be $25 + 3$, or 28 mph. Its rate traveling upstream against the current will be $25 - 3$, or 22 mph.

The rate of speed of an airplane is similarly affected by an air current. Suppose the motors of the plane supply enough power for it to travel at the rate of 400 mph in still air, and there is a wind blowing at the rate of 20 mph. When flying with the wind, the plane will be traveling at the rate of $400 + 20$, or 420 mph. When flying against the wind, the plane will be traveling at the rate of $400 - 20$, or 380 mph.

KEEP IN MIND

If $r =$ the rate in still water or still air,
and $c =$ the rate of the water current or air current,
then $r + c =$ the rate traveling with the current,
and $r - c =$ the rate traveling against the current.

SOLVING STREAM AND WIND MOTION PROBLEMS

～～～～～～～ MODEL PROBLEM ～～～～～～～

A ship can travel at the rate of 15 mph in still water. To travel 60 miles downstream in a river, the ship requires $\frac{2}{3}$ of the time that it requires to travel the same distance upstream in the same river. Find the rate of the river's current.

Solution:

Note. When a ship can sail r mph in still water and it is sailing in a stream whose current is flowing at the rate of c mph, then:

1. The ship can sail *downstream*, with the current, at the rate of $(r + c)$ mph.

2. The ship can sail *upstream*, against the current, at the rate of $(r - c)$ mph.

Let $c =$ the rate of the current.
Then $15 + c =$ the rate of the ship downstream.
And $15 - c =$ the rate of the ship upstream.

	(mi.) Distance	(mph) ÷ Rate	(hr.) = Time
Upstream	60	$15 - c$	$\dfrac{60}{15 - c}$
Downstream	60	$15 + c$	$\dfrac{60}{15 + c}$

$$\text{Time} = \frac{\text{Distance}}{\text{Rate}}$$

Time downstream equals $\frac{2}{3}$ of the time upstream.

$$\frac{60}{15 + c} = \frac{2}{3}\left(\frac{60}{15 - c}\right)$$

$$\frac{60}{15 + c} = \frac{40}{15 - c} \qquad (\tfrac{2}{3} \times 60 = 40)$$

Multiply both members of the equation by $(15 + c)(15 - c)$.

$$(15 - c)60 = (15 + c)40$$
$$900 - 60c = 600 + 40c$$
$$300 = 100c$$
$$3 = c$$

Check: Is the time downstream equal to $\frac{2}{3}$ of the time upstream?

Time downstream is $\dfrac{60}{15 + 3} = \dfrac{60}{18} = \dfrac{10}{3}$ hours.

Time upstream is $\dfrac{60}{15 - 3} = \dfrac{60}{12} = 5$ hours.

Since $\frac{2}{3} \times 5 = \frac{10}{3}$, then $\frac{10}{3}$ is $\frac{2}{3}$ of 5. (true)

Answer: The rate of the river's current is 3 mph.

Exercises

1. Mr. Sweeney has a motorboat that can travel 14 mph in still water. He wishes to make a trip on a river whose current flows at a rate of 2 mph. If he has 7 hours at his disposal, how far can he travel up the river and return?
2. Gayle traveled 20 miles upstream in a river, whose current flowed at 3 miles per hour, in $2\frac{1}{2}$ times as much time as she required to return downstream. Find Gayle's rate in still water.

3. A boat can travel 8 miles an hour in still water. If it can travel 15 miles down a stream in the same time that it can travel 9 miles up the stream, what is the rate of the stream?

4. A plane which can fly 100 miles an hour in still air can fly 500 miles with a wind which is blowing at a certain rate in $\frac{5}{8}$ of the time it would require to fly 500 miles against a wind blowing at the same rate. What was the rate of the wind?

5. A light plane can fly 120 mph in still air. Flying with the wind, it can fly 700 miles in a certain time. Flying against the wind, it can fly only $\frac{5}{7}$ of this distance in the same time. Find the rate of the wind.

PER CENT AND PERCENTAGE PROBLEMS

PREPARING TO SOLVE PER CENT AND PERCENTAGE PROBLEMS

Per cent means *per hundred* or *hundredths*. For example, 15% means $\frac{15}{100}$ or .15; 6% means $\frac{6}{100}$ or .06; $2\frac{1}{2}$% means $\frac{2\frac{1}{2}}{100}$, or .025; 100% means $\frac{100}{100}$ or 1; 125% $=\frac{125}{100}$ or 1.25.

Various types of business problems frequently involve per cents. For example, to find the amount of tax when $80 is taxed at the rate of 5%, we multiply $80 by 5%. Since $.05 \times 80 = 4$, we get $4 as the result. In this case, the three related quantities are:

1. The sum of money being taxed, the **base**, which is $80.
2. The rate of tax, the **rate**, which is 5% or .05.
3. The amount of tax, the **percentage**, which is $4.

The relation involving base, B, rate, R, and percentage, P, may be expressed as $P = BR$.

Note. In $P = BR$, P and B must be expressed in the same unit of money; that is, if B is in dollars, then P must be in dollars.

$$\boxed{\begin{array}{c} \textit{\textbf{KEEP IN MIND}} \\[6pt] P = BR \qquad R = \frac{P}{B} \qquad B = \frac{P}{R} \end{array}}$$

SOLVING PER CENT AND PERCENTAGE PROBLEMS

~~~~~~~~~~~~~ *MODEL PROBLEM* ~~~~~~~~~~~~~

Of 150 planes at an airbase, 135 took off on a mission. What per cent of the planes took off?

Solution:

Let $\dfrac{x}{100}$ = the per cent of the planes that took off.

$P = BR$ where $P = 135$, $B = 150$

$$135 = 150\left(\frac{x}{100}\right)$$

M_{100}: $100 \cdot 135 = 100 \cdot 150\left(\dfrac{x}{100}\right)$

$$13{,}500 = 150x$$

$$90 = x$$

$$\frac{x}{100} = \frac{90}{100} = 90\%$$

Check in the original problem:
Is 90% of 150 equal to 135?
$.90(150) = 135$ (true)

Answer: 90% of the planes took off.

~~~~~~

Exercises

1. The price of a new car is $16,800. Mr. Sawyer made a down payment of 25% of the price of the car when he bought it. How much was his down payment?
2. How much pure copper is in 80 pounds of an alloy which is 8% copper?
3. How much pure iodine is in 12 ounces of a solution which is $2\frac{1}{2}\%$ pure iodine?
4. Sally bought a dress at a "20% off" sale and saved $13. What was the marked price of the dress originally?
5. A businessman is required to collect a 6% sales tax. One day he collected $144 in taxes. Find the total amount of sales he made that day.
6. A merchant sold a radio for $60, which was 25% above its cost to him. Find the cost of the radio set to the merchant.
7. After Mr. Karp lost 12% of his investment, he had $4400 left. How much did he invest originally?
8. Mr. Crowley bought a $90 sport jacket. He had to pay $4.50 as a sales tax. What per cent of the price of the jacket is the tax?
9. The marked price of a coat is $119. What was the cost of the coat if the marked price represents a profit of 40%?

10. A dealer paid \$24 for a chair. At what price should he mark it for sale if he expects to allow a discount of 20% and wishes to make a profit of $33\frac{1}{3}\%$ on the cost price?

MIXTURE PROBLEMS

PREPARING TO SOLVE MIXTURE PROBLEMS

Many problems deal with the mixing of ingredients which have different costs. In solving these problems, it is helpful to express the total value of each ingredient in the same unit of money, such as cents. For example:

The value of 5 pounds of apples at 70 cents per pound is 5(70), or 350 cents.

The value of x pounds of sugar at 90 cents per pound is $x(90)$, or $90x$ cents.

The value of $(40 - x)$ gallons of water at 75 cents per gallon is $(40 - x)75$, or $75(40 - x)$ cents.

The relationship involving the number of units, N, each having the same value, V, and the total value of all the units, T, can be expressed as $NV = T$.

── *KEEP IN MIND* ──

Number of units of $\times$	Value of	=	Total value of all
the same kind	each unit		the units
(N)	(V)		(T)

$$NV = T \qquad V = \frac{T}{N} \qquad N = \frac{T}{V}$$

SOLVING MIXTURE PROBLEMS

In solving mixture problems, we can use a chart to organize the facts of the problem compactly.

〜〜〜〜〜 *MODEL PROBLEM* 〜〜〜〜〜

A dealer wishes to mix beans worth \$1.20 per pound with beans worth \$1.60 per pound to produce a mixture of 20 pounds of beans which he can sell for \$1.50 per pound. How many pounds of each type should he use?

Solution:

Let n = the number of pounds of \$1.60 beans.

Then $20 - n$ = the number of pounds of \$1.20 beans.

	(lb.) Number	($) × Price per pound =	($) Total value
$1.60	n	1.60	$1.60n$
$1.20	$20 - n$	1.20	$1.20(20 - n)$
Mixture	20	1.50	$1.50(20)$

The total value of the $1.60 beans and the $1.20 beans equals the value of the mixture.

$$1.60n + 1.20(20 - n) = 1.50(20)$$
$$1.60n + 24 - 1.20n = 30$$
$$0.40n + 24 = 30$$
$$0.40n = 6$$
$$n = 15$$
$$20 - n = 5$$

Check in the original problem.
Is the total number of pounds in the
   mixture 20?
$15 + 5 = 20$ (true)
Does the total value of the ingredients
   equal the value of the mixture?
Value of 15 lb. at $1.60 per lb. = $24.00
Value of 5 lb. at $1.20 per lb. =   6.00
Total value   = $30.00
Value of 20 lb. at $1.50 per lb. = $30.00

Answer: 15 lb. of the $1.60 beans; 5 lb. of the $1.20 beans.

Exercises

1. A merchant mixed nuts worth $3.50 per pound with nuts worth $5.60 per pound. If he wishes to make a mixture of 30 pounds to sell at $5.25 per pound, how many pounds of each should he use?
2. How many pounds of $5.60 nuts and how many pounds of $3.50 nuts must a dealer use to produce a mixture of 90 pounds to sell at $5.25 per pound?
3. A garden store has seeds worth $1.40 per pound and seeds worth $1.80 per pound. How many pounds of each must be used to make 300 pounds to sell at $1.50 per pound?
4. How many pounds of nuts worth $4.20 per pound must be mixed with 12 pounds of nuts worth $3.00 per pound to produce a mixture which can be sold for $3.90 per pound?

5. How many pounds of candy worth $4.40 per pound must be mixed with 36 pounds of candy worth $6.40 per pound to produce a mixture worth $6.00 per pound?

6. At a matinee performance, orchestra seats are sold for $5 each and balcony seats for $3 each. Explain why it is impossible to sell 700 tickets for this performance with the total receipts amounting to $2975.

PER CENT MIXTURE PROBLEMS

PREPARING TO SOLVE PER CENT MIXTURE PROBLEMS

The quantity of salt in 50 oz. of a 10% solution of salt and water is 10% of 50, which is equal to .10(50), or 5 oz.

The quantity of butterfat in x lb. of milk containing 6% butterfat can be represented by .06(x), or .06x lb.

The quantity of pure acid in $(30 - x)$ oz. of a 20% solution of acid in water may be represented by .20$(30 - x)$ oz.

The number of units, S, of a solution (mixture) that contains a given pure substance × the part, R, of the solution (mixture) which is that pure substance = the quantity, Q, of that pure substance in the solution (mixture).

This relationship can be expressed by the formula $SR = Q$.

KEEP IN MIND

$$SR = Q \qquad S = \frac{Q}{R} \qquad R = \frac{Q}{S}$$

SOLVING PER CENT MIXTURE PROBLEMS

~~~~~~~ *MODEL PROBLEMS* ~~~~~~~

1. How much pure acid must be added to 30 ounces of an acid solution which is 40% acid in order to produce a solution which is 50% acid?

Solution:

Let $n =$ the number of ounces of pure acid to be added.

Then $30 + n =$ the number of ounces that the new solution weighs.

	(oz.) Solution	(%) × Part pure acid	(oz.) = Quantity of pure acid
Original solution	30	.40	.40(30)
Pure acid to be added	n	100% = 1.00	1(n)
New solution	$30 + n$	.50	.50(30 + n)

The amount of pure acid in the original solution
+ the amount of pure acid added = the amount
of pure acid in the new solution.

$.40(30) + 1(n) = .50(30 + n)$
$12 + n = 15 + .50n$
$n - .50n = 15 - 12$
$.50n = 3$
$n = 6$

Check in the original problem:
If 6 oz. of pure acid are added to the original solution, will the new solution be 50% acid?
40% of 30 = .40(30) = 12
12 + 6 = 18
50% of (30 + 6) = .50(36) = 18 (true)

Answer: 6 ounces of pure acid must be added.

2. A chemist has 160 pints of a solution which is 20% acid. How much water must be evaporated to make a solution which is 40% acid?

Solution:
Let n = the number of pints of water to be evaporated.
Then $160 - n$ = the number of pints in the new solution.

	(pt.) Solution	(%) × Part pure acid	(pt.) = Quantity of pure acid
Original solution	160	.20	.20(160)
Water evaporated	n	0	0(n)
New solution	$160 - n$	.40	.40(160 - n)

The amount of pure acid in the original solution — the amount of pure acid in the water that was evaporated = the amount of pure acid in the new solution.

$$.20(160) - 0(n) = .40(160 - n)$$
$$32 - 0 = 64 - .40n$$
$$.40n = 64 - 32$$
$$.40n = 32$$
$$n = 80$$

Check in the original problem:
If 80 pints of water are evaporated from the original solution, will the resulting solution be 40% acid?
$$20\% \text{ of } 160 = .20(160) = 32$$
$$32 - 0 = 32$$
$$160 - 80 = 80$$
$$40\% \text{ of } 80 = .40(80) = 32(\text{ true})$$

Answer: 80 pints of water must be evaporated.

Exercises

1. A chemist has one solution that is 30% pure salt and another solution that is 60% pure salt. How many ounces of each solution must she use to produce 60 ounces of a solution that is 50% pure salt?

2. A farmer has some cream that is 24% butterfat and some cream that is 18% butterfat. How many quarts of each must he use to produce 90 quarts of cream that is 22% butterfat?

3. A chemist has one solution that is 40% pure acid and another solution that is 75% pure acid. How many pints of each solution must she use to produce 60 pints of a solution that is 50% pure acid?

4. How many pints of a solution that is 30% alcohol must be mixed with 21 pints of a solution that is 80% alcohol to produce a mixture that is 60% alcohol?

5. How many ounces of a silver alloy that is 30% silver must be mixed with 18 ounces of a silver alloy that is 12% silver to produce a new alloy that is 18% silver?

6. How many quarts of a solution that is 75% acid must be mixed with 16 quarts of a solution that is 30% acid to produce a solution that is 55% acid?

7. How many pounds of pure salt must be added to 60 pounds of a 2% solution of salt and water to increase it to a 10% solution?

8. An alloy of copper and tin is 20% copper. How many pounds of copper must be added to 80 pounds of the alloy in order that the resulting alloy be 50% copper?

9. A certain grade of gun metal that is a mixture of tin and copper contains 16% tin. How much tin must be added to 820 pounds of this gun metal to make a mixture that is 18% tin?

10. Fifteen gallons of alcohol are mixed with 60 gallons of water. How much alcohol must be added to make a solution that is 55% alcohol?

11. How much water must be added to 40 ounces of a 10% solution of boric acid to reduce it to a 4% solution?

12. How many pounds of water must be added to 24 pounds of a 10% solution of salt to reduce it to a 6% solution?

13. How much water must be evaporated from 32 pounds of a 4% solution of salt and water to make the result a solution that is 6% salt?

14. How much water must be evaporated from 240 pounds of a solution that is 3% salt to make a solution that is 5% salt?

SOLVING PER CENT MIXTURE PROBLEMS USING FRACTIONAL EQUATIONS

If 20 ounces of a solution of salt in water contains 5 ounces of pure salt, then $\frac{5}{20}$ or $\frac{1}{4}$ is the fractional part of the mixture which is pure salt. In general,

$$\frac{\text{number of units of pure substance}}{\text{number of units in the mixture}} = \frac{\text{fractional part of the mixture}}{\text{that is pure substance}}$$

~~~~~~~~~~ *MODEL PROBLEM* ~~~~~~~~~~

How much pure acid must be added to 30 ounces of an acid solution which is 40% acid in order to produce a solution which is 50% acid?

Solution:

$$\textit{Part pure acid} \times \textit{Weight of solution} = \textit{Weight of pure acid}$$

The number of ounces of pure acid in the given solution is .40(30) = 12.

Let $n =$ the number of ounces of pure acid to be added.

Then $12 + n =$ the number of ounces of pure acid in the new solution.

And $30 + n =$ the number of ounces that the new solution weighs.

Also $\dfrac{12 + n}{30 + n} =$ the fractional part of the new solution which is pure acid.

The fractional part of the new solution that is pure acid is $\frac{50}{100}$ or $\frac{1}{2}$.

$$\mathrm{M}_{2(30+n)}: \frac{12 + n}{30 + n} = \frac{1}{2}$$

$$2(12 + n) = 1(30 + n)$$
$$24 + 2n = 30 + n$$
$$n = 6$$

Answer: 6 ounces of pure acid must be added.

Note. See alternate solution and check on pages 145–146.

~~~~~~

Exercises

1. Of 24 pounds of salt water, 8% is salt. Of another mixture, 4% is salt. How many pounds of the second mixture should be added to the first mixture in order to get a mixture that is 5% salt?
2. Thirteen ounces of a solution of iodine in alcohol contain 1 ounce of pure iodine. How many ounces of pure iodine must be added to make a solution that is 25% iodine?
3. How much salt must be added to 40 pounds of a 5% salt solution to make a 24% salt solution?
4. A chemist has 20 pints of a solution of iodine and alcohol that is 15% iodine. How much iodine must be added to make a solution that is 20% iodine?
5. A certain alloy of copper and silver weighs 50 pounds and is 10% silver. How much silver must be added to produce a metal that is 25% silver?
6. One solution is 20% pure salt and another solution is 13% pure salt. How many ounces of each solution must be used to produce 35 ounces of a solution that is 15% pure salt?
7. A beaker contains 40 cubic centimeters (cc.) of a 20% solution of acid. How many cc. of this solution must be drawn off and replaced by pure acid so that the resulting solution will be 60% pure acid?

INVESTMENT PROBLEMS

PREPARING TO SOLVE INVESTMENT PROBLEMS

Mr. Field invests $700 at 8%. His annual income is 8% of $700, which equals .08(700), or $56. In finding the annual income, we make use of the *annual interest formula* $I = PR$. In this formula, P represents the **principal**, or amount invested, $700; R represents the **annual rate of interest**, 8%; and I represents the **annual interest**, or **annual income**, $56.

$$\boxed{\begin{array}{ccc} \textbf{\textit{KEEP IN MIND}} \\[4pt] I = PR \qquad R = \dfrac{I}{P} \qquad P = \dfrac{I}{R} \end{array}}$$

SOLVING INVESTMENT PROBLEMS

In solving investment problems, we can use a chart to organize the facts of the problem compactly.

~~~~~~~~~~~~~~~~ *MODEL PROBLEM* ~~~~~~~~~~~~~~~~

Ms. Rabin invested a part of $2000 at 6% and the remainder at $4\frac{1}{2}\%$. The annual income from the $4\frac{1}{2}\%$ investment exceeded the annual income on the 6% investment by $27. Find the amount she invested at each rate.

Solution:

Let $p =$ the amount invested at $4\frac{1}{2}\%$.

Then $2000 - p =$ the amount invested at 6%.

First fill in the principal and rate of interest for each investment. Then represent the annual income for each investment.

	($) Principal $\times$ Annual rate of interest $=$ Annual income		($)
$4\frac{1}{2}\%$ investment	p	.045	.045p
6% investment	$2000 - p$	.06	$.06(2000 - p)$

Principal $\times$ Annual rate of interest $=$ Annual income

The annual income from the $4\frac{1}{2}\%$ investment is $27 more than the annual income from the 6% investment.

$.045p = .06(2000 - p) + 27$
$.045p = 120 - .06p + 27$
$.045p + .06p = 147$
$.105p = 147$
$\text{M}_{1000}: \ 105p = 147{,}000$
$p = 1400$
$2000 - p = 600$

Check in the original problem:
Is the sum of both investments $2000?
$1400 + $600 = $2000 (true)
Does the annual income from the $4\frac{1}{2}\%$ investment exceed the annual income from the 6% investment by $27?
$.045(1400) = \$63$
$.06(600) = \$36$
$63 exceeds $36 by $27. (true)

Answer: $1400 was invested at $4\frac{1}{2}\%$; $600 was invested at 6%.

Exercises

1. Mr. Carter invested a sum of money at 4%. He invested a second sum, $250 more than the first sum, at 6%. If his total annual income was $90, how much did he invest at each rate?

2. Ms. Hernandez has invested a sum of money at 6%. A second sum, $400 more than the first sum, is invested at 3%. A third sum, 4 times as much as the first sum, is invested at 4%. The total annual income is $237. How much has she invested at each rate?

3. Ms. Baum invested $8000, part at 4% and the rest at 9%. The total annual income is $495. Find the amount invested at each rate.

4. Mr. Banks invested $\frac{1}{3}$ of his capital at 5%, $\frac{1}{4}$ of his capital at 6%, and the remainder at 3%. The total annual income was $530. Find his capital.

5. Mr. Harvey invested a sum of money at 4%. He invested a second sum, $1000 greater than the first sum, at 3%. The annual income on the first investment was equal to the annual income on the second investment. Find the amount invested at each rate.

6. Mr. Rose invested $3500 in two business enterprises. In one enterprise, he made a profit of 6%; in the other, he suffered a loss of 4%. His net profit for the year was $160. Find the amount invested at each rate.

7. A sum of $2200 is invested, part at 5% and the remainder at 3%. The annual income on the 3% investment is $46 less than the annual income on the 5% investment. How much was invested at each rate?

8. A sum of $3800 is invested, part at 4% and the remainder at 6%. The annual interest on the 6% investment is $98 more than the annual income on the 4% investment. Find the sum invested at each rate.

9. Mr. Roscoe invested $9000, part at 4% and the remainder at 7%. If his total annual income was 5% of his total investment, how much did he invest at each rate?

10. Ms. Green invested $1000 at 4% and $6000 at 6%. How much must she invest at 10% to make her total annual income $7\frac{1}{2}$% of her total investment?

11. Mr. Sable invested $5000 in a mortgage that pays 9% interest annually. He bought bonds paying $4\frac{1}{2}$% interest annually. His total annual income is 6% of his total investment. How much did he invest in the bonds?

GEOMETRIC PROBLEMS

PREPARING TO SOLVE GEOMETRIC PROBLEMS

Recall the following geometric relationships:

1. The perimeter of a geometric plane figure is the sum of the lengths of all its sides.
2. The area of a rectangle = base × altitude.
3. The area of a triangle = $\frac{1}{2}$ × base × altitude.

Note. In referring to rectangles, the words *length* and *width* are commonly used in place of the words *base* and *altitude*.

SOLVING GEOMETRIC PROBLEMS

In solving problems dealing with geometric figures, it is helpful to draw the figures.

~~~~~~~~~~ *MODEL PROBLEM* ~~~~~~~~~~

The base of a rectangle exceeds its altitude by 11 feet. If the base is decreased by 5 feet and the altitude is increased by 2 feet, a new rectangle is formed whose area is equal to the area of the original rectangle. Find the dimensions of the original rectangle.

Solution:

Let x = the altitude if the original rectangle in feet.

Then $x + 11$ = the base of the original rectangle in feet.

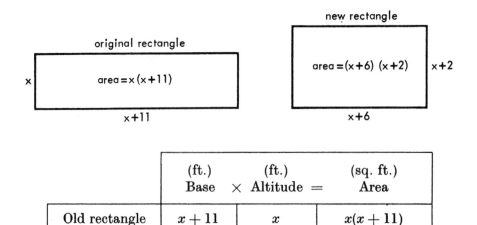

	(ft.) Base	(ft.) × Altitude =	(sq. ft.) Area
Old rectangle	$x + 11$	x	$x(x + 11)$
New rectangle	$x + 6$	$x + 2$	$(x + 6)(x + 2)$

Base × Altitude = Area of rectangle

The areas of the rectangles are equal.

$$x(x+11) = (x+6)(x+2)$$
$$x^2 + 11x = x^2 + 8x + 12$$
$$3x = 12$$
$$x = 4$$
$$x + 11 = 15$$

Check in the original problem:

Does the area of the original rectangle = the area of the new rectangle?

Area of original rectangle = 15(4) = 60 sq. ft.

In new rectangle, base = 10 ft., altitude = 6 ft.

Area of new rectangle = 10(6) = 60 sq. ft.

The areas of the rectangles are equal.

Answer: The base of the original rectangle is 15 feet; the altitude is 4 feet.

Exercises

1. The length of a rectangle exceeds twice its width by 1 inch. The perimeter of the rectangle is 32 inches. Find the dimensions of the rectangle.

2. The base of a rectangle is 5 feet less than $\frac{2}{3}$ of its width. The perimeter of the rectangle is 80 feet. Find the dimensions of the rectangle.

3. The length of a rectangle exceeds its width by 4 inches. If the width is doubled and its length is diminished by 2 inches, a new rectangle is formed whose perimeter is 8 feet more than the perimeter of the original rectangle. Find the dimensions of the original rectangle.

4. If the length of one side of a square is increased by 3 feet and the length of the adjacent side is decreased by 2 feet, a rectangle is formed whose area is equal to the area of the square. Find the length of a side of the square.

5. The base of a rectangle exceeds 3 times its altitude by 1 foot. If the base is decreased by 5 feet and the altitude is increased by 2 feet, a new rectangle is formed whose area is the same as the area of the original rectangle. Find the dimensions of the original rectangle.

6. The width of a rectangle is 2 inches less than its length. If its width is increased by 4 inches and its length is decreased by 2 inches, the area of the rectangle will be increased by 8 square inches. Find the length and width of the original rectangle.

7. If the length of one side of a square is increased by 2 feet and the length of an adjacent side is decreased by 3 feet, a rectangle is formed whose area is 14 sq. ft. less than the area of the square. Find the length of a side of the square.

8. The base of a rectangle exceeds twice its altitude by 2 inches. If the base is increased by 5 inches and the altitude is decreased by 1 inch, a new rectangle is formed whose area exceeds the area of the original rectangle by 20 square inches. Find the dimensions of the original rectangle.

9. The base of a triangle is 10 inches more than its altitude. If the base is increased by 5 inches and the altitude is decreased by 2 inches, the area of

the new triangle is the same as the area of the original triangle. Find the base and altitude of the original triangle.

WORK PROBLEMS

PREPARING TO SOLVE WORK PROBLEMS

If Kim can mow a lawn in 4 hours, then in 1 hour she will complete $\frac{1}{4}$ of the job. The part of the job that can be completed in 1 unit of time is called the ***rate of work.*** In general, if the job can be completed in x units of time, the rate of work is $\frac{1}{x}$. Notice that the rate of work is the reciprocal of the number of units of time required to complete the job. Therefore, Kim's rate of work is $\frac{1}{4}$. In 2 hours, Kim will complete $2(\frac{1}{4})$ or $\frac{2}{4}$ of the job; in x hours, she will complete $x(\frac{1}{4})$ or $\frac{x}{4}$ of the job. Thus we see that:

Rate of work, R, $\times$ amount of time worked, T, = part of the work done, W.

The relation involving R, T, and W may be expressed as $RT = W$.

Kim and Sam work on a job together. If Kim finishes $\frac{1}{4}$ of the job while Sam finishes $\frac{3}{4}$ of the job, then together they finish $\frac{1}{4} + \frac{3}{4}$ or $\frac{4}{4}$ of the job. Notice that in order for Kim and Sam to complete the whole job, the sum of the fractional part of the job that Kim finished and the fractional part of the job that Sam finished must be a fraction whose value is 1.

KEEP IN MIND

$$RT = W \qquad T = \frac{W}{R} \qquad R = \frac{W}{T}$$

SOLVING WORK PROBLEMS

~~~~~~~~~ *MODEL PROBLEM* ~~~~~~~~~

It takes Fred twice as long to paint a fence as it takes Harry to paint the same fence. If the two boys work together, they can paint the fence in 2 hours. How many hours would each boy, working alone, need to paint the fence?

Solution:

Let $x =$ number of hours Harry needs to do the job alone.

Then $2x =$ number of hours Fred needs to do the job alone.

	(part of job per hr.) Rate of work	$\times$	(hr.) Time of work	$=$	(part of job) Work done
Harry	$\dfrac{1}{x}$		2		$\dfrac{2}{x}$
Fred	$\dfrac{1}{2x}$		2		$\dfrac{2}{2x}$

Rate of work $\times$ Amount of time worked $=$ Part of work done

If the job is finished, the sum of the fractional part of
the job finished by Harry and the fractional part finished
by Fred must equal 1.

$$\frac{2}{x} + \frac{2}{2x} = 1$$

$$\mathrm{M}_{2x}: \ 2x\left(\frac{2}{x} + \frac{2}{2x}\right) = 2x(1)$$

$$4 + 2 = 2x$$
$$6 = 2x$$
$$3 = x$$
$$2x = 6$$

Check in the original problem:
Will the boys paint the whole fence if they work
together for 2 hours?
In 2 hours Harry will paint $2(\frac{1}{3})$ or $\frac{2}{3}$ of the fence.
In 2 hours Fred will paint $2(\frac{1}{6})$ or $\frac{1}{3}$ of the
fence.
In 2 hours, together, they will paint $\frac{2}{3} + \frac{1}{3} = \frac{3}{3}$,
or the whole fence.

Answer: Harry requires 3 hours; Fred requires 6 hours.

Exercises

1. Sam can mow a lawn in 20 minutes and Robert can mow the same lawn in 30 minutes. If they worked together, how long would it take them to do the job?
2. A printing press can print 300,000 cards in 12 hours. An older press requires 18 hours to print 300,000 copies of a card. How long would it take for both presses working together to print 300,000 copies?
3. A farmer working together with his son needs 3 hours to plow a field. The farmer working alone can plow the field in 4 hours. How long would it take the son working alone to plow the field?
4. One pipe can fill a tank in 20 minutes. A second pipe can fill the tank in 30 minutes. If the tank is empty, how long would be required for the two pipes operating together to fill it?

5. A farmer working with his son needs 3 hours to plow a field. Working alone, the farmer can plow the field in 12 hours. How long would it take the son working alone to plow the field?

6. A farmer can milk her cows in $\frac{2}{3}$ of an hour, and her son can do the job in 1 hour. How long would it take them to milk the cows together?

7. An inlet pipe can fill a tank in 3 hours. An outlet pipe can empty the tank in 6 hours. If the tank is empty and both pipes are opened, how many hours will it take to fill the tank?

8. A tank can be filled by one pipe in 20 minutes, and it can be filled by a second pipe in 10 minutes. It can be drained by a third pipe in 12 minutes. If the tank is empty and the inlet pipes and the drain are opened, in how many hours will the tank be filled?

9. Lori working alone can mow a lawn in 20 minutes; Cary can mow the same lawn in 30 minutes. They mow the lawn together for 10 minutes, after which Lori leaves the job. How much longer will it take Cary to finish the job?

10. A clerk is assigned a job that she can complete in 8 hours. After she has been working 2 hours, another clerk, who is able to do this job in 10 hours, is assigned to help her. In how many hours will the clerks working together complete the job?

11. A man can do a piece of work in 9 hours. After he worked 3 hours alone, he was joined by his son, and they finished the job in $4\frac{1}{2}$ hours. How many hours would it take the son to do the job alone?

12. Leo finished $\frac{2}{3}$ of a job in 12 hours. When he was joined by Ed, they completed the job in 2 hours. How many hours would it take Ed to do the job alone?

13. A mechanic's helper requires twice as long as the mechanic to do the same amount of work. On a particular job, they work together for 2 hours. Then the mechanic is called away, and the helper finishes it in 1 hour. How many hours would it take the helper to do the entire job?

14. It takes a human 8 hours to do a certain job and a robot 12 hours to do the same job. If 2 humans and 3 robots work on the job, how many hours does it take them to finish it?

NUMBER PROBLEMS INVOLVING FRACTIONS

~~~~~~~~~~~~ *MODEL PROBLEMS* ~~~~~~~~~~~~

1. Two numbers are in the ratio 4 : 5. If 10 is added to each number, the resulting numbers will be in the ratio 5 : 6. Find the original numbers.

Solution:

Let $4x =$ the first number.

Then $5x =$ the second number.

Then $4x + 10 =$ the first number increased by 10.
And $5x + 10 =$ the second number increased by 10.

The new numbers are in the ratio 5 : 6.

$$\frac{4x + 10}{5x + 10} = \frac{5}{6}$$

Multiply both members
of the equation
by $6(5x + 10)$.

$6(4x + 10) = 5(5x + 10)$
$24x + 60 = 25x + 50$
$10 = x$
$4x = 40$
$5x = 50$

Check in the original problem:
Is the ratio of the two numbers 4 : 5?

$\frac{40}{50} = \frac{4}{5}$ (true)

If 10 is added to each number, will the ratio of
the resulting numbers be 5 : 6?

$40 + 10 = 50$
$50 + 10 = 60$

$\frac{50}{60} = \frac{5}{6}$ (true)

Answer: The first number is 40; the second number is 50.

2. The larger of two numbers exceeds twice the smaller by 17. When the larger number is divided by the smaller number, the quotient is 3 and the remainder is 7. Find the two numbers.

Solution:

Note. When 23 is divided by 5, the quotient is 4 and the remainder is 3.
This may be written as follows: $\frac{23}{5} = 4 + \frac{3}{5}$. Similarly, "when D is divided

by d, the quotient is Q and the remainder is R" may be written as $\frac{D}{d} = Q + \frac{R}{d}$.

Let $x =$ the smaller number.
Then $2x + 17 =$ the larger number.

When the larger number is divided by the smaller
number, the quotient is 3 and the remainder is 7.

$$\frac{2x + 17}{x} = 3 + \frac{7}{x}$$

M_x :

$2x + 17 = 3x + 7$
$10 = x$
$2x + 17 = 37$

Check in the original problem:
The larger number, 37, is 17 more than 2 times the
smaller number, 10. When 37 is divided by 10,
the quotient is 3 and the remainder is 7.

Answer: The larger number is 37; the smaller number is 10.

3. The denominator of a fraction is 1 less than 4 times the numerator. If the numerator is doubled and the denominator is increased by 6, the value of the resulting fraction is $\frac{2}{5}$. Find the original fraction.

Solution:

Let $n =$ the numerator of the original fraction.

Then $4n - 1 =$ the denominator of the original fraction.

Then $\dfrac{2n}{(4n - 1) + 6} =$ the resulting fraction.

The value of the resulting fraction is $\frac{2}{5}$.

$$\frac{2n}{(4n - 1) + 6} = \frac{2}{5}$$

$$\frac{2n}{4n + 5} = \frac{2}{5}$$

$\text{M}_{5(4n+5)}$:

$$5(2n) = 2(4n + 5)$$
$$10n = 8n + 10$$
$$2n = 10$$
$$n = 5$$
$$4n - 1 = 19$$

Check in the original problem:
Is the denominator of the fraction 1 less than 4 times the numerator?
19 is 1 less than 4 times 5. (true)
Is the value of the resulting fraction $\frac{2}{5}$?

The resulting fraction is $\dfrac{2(5)}{19 + 6} = \dfrac{10}{25} = \dfrac{2}{5}$ (true)

Answer: The original fraction is $\dfrac{5}{19}$.

~~~~~~

Exercises

1. If one-half of a number is subtracted from three-fifths of that number, the difference is 10. Find the number.

2. The larger of two numbers is 12 less than 5 times the smaller. If the smaller number is equal to $\frac{1}{3}$ of the larger number, find the numbers.

3. One-fifth of the result obtained when a number is increased by 12, is equal to one-fourth of the result that is obtained when that number is increased by 2. Find the number.

4. One of two positive numbers exceeds 4 times the other number by 4. One-third of the larger exceeds one-half of the smaller by 8. Find both numbers.

5. Find three consecutive even numbers such that the sum of the first and the third exceeds one-half of the second by 54.

6. The numerator of a fraction exceeds 8 times the denominator by 2. The value of the fraction is $\frac{17}{2}$. Find the fraction.

7. The numerator of a fraction is 7 less than the denominator. If 3 is added to the numerator and 9 is subtracted from the denominator, the resulting fraction is equal to $\frac{3}{2}$. Find the original fraction.

8. What number must be added to both the numerator and denominator of the fraction $\frac{5}{23}$ to give a fraction equal to $\frac{1}{3}$.

9. The larger of two numbers exceeds the smaller by 8. When the smaller is divided by the larger, the quotient is equal to $\frac{5}{7}$. Find the numbers.

10. The difference between two numbers is 24. If the larger is divided by the smaller, the quotient is 4 and the remainder is 3. Find the numbers.

11. Two numbers are in the ratio of 3 : 5. One-half of the smaller exceeds $\frac{1}{5}$ of the larger by 6. Find the numbers.

12. What number must be added to 12 and 18 respectively so that the new numbers will be in the ratio of 3 : 4?

13. The numerator and the denominator of a fraction are in the ratio of 5 : 4. If 5 is added to the numerator and 12 is subtracted from the denominator, the value of the resulting fraction is $\frac{5}{2}$. Find the original fraction.

14. The larger of two numbers exceeds twice the smaller by 1. If 2 is added to the larger and 1 is subtracted from the smaller, the resulting numbers are in the ratio of 3 : 1. Find the numbers.

15. One-fourth of the reciprocal of a number exceeds one-fifth of the reciprocal of the number by 1. Find the number.

9. Solving Equations Involving More Than One Variable

An equation may contain more than one variable. Examples of such equations are $ax = b$, $x + y = 7$, and $ax + bx = a^2 - b^2$.

To solve such an equation for one of its variables means to express this particular variable in terms of the other variables. In order to plan the steps in the solution, it may be helpful to compare the equation with a similar equation which contains only the variable being solved for. The same operations are used in solving both equations. Note how these ideas are applied in the following model problems:

〰〰〰〰〰〰 *MODEL PROBLEMS* 〰〰〰〰〰〰

1. Solve for x: $a(x - a) = b(x - b)$ [Compare with $7(x - 7) = 5(x - 5)$.]

How To Proceed	*Solution*
1. Transform the equation into an equivalent equation in which all terms involving the variables being solved for are collected on one side of the equation and all the other terms are collected on the other side.	$a(x-a)=b(x-b)$ $ax-a^2=bx-b^2$ $ax-bx=a^2-b^2$
2. Find the coefficient of the variable being solved for.	$(a-b)x=a^2-b^2$
3. Divide by the coefficient of the variable being solved for.	$\dfrac{(a-b)x}{(a-b)}=\dfrac{a^2-b^2}{a-b}$
4. Simplify the equation obtained in step 3.	$x=\dfrac{(a-b)(a+b)}{(a-b)}$ $x=a+b\quad Ans.$

Check by substituting $a+b$ for x in the given equation.

2. Solve for x: $\dfrac{1}{a}+\dfrac{1}{b}=\dfrac{1}{x}$ $\left[\text{Compare with }\dfrac{1}{2}+\dfrac{1}{3}=\dfrac{1}{x}\right]$

Solution:

$$\frac{1}{a}+\frac{1}{b}=\frac{1}{x}$$

L.C.D. $=abx$

$$\mathbf{M}_{abx}:\ abx\left(\frac{1}{a}+\frac{1}{b}\right)=abx\left(\frac{1}{x}\right)$$

$$abx\left(\frac{1}{a}\right)+abx\left(\frac{1}{b}\right)=abx\left(\frac{1}{x}\right)$$

$$bx+ax=ab$$
$$(b+a)x=ab$$

$$\mathbf{D}_{b+a}:\ \frac{(b+a)x}{b+a}=\frac{ab}{b+a}$$

$$x=\frac{ab}{b+a}\quad Ans.$$

Check in the original problem:

$$\frac{1}{a}+\frac{1}{b}=\frac{1}{x}$$

Let $x=\dfrac{ab}{b+a}$

$$\frac{1}{a}+\frac{1}{b}\overset{?}{=}\frac{1}{\dfrac{ab}{b+a}}$$

$$\frac{b+a}{ab}=\frac{b+a}{ab}\ \text{(true)}$$

Note. Division by $b+a$ is undefined if $b+a=0$. Hence, $b\neq(-a)$.

Exercises

In 1–39, solve for x or y in terms of the other variables.

1. $3x = a$

2. $rx = 10$

3. $ax = b$

4. $x + 2 = a$

5. $x + a = b$

6. $y - 4 = m$

7. $y + r = s$

8. $\dfrac{y}{10} = a$

9. $\dfrac{y}{b} = 7$

10. $3x - 4e = 5e$

11. $bx + b^2 = 5b^2 - 3bx$

12. $\dfrac{ax}{b} = \dfrac{c}{d}$

13. $\dfrac{m}{x} = \dfrac{m^2}{r}$

14. $\dfrac{g}{h} = \dfrac{g^2}{x}$

15. $cy - 3 = d$

16. $tx + r = s$

17. $\dfrac{sx}{t} + a = b$

18. $m = \dfrac{ny}{r} - s$

19. $\dfrac{5}{x} - \dfrac{4}{x} = t$

20. $\dfrac{a}{x} + c = \dfrac{b}{x}$

21. $\dfrac{m}{y} = \dfrac{n}{y} + p$

22. $3x + 2a = x + 4a$

23. $my - m^2 = 5m^2 - 2my$

24. $y - 5 = 5r - ry$

25. $ey + f^2 = e^2 + fy$

26. $(x - r)(x - s) = x^2$

27. $(c - d)x = c^2 - cd$

28. $c(c + x) = b(b + x)$

29. $3cx + 8d^2 = 2d(2x + 3c)$

30. $\dfrac{x}{a} + \dfrac{x}{b} = 1$

31. $\dfrac{c}{x} - c^2 = \dfrac{b}{x} - b^2$

32. $\dfrac{x}{r} + \dfrac{x}{s} + \dfrac{x}{t} = 1$

33. $\dfrac{3}{x - c} = \dfrac{4}{x - d}$

34. $\dfrac{y - r}{y - s} = \dfrac{e}{f}$

35. $\dfrac{y - m}{y + n} = \dfrac{y + n}{y - m}$

36. $\dfrac{y}{y - r} - \dfrac{y}{y + r} = \dfrac{6r}{y^2 - r^2}$

37. $\dfrac{x}{a^2 - 4} = \dfrac{b}{a - 2} - \dfrac{b}{a + 2}$

38. $\dfrac{x}{a^2 - a} = \dfrac{b}{a - 1} - \dfrac{b}{a}$

39. $\dfrac{y - 2}{a^2 - 5a + 6} = \dfrac{a + 1}{a - 3} - \dfrac{a + 2}{a - 2}$

10. Problems Involving Literal Numbers

〰〰〰〰〰〰〰 *MODEL PROBLEM* 〰〰〰〰〰〰〰

Marie and Anne start from the same place at the same time and travel in opposite directions. Marie travels r miles per hour and Anne travels s miles per hour. In how many hours will the girls be t miles apart?

Solution:
   Let $x =$ the required number of hours.
Then $rx =$ the distance Marie travels.
 And $sx =$ the distance Anne travels.

The total distance is t miles.

$$rx + sx = t$$

$$(r + s)x = t$$

$$x = \frac{t}{r + s}$$

Answer: The time required is $\dfrac{t}{r + s}$ hours.

~~~~~~

Exercises

1. It took m army engineers h hours to build a pontoon bridge. If r more engineers had been assigned to the work and they had worked at the same rate, how many hours would it have taken?

2. A woman travels m miles per hour for t hours and then changes her rate to s miles per hour for h hours. Find her average rate.

3. If a pounds of coffee worth c dollars a pound are mixed with b pounds of coffee worth d dollars a pound, find how much the resulting mixture is worth per pound.

4. If a piece of work can be done by one machine in p hours and by another machine in q hours, find how many hours both machines working together require to do the job.

5. A woman is r times as old as her daughter. In s years she will be t times as old as her daughter will be then. Find the daughter's present age.

6. A boy is c years old. His father is m times as old as he is. How many years ago was the father d times as old as the boy was then?

7. In a purse which contains quarters and dimes, there are w coins whose value is c cents. Find the number of quarters.

8. A man invested p dollars, part at $c\%$ and part at $d\%$. If his annual income is I dollars, find the amount he invested at $c\%$.

9. How far can a man ride out into the country at the rate of w miles per hour and return at the rate of z miles per hour if he travels h hours on the entire trip?

10. Traveling at the rate of r miles per hour, a motorist leaves a town. Another

motorist leaves the same town h hours later and travels on the same road at the rate of s miles per hour. In how many hours will the second motorist overtake the first motorist?

11. A patrol plane flies at r miles an hour in still air. On a certain day, the plane flew a distance of d miles from its base against a headwind of w miles an hour. It returned immediately over the same route, the round trip taking t hours. The rate and the direction of the wind remained the same during the entire trip.
    a. Express in terms of r, d, and w the time required for the flight (1) away from the base (2) back to the base.
    b. Write an equation that can be used to solve for d in terms of r, t, and w.
    c. Solve for d the equation written in answer to b.

12. a. A solution of water and alcohol is 90% alcohol. Express in terms of n the amount of water that must be added to n gallons of this solution to make it 80% alcohol.
    b. A solution of water and alcohol is 80% alcohol. Express in terms of n the amount of alcohol that must be added to n gallons of this solution to make it 90% alcohol.

13. A mixture of m pounds of sand and cement is c per cent cement.
    a. Express the number of pounds of cement in the mixture in terms of c and m.
    b. If x pounds of sand are added to this mixture, it will then be d per cent cement. Express the number of pounds of cement in the new mixture in terms of d, m, and x.
    c. Find x in terms of c, d, and m.

14. A motorist finds that if he travels r miles per hour, he can cover a certain distance in h hours. By how many miles must he increase his rate per hour if he is to cover the same distance in one hour less time?

15. In a certain solution, there are a pints of acid and w pints of water. How many pints of pure acid must be added to the solution to make a mixture which is $p\%$ pure acid?

11. Finding the Value of a Variable in a Formula by Solving an Equation

Procedure. To find the value of an indicated variable in a formula by solving an equation:
1. Replace all variables in the formula except the one that is being evaluated by their specified values.
2. Solve the resulting equation.

~~~~~~~~~~~ *MODEL PROBLEM* ~~~~~~~~~~~

If $V = \frac{1}{3}BH$, find H when $V = 24$ and $B = 8$.

Solution:

Substitute the given values.

$V = \frac{1}{3}BH$	*Check*
$24 = \frac{1}{3}(8)H$	$V = \frac{1}{3}BH$
$24 = \frac{8}{3}H$	$24 \overset{?}{=} \frac{1}{3}(8)(9)$
$9 = H$ *Ans.*	$24 = 24$ (true)

Exercises

1. If $A = \frac{1}{2}bh$, find h when $A = 30$ and $b = 15$.

2. If $F = \frac{9}{5}C + 32$, find C when $F = -13$.

3. If $S = \dfrac{n}{2}(a + l)$, find n when $S = 40$, $a = 7$, and $l = 13$.

4. If $C = \frac{5}{9}(F - 32)$, find F when $C = 15$.

5. If $T = 2\pi r(r + h)$, find h when $T = 1408$, $\pi = \dfrac{22}{7}$, and $r = 14$.

6. If $S = \dfrac{a}{1 - r}$, find r when $a = 2$ and $S = 4$.

7. In the formula $F = \frac{9}{5}C + 32$, find the temperature reading at which $F = C$.

8. If $\dfrac{1}{f} = \dfrac{1}{D} + \dfrac{1}{d}$, find d when $f = 4$ and $D = 12$.

12. Transforming Formulas

A formula may be expressed in more than one form. Sometimes it is desirable to solve a formula for a variable which is different from the one for which it is solved. This is called **transforming** the formula or **changing the subject** of the formula. For example, the formula $D = RT$ can be transformed into the formula $\dfrac{D}{R} = T$ by dividing both of its members by R. Originally D was expressed in terms of R and T; now T is expressed in terms of D and R.

Procedure. To transform a formula:
1. Consider the formula as an equation with several variables.
2. Solve the equation for the indicated variable in terms of the other variables.

The value of a variable in a formula may be found by first transforming the equation so that it is solved for that variable and then by substituting the given values for the other variables.

Thus, to find the value of K in the formula $\dfrac{K}{d^2} = F$ when $F = 120$ and $d = \frac{1}{2}$:

1. Transform $\dfrac{K}{d^2} = F$ to $K = Fd^2$.

2. Substitute and solve $K = 120(\frac{1}{2})^2 = 120(\frac{1}{4}) = 30$.

〰〰〰〰〰 *MODEL PROBLEMS* 〰〰〰〰〰

1. Solve the formula $a = p(1 + rt)$ for t in terms of a, p, and r.

How To Proceed	*Solution*
1. Consider the formula as an equation with several variables.	$a = p(1 + rt)$
2. Solve for the indicated variable.	$a = p + prt$
	$S_p\colon\ a - p = prt$
	$D_{pr}\colon\ \dfrac{a - p}{pr} = \dfrac{\cancel{prt}}{\cancel{pr}}$
	$\dfrac{a - p}{pr} = t \quad Ans.$

2. If $u = \dfrac{MV}{M + V}$, express M in terms of u and V.

Solution:

$$u = \frac{MV}{M + V}$$

$$M_{M+V}\colon\ (M + V)u = (M + V) \cdot \frac{MV}{M + V}$$

$$uM + uV = MV$$

$$S_{uM}\colon\ uV = MV - uM$$

$$uV = (V - u)M$$

$$D_{V-u}\colon\ \frac{uV}{(V - u)} = \frac{(\cancel{V - u})M}{(\cancel{V - u})}$$

$$\frac{uV}{V - u} = M \quad Ans.$$

Exercises

In 1–24, transform the formula by solving for the indicated variable.

1. $A = bh$ for h **2.** $C = \pi d$ for d **3.** $E = ir$ for i

4. $A = lw$ for l **5.** $V = lwh$ for h **6.** $i = prt$ for p

7. $A = \frac{1}{2}bh$ for h **8.** $C = \dfrac{360}{N}$ for N **9.** $I = \dfrac{E}{R}$ for E

10. $v = \dfrac{s}{t}$ for t **11.** $F = \dfrac{mv^2}{gr}$ for r **12.** $S = \frac{1}{2}at^2$ for a

13. $S = c + g$ for g **14.** $p = 2l + 2w$ for w **15.** $F = \frac{9}{5}C + 32$ for C

16. $A = \frac{1}{2}h(b + c)$ for b **17.** $A = p + prt$ for t **18.** $\dfrac{D}{d} = q + \dfrac{r}{d}$ for d

19. $S = vt - \frac{1}{2}gt^2$ for v **20.** $A = p + prt$ for p **21.** $\dfrac{1}{f} = \dfrac{1}{p} + \dfrac{1}{q}$ for p

22. $R = \dfrac{gs}{g + s}$ for s **23.** $S = \dfrac{a}{1 - r}$ for r **24.** $C = \dfrac{nE}{R + nr}$ for n

25. Solve for h: $V = \dfrac{\pi r^2 h}{3}$ **26.** Solve for k: $n = \dfrac{a - k}{5k}$

27. Using the formula $A = P(1 + rt)$, express r in terms of A, P, and t.

In 28–33, (*a*) transform the given formula by solving for the variable to be evaluated, and (*b*) substitute the given values in the result obtained in part (*a*) to evaluate this variable.

28. Find h if $V = 250$, $r = 5$, $\pi = \frac{22}{7}$: $V = \dfrac{\pi r^2 h}{3}$.

29. Find W if $A = 27$, $t = 3$: $t = \dfrac{6W}{A - W}$.

30. Find r if $A = 2000$, $P = 1000$, $t = 10$: $A = P(1 + rt)$.

31. Find I if $E = 42$, $R = 4$, $r = 3$: $E = IR + Ir$.

32. Find t if $V = 1200$, $k = 850$, $g = 7$: $V = k + gt$.

33. Find F if $g = \frac{1}{2}$, $h = \frac{2}{5}$: $\dfrac{1}{F} = \dfrac{1}{g} + \dfrac{1}{h}$.

34. The formula for the area of a triangle is $A = \frac{1}{2}bh$. Rewrite this formula if $h = 8b$; that is, express A in terms of b.

35. The formula for the total surface of a cylinder is $S = 2\pi R(R + H)$. Express S in terms of H if $R = 2H$.

13. Solving Equations Involving Absolute Values

The absolute value of a, $|a|$, has been defined as follows:

$$|a| = a \text{ when } a \geq 0, |a| = -a \text{ when } a < 0.$$

If we apply the absolute value definition to $|x + 1| = 5$, we see that $x + 1 = 5$ or $x + 1 = -5$. Hence, the roots of $|x+1| = 5$ are $x = 4$, the root of $x + 1 = 5$, together with $x = -6$, the root of $x + 1 = -5$. Therefore, the solution set of $|x + 1| = 5$ is $\{4, -6\}$.

Check in the original equation:

$\|x + 1\| = 5$	$\|x + 1\| = 5$
Let $x = 4$: $\|4 + 1\| \stackrel{?}{=} 5$	Let $x = -6$: $\|-6 + 1\| \stackrel{?}{=} 5$
$\|5\| \stackrel{?}{=} 5$	$\|-5\| \stackrel{?}{=} 5$
$5 = 5$	$5 = 5$

We refer to $x + 1 = 5$ and $x + 1 = -5$ as the *derived equations* of $|x + 1| = 5$. In this case, these two equations together are equivalent to $|x + 1| = 5$.

It is most important to check the roots of the derived equations in the original absolute value equation because a root of a derived equation may not satisfy the absolute value equation. For example, in the case of $|x| = -5$, the roots of the derived equations $x = 5$ and $x = -5$ do not satisfy $|x| = -5$ since the absolute value of a number cannot be negative. Hence, the solution set of $|x| = -5$ is the empty set, $\varnothing$.

Likewise, in the case of the equation $|2x + 5| = x + 1$, the derived equations are $2x + 5 = x + 1$ and $2x + 5 = -(x + 1)$. Neither $x = -4$, which is the root of $2x + 5 = x + 1$, nor $x = -2$, which is the root of $2x + 5 = -(x + 1)$, satisfies $|2x + 5| = x + 1$. Therefore, the solution set of $|2x + 5| = x + 1$ is the empty set, $\varnothing$.

Procedure. To solve equations involving absolute values:
1. Write the derived equations.
2. Solve each equation.

KEEP IN MIND

When solving an absolute value equation, check to see whether the roots of the derived equations satisfy the given absolute value equation.

~~~~~~~~~~~~~~~~ *MODEL PROBLEM* ~~~~~~~~~~~~~~~~

a. Solve and check: $|3x + 2| = 4x + 5$

b. Graph the solution set of $|3x + 2| = 4x + 5$

a.   *How To Proceed*

                                        Solution

1. Write the derived equations.
2. Solve each equation.

$$|3x + 2| = 4x + 5$$

$3x + 2 = 4x + 5$	$3x + 2 = -(4x + 5)$
$3x - 4x = 5 - 2$	$3x + 2 = -4x - 5$
$-x = 3$	$7x = -7$
$x = -3$	$x = -1$

Check in the original equation:

$|3x + 2| = 4x + 5$	$|3x + 2| = 4x + 5$
Let $x = -3:\ |-9 + 2| \overset{?}{=} -12 + 5$	Let $x = -1:\ |-3 + 2| \overset{?}{=} -4 + 5$
$|-7| \overset{?}{=} -7$	$|-1| \overset{?}{=} 1$
$7 \neq -7$	$1 = 1$ (true)

Answer: $x = -1$, or the solution set is $\{-1\}$.

b. Graph the solution set $\{-1\}$.

Answer:

~~~~~~

Exercises

In 1–8, write two derived equations that do not contain the absolute value symbol and that together are equivalent to the given equation.

1. $|x| = 8$ **2.** $|4x| = 16$ **3.** $|m + 5| = 7$

4. $|2y - 4| = 6$ **5.** $|5 - 3x| = 11$ **6.** $\left|\dfrac{3t + 1}{4}\right| = 7$

7. $|3x - 20| = 2x$ **8.** $|3x - 7| = 2x - 3$

In 9–27, find the solution set of the equation.

9. $|x| = 12$ **10.** $|y| = -6$ **11.** $|2y| = 24$

12. $3|x| = 27$ **13.** $|x + 9| = 25$ **14.** $|4t - 1| = 27$

15. $|6 - 3y| = -9$ **16.** $\left|\dfrac{x}{4}\right| = 8$ **17.** $3\left|\dfrac{d}{4}\right| = 20$

18. $\left|\dfrac{2x}{3}\right| = -12$ **19.** $\left|\dfrac{4x - 1}{3}\right| = 15$ **20.** $\left|\dfrac{2(y + 1)}{7}\right| = 6$

21. $\left|\dfrac{4(3x - 1)}{5}\right| = 4$ **22.** $|5x - 4| = 3x$ **23.** $|3x + 1| = 4x$

24. $|4x + 8| = 2x$ **25.** $|4x - 2| = 3x + 2$ **26.** $|6x - 8| = 5x - 2$
27. $|8x + 20| = 7x + 10$

14. Properties of Inequalities

THE ORDER PROPERTY OF NUMBER

If x and y are real numbers, then one and only one of the following sentences is true:

$$x < y \qquad x = y \qquad x > y$$

The following graphs illustrate this *order property of number:*

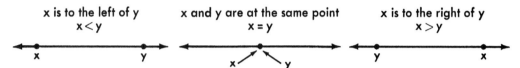

THE TRANSITIVE PROPERTY OF INEQUALITIES

If x, y, and z are real numbers:

$$\text{if } x < y \text{ and } y < z, \text{ then } x < z$$
$$\text{if } z > y \text{ and } y > x, \text{ then } z > x$$

The following graph illustrates the *transitive property of inequalities*:

THE ADDITION PROPERTY OF INEQUALITIES

If x, y, and z are real numbers:

$$\text{if } x > y, \text{ then } x + z > y + z$$
$$\text{if } x < y, \text{ then } x + z < y + z$$

The following examples illustrate the *addition property of inequalities*:

$$\text{if } 10 > 7, \text{ then } 10 + 5 > 7 + 5$$
$$\text{if } 6 < 8, \text{ then } 6 + (-4) < 8 + (-4)$$

Since subtracting a real number from both members of an inequality means adding its opposite to both members of the inequality, we can say:

When the same number is added to or subtracted from both members of an inequality, the order of the inequality remains unchanged.

THE MULTIPLICATION PROPERTY OF INEQUALITIES

If x, y, and z are real numbers, then:

if $x > y$, then $xz > yz$ when z is positive ($z > 0$)

if $x < y$, then $xz < yz$ when z is positive ($z > 0$)

if $x > y$, then $xz < yz$ when z is negative ($z < 0$)

if $x < y$, then $xz > yz$ when z is negative ($z < 0$)

The following examples illustrate the **multiplication property of inequalities**:

if $5 > 3$, then $5(7) > 3(7)$

if $6 < 9$, then $6(\frac{1}{3}) < 9(\frac{1}{3})$

if $5 > 3$, then $5(-7) < 3(-7)$

if $6 < 9$, then $6(-\frac{1}{3}) > 9(-\frac{1}{3})$

Note that the order of the inequality is reversed in the last two examples.

Since dividing both members of an inequality by a nonzero number means to multiply by the reciprocal of the number, we can say:

When both members of an inequality are multiplied or divided by a positive number, the order of the inequality remains unchanged; when both members are multiplied or divided by a negative number, the order of the inequality is reversed.

Exercises

In 1–16, replace the question mark with the symbol $>$ or the symbol $<$ so that the resulting sentence will be true. All variables are nonzero real numbers.

1. If $x < 7$ and $7 < y$, then x ? y.
2. If $r > -5$ and $-5 > q$, then r ? q.
3. If $y < 8$ and $z > 8$, then y ? z.
4. If $d > e$ and $g < e$, then d ? g.
5. If $5 < 7$, then $5 - 12$? $7 - 12$.
6. If $12 > 9$, then $12(4)$? $9(4)$.
7. If $15 > 10$, then $15(-2)$? $10(-2)$.
8. If $36 < 42$, then $36 \div 6$? $42 \div 6$.
9. If $24 > 18$, then $24 \div (-2)$? $18 \div (-2)$.
10. If $35 > 20$, then $35 \div 5$? $20 \div 5$.
11. If $x + 5 > 7$, then $x + 5 + (-5)$? $7 + (-5)$ or x ? 2.
12. If $x - 3 < 15$, then $x - 3 + 3$? $15 + 3$ or x ? 18.
13. If $4x > 20$, then $\frac{4x}{4}$? $\frac{20}{4}$ or x ? 5.
14. If $\frac{1}{2}y > 5$, then $2 \times \frac{1}{2}y$? 2×5 or y ? 10.

15. If $-3x < 27$, then $\dfrac{-3x}{-3}$? $\dfrac{27}{-3}$ or x ? -9.

16. If $\dfrac{-x}{4} > 16$, then $-4\left(\dfrac{-x}{4}\right)$? $-4(16)$ or x ? -64.

15. Solving First-Degree Inequalities

To *solve an inequality* means to find its solution set. Consider the inequality $3x > 6$, the domain of the variable being the set of real numbers.

If $x = 1.9$, then $3(1.9) > 6$, or $5.7 > 6$ is a false sentence.
If $x = 2$, then $3(2) > 6$, or $6 > 6$ is a false sentence.
If $x = 2.1$, then $3(2.1) > 6$, or $6.3 > 6$ is a *true* sentence.

Notice that if x is replaced by any real number greater than 2, the resulting sentence is true. Therefore, the solution set of $3x > 6$ is the set of all real numbers greater than 2, which may be represented by $x > 2$ or $\{x \mid x > 2\}$.

Equivalent inequalities are inequalities that have the same solution set. Therefore, $3x > 6$ and $x > 2$ are equivalent inequalities because they have the same solution set.

If the domain of x is the set of real numbers, the inequality $3x > 6$ is called a *conditional inequality* because it is true for at least one, but not all members of the domain. Other examples of conditional inequalities are $x + 7 > 9$ and $3x - 2 < 7$, $x \in \{\text{real numbers}\}$.

If $x \in \{\text{real numbers}\}$, the inequality $x + 8 > x$ is true for every element of the domain. Such an inequality is called an *absolute inequality*, or an *unconditional inequality*. Other examples of absolute inequalities are $5x + 8 > 5x$ and $x - 4 < x$.

To find the solution set of an inequality, we will solve the inequality using methods similar to those used in solving an equation. We will transform the inequality into a simpler equivalent inequality of the form $x > a$, or $x < a$, making use of the following four postulates, which correspond to the addition, subtraction, multiplication, and division properties of inequalities.

POSTULATES INVOLVING EQUIVALENT INEQUALITIES

Postulate 1. If an equivalent expression is substituted for an expression in an inequality, then the result is an inequality that is equivalent to the original inequality. Thus, $6x + 2x > 4$ and $8x > 4$ are equivalent inequalities.

Postulate 2. If the same number is added to or subtracted from both members of an inequality, the resulting inequality is equivalent to the original inequality.

Thus, if $x - 5 > 2$, then $(x - 5) + 5 > 2 + 5$, and $x > 7$. Therefore, the solution set of $x - 5 > 2$ is $\{x \mid x > 7\}$.

Also, if $x + 3 < 9$, then $(x + 3) - 3 < 9 - 3$, and $x < 6$. Therefore, the solution set of $x + 3 < 9$ is $\{x \mid x < 6\}$.

Postulate 3. If both members of an inequality are multiplied by or divided by the same positive number, the resulting inequality is equivalent to the original inequality.

Thus, if $\dfrac{x}{5} < 3$, then $5\left(\dfrac{x}{5}\right) < 5(3)$, and $x < 15$. Therefore, the solution set of $\dfrac{x}{5} < 3$ is $\{x \mid x < 15\}$.

Also, if $4x > 12$, then $\dfrac{4x}{4} > \dfrac{12}{4}$, and $x > 3$. Therefore, the solution set of $4x > 12$ is $\{x \mid x > 3\}$.

Postulate 4. If both members of an inequality are multiplied by or divided by the same negative number, and the order of the inequality is reversed, the resulting inequality is equivalent to the original inequality.

Thus, if $\dfrac{x}{-4} > 3$, then $(-4)\left(\dfrac{x}{-4}\right) < (-4)(3)$, and $x < -12$. Therefore, the solution set of $\dfrac{x}{-4} > 3$ is $\{x \mid x < -12\}$.

Also, if $-3x < 6$, then $\dfrac{-3x}{-3} > \dfrac{6}{-3}$, and $x > -2$. Therefore, the solution set of $-3x < 6$ is $\{x \mid x > -2\}$.

～～～～～～ MODEL PROBLEMS ～～～～～～

In 1 and 2 (*a*) solve and (*b*) graph the solution sets of the inequalities.
1. $8x - 3 < 5x + 12$

Solution:

$$8x - 3 < 5x + 12$$
$$8x - 3 + 3 < 5x + 12 + 3$$
$$8x < 5x + 15$$
$$8x - 5x < 5x + 15 - 5x$$
$$3x < 15$$
$$\frac{3x}{3} < \frac{15}{3}$$
$$x < 5$$

Answer: a. $\{x \mid x < 5\}$

b.

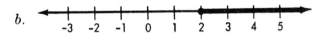

2. $4(2x - 6) - 10x \le -28$

Solution:

$$4(2x - 6) - 10x \le -28$$
$$8x - 24 - 10x \le -28$$
$$-2x - 24 \le -28$$
$$-2x - 24 + 24 \le -28 + 24$$
$$-2x \le -4$$
$$\frac{-2x}{-2} \ge \frac{-4}{-2}$$
$$x \ge 2$$

Answer: a. $\{x \mid x \ge 2\}$.

b.

Exercises

In 1–24, find and graph the solution set of the inequality.

1. $x - 3 > 4$ **2.** $10 \le 1 + y$ **3.** $4y > 20$

4. $12 \ge 2x$ **5.** $-3x > -12$ **6.** $-4c \le -8$

7. $\dfrac{x}{2} > 4$ **8.** $8 \ge \dfrac{d}{4}$ **9.** $\dfrac{x}{-2} < 5$

10. $-4 \le \dfrac{x}{-3}$ **11.** $4t - 3 < 17$ **12.** $-8 \le 3y - 2$

13. $2x + 6x - 12 > 4$ **14.** $3x + 5 > 4x$ **15.** $3x + 2 > 2x + 7$

16. $\dfrac{x}{3} - 2 < \dfrac{x}{2} - 1$ **17.** $3(2x - 1) < 21$ **18.** $7y > 2(2y + 3)$

19. $4\left(\frac{1}{2} - \dfrac{3y}{4}\right) \ge -7$ **20.** $10x - 4(2x + 1) \ge 0$

21. $2y^2 - 24 \le 2y(6 + y)$ **22.** $5t \le 8 + 3(2t - 2)$
23. $2(2x + 3) > -3(4x - 8)$ **24.** $-2(2c - 9) \ge 4 - 5(c - 2)$

In 25–28, state whether the inequality is a conditional inequality or an absolute inequality. $x \in \{$real numbers$\}$.

25. $5x + 6 > 5x$ **26.** $5x + 6 > 11$
27. $2x + 6x - 10 < 8x$ **28.** $2(3x + 5) > 4x - (6 - x)$

In 29–32, find the set of numbers that satisfy the condition.
29. A number decreased by 4 is greater than 6.
30. Twice a number increased by 8 is less than —4.
31. Three times a number decreased by 10 is less than 2 times the number.
32. Five times a number decreased by 12 is greater than 2 times the number increased by 6.

In 33 and 34, x and $y \in$ {real numbers}.
33. *a.* If $x > y$, prove that $x - y > 0$.   *b.* If $x - y > 0$, prove that $x > y$.
34. *a.* If $x - y < 0$, prove that $x < y$.   *b.* If $x < y$, prove that $x - y < 0$.

In 35 and 36, a, b, x, and $y \in$ {real numbers}.
35. If $a < b$ and $x < y$, prove that $a + x < b + y$.
36. If a, b, x, and y are positive numbers, and $a > b$ and $x > y$, prove that $ax > by$.

In 37–39, a and $b \in$ {real numbers}.
37. If $a > 0$ and $b > 0$, prove that $ab > 0$.
38. If $a > 0$ and $b < 0$, prove that $ab < 0$.
39. If $a < 0$ and $b < 0$, prove that $ab > 0$.

16. Solving Inequalities Involving Absolute Values

On a number line, the absolute value of a real number may represent the non-directed distance between the graph of the number and the origin.

To find the solution set of $|x| < 3$, we must find the set of points such that, on a number line, the distance between the graph of every one of these points and the 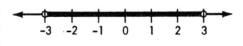 origin is less than 3. The graph at the right shows us that the distance between the origin and the graph of any number x between —3 and 3 is less than 3. Hence, x is any real number greater than —3 and less than 3; that is, $-3 < x$ and $x < 3$. These two inequalities may be written as $-3 < x < 3$. Therefore, the solution set of $|x| < 3$ is $\{x \mid -3 < x < 3\}$. This example illustrates the truth of the following:

Principle 1. If a is a positive number ($a > 0$), then the solution set of $|x| < a$ is $\{x \mid -a < x < a\}$.

Note. As shown in the figure at the right, the graph of the solution set of $|x| < a$ is the heavy line 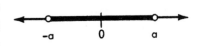 segment between —a and a, not including its endpoints.

If we wish to find the solution set of $|x| > 3$, we must find the set of points such that, on a number line, the distance between the graph of every one of these points and the origin is greater than 3.

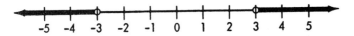

The preceding graph shows us that the distance between the origin and the graph of any number x which is greater than 3, also, the distance between the origin and the graph of any number x which is less than -3, must be greater than 3. Therefore, $|x| > 3$ for $x < -3$ or $x > 3$. Hence, the solution set of $|x| > 3$ is $\{x \mid x < -3 \text{ or } x > 3\}$. This example illustrates the following principle:

Principle 2. If a is a positive number $(a > 0)$, then the solution set of $|x| > a$ is $\{x \mid x < -a \text{ or } x > a\}$.

Note. As shown in the figure at the right, the graph of the solution set of $|x| > a$ is the 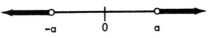 union of the heavy ray to the right of a and the heavy ray to the left of $-a$, not including the endpoints a and $-a$.

~~~~~~~~~~ **MODEL PROBLEMS** ~~~~~~~~~~

1. Solve $|4x - 8| < 12$ and graph the solution set.

How To Proceed	*Solution*
1. Write the inequality.	$|4x - 8| < 12$
2. Apply Principle 1.	$-12 < 4x - 8 < 12$
3. Add 8 to each member of the inequality.	$-4 < 4x < 20$
4. Divide each member of the inequality by 4.	$-1 < x < 5$

Answer: Solution set is $\{x \mid -1 < x < 5\}$.

2. Solve $\left| \dfrac{x}{3} - 2 \right| \geq 1$ and graph the solution set.

How To Proceed	*Solution*
1. Write the inequality.	$\left| \dfrac{x}{3} - 2 \right| \geq 1$

2. Apply Principle 2.

$$\frac{x}{3} - 2 \leq -1 \text{ or } \frac{x}{3} - 2 \geq 1$$

3. Add 2 to each member of the inequality.

$$\frac{x}{3} \leq 1 \qquad \frac{x}{3} \geq 3$$

4. Multiply each member of the inequality by 3.

$$x \leq 3 \qquad x \geq 9$$

Answer: Solution set is $\{x \mid x \leq 3 \text{ or } x \geq 9\}$.

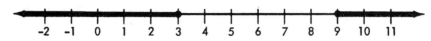

Exercises

In 1–12, solve the inequality and graph the solution set.

1. $|x| > 5$

2. $\left|\dfrac{x}{3}\right| \geq 2$

3. $|3x| < 9$

4. $|-2r| \leq 10$

5. $|x - 2| > 4$

6. $\left|\dfrac{3 + d}{2}\right| \leq 2$

7. $|2x - 1| < 7$

8. $\left|\dfrac{3x}{4} - 1\right| \geq 2$

9. $|2y + 5| \geq 9$

10. $13 \geq |6x + 1|$

11. $|6 - x| \leq 18$

12. $\left|\dfrac{4 - 2x}{3}\right| \geq 4$

In 13–16, select the inequality whose solution set is pictured in the graph.

13. (1) $|x| > 4$ (2) $|x| < 4$ (3) $|x| \geq 4$ (4) $|x| \leq 4$

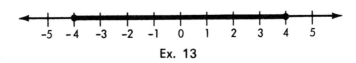

Ex. 13

14. (1) $|x| \leq 2$ (2) $|x| > 2$ (3) $|x| < 2$ (4) $|x| \geq 2$

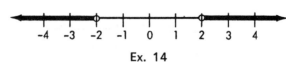

Ex. 14

15. (1) $|x - 2| > 3$ (2) $|x + 2| \leq 3$ (3) $|x + 2| \geq 3$ (4) $|x - 2| < 3$

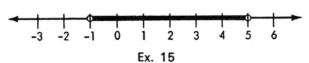

Ex. 15

16. (1) $|x + 3| < 4$ (2) $|x - 3| < 4$ (3) $|x - 3| \leq 2$ (4) $|x + 3| \geq 4$

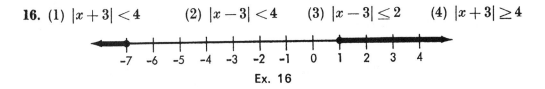

Ex. 16

17. Solving Problems Involving Inequalities

PREPARING TO SOLVE PROBLEMS INVOLVING INEQUALITIES

The following examples illustrate how to represent algebraically sentences that involve relationships of inequality.

Sentence	Meaning	Representation
1. x is at least 25.	x is equal to 25, or x is greater than 25.	$x \geq 25$
2. The minimum value of x is 25.	x is equal to 25, or x is greater than 25.	$x \geq 25$
3. x is at most 25.	x is equal to 25, or x is less than 25.	$x \leq 25$
4. The maximum value of x is 25.	x is equal to 25, or x is less than 25.	$x \leq 25$
5. x is at least 25 and at most 30.	x is equal to 25 or x is greater than 25 and x is equal to 30 or x is less than 30.	$x \geq 25$ and $x \leq 30$; or $25 \leq x \leq 30$

Exercises

In 1–10, represent the sentence as an algebraic inequality.

1. y is at least 100.
2. The least possible value of x is 18.
3. x is at most 69.
4. The greatest possible value of $5y$ is 150.
5. The sum of $5x$ and $2x$ is at least 77.
6. 8 more than $2y$ is at most 24.
7. The maximum value of $6x - 4$ is 48.
8. The minimum value of the sum of y and $y + 2$ is 62.
9. y is at least 30 and at most 40.
10. The sum of twice x and three times $(x - 5)$ is at least 70 and at most 85.

SOLVING PROBLEMS INVOLVING INEQUALITIES

〜〜〜〜〜〜〜〜〜 *MODEL PROBLEMS* 〜〜〜〜〜〜〜〜〜

1. In a town, the number of adults exceeds 3 times the number of children by 80. If the town has at most 3560 persons, find the greatest possible number of adults and the greatest possible number of children there can be in the town.

   Solution: If the town has at most 3560 persons, the sum of the number of adults and the number of children either is equal to 3560 or is less than 3560.

   Let $x =$ the possible number of children in the town.
   Then $3x + 80 =$ the possible number of adults in the town.

   The possible number of children plus the possible number of adults is equal to 3560 or is less than 3560.

   $$x + 3x + 80 \leq 3560$$
   $$4x + 80 \leq 3560$$
   $$4x \leq 3480$$
   $$x \leq 870 \qquad \text{Hence, there are at most 870 children.}$$
   $$3x + 80 \leq 2690 \qquad \text{Hence, there are at most 2690 adults.}$$

   Answer: There are at most 2690 adults and at most 870 children in the town.

2. Elliot is 5 times as old as Sid. Ten years from now, Elliot will be at least 3 times as old as Sid will be then. At least how old is each person now?

   Solution: If 10 years from now Elliot will be at least 3 times as old as Sid will be then, Elliot's age at that time either will be equal to or greater than 3 times Sid's age at that time.

   Let $x =$ the possible number of years in Sid's age now.

	Age now	Age 10 years hence
Sid	x	$x + 10$
Elliot	$5x$	$5x + 10$

   *Ten years from now, Elliot's age will be equal to or
   greater than 3 times Sid's age then.*

$$5x + 10 \geq 3(x + 10)$$
$$5x + 10 \geq 3x + 30$$
$$2x \geq 20$$

$x \geq 10$ Hence, Sid is at least 10 years old now.

$5x \geq 50$ Hence, Elliot is at least 50 years old now.

Answer: Sid is at least 10 years old now; Elliot is at least 50 years old now.

Exercises

1. Ms. Freed and her helper, Mr. Standish, work together on a job. Ms. Freed earns twice as much as Mr. Standish. If they receive at least $840 for doing the job, find the least amount that each would receive.
2. Three times a number increased by 8 is at most 40 more than the number. Find the greatest value of the number.
3. The length of a rectangle is 8 inches less than 5 times its width. If the perimeter of the rectangle is at most 104 inches, find the maximum length of the rectangle.
4. Mr. Gold is 8 years younger than Mr. Breen, and Mr. Carney is half as old as Mr. Breen. The sum of Mr. Gold's age and Mr. Breen's age is at least 58 years more than Mr. Carney's age. At least, how old is Mr. Breen now?
5. Mr. Taylor has $15,000. He wishes to invest part of his money in bonds which pay 5% interest and the rest of his money in a business loan which pays 8% interest. Find the maximum amount that he may invest at 5% if his annual income from these investments is to be at least $1050.
6. The smaller of two consecutive even integers is greater than 49 more than one-fourth of the larger. Find the least possible values for the integers.
7. Ms. Gray can drive from her home out into the country traveling at the rate of 30 mph and return traveling at the rate of 40 mph. What is the greatest distance that she can drive out and then return to her home if she has at most 7 hours to spend on the trip?
8. A chemist has 40 quarts of an acid solution which is 20% pure acid. At least, how many quarts of pure acid must she add to this solution to obtain a solution which is at least 50% pure acid?
9. A shoemaker can fix a pair of heels in 10 minutes and a pair of soles in 20 minutes. The number of pairs of heels he fixes is 5 times the number of pairs of soles he fixes. In a week, he works a minimum of 42 hours and a maximum of 49 hours. Find the minimum and the maximum number of pairs of heels that he can fix in a week.
10. Mr. Green would like to save between $250 and $310 a month. He can do this if in a month he will save $50 less than 3 times the amount that he saves a month now. In what range is the amount he saves a month now?

~~~~~~~~ *CALCULATOR APPLICATIONS* ~~~~~~~~

There are several ways in which a graphing calculator can be used to find solution sets.

~~~~~~~~~~~~ *MODEL PROBLEMS* ~~~~~~~~~~~~

1. Find the solution set of $6 - 2x = 8$ if the domain is $\{-3, -2, -1, 0, 1, 2, 3\}$.

   Solution: Enter the equation to be tested as Y_1.

   Enter: $\boxed{Y=}$ 6 $\boxed{-}$ 2 $\boxed{X,T,\theta,n}$ $\boxed{\text{2nd}}$ $\boxed{\text{TEST}}$ $\boxed{\text{ENTER}}$ 8 $\boxed{\text{ENTER}}$

   Store -3 as x on the home screen.

   Enter: $\boxed{(-)}$ 3 $\boxed{\text{STO►}}$ $\boxed{X,T,\theta,n}$ $\boxed{\text{ENTER}}$

   Then test Y_1.

   Enter: $\boxed{\text{VARS}}$ $\boxed{►}$ 1 $\boxed{\text{ENTER}}$ $\boxed{\text{ENTER}}$

   Display: $\boxed{0}$

   Since the calculator returns 0, the equation is false when $x = -3$, and -3 is not a root.

   Continue by storing and testing each value until all the elements of the domain have been tested. The only value for which the display is 1, indicating the equation is true, is -1.

   Answer: $\{-1\}$

   +--------------------- *KEEP IN MIND* ---------------------+

   Tables make it possible to evaluate an expression for a range of values. The user enters the first value and the increment (or amount added to each value) to obtain the next. If the expression being evaluated is entered as an equation, the calculator returns a value of 1 when the equation is true and a value of 0 when the equation is false.

2. Find the solution set of $|x + 2| = 5$ if the domain is all integers between -10 and 10.

   Solution: Enter the equation to be tested as Y_1.

   Enter: $\boxed{Y=}$ $\boxed{\text{MATH}}$ $\boxed{►}$ 1 $\boxed{X,T,\theta,n}$ $\boxed{+}$ 2 $\boxed{)}$ $\boxed{\text{2nd}}$ $\boxed{\text{TEST}}$ $\boxed{\text{ENTER}}$ 5 $\boxed{\text{ENTER}}$

Enter the elements of the domain in a table. Since the smallest value in the domain is −10, enter this value as TblStart, the starting value. Since the integers increase by 1, the table increment, ΔTbl, should be 1.

Enter: [2nd] [TBLSET] 10 [ENTER] 1 [2nd] [TABLE]

Read the results from the table. The values of the variable are given in the first column, and 1 for true or 0 for false in the second column. Use the arrow keys to scroll down the table and examine all the domain values.

The result is 1 (true) for $x = -7$ and $x = 2$. The result is 0 (false) for all other values.

Answer: {−7, 2}

KEEP IN MIND

Some first-degree equations can be solved by graphing the right and left sides separately and then determining where the graphs intersect.

---

3. Solve $2x + 3 = 5$.

Solution: Use the standard window. Enter $2x + 3$ as Y_1 and −5 as Y_2.

Enter: [Y=] 2 [X,T,θ,n] [+] 3 [ENTER] [(−)] 5 [GRAPH]

Display:

Use the [CALC] feature to determine the intersection of the two lines.

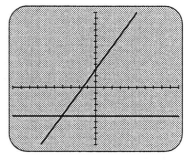

Enter: [2nd] [CALC] 5

Use the arrow keys to move the flashing cursor until it is approximately at the intersection point. The prompts *First curve?*, *Second curve?*, and *Guess?* will appear; press [ENTER] in response to each. The calculator will determine the x- and y-values at the intersection. The x-value is the solution to the equation.

Answer: $x = -4$

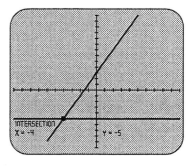

CHAPTER V

Linear Relations and Functions

1. Finding Solution Sets of Open Sentences in Two Variables

An example of an open sentence in two variables is $x + 2y = 6$, which is neither true nor false. If x is replaced by 1 and y is replaced by 4, the resulting sentence $1 + 8 = 6$ is *false*. If x is replaced by 4 and y is replaced by 1, the resulting sentence $4 + 2 = 6$ is *true*. Therefore, the pair of numbers $x = 4$ and $y = 1$ satisfies the open sentence $x + 2y = 6$. Such a pair of numbers is called a ***root*** or ***solution*** of $x + 2y = 6$. We can write the solution $x = 4$ and $y = 1$ as an ***ordered pair*** of numbers using the symbol (4, 1) if we agree that the first number in the pair, called the ***first coordinate*** or the ***first component***, represents a value of the variable x, and the second number of the pair, called the ***second coordinate*** or the ***second component***, represents a value of the variable y. We saw that the pair of numbers $x = 1$, $y = 4$ does not satisfy the sentence $x + 2y = 6$. Therefore, the ordered pair (1, 4) is not a solution of $x + 2y = 6$.

Observe that the ordered pair (4, 1) is not the same as the ordered pair (1, 4); that is, $(4, 1) \neq (1, 4)$. When dealing with ordered pairs of numbers, we must be careful not to interchange their coordinates.

Two ordered pairs of numbers are equal if and only if their first coordinates are equal and their second coordinates are equal. For example, $(4, 1) = (\frac{8}{2}, \frac{3}{3})$ because $4 = \frac{8}{2}$ and $1 = \frac{3}{3}$.

In general, for all real numbers a, b, c, d,

$$(a, b) = (c, d) \text{ when } a = c \text{ and } b = d$$

When the domain, or replacement set for x and for y, is {1, 2, 3, 4, 5, 6, 7, 8, 9}, the solutions of $x + 2y = 6$ are the ordered pairs (2, 2) and (4, 1). We call this set of ordered pairs {(2, 2), (4, 1)} the *solution set* of the sentence $x + 2y = 6$. The ***solution set of an open sentence in two variables*** is the set of ordered pairs of numbers that are members of the replacement set of the variables and that also satisfy the open sentence. If there are no ordered pairs that are solutions of the sentence, we say that the solution set is the empty set, $\emptyset$.

If the replacement set for both x and y is the set of real numbers, the solution set of $x + 2y = 6$ has an infinite number of members, some of which are (1, 2.5),

(2, 2), (0, 3), (−1, 3.5). It is impossible to list all the members of the solution set. In such a case, we can describe the solution set as $\{(x, y) \mid x + 2y = 6\}$, which is read "the set of all ordered pairs (x, y) such that $x + 2y = 6$."

When the replacement set of both x and y is $\{1, 2, 3, 4\}$, the solution set of the open sentence $x + 2y < 6$ is the set of ordered pairs $\{(1, 1), (1, 2), (2, 1), (3, 1)\}$. To verify, for example, that (1, 2) is a solution of $x + 2y < 6$, we replace x by 1 and y by 2. We obtain $1 + 4 < 6$, which is a true sentence.

If the replacement set of both x and y is the set of real numbers, the solution set of $x + 2y < 6$ has an infinite number of members, some of which are (5, 0), ($\frac{1}{2}$, 2), (0, 1), (−1, 3). It is impossible to list all the members of the solution set. In such a case, we can describe the solution set as $\{(x, y) \mid x + 2y < 6\}$, which is read "the set of all ordered pairs (x, y) such that $x + 2y < 6$."

〜〜〜〜〜〜〜 *MODEL PROBLEMS* 〜〜〜〜〜〜〜

1. Find the solution set of $y - 3x = 2$ when the replacement set of x is $R = \{1, 2, 3, 4, 5\}$ and the replacement set of y is $S = \{8, 9, 10, 11, 12\}$.

How To Proceed	*Solution*
1. Transform the sentence into an equivalent sentence which expresses y in terms of x.	1. $y - 3x = 2$ $\quad y = 3x + 2$
2. Replace x by each member of R, the replacement set of x. Then compute each of the corresponding y-values.	
3. Determine whether or not each y-value in step 2 is a member of S, the replacement set of y.	
4. List the ordered pairs (x, y), $x \in R$, and $y \in S$, which are solutions of the given open sentence.	

Step 2   Step 3

$x \in R$	$3x + 2 = y$	Does $y \in S$?
1	$3(1) + 2 = 5$	$5 \notin S$
2	$3(2) + 2 = 8$	$8 \in S$
3	$3(3) + 2 = 11$	$11 \in S$
4	$3(4) + 2 = 14$	$14 \notin S$
5	$3(5) + 2 = 17$	$17 \notin S$

4. (2, 8), (3, 11)

Answer: Solution set is $\{(2, 8), (3, 11)\}$.

2. Find the solution set of $y + 1 \geq 2x$ when the domain of x is $R = \{1, 2, 3, 4\}$ and the domain of y is $S = \{2, 4, 6\}$.

How To Proceed	*Solution*
1. Transform the sentence into an equivalent sentence which expresses y in terms of x.	1. $y + 1 \geq 2x$ $\quad y \geq 2x - 1$

184 *Algebra II*

2. Replace x in the expression found in step 1 by each member of R, the domain of x. Then compute each corresponding value of the expression.

Step 2

$x \in R$	$2x - 1$
1	$2(1) - 1 = 1$
2	$2(2) - 1 = 3$
3	$2(3) - 1 = 5$
4	$2(4) - 1 = 7$

3. Determine whether or not the y-value(s), found by making use of the results obtained in step 2 are members of S, the domain of y.

Step 3

$y \geq 2x - 1$	$y \in S$
$y \geq 1$	2, 4, 6
$y \geq 3$	4, 6
$y \geq 5$	6
$y \geq 7$	no values

4. List the ordered pairs (x, y), $x \in R$, and $y \in S$, which are solutions of the given open sentence.

4. $(1, 2), (1, 4), (1, 6), (2, 4), (2, 6), (3, 6)$

Answer: Solution set is $\{(1, 2), (1, 4), (1, 6), (2, 4), (2, 6), (3, 6)\}$.

Exercises

In 1–4, find the value for x and the value for y for which the ordered pairs of numbers are equal.

1. $(4, y)$ and $(x, 7)$

2. $(2x, -9)$ and $(6, 3y)$

3. $(\frac{1}{3}x, 8)$ and $(6, \frac{3}{4}y)$

4. $(2x - 6, y + 2)$ and $(6x, 3y + 5)$

In 5–8, state whether or not the given ordered pair of numbers is a solution of the sentence. The replacement set for x and for y is the set of integers.

5. $y = 2x + 1$, $(3, 7)$

6. $5x - 3y = 0$, $(3, 5)$

7. $3y > 2x$, $(3, 4)$

8. $5y - 2x \leq 18$, $(5, 1)$

In 9–12, state whether or not the given ordered pair of numbers is a solution of the sentence. The replacement set for x and for y is the set of real numbers.

9. $4x + 5y = 2$, $(\frac{1}{4}, \frac{1}{5})$

10. $3x < 4y$, $(-4, -2)$

11. $y \geq 4 - 3x$, $(-1, \frac{1}{3})$

12. $4y - 3x \leq 13$, $(-2, -\frac{1}{2})$

In 13–15, find the solution set of the sentence.

13. $y = 3x$ when the replacement set of x is $\{1, 2, 3\}$ and the replacement set of y is $\{5, 6, 7, 8, 9\}$.

14. $2x + 3y = 11$ when the domain of x is $\{1, 2, 3, 4, 5, 6\}$ and the domain of y is $\{0, 1, 2, 3, 4\}$.

15. $x + y \geq 8$ when the domain of x is $\{-6, 8, 10\}$ and the domain of y is $\{-2, 2, 6, 10\}$.

16. If the replacement set of x and of y is $\{$natural numbers$\}$, find three ordered pairs that satisfy the sentence $x + 2y = 20$.

17. If the domain of x and of y is the set of odd integers, find three ordered pairs that are members of the solution set of the sentence $y + 3x > 8$.

In 18–21, use set notation to describe the solution set when the replacement set of x and of y is $\{$real numbers$\}$.

18. $y = 8x$ **19.** $4x + y = 8$ **20.** $y < 2x - 3$ **21.** $y - 2x \leq 6$

2. Graphing Ordered Number Pairs in a Plane

An ordered pair of numbers can be associated with a particular point in a plane. We begin with two lines, called **coordinate axes,** drawn at right angles to each other. The horizontal line is called the **x-axis.** The vertical line is called the **y-axis.** In a **coordinate plane**, or **Cartesian plane**, the point O at which the two axes intersect is called the **origin.**

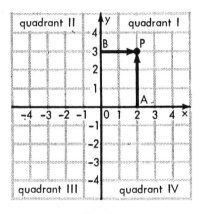

Fig. 1

The x-axis and the y-axis divide the plane into four regions called **quadrants,** which are numbered I, II, III, and IV in a counterclockwise order, as shown in Fig. 1. To determine the ordered pair of numbers that is associated with point P, we draw, through P, a line that is perpendicular to the x-axis and that intersects it at 2. This number 2 is called the **x-coordinate**, or **abscissa**, of point P. We also draw, through P, a line that is perpendicular to the y-axis and that intersects it at 3. This number 3 is called the **y-coordinate,** or **ordinate,** of point P. The x-coordinate, 2, and the y-coordinate, 3, are called the **coordinates** of point P. The coordinates of a point may be written as an ordered pair of numbers in which the first number is always the x-coordinate, and the second number is always the y-coordinate. Thus, the coordinates of point P in Fig. 1 may be written $(2, 3)$. In general, the coordinates of a point may be represented by (x, y).

To find a point, given an ordered pair, we merely reverse the previous procedure. For example, to find the point associated with the ordered pair (3, 2), we draw a line perpendicular to the x-axis at 3; we draw a line perpendicular to the y-axis at 2. The point, P_1, at which these lines intersect (see Fig. 2), is the point associated with the ordered pair (3, 2). Point P_1 is called the **graph** of the ordered number pair (3, 2). The point associated with (−2, 4) is point P_2 in quadrant II. The point associated with (−3, −4) is point P_3 in quadrant III. The point associated with (4, −2) is point P_4 in quadrant IV. When we graph the

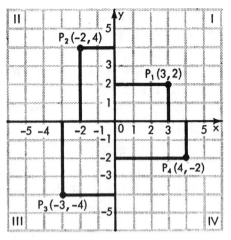

Fig. 2

point associated with an ordered number pair, we are *plotting the point.*

Each ordered pair of real numbers corresponds to one and only one point in the plane. Also, each point in the plane corresponds to one and only one ordered pair of real numbers. This system, which sets up a one-to-one correspondence between the set of all ordered pairs of real numbers and the set of all points in a plane, is called the **Cartesian coordinate system,** or **rectangular coordinate system.**

If one of the coordinates of a point is zero, the point lies on one of the axes. In Fig. 3, points such as $A(3, 0)$ and $C(−4, 0)$, whose ordinate is 0, lie on the x-axis. Points such as $B(0, 2)$ and $D(0, −3)$, whose abscissa is 0, lie on the y-axis. The origin, $O(0, 0)$, lies on both axes.

The signs and zero values of the co-ordinates of points in the four quadrants and between the quadrants may be summarized as follows:

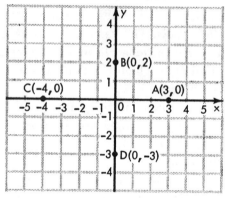

Fig. 3

	Quadrant				Between Quadrants			
	I	*II*	*III*	*IV*	*IV & I*	*I & II*	*II & III*	*III & IV*
Abscissa, (x)	+	−	−	+	+	0	−	0
Ordinate, (y)	+	+	−	−	0	+	0	−

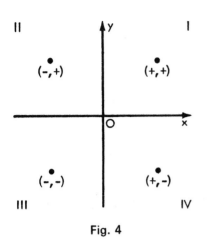

Fig. 4

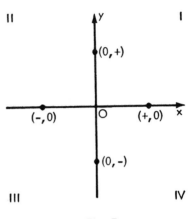

Fig. 5

Exercises

1. Write as ordered number pairs the coordinates of points $A, B, C, D, E, F, G, H, K, L, M, N,$ and O in the graph at the right.

In 2–13, draw a pair of coordinate axes on a sheet of graph paper and plot the point associated with the ordered number pair.

2. $(3, 5)$ 3. $(-4, 2)$ 4. $(3, -5)$ 5. $(-5, -4)$

6. $(3, 6\frac{1}{2})$ 7. $(-2\frac{1}{2}, 5)$ 8. $(4, 0)$ 9. $(-2, 0)$

10. $(-9, 0)$ 11. $(0, 6)$ 12. $(0, -4)$ 13. $(|3|, |5|)$

In 14–16, graph each member of the set.

14. $\{(1, 3), (2, 5), (3, 7)\}$
15. $\{(0, 5), (2, 4), (4, 3)\}$
16. $\{(0, 0), (-1, 1), (-3, 3)\}$

17. Graph several points on the x-axis. What is the value of the ordinate for every point in the set of points on the x-axis?
18. Graph several points on the y-axis. What is the value of the abscissa for every point in the set of points on the y-axis?
19. Name the quadrant in which the graph of $P(x, y)$ lies when:
    a. $x > 0, y > 0$ b. $x > 0, y < 0$ c. $x < 0, y > 0$ d. $x < 0, y < 0$
20. Name the quadrant in whose interior the graph of the point $P(|x|, |y|)$ lies when x and y are members of the set of nonzero real numbers.

3. Graphing a Linear Equation in Two Variables by Using Its Solutions

If we wish to discover pairs of numbers which satisfy, that is, are solutions of, the equation $x + 2y = 6$, we can proceed in either of the following two ways:

Method 1	*Method* 2
In the equation $x + 2y = 6$ replace one variable, for example x, by a convenient value. Then solve the resulting equation for the corresponding value of the other variable, y.	Transform $x + 2y = 6$ into an equivalent equation which has y as one of its members. Then we can assign a value to x and find the corresponding y-value.

Method 1:

Let $x = 4$. Then
$$4 + 2y = 6$$
$$2y = 2$$
$$y = 1$$
$(4, 1)$ is a solution of $x + 2y = 6$.

Method 2:

For example,
$$x + 2y = 6$$
$$2y = 6 - x$$
$$y = 3 - \tfrac{1}{2}x$$
Let $x = 4$. Then
$$y = 3 - \tfrac{1}{2}(4)$$
$$y = 3 - 2$$
$$y = 1$$
$(4, 1)$ is a solution of $x + 2y = 6$.

Note. When several ordered pairs are to be found, method 2 is preferable since, in this method, y is expressed in terms of x.

We have seen that if the domain of both x and y is the set of real numbers, there is an infinite number of ordered pairs that are solutions of $x + 2y = 6$. Some of these solutions are shown in the following table:

x	-2	0	2	4	6	8
y	4	3	2	1	0	-1

When we graph the number pairs in the table, as shown in the figure below, the points that are associated with these number pairs lie on a straight line; that is, they are called *collinear points*. In fact, any point that is the graph of an ordered pair that is a solution of $x + 2y = 6$ must lie on this same line. Furthermore, any point that is the graph of an ordered pair that is not a solution of $x + 2y = 6$ does not lie on this same line.

The graph of $x + 2y = 6$ is the line which is the set of all those points and only those points whose coordinates satisfy $x + 2y = 6$. In turn, $x + 2y = 6$ is

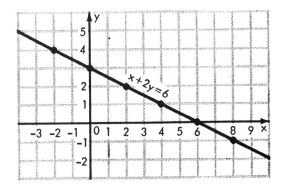

called the **equation of the line.** This line is also the graph of every equation which is equivalent to $x + 2y = 6$. Examples of such equations are $x = 6 - 2y$ and $y = \dfrac{6 - x}{2}$.

The equation $x + 2y = 6$ is in the form $Ax + By = C$.

Any first-degree equation in two variables such as $x = 6 - 2y$ or $y = \dfrac{6 - x}{2}$ may be transformed into the form $Ax + By = C$.

The following theorems, which can be proved, apply to first-degree equations transformable into $Ax + By = C$ where the replacement set of x, y, A, B, and C is the set of real numbers and A and B are not both 0:

Theorem 1. The graph of every first-degree equation is a straight line. For this reason, we call such an equation a **linear equation.**

Theorem 2. Any straight line in a Cartesian plane is the graph of a first-degree equation.

Note. When we graph a linear equation, unless otherwise indicated, the replacement set of the variables is understood to be the set of real numbers.

KEEP IN MIND

1. The graph of an open sentence in two variables is the set of those points that are the graphs of the ordered pairs of real numbers (x, y) for which the sentence becomes a true statement.
2. Each ordered pair of real numbers that satisfies an equation represents the coordinates of a point on the graph of the equation.
3. Each point on the graph of an equation has as its coordinates an ordered pair of real numbers that satisfies the equation.

~~~~~~~~~~~~ **MODEL PROBLEMS** ~~~~~~~~~~~~

1. *a.* Write the following verbal sentence as an equation: "The sum of twice
   the abscissa of a point and 3 times the ordinate of that point is 6."
   b. Graph the equation written in part *a.*

Solution:

a. Let $x =$ the abscissa of the point.
Let $y =$ the ordinate of the point.
Then $2x + 3y = 6$ *Ans.*

b.   *How To Proceed* | *Solution*

1. Transform the equation into an
   equivalent equation that expresses
   y in terms of x.

   $$2x + 3y = 6$$
   $$3y = 6 - 2x$$
   $$y = 2 - \tfrac{2}{3}x$$

2. Determine three solutions of this
   equation by assuming convenient
   values for x and computing the corresponding values for y. (In this case,
   if we choose multiples of 3 as our
   x-values, we will avoid y-values that
   are fractions.)

x	$2 - \tfrac{2}{3}x$	$= y$
-3	$2 - \tfrac{2}{3}(-3)$	$= 4$
0	$2 - \tfrac{2}{3}(0)$	$= 2$
3	$2 - \tfrac{2}{3}(3)$	$= 0$

3. Plot the points which are associated
   with the three solutions found in step
   2.

4. Draw a straight line which passes
   through the three points. This is the
   required graph.

Steps 3 and 4

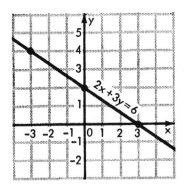

2. If the graph of $4x = 3y + 1$ passes through a point whose ordinate is -3,
   find the abscissa of the point.

Solution:
Let $y = -3$ in the equation $4x = 3y + 1$. Then, $4x = 3(-3) + 1$.
$4x = -8$, or $x = -2$. Hence, the abscissa is -2.   *Answer:* -2

~~~~~~

Exercises

In 1–6, solve the equation for y in terms of x.

1. $2x + y = 3$
2. $3x - y = 9$
3. $2x + 5y = 10$
4. $3x + 4y = 12$
5. $4x - 2y = 8$
6. $2x + 3y - 12 = 0$

In 7–15, graph the equation.

7. $y = 3x$
8. $3y = 4x$
9. $y = -2x$
10. $2x = -5y$
11. $x + y = 5$
12. $y - x = 6$
13. $2x + y = 8$
14. $3x - 4y = 12$
15. $5y - 3x + 15 = 0$

In 16–21, state whether or not the point whose coordinates are given is on the graph of the given equation.

16. $x + y = 8$, $(6, 2)$
17. $x - y = 4$, $(8, -4)$
18. $2x + 3y = -2$, $(5, -4)$
19. $2x = -5y + 2$, $(6, -2)$
20. $y = 5$, $(1, 5)$
21. $x = -2$, $(-2, -1)$

In 22–27, a point is to lie on the graph of the equation. Find its abscissa if its ordinate is the number indicated in the parentheses.

22. $x + 3y = 5$, (2)
23. $3x - y = 8$, (-1)
24. $2x + 3y = 8$, (4)
25. $2y + x = -10$, (6)
26. $4x - 3y = -9$, (0)
27. $y = -2x + 5$, (-3)

In 28–33, a point is to lie on the graph of the equation. Find its ordinate if its abscissa is the number indicated in the parentheses.

28. $x + y = 7$, (5)
29. $2x - y = 11$, (-1)
30. $3x + 2y = 5$, (-5)
31. $4y + x = 10$, (2)
32. $5x - 4y = 8$, (0)
33. $y = -3x - 5$, (4)

In 34–38, find a value that can replace k so that the graph of the resulting equation will pass through the point whose coordinates are given.

34. $x + y = k$, $(1, 3)$
35. $x - y = k$, $(2, -4)$
36. $3x + y = k$, $(-2, -4)$
37. $2x + 3y = k$, $(5, 3)$
38. $3x + 4y = k$, $(4, -3)$

39. If the graph of $y = -3x + b$ passes through the point $(-2, 3)$, find the value of b.

40. If the graph of the equation $Ax + By = 12$ passes through the point $(4, 0)$, find the value of A.

41. Without drawing a graph, find the coordinates of a point on the graph of $x + 4y = 18$ such that the ordinate of the point is twice its abscissa.

In 42–44, (a) write the verbal sentence as an equation, and (b) graph the equation.

42. The abscissa of a point is equal to the ordinate of the point.

43. The ordinate of a point exceeds twice its abscissa by 3.

44. The sum of the abscissa of a point and the ordinate of the point is 7.

4. Graphing a Linear Equation in Two Variables by the Intercept Method

The x-intercept of a line is the x-coordinate of the point at which the line intersects the x-axis. The graph of $2x - 3y = 6$, shown in the figure, intersects the x-axis at $A(3, 0)$. Therefore, the x-intercept of the graph of $2x - 3y = 6$ is 3. Notice that the value of y at point A must be zero, since A is on the x-axis.

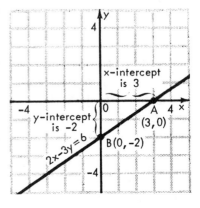

The y-intercept of a line is the y-coordinate of the point at which the line intersects the y-axis. The graph of $2x - 3y = 6$ intersects the y-axis at $B(0, -2)$. Therefore, the y-intercept of the graph of $2x - 3y = 6$ is -2. Notice that the value of x at point B must be zero, since B is on the y-axis.

Procedure. To find the x- and y-intercepts of a line:

1. **To find the x-intercept of a line, substitute 0 for y in the equation of the line. Then solve the resulting equation for x.**

2. **To find the y-intercept of a line, substitute 0 for x in the equation of the line. Then solve the resulting equation for y.**

〜〜〜〜〜〜〜〜 *MODEL PROBLEM* 〜〜〜〜〜〜〜〜

a. Find the x- and y-intercepts of the line $x - 2y = 4$.

b. Graph the equation using the intercepts.

How To Proceed	*Solution*

1. Find the x-intercept by substituting 0 for y; find the y-intercept by substituting 0 for x.

Find the x-intercept:	Find the y-intercept:
$x - 2y = 4$	$x - 2y = 4$
$x - 2(0) = 4$	$0 - 2y = 4$
$x = 4$	$y = -2$

The points $(4, 0)$ and $(0, -2)$ lie on the graph.

2. On the x-axis, graph a point whose abscissa is the x-intercept. On the y-axis, graph a point whose ordinate is the y-intercept.

3. Draw a line through these points.

Steps 2 and 3

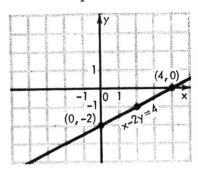

4. To check, find a third ordered pair of numbers that satisfies the given equation. The graph of this ordered pair must be a point in the line.

4. *Check* in the original equation:
$$x - 2y = 4$$
Let $x = 2$: $2 - 2y = 4$
$$-2y = 2$$
$$y = -1$$
The point $(2, -1)$ is on the graph of the line $x - 2y = 4$.

~~~~~~

Exercises

In 1–6, find the x- and y-intercepts of the line that is the graph of the equation.

1. $x + y = 6$ **2.** $x - 4y = 8$ **3.** $y = 3x - 12$
4. $y = 4x$ **5.** $3x - 5y = 15$ **6.** $3x + 4y = 6$

In 7–12, draw the graph of the equation by the intercept method.

7. $x + y = 4$ **8.** $x - 3y = 12$ **9.** $y = 3x + 2$
10. $x = 4y - 8$ **11.** $2x + 5y = 10$ **12.** $4x - 6y = 6$

13. *a.* Find the x-intercept and y-intercept for the graph of the equation $y = 2x$.
    b. Is it possible to graph the equation $y = 2x$ by the intercept method?
    c. Draw the graph of $y = 2x$ using another method.

5. Graphing Lines Parallel to the *x*-Axis or *y*-Axis

LINES PARALLEL TO THE *x*-AXIS

If we wish to graph the equation $y = 2$ in a Cartesian plane, we can write this equation in the form $0x + y = 2$.

Every ordered pair in the solution set of this equation has an x-coordinate which is any real number and a y-coordinate which must be 2. Note in Fig. 1 the points $(-2, 2)$, $(0, 2)$, $(2, 2)$, etc. When the graph of $y = 2$ is drawn, it consists of the set of all points and only those points whose ordinate is 2. This set of points is a line parallel to the x-axis and whose y-intercept is 2. Similarly, $y = -2$ represents an equation of a line whose graph is parallel to the x-axis and whose y-intercept is -2.

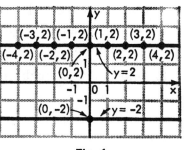

Fig. 1

In general, for any nonzero real number a, the graph of the equation $y = a$ in a Cartesian plane is a line parallel to the x-axis and whose y-intercept is a. Note that when $a = 0$, the graph of $y = a$ in a Cartesian plane is the x-axis itself. Hence, an equation of the x-axis is $y = 0$.

LINES PARALLEL TO THE y-AXIS

If we wish to graph the equation $x = 2$ in a Cartesian plane, we can write this equation in the form $x + 0y = 2$.

The solution set of this equation will be an infinite number of ordered pairs whose x-coordinate is 2 and whose y-coordinate is any real number. Note in Fig. 2 the points $(2, -1)$, $(2, 0)$, $(2, 1)$. The graph of the solution set of $x = 2$ is a line parallel to the y-axis and whose x-intercept is 2. Similarly, $x = -2$ represents the equation of a line whose graph is parallel to the y-axis and whose x-intercept is -2.

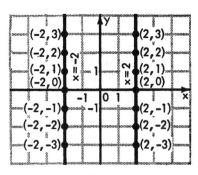

Fig. 2

In general, for any nonzero real number a, the graph of the equation $x = a$ in a Cartesian plane is a line parallel to the y-axis and whose x-intercept is a. Note that when $a = 0$, the graph of the equation $x = a$ in a Cartesian plane is the y-axis itself. Hence, an equation for the y-axis is $x = 0$.

On a number line, there is only one point that has the coordinate a. Therefore, on a number line, the graph of $\{x \mid x = a\}$ is a single point, a.

However, in a Cartesian plane, there is an infinite number of collinear points whose x-coordinate is equal to a. Hence, the graph of $x = a$ in a Cartesian plane is a line, each of whose points has a as its x-coordinate. Therefore, in a Cartesian plane, the graph of $\{(x, y) \mid x = a\}$ is a line, each of whose points is the graph of an ordered pair (x, y) whose x-coordinate is a.

Exercises

In 1–10, draw the graph of the equation in a Cartesian plane.

1. $x = 1$ **2.** $x = \frac{5}{2}$ **3.** $x = 0$ **4.** $x = -4$ **5.** $x + 6 = 0$

6. $y = 3$ **7.** $y = 1.5$ **8.** $y = 0$ **9.** $y = -5$ **10.** $y + 6 = 0$

In 11–13, draw the graph of the set in a Cartesian plane.

11. $\{(x, y) \mid x = 3\}$ **12.** $\{(x, y) \mid y = 1\}$ **13.** $\{(x, y) \mid y + 3 = 0\}$

14. Write an equation of a line that is parallel to the x-axis and whose y-intercept is (a) 4 and (b) -8.

15. Write an equation of a line that passes through all points whose:

a. abscissa is 4. b. ordinate is 5.

c. abscissa is -9. d. ordinate is -6.

6. The Slope of a Line

LINE REPRESENTATION

In discussions that are to come, we will be dealing with lines. The meanings of the symbols that we will use to represent lines are as follows:

In Fig. 1, $\overleftrightarrow{AB}$ represents a straight line passing through points A and B, extending infinitely far in both directions.

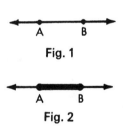

Fig. 1

In Fig. 2. $\overline{AB}$ represents line segment AB, or simply segment AB. $\overline{AB}$ is the set of points containing A, B, and all points lying between A and B.

Fig. 2

AB represents the length of $\overline{AB}$. AB is a number, the distance between A and B.

MEANING OF THE SLOPE OF A LINE

In triangle ABC (Fig. 3), $\overline{AB}$ represents a section of a straight road. Using $\overline{AC}$ and $\overline{BC}$, we can tell that as a person moves along the road, he will go up a vertical distance

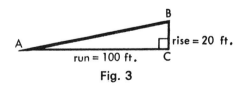

Fig. 3

of 20 feet whenever he travels a horizontal distance of 100 feet. The change in the vertical distance, BC, is called the **rise** and the change in the horizontal distance, AC, is called the **run.**

We may measure the steepness of the road by finding the *slope* of $\overline{AB}$, which is defined as follows:

$$\text{Slope of } \overline{AB} = \frac{\text{change in the vertical distance}}{\text{change in the horizontal distance}} = \frac{\text{rise}}{\text{run}} = \frac{CB}{AC} = \frac{20 \text{ feet}}{100 \text{ feet}} = \frac{1}{5}$$

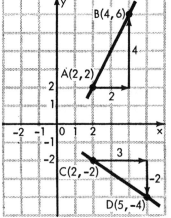

Hence, when we say that the slope of a road is $\frac{1}{5}$, we mean that the road rises 1 foot in each run of 5 feet, or the road rises $\frac{1}{5}$ foot in each run of 1 foot.

Procedure. To find the slope of a line that passes through two points:
1. **Find the horizontal change, the difference in the *x*-values, in going from the point on the left to the point on the right.**
2. **Find the vertical change, the difference in the *y*-values, in going from the point on the left to the point on the right.**
3. **Divide the vertical change by the horizontal change.**

Fig. 4

For example, in Fig. 4:

$$\text{slope of } \overleftrightarrow{AB} = \frac{\text{vertical change}}{\text{horizontal change}} = \frac{6-2}{4-2} = \frac{4}{2} = 2$$

$$\text{slope of } \overleftrightarrow{CD} = \frac{\text{vertical change}}{\text{horizontal change}} = \frac{(-4)-(-2)}{5-2} = \frac{-2}{3} = -\frac{2}{3}$$

In general, the slope m of a line that passes through two points $P_1(x_1, y_1)$ and $P_2(x_2, y_2)$, $x_2 \neq x_1$, is the ratio of the difference of the *y*-values, $y_2 - y_1$, to the difference of the corresponding *x*-values, $x_2 - x_1$. Thus, as shown in Fig. 5, we may indicate the slope of a line passing through (x_1, y_1) and (x_2, y_2) as

$$m = \frac{y_2 - y_1}{x_2 - x_1}$$

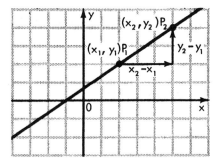

Fig. 5

The expression "difference in *x*-values, $x_2 - x_1$" may be represented by the symbol Δx, read "delta *x*." Similarly, "the difference in *y*-values, $y_2 - y_1$" may be represented by Δy, read "delta *y*." Therefore, we may indicate the slope of the line passing through (x_1, y_1) and (x_2, y_2) as

$$m = \frac{\Delta y}{\Delta x}$$

Note that in finding the slope of a line passing through two points, it does not matter which point is represented by (x_1, y_1) and which point is represented by (x_2, y_2) since $\dfrac{y_2 - y_1}{x_2 - x_1} = \dfrac{y_1 - y_2}{x_1 - x_2}$. Hence, we may write that slope $m = \dfrac{y_1 - y_2}{x_1 - x_2}$.

The slope of a line that passes through points $A(3, 1)$ and $B(5, 4)$, as in Fig 6, can be found as follows:

$$m = \frac{\Delta y}{\Delta x} = \frac{y_2 - y_1}{x_2 - x_1} = \frac{4 - 1}{5 - 3} = \frac{3}{2}$$

Since this line also passes through points $B(5, 4)$ and $C(9, 10)$, its slope can also be found as follows:

$$m = \frac{\Delta y}{\Delta x} = \frac{10 - 4}{9 - 5} = \frac{6}{4} = \frac{3}{2}$$

Notice that in both cases, the slope of the line was found to be $\frac{3}{2}$. This example illustrates the following:

Principle 1. The slope of a line can be found by using the coordinates of any two of its points.

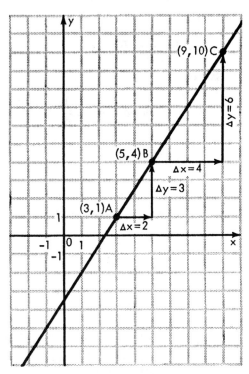

Fig. 6

In Fig. 7, points A, B, C, and D are on the given line. Through A and B, draw lines that are parallel to the x-axis and y-axis respectively, meeting in F. Draw corresponding lines through C and D, meeting in E. Since $\overline{AF}$ and $\overline{CE}$ are both horizontal segments, they are parallel, making $\angle A \cong \angle C$. Therefore, right $\triangle AFB \sim$ right $\triangle CED$ and $\dfrac{BF}{AF} = \dfrac{DE}{CE}$. Since the slope of the line, $m = \dfrac{BF}{AF} = \dfrac{DE}{CE}$, we see that the slope of the line can be found by using any two points.

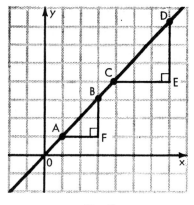

Fig. 7

POSITIVE SLOPES

As a point moves along $\overleftrightarrow{LM}$ from left to right, for example from A to B in Fig. 8, the line is "rising." As the x-values increase, the y-values also increase. Between point A and point B, $y_2 - y_1$, or $\varDelta y$ is a positive number $(+4)$; $x_2 - x_1$ or $\varDelta x$ is also a positive number $(+2)$. Since both $\varDelta y$ and $\varDelta x$ are positive numbers, the slope of $\overleftrightarrow{LM}$, $\dfrac{\varDelta y}{\varDelta x}$, must be a positive number.

In this case,

$$\text{slope} = m = \frac{\varDelta y}{\varDelta x} = \frac{+4}{+2} = 2$$

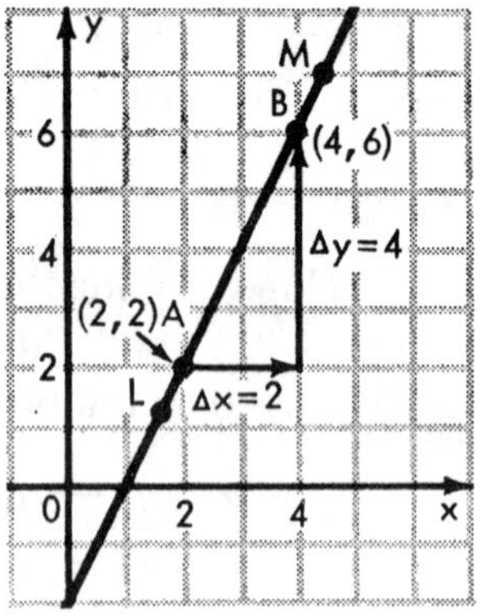

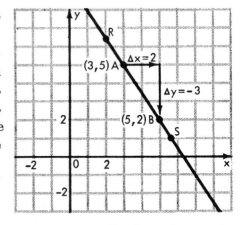

Fig. 8

Thus, we have illustrated the truth of the following:

Principle 2. As a point moves from left to right along a line that is rising, y increases as x increases, and the slope of the line is positive.

NEGATIVE SLOPES

As a point moves along $\overleftrightarrow{RS}$ from left to right, for example from A to B in Fig. 9, the line is "falling." As the x-values increase, the y-values decrease. Between point A and point B, $y_2 - y_1$ or $\varDelta y$ is a negative number (-3); $x_2 - x_1$ or $\varDelta x$ is a positive number (2). Since $\varDelta y$ is a negative number and $\varDelta x$ is a positive number, the slope of $\overleftrightarrow{RS}$, $\dfrac{\varDelta y}{\varDelta x}$, must be a negative number.

In this case,

$$\text{slope} = m = \frac{\varDelta y}{\varDelta x} = \frac{-3}{2} = -\frac{3}{2}$$

Fig. 9

Thus, we have illustrated the truth of the following:

Principle 3. As a point moves from left to right along a line that is falling, y decreases as x increases, and the slope of the line is negative.

ZERO SLOPE

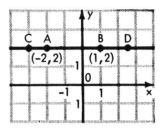

Fig. 10

$\overleftrightarrow{CD}$ is parallel to the x-axis. Consider a point moving along $\overleftrightarrow{CD}$ from left to right, for example from A to B in Fig. 10. As the x-values increase, the y-values are unchanged. Between A and B, $y_2 - y_1$ or Δy is 0; and $x_2 - x_1$ or Δx is a positive number $(+3)$. Since Δy is 0 and Δx is a positive number, the slope of $\overleftrightarrow{CD}, \dfrac{\Delta y}{\Delta x}$, must be 0. In this case,

$$\text{slope} = m = \frac{\Delta y}{\Delta x} = \frac{0}{+3} = 0$$

Thus, we have illustrated the truth of the following:

Principle 4. The slope of a line parallel to the x-axis is 0. (*Note.* The slope of the x-axis itself is also 0.)

NO SLOPE

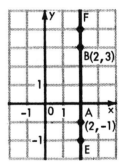

Fig. 11

$\overleftrightarrow{EF}$ is parallel to the y-axis. Consider a point moving upward along $\overleftrightarrow{EF}$, for example from A to B in Fig. 11. The x-values are unchanged, but the y-values increase. Between point A and point B, $y_2 - y_1$ or Δy is a positive number (4) and $x_2 - x_1$ or Δx is 0. Since the slope of $\overleftrightarrow{EF} = \dfrac{\Delta y}{\Delta x}$, and $\Delta x = 0$, $\overleftrightarrow{EF}$ has no defined slope because a number cannot be divided by 0.

Thus, we have illustrated the truth of the following:

Principle 5. If a line is parallel to the y-axis, it has no defined slope. (*Note.* The y-axis itself has no defined slope.)

SLOPES OF PARALLEL LINES

If two lines are vertical, they are parallel to the y-axis and as a result they have no slope.

If two lines are horizontal, they are parallel to the x-axis. Therefore, the slopes of these lines are equal, the slope of each line being 0. Also, if the slope of each of two lines is 0, each of the lines is horizontal and the two lines must be parallel.

By making use of similar triangles, we can prove the truth of the following:

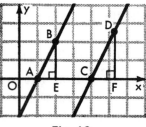

Fig. 12

Principle 6. If two nonvertical lines are parallel, their slopes are equal.

Thus, if $\overleftrightarrow{AB} \| \overleftrightarrow{CD}$, then slope of $\overleftrightarrow{AB}$ = slope of $\overleftrightarrow{CD}$.

Principle 7. If two nonvertical lines have the same slope they are parallel lines.

Thus, if the slope of $\overleftrightarrow{AB}$ = slope of $\overleftrightarrow{CD}$, then $\overleftrightarrow{AB} \| \overleftrightarrow{CD}$.

Note. Use Fig. 12 to prove principle 6 and principle 7. First prove that right triangle $AEB \sim$ right triangle CFD.

For example (principle 6), if $\overleftrightarrow{AB} \| \overleftrightarrow{CD}$ and the slope of $\overleftrightarrow{AB} = \frac{5}{2}$, then the slope of $\overleftrightarrow{CD}$ is also $\frac{5}{2}$. Also (principle 7), if the slope of $\overleftrightarrow{AB}$ is $\frac{5}{2}$ and the slope of $\overleftrightarrow{CD}$ is also $\frac{5}{2}$, then $\overleftrightarrow{AB} \| \overleftrightarrow{CD}$.

SLOPES OF PERPENDICULAR LINES

By using the theorem of Pythagoras and its converse, or using principles that we will develop in later chapters, we can prove the truth of the following:

Principle 8. If two nonvertical lines are perpendicular, then the product of their slopes is -1. That is, the slope of one line is the *negative reciprocal* of the slope of the other.

Thus, if $\overleftrightarrow{AB} \perp \overleftrightarrow{CD}$ and the slope of $\overleftrightarrow{AB}$ is m_1 and the slope of $\overleftrightarrow{CD}$ is m_2, then

$$m_1 m_2 = -1, \text{ or } m_2 = -\frac{1}{m_1}.$$

Principle 9. If the product of the slopes of two nonvertical lines is -1 (that is, the slope of one line is the negative reciprocal of the slope of the other line), then the lines are perpendicular to each other.

Thus, if the slope of $\overleftrightarrow{AB}$ is m_1 and the slope of $\overleftrightarrow{CD}$ is m_2, and $m_1 m_2 = -1$, then $\overleftrightarrow{AB} \perp \overleftrightarrow{CD}$.

Study the following:

Example 1. If $\overleftrightarrow{AB} \perp \overleftrightarrow{CD}$, and the slope of $\overleftrightarrow{AB} = m_1 = \frac{2}{3}$, find the slope of $\overleftrightarrow{CD}$, m_2. Find the slope of $\overleftrightarrow{CD}$ by applying principle 8: Since $\overleftrightarrow{AB} \perp \overleftrightarrow{CD}$, the slope of $\overleftrightarrow{CD}$ is the negative reciprocal of $\frac{2}{3}$, the slope of $\overleftrightarrow{AB}$. Hence, the slope of $\overleftrightarrow{CD} = -\frac{3}{2}$.

Example 2. If the slope of $\overleftrightarrow{AB} = m_1 = 5$ and the slope of $\overleftrightarrow{CD} = m_2 = -\frac{1}{5}$, show that $\overleftrightarrow{AB} \perp \overleftrightarrow{CD}$. Reason as follows: Since the slope of $\overleftrightarrow{AB} = m_1 = 5$ and the slope of $\overleftrightarrow{CD} = m_2 = -\frac{1}{5}$, then $m_1 m_2 = 5(-\frac{1}{5}) = -1$. Applying principle 9, since $m_1 m_2 = -1$, then $\overleftrightarrow{AB} \perp \overleftrightarrow{CD}$.

~~~~~~~~~~~~~~~~~ **MODEL PROBLEMS** ~~~~~~~~~~~~~~~~~

1. *a.* Graph the straight line that passes through the points $(-2, 3)$ and $(4, 6)$.
   b. Find the slope of the line.

Solution:

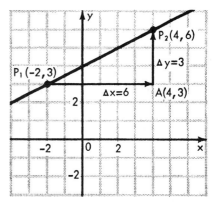

a. Plot $P_1(-2, 3)$ and $P_2(4, 6)$ as shown. Draw $\overleftrightarrow{P_1 P_2}$.

b. Complete right triangle $P_1 A P_2$ as shown.

At $P_2(4, 6) \rightarrow x_2 = 4, y_2 = 6$
At $P_1(-2, 3) \rightarrow x_1 = -2, y_1 = 3$

$\overleftrightarrow{P_1 P_2}$ passes through $P_1(x_1, y_1)$ and $P_2(x_2, y_2)$.

Slope of $\overleftrightarrow{P_1 P_2} = \dfrac{y_2 - y_1}{x_2 - x_1}$

$= \dfrac{6 - 3}{4 - (-2)}$

$= \frac{3}{6} = \frac{1}{2}$ *Ans.*

2. Show that the line which joins the point $A(1, 1)$ and the point $B(2, 4)$ is parallel to the line which connects the point $C(3, -2)$ and the point $D(4, 1)$.

Plan: We can prove that $\overleftrightarrow{AB}$ is parallel to $\overleftrightarrow{CD}$ if we can show that the slope of line $\overleftrightarrow{AB}$ is equal to the slope of line $\overleftrightarrow{CD}$.

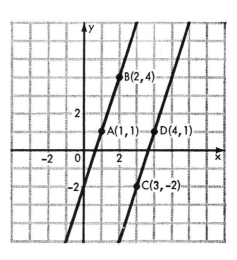

Solution:

Let A be $P_1(x_1, y_1)$.

Let B be $P_2(x_2, y_2)$.

At $B(2, 4) \rightarrow x_2 = 2,\ y_2 = 4$

At $A(1, 1) \rightarrow x_1 = 1,\ y_1 = 1$

Let C be $P_1(x_1, y_1)$.

Let D be $P_2(x_2, y_2)$.

At $D(4, 1) \rightarrow x_2 = 4,\ y_2 = 1$

At $C(3, -2) \rightarrow x_1 = 3,\ y_1 = -2$

Slope of a line passing through (x_1, y_1) and $(x_2, y_2) = \dfrac{y_2 - y_1}{x_2 - x_1}$.

Slope of $\overleftrightarrow{AB} = \dfrac{4 - 1}{2 - 1}$

$= \dfrac{3}{1}$

Slope of $\overleftrightarrow{CD} = \dfrac{1 - (-2)}{4 - 3}$

$= \dfrac{3}{1}$

Since slope of $\overleftrightarrow{AB}$ = slope of $\overleftrightarrow{CD}$, then $\overleftrightarrow{AB}$ is parallel to $\overleftrightarrow{CD}$.

3. Show that the line which passes through the points $A(-1, 4)$ and $B(2, -2)$ is perpendicular to the line which passes through the points $C(-1, 0)$ and $D(3, 2)$.

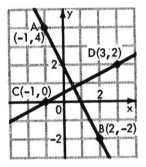

Plan: We can prove that $\overleftrightarrow{AB}$ is perpendicular to $\overleftrightarrow{CD}$ if we can show that the product of the slopes of these lines is -1.

Solution:

Let A be $P_1(x_1, y_1)$.

Let B be $P_2(x_2, y_2)$.

At $B(2, -2) \rightarrow x_2 = 2,\ y_2 = -2$

At $A(-1, 4) \rightarrow x_1 = -1,\ y_1 = 4$

Let C be $P_1(x_1, y_1)$.

Let D be $P_2(x_2, y_2)$.

At $D(3, 2) \rightarrow x_2 = 3,\ y_2 = 2$

At $C(-1, 0) \rightarrow x_1 = -1,\ y_1 = 0$

Slope of a line passing through (x_1, y_1) and $(x_2, y_2) = \dfrac{y_2 - y_1}{x_2 - x_1}$.

Slope of $\overleftrightarrow{AB} = \dfrac{(-2) - 4}{2 - (-1)}$

$= \dfrac{-6}{3} = \dfrac{-2}{1}$

Slope of $\overleftrightarrow{CD} = \dfrac{2 - 0}{3 - (-1)}$

$= \dfrac{2}{4} = \dfrac{1}{2}$

Since the slopes of $\overleftrightarrow{AB}$ and $\overleftrightarrow{CD}$ are negative reciprocals of each other, then $\overleftrightarrow{AB}$ is perpendicular to $\overleftrightarrow{CD}$.

Exercises

1. In *a-f*, tell whether the line has a positive slope, a negative slope, a slope of 0, or no slope.

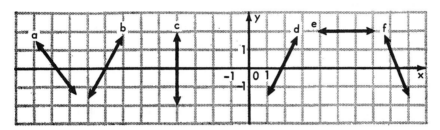

In 2–7, find the slope of the line.

2.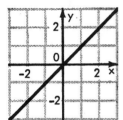

3.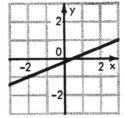

4.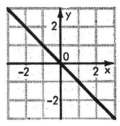

5.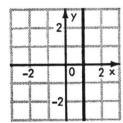

6.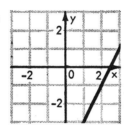

7.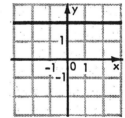

In 8–16, plot both points, draw the straight line that joins them, and find the slope of this line.

8. (0, 0) and (3, 3) 9. (0, 0) and (2, 5) 10. (0, 0) and (10, 5)
11. (0, 2) and (0, 5) 12. (2, 3) and (4, 15) 13. (4, 6) and (10, 4)
14. (−3, 8) and (−1, 12) 15. (3, −4) and (8, −4) 16. (−1, 2) and (6, −3)

In 17–25, through the given point, draw a line with the given slope *m*.
17. $(0, 0), m = \frac{1}{2}$ 18. $(-2, 4), m = \frac{4}{3}$ 19. $(0, 0), m = 2$
20. $(-2, -3), m = 1$ 21. $(0, 0), m = -\frac{2}{3}$ 22. $(-3, -4), m = -\frac{1}{4}$
23. $(0, 0), m = -3$ 24. $(0, 5), m = -2$ 25. $(0, -2), m = -1$

26. Find the value of *y* so that the slope of the line passing through the points (4, *y*) and (8, 12) will be (*a*) 1, (*b*) 3, (*c*) $-\frac{1}{2}$, and (*d*) 0.

27. The straight line which passes through the points (5, 4) and (2, 4) (1) has a slope of 3 (2) has a slope of zero (3) has no slope

28. Show that the line which joins the points (2, 4) and (6, 7) is parallel to the line which joins the point (6, 2) and (10, 5).

29. Show that the line whose y-intercept is 2 and which passes through the point $(-2, 1)$ is parallel to the line whose y-intercept is -1 and which passes through the point $(4, 1)$.

In 30–33, use the definition of the slope of a line to determine whether or not the points lie on the same straight line (are collinear).

30. (1, 2), (4, 5), (6, 7)   **31.** $(-2, 6)$, (0, 2), (1, 0)

32. (0, 0), $(-8, -2)$, (4, 1)   **33.** $(-3, 4)$, $(-1, 1)$, $(1, -3)$

In 34–39, find the slope of a line which is perpendicular to the line with the given slope.

34. $m = \frac{3}{4}$ **35.** $m = 2$ **36.** $m = \frac{1}{3}$

37. $m = -\frac{3}{5}$ **38.** $m = -4$ **39.** $m = -1$

In 40–42, show that the line which passes through the first pair of points is perpendicular to the line which passes through the second pair of points.

40. (7, 3) and (4, 1); (8, 12) and (6, 15)

41. $(-2, 6)$ and (1, 1); (9, 2) and $(4, -1)$

42. the origin and (4, 4); $(-1, 1)$ and $(-4, 4)$

43. The vertices of parallelogram $ABCD$ have the following coordinates: $A(-2, 4)$, $B(2, 6)$, $C(7, 2)$, $D(x, 0)$.

    a. Find the slope of $\overline{AB}$.

    b. Express the slope of $\overline{DC}$ in terms of x.

    c. Using the results found in answer to a and b, find the value of x.

44. Show that the triangle whose vertices are the points (8, 6), (4, 2), and (6, 0) is a right triangle.

7. The Slope-Intercept Form of a Linear Equation

Let us determine the equation of a line whose slope is m and whose y-intercept is b.

Since the y-intercept is b, the line passes through $A(0, b)$, as shown in the figure at the right. Let $P(x, y)$ be any other point on this line ($x \neq 0$) whose slope is m.

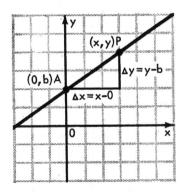

Since $\dfrac{\Delta y}{\Delta x} =$ slope of the line $= m$,

$$\frac{y - b}{x - 0} = m$$

Multiply both members of the equation by x, $x \neq 0$.

$y - b = mx$

Hence, $y = mx + b$.

Therefore, if a line has a slope m and a y-intercept b, the equation of the line is

$$y = mx + b$$

It can also be shown that, conversely, if a line has a slope and its equation is in the form $y = mx + b$, then m represents the slope of the line and b represents the y-intercept of the line.

～～～～～～～～ MODEL PROBLEMS ～～～～～～～～

1. Write an equation of a line whose slope is $\frac{1}{2}$ and whose y-intercept is -1.
   Solution: Substitute $\frac{1}{2}$ for m and -1 for b in the slope-intercept form.

$$y = mx + b \quad m = \tfrac{1}{2}, \ b = -1$$
$$y = \tfrac{1}{2}x - 1, \text{ or } 2y = x - 2 \quad Ans.$$

2. Find the slope and y-intercept of the line whose equation is $6x + 2y = 8$.
   Solution: Transform the equation into the slope-intercept form $y = mx + b$.

$$6x + 2y = 8$$
$$2y = -6x + 8$$
$$y = -3x + 4$$

Hence, the slope of the line, $m = -3$; the y-intercept, $b = 4$.
Answer: Slope is -3; y-intercept is $+4$.

3. Show that the graph of $y = -2x + 3$ is parallel to the graph of $4x + 2y = 5$.
   Solution: The equation $y = -2x + 3$ is in the slope-intercept form
   $y = mx + b$. Hence, the slope of $y = -2x + 3$ is -2. Transform $4x + 2y = 5$
   into the slope-intercept form $y = mx + b$.
   Since $4x + 2y = 5$, $2y = -4x + 5$ and $y = -2x + \frac{5}{2}$.
   Therefore, the slope of $4x + 2y = 5$ is -2. Since the slopes of the two lines are equal and their y-intercepts are not equal, the lines are parallel.

Exercises

In 1–8, write an equation of a straight line, with integral coefficients, whose slope and y-intercept are respectively:

1. 3 and 4	**2.** -5 and 1	**3.** -7 and -5	**4.** 0 and 2
5. 6 and 0	**6.** 0 and 0	**7.** $\frac{3}{4}$ and 4	**8.** $-\frac{5}{6}$ and $-\frac{2}{3}$

In 9–20, find the slope and y-intercept of the graph of the equation.

9. $y = 2x + 1$ **10.** $y = -4x + 2$ **11.** $y = -x - 7$ **12.** $y = 5x$

13. $y = \frac{1}{4}x + 5$ **14.** $y = -\frac{3}{5}x + 2$ **15.** $y = -\frac{2}{3}x$ **16.** $y - 4x = 7$

17. $3y = 6x + 4$ **18.** $2y = -x + 6$ **19.** $3x + 5y = 15$ **20.** $\frac{1}{3}x + \frac{3}{4} = \frac{1}{2}y$

21. The slope of the graph of $2x + 3y = 5$ is   (1) -2 (2) $\frac{5}{3}$ (3) $-\frac{2}{3}$ (4) $\frac{2}{3}$

In 22 and 23, state whether or not the graphs of the two lines are parallel. Give a reason for your answer.

22. $y = 2x + 5$, $y = 2x + 9$ **23.** $2x - y = 6$, $y - 2x = 6$

In 24 and 25, state whether or not the graphs of the two lines are perpendicular to each other. Give a reason for your answer.

24. $2x + 3y = 7$, $3x - 2y = 9$ **25.** $2x = 7y + 2$, $4 - 2y = 7x$

26. Write an equation of a line whose slope is 2 and which passes through the point $(0, -3)$.

27. Write an equation of a line which passes through the origin and has a slope of -4.

28. How are the graphs of $y = mx + b$ affected when m is always replaced by the same number and b is replaced by different numbers?

29. How are the graphs of $y = mx + b$ affected when b is always replaced by the same number and m is replaced by different numbers?

30. Find the value of a if the graph of $6x + 9y = 13$ is perpendicular to the graph of $ax + 2y = 8$.

In 31 and 32: Is the line which is the graph of the first set of points parallel to the line which is the graph of the second set of points? Give a reason for your answer.

31. $\{(x, y) \mid 2x + 4y = 7\}$; $\{(x, y) \mid 3x + 9y = 8\}$

32. $\{(x, y) \mid 6x + 9y = 4\}$; $\{(x, y) \mid 3y = 5 - 2x\}$

In 33 and 34: Is the line which is the graph of the first set of points perpendicular to the line which is the graph of the second set of points? Give a reason for your answer.

33. $\{(x, y) \mid 4x + 3y = 24\}$; $\{(x, y) \mid 3x - 4y = 18\}$

34. $\{(x, y) \mid 2x + 3y = 15\}$; $\{(x, y) \mid 4y - 3 = -6x\}$

35. Write an equation of a line which passes through the point $(d, 4d)$ and has $\frac{3}{4}$ as its slope.

36. Find the value of d for which the graph of $\{(x, y) \mid x - 4y = 9\}$ will be perpendicular to the graph of $\{(x, y) \mid dx - 12y = 14\}$.

GRAPHING A LINEAR EQUATION BY THE SLOPE-INTERCEPT METHOD

Procedure. To graph a first-degree equation in two variables by the slope-intercept method:
1. **Transform the equation into an equivalent equation in the slope-intercept form, $y = mx + b$.**
2. **Then graph this resulting equation, making use of the slope m and y-intercept b.**

Exercises

In 1–12, using the slope-intercept method, graph the line described by the equation.

1. $y = 2x + 3$	**2.** $y = 2x$	**3.** $y = -3x$
4. $y = -x - 4$	**5.** $y = \frac{2}{3}x$	**6.** $y = -\frac{4}{3}x - 2$
7. $y - 2x = 4$	**8.** $2y = 3x + 2$	**9.** $5x + 2y = 10$
10. $4x - y = 5$	**11.** $2x = 3y + 8$	**12.** $2x - 3y - 12 = 0$

In 13–15, using the slope-intercept method, graph the line described by the set of points.

13. $\{(x, y) \mid y = 3x - 2\}$ **14.** $\{(x, y) \mid 2y = x + 4\}$ **15.** $\{(x, y) \mid 3x - 2y = 6\}$

WRITING A LINEAR EQUATION BY USING THE SLOPE-INTERCEPT FORM

Procedure. To write an equation of a line that satisfies a given set of conditions:
1. **Determine the slope of the line, m, and its y-intercept, b.**
2. **Then use the values found for m and b in the slope-intercept formula $y = mx + b$.**

〜〜〜〜〜〜〜〜 *MODEL PROBLEMS* 〜〜〜〜〜〜〜〜

1. Write an equation of a line whose slope is 4 and which passes through the point $(-1, 2)$.

How To Proceed

1. In the slope-intercept form, $y = mx + b$, find b by substituting the given values for m, x, and y.

Solution

1. It is given that the slope, $m = 4$. Since the line passes through the point $(-1, 2)$, $x = -1$ and $y = 2$ satisfy the equation of the line. Substitute in $y = mx + b$.
$$2 = 4(-1) + b$$
$$2 = -4 + b$$
$$6 = b$$

2. In $y = mx + b$, substitute the value given for m and the value found for b in step 1.

2. $y = mx + b$
   $y = 4x + 6$ *Ans.*

2. Using integral coefficients, write an equation of a line that is parallel to the graph of $2y - 3x = 8$ and whose y-intercept is 5.

Solution:

1. Transform the equation $2y - 3x = 8$ into the slope-intercept form, $y = mx + b$.

$$2y - 3x = 8$$
$$2y = 3x + 8$$
$$y = \tfrac{3}{2}x + 4$$

The slope of the line $2y - 3x = 8$ is $\tfrac{3}{2}$.

2. Since parallel lines have the same slope, the slope of a line parallel to the line $2y - 3x = 8$ is also $\tfrac{3}{2}$.

3. The y-intercept of the required line is 5. Substitute $\tfrac{3}{2}$ for m and 5 for b in $y = mx + b$.

$$y = mx + b$$
$$y = \tfrac{3}{2}x + 5$$
$$2y = 3x + 10 \ \ Ans.$$

Exercises

In 1–6, using integral coefficients, write an equation of a line which has the given slope m and which passes through the given point.

1. $m = 2$, $(1, 4)$ **2.** $m = 3$, $(-1, 5)$ **3.** $m = -2$, $(-1, -4)$

4. $m = \tfrac{1}{2}$, $(-2, 3)$ **5.** $m = -\tfrac{3}{4}$, $(0, 0)$ **6.** $m = -\tfrac{2}{3}$, $(-2, -3)$

In 7–12, write an equation of the line that passes through the given points.

7. $(2, 4)$, $(4, 8)$ **8.** $(2, 1)$, $(5, 4)$ **9.** $(2, 4)$, $(10, 9)$

10. $(4, 7)$, $(2, 10)$ **11.** $(-2, 7)$, $(2, 10)$ **12.** $(0, 0)$, $(4, -2)$

In 13–16, write a linear equation that expresses the relation between the two variables in the table. Use this equation to find the replacements for the question marks.

13.

x	0	1	2	3	?
y	0	2	4	?	10

14.

x	1	3	5	7	?
y	5	9	13	?	23

15.

a	0	2	3	5	?
b	3	9	12	?	24

16.

c	0	−1	−2	4	?
w	5	7	9	?	−5

In 17–23, write an equation of a line that is:

17. parallel to $y = -\frac{2}{3}x + 5$ and whose y-intercept is -3.

18. parallel to $y = 3x - 1$ and passes through the point $(2, 4)$.

19. parallel to $y = 2x + 3$ and has the same y-intercept as $y + 5 = 4x$.

20. perpendicular to $y = \frac{2}{3}x + 5$ and whose y-intercept is -4.

21. perpendicular to $y = 2x - 3$ and passes through the point $(5, 7)$.

22. perpendicular to $x + 3y = 9$ and passes through the origin.

23. perpendicular to the line which passes through the points $(0, 0)$ and $(-2, 4)$ and whose y-intercept is the same as the y-intercept of $3x - 4y = 24$.

24. Show that an equation of a line whose nonzero x-intercept is a and whose nonzero y-intercept is b is:

$$\frac{x}{a} + \frac{y}{b} = 1$$

(This is called the **intercept form** of the equation.)

In 25–28, use the intercept form to write an equation of a line whose x- and y-intercepts are, respectively, the given numbers.

25. 3 and 4   **26.** 5 and -2 **27.** -4 and 6   **28.** -3 and -5

8. The Point-Slope Form of a Linear Equation

Let us discover an equation of a line which passes through a point P_1, with given coordinates (x_1, y_1), and whose slope is m.

Let $P(x, y)$ be any other point on the given line which passes through P_1, as shown in the figure at the right. Hence, $x \neq x_1$.

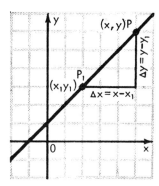

The slope of the line $= m = \dfrac{\Delta y}{\Delta x}$, or

$$m = \frac{y - y_1}{x - x_1}$$

Since $x \neq x_1$, $x - x_1 \neq 0$. Multiply both members of the equation by $x - x_1$ to obtain

$$y - y_1 = m(x - x_1)$$

The equation $y - y_1 = m(x - x_1)$ is called the **point-slope form** of a linear equation.

~~~~~~~~~~~~ **MODEL PROBLEM** ~~~~~~~~~~~~

Write, in the form $y = mx + b$, an equation of a line whose slope is 2 and which
  passes through the point $(-1, 3)$.

How To Proceed	*Solution*
1. In the point-slope form of the equation, $y - y_1 = m(x - x_1)$, substitute the coordinates of the given point for x_1 and y_1 and the value of the slope for m.	1. At $P_1(-1, 3) \to x_1 = -1, y_1 = 3$. The slope $m = 2$. Substitute: $y - y_1 = m(x - x_1)$ $y - 3 = 2[x - (-1)]$
2. Simplify the resulting equation.	2. $y - 3 = 2(x + 1)$ $y - 3 = 2x + 2$ $y = 2x + 5$

Answer: $y = 2x + 5$

~~~~~~

Exercises

In 1–9, write an equation of a line that has the given slope, m, and passes
through the given point.

1. $m = 4$, $(2, 3)$ 2. $m = 3$, $(0, 0)$ 3. $m = -2$, $(6, 1)$
4. $m = 2$, $(4, -1)$ 5. $m = -1$, $(-2, -3)$ 6. $m = \frac{1}{2}$, $(8, 6)$
7. $m = \frac{2}{3}$, $(4, 2)$ 8. $m = -\frac{1}{2}$, $(-1, 5)$ 9. $m = -\frac{3}{4}$, $(-2, -4)$

In 10–12, write an equation of a line that is:
10. parallel to $y = 4x + 5$ and passes through the point $(4, 2)$.
11. parallel to $2x + 3y = 6$ and passes through the point $(0, -3)$.
12. perpendicular to $y = -\frac{2}{3}x + 4$ and passes through the point $(3, 5)$.

In 13–18, write an equation of a line which passes through the points:

13. $(0, 0)$ and $(4, 8)$ 14. $(1, 3)$ and $(4, 12)$ 15. $(-1, 2)$ and $(2, 5)$
16. $(-3, -2)$ and $(5, 2)$ 17. $(-5, 1)$ and $(-2, 5)$ 18. $(-1, 3)$ and $(-4, 9)$

19. Write an equation of a line which passes through the point $(-3, 7)$ and is
    parallel to the graph of $\{(x, y) | x - 2y = 5\}$.
20. Write an equation of a line which passes through the point $(2, -5)$ and is
    perpendicular to the graph of $\{(x, y) | y + 3x = 8\}$.
21. Show that an equation of a line which passes through two different points
    $P_1(x_1, y_1)$ and $P_2(x_2, y_2)$, $(x_1 \neq x_2)$ is:

$$y - y_1 = \frac{y_2 - y_1}{x_2 - x_1}(x - x_1)$$

(This is called the ***two-point form*** of the equation.)

In 22–25, use the two-point form to write an equation of a line which passes through the two given points.

22. $(3, 5)$, $(5, 12)$ **23.** $(6, 2)$, $(4, 10)$

24. $(-1, 3)$, $(2, -6)$ **25.** $(0, -5)$, $(-2, -7)$

9. Graphing a First-Degree Inequality in the Cartesian Plane

The line which is the graph of the first-degree equation $y = 2x - 1$ (Fig. 1) is the set of all points whose ordinate y, *equals* "1 less than twice its abscissa, x." For example, at point $A(1, 1)$, the ordinate, 1, is 1 less than twice the abscissa, 1. That is, $1 = 2(1) - 1$. This line divides the Cartesian plane into two regions called *half-planes*.

Graph of $y = 2x - 1$

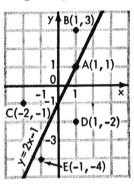

Fig. 1

The half-plane above the line $y = 2x - 1$ (Fig. 1) is the set of all points whose ordinate, y, *is greater than* "1 less than twice its abscissa, x"; that is, $y > 2x - 1$. For example, at point $B(1, 3)$, the ordinate, 3, is greater than "1 less than twice the abscissa, 1." That is, $3 > 2(1) - 1$ is a true statement. The coordinates of point $C(-2, -1)$ also satisfy $y > 2x - 1$ because $-1 > 2(-2) - 1$ is a true sentence. The graph of $y > 2x - 1$ is the shaded half-plane above the line $y = 2x - 1$ (Fig. 2). Notice that the line which is the plane-divider is drawn dashed to indicate that it is not part of the graph. To graph $y \geq 2x - 1$ (Fig. 3), we draw the plane-divider as a solid line to indicate that the line $y = 2x - 1$ is part of the graph.

Graph of $y > 2x - 1$

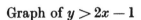
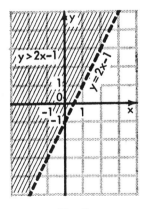

Fig. 2

Graph of $y \geq 2x - 1$

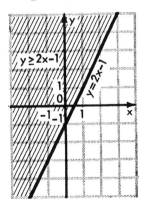

Fig. 3

The half-plane below the line $y = 2x - 1$ (Fig. 1) is the set of all points at which the ordinate, y, *is less than* "1 less than twice the abscissa, x"; that is, $y < 2x - 1$. For example, at point $D(1, -2)$, the ordinate, -2, is less than "1 less than twice the abscissa, 1." That is, $-2 < 2(1) - 1$ is a true sentence. The coordinates of point $E(-1, -4)$ also satisfy $y < 2x - 1$ because the sentence $-4 < 2(-1) - 1$ is true. The graph of $y < 2x - 1$ is the shaded half-plane below the line $y = 2x - 1$ (Fig. 4). Notice that the line which is the plane-divider is drawn dashed to indicate that it is not part of the graph. To graph $y \leq 2x - 1$ (Fig. 5), we draw the plane-divider as a solid line to indicate that the line $y = 2x - 1$ is part of the graph.

Graph of $y < 2x - 1$ Graph of $y \leq 2x - 1$

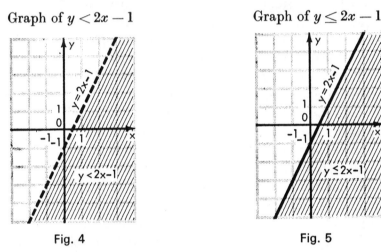

Fig. 4   Fig. 5

From the study of the graphs of $y = 2x - 1$, $y > 2x - 1$, and $y < 2x - 1$ we see that the line which is the graph of $y = 2x - 1$ acts as a ***plane-divider***. It divides the Cartesian plane into two regions. The region above the line is the graph of $y > 2x - 1$; the region below it is the graph of $y < 2x - 1$.

In general, the graph of $y = mx + b$ is a line which divides the coordinate plane into three sets of points:

1. *The set of points on the line.* Each ordered pair that describes a member of this set of points is a solution of $y = mx + b$. The line is the graph of the equation $y = mx + b$; also, the line is the graph of the solution set of $y = mx + b$.

2. *The set of points in the half-plane above the line.* Each ordered pair that describes a member of this set of points is a solution of $y > mx + b$. The half-plane is the graph of the inequality $y > mx + b$; also, the half-plane is the graph of the solution set of the inequality $y > mx + b$.

3. *The set of points in the half-plane below the line.* Each ordered pair that describes a member of this set of points is a solution of $y < mx + b$. The half-plane is the graph of the inequality $y < mx + b$; also, the half-plane is the graph of the solution set of the inequality $y < mx + b$.

KEEP IN MIND

To graph $Ax + By > C$ or $Ax + By < C$, $B \neq 0$, transform the inequality into an equivalent inequality whose left member is y alone: $y > mx + b$ or $y < mx + b$. This enables us first to graph the plane divider $y = mx + b$.

~~~~~~~~~~ **MODEL PROBLEMS** ~~~~~~~~~~

1. Graph the inequality $2x - y \geq -2$ in the coordinate plane.

How To Proceed	*Solution*
1. Transform the inequality into an equivalent inequality having y as its left member. Remember that when we multiply by -1 or any negative number, we reverse the order of the inequality.	$2x - y \geq -2$ $-y \geq -2x - 2$ $y \leq 2x + 2$
2. Graph the resulting inequality by first graphing the plane-divider $y = 2x + 2$.	$y = 2x + 2$

	x	y
A	-1	0
B	0	2
C	1	4

3. Shade the half-plane below the plane-divider. The required graph is the union of the shaded half-plane, which is the graph of $2x - y > -2$, and the line, which is the graph of $2x - y = -2$. Note that the line is drawn solid to show that it is part of the graph.

Graph of $2x - y \geq -2$

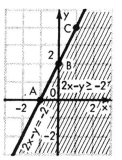

2. Graph each of the following in the coordinate plane: *a.* $x > 1$ *b.* $y \le 1$

How To Proceed	*Solution*
1. Graph the plane-divider, using either a dotted line or a solid line, whichever the question requires.	*a.* Graph of $x > 1$ *b.* Graph of $y \le 1$
2. Shade the proper half-plane.	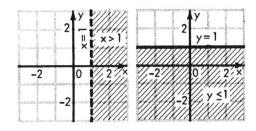

3. Graph $\{(x, y) \mid x < 1\}$.

Solution: $\{(x, y) \mid x < 1\}$ is the set of points whose abscissa is less than 1 and whose ordinate is any real number. Examples of such points are:

$(0, 1)$, $(0, -3)$, $(-1, 5)$, $(-2, -4)$.

The graph of $\{(x, y) \mid x < 1\}$ is the half-plane to the left of the plane-divider $x = 1$.

Graph of $\{(x, y) \mid x < 1\}$

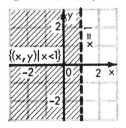

Exercises

In 1–6, transform the inequality into an equivalent inequality whose left member is y.

1. $y - x < 0$ **2.** $5x \ge 2y$ **3.** $y - x \le 5$

4. $3x - y \le 6$ **5.** $2x > 6 + 2y$ **6.** $8 \le 2x + 3y$

In 7–18, graph the inequality in the coordinate plane.

7. $y > 2x$ **8.** $y < 4x + 2$ **9.** $y > \frac{1}{2}x - 1$

10. $x + y > 6$ **11.** $y - x \ge 4$ **12.** $x - y > 4$

13. $x - y \le -2$ **14.** $x - 2y \le 3$ **15.** $3x + y - 4 \le 0$

16. $2x + 3y - 6 \le 0$ **17.** $6 \le 3x + 2y$ **18.** $x \ge -3$

In 19–24, graph the set in the coordinate plane.

19. $\{(x, y) \mid y \ge 2x + 4\}$ **20.** $\{(x, y) \mid 3x + y \le -2\}$ **21.** $\{(x, y) \mid 2x \ge 4 - y\}$

22. $\{(x, y) \mid 4x < 8 + y\}$ **23.** $\{(x, y) \mid x \ge 6\}$ **24.** $\{(x, y) \mid y \le -4\}$

In 25–27, (*a*) write the verbal sentence as an open sentence involving x and y and (*b*) graph the open sentence in the coordinate plane.

25. The ordinate of a point is less than 3 times its abscissa.

26. The sum of the abscissa and ordinate of a point is greater than or equal to 5.

27. The abscissa of a point decreased by the ordinate of the point is less than 2.

10. Comparing the Graphs of Open Sentences in Two Variables When the Replacement Set Changes

The graph of an open sentence in two variables is the set of points whose coordinates must have the following characteristics:

1. Both coordinates must be members of the replacement sets of the variables.
2. Both coordinates must satisfy the open sentence.

For example: If the replacement set of both x and y is the set of real numbers, the ordered number pair $(5, -6)$ is a member of the solution set of $x - y = 11$ and the graph of $(5, -6)$ is a point in the graph of $x - y = 11$ because (1) both 5 and -6 are members of the replacement set of the variables and (2) $(5, -6)$ satisfies $x - y = 11$. However, should the replacement set of both x and y be the set of positive integers, the point which is the graph of $(5, -6)$ would not be a point in the graph of $x - y = 11$. This would be so because in spite of the fact that $(5, -6)$ satisfies $x - y = 11$, the coordinate -6 is not a member of the replacement set of y.

In Figs. 1–4, we see how the graph of an open sentence in two variables changes when the replacement sets of the variables change.

Graph of $y = x - 1$.
Replacement set for x
and y: {integers}

Graph of $y = x - 1$.
Replacement set for x
and y: {real numbers}

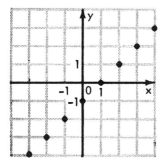

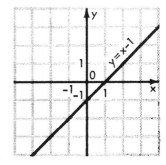

Fig. 1   Fig. 2

In Fig. 1, we see the graph of $y = x - 1$ when the replacement set for both x and y is the set of integers. The graph is a set of distinct, individual points, all of which lie on a straight line whose equation is $y = x - 1$, but the graph *does not include all the points on that line.*

In Fig. 2, we see the graph of $y = x - 1$ when the replacement set for both x and y is the set of real numbers. Now the graph is a line, the same line that was referred to in the discussion of Fig. 1. Observe that the graph *includes all the points on this line.*

Graph of $y < x - 1$.
Replacement set for x
and y: {integers}

Graph of $y < x - 1$.
Replacement set for x
and y: {real numbers}

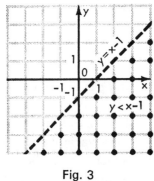

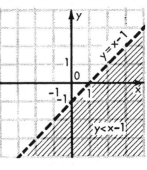

Fig. 3   Fig. 4

In Fig. 3, we see the graph of $y < x - 1$ when the replacement set of both x and y is the set of integers. The graph is a set of distinct individual points, all of which lie in the half-plane below the plane-divider $y = x - 1$, but the graph *does not include all the points in this half-plane.*

In Fig. 4, we see the graph of $y < x - 1$ when the replacement set of both x and y is the set of real numbers. Now the graph is the half-plane below the plane-divider $y = x - 1$, the same half-plane that was referred to in the discussion of Fig. 3. Observe that the graph *includes all the points in the half-plane.*

Exercises

In 1–15, draw the graph of the open sentence when the replacement set of both variables is (*a*) the set of real numbers, (*b*) the set of integers, and (*c*) the set of positive integers.

1. $y = 2x$	**2.** $y = x + 2$	**3.** $y = 2x - 1$
4. $y > 2x$	**5.** $y < x + 2$	**6.** $y > 2x - 1$
7. $y \leq 2x$	**8.** $y \geq x + 2$	**9.** $y \leq 2x - 1$
10. $y = \frac{1}{2}x + 4$	**11.** $y < \frac{1}{2}x + 4$	**12.** $y \geq \frac{1}{2}x + 4$
13. $x + 2y = 6$	**14.** $3x + y > 4$	**15.** $2x - y \leq 0$

11. Understanding Relations and Functions

RELATIONS

DOMAIN AND RANGE OF A RELATION

Mr. Dale is a teacher. In the table at the right, we find the amounts he earned and saved during each of the first four weeks of January. The pairings in this table may be written as a set of ordered pairs in which the first coordinate represents the amount earned during a week and the second coordinate represents the amount saved during that week:

($) Earnings	($) Savings
550	30
565	32
540	35
550	25

$$A = \{(550, 30), (565, 32), (540, 35), (550, 25)\}$$

A set of ordered pairs is called a *relation.* The set consisting of the first members of the ordered pairs is called the *domain* of the relation. Thus, the domain of relation A is $\{550, 565, 540\}$.

The set consisting of the second members of the ordered pairs is called the *range* of the relation. Thus, the range of relation A is $\{30, 32, 35, 25\}$.

In general, a relation involves:
1. A set of first members, the domain.
2. A set of second members, the range.
3. A pairing of a member of the domain with a member of the range to define the relation.

The pairing can be shown in several ways.

METHODS OF DEFINING A RELATION

1. *Using a mapping diagram to define a relation.*
On a block, there is a set of children, $C = \{$Alice, Robert, Martha, Walter$\}$ and there is a set of houses, $H = \{X, Y, Z\}$. Fig. 1 at the right shows how the members of the set of children, C, are paired with the members of the set of houses, H, to define a relation R.

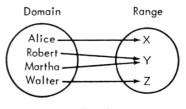

Fig. 1

$$R = \{(\text{Alice}, X), (\text{Robert}, Y), (\text{Martha}, Y), (\text{Walter}, Z)\}$$

The arrow is used to show the element of the range that is paired with a particular element of the domain. We call this a ***mapping diagram*** and say that the domain is mapped into the range.

2. *Using a table to define a relation.*

The table at the right shows how the numbers x and y are paired to define the relation $R = \{(1, 4), (2, 8), (3, 12)\}$.

x	y		(x, y)
1	4		(1, 4)
2	8	or	(2, 8)
3	12		(3, 12)

3. *Using a rule to define a relation.*

A relation may be defined by a rule which makes it possible to determine how the member or members of the range are to be paired with each member of the domain. For example, the solution set of an open sentence such as $y = 4x$ is a relation in which the second member of each ordered pair is 4 times the first member. We can write the rule as $x \rightarrow 4x$ or define this relation as follows:

$$R = \{(x, y) \,|\, y = 4x, \ x \text{ and } y \text{ are real numbers}\}$$

We can omit "x and y are real numbers" from the rule defining the relation by assuming that the domain and range include only real numbers. In the case of the relation $\{(x, y) \,|\, y = \dfrac{1}{x}\}$ we assume that the domain is the set of real numbers, and also that $x \neq 0$, since $\dfrac{1}{x}$ is not defined if $x = 0$.

4. *Using graphs to define a relation by means of a picture.*

When we draw the graphs of first-degree equalities and first-degree inequalities, we are picturing sets of ordered pairs, or relations.

Observe that $\{(x, y) \,|\, x + y = 2, \ x \text{ and } y \text{ are integers}\}$, whose graph is shown in Fig. 2, and $\{(x, y) \,|\, x + y = 2, \ x \text{ and } y \text{ are real numbers}\}$, whose graph is shown in Fig. 3, are different relations.

Graph of $\{(x, y) \,|\, x + y = 2\}$
Domain: {integers}

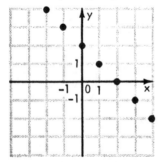

Fig. 2

Graph of $\{(x, y) \,|\, x + y = 2\}$
Domain: {real numbers}

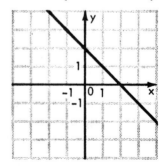

Fig. 3

FUNCTIONS

There is an important difference between the following two relations:

$$A = \{(-2, 2), (-1, 1), (0, 0), (1, 1), (2, 2)\}$$

$$B = \{(0, 0), (1, 1), (1, -1), (2, 2), (2, -2)\}$$

In relation A, observe that every member of the domain $\{-2, -1, 0, 1, 2\}$ is paired with one and only one member of the range $\{0, 1, 2\}$. For example, the member of the domain, -1, is paired with a unique member of the range, 1.

In relation B, however, some members of the domain $\{0, 1, 2\}$ are paired with two members of the range $\{-2, -1, 0, 1, 2\}$. For example, one member of the domain, 2, is paired with two members of the range, 2 and -2. A relation such as set A is called a *function*. A relation such as set B is not a function, in accordance with the following definition:

A *function* is a relation (a set of ordered pairs) in which every member of the domain is paired with one and only one member of the range.

VERTICAL LINE TEST FOR A FUNCTION

If a relation is graphed, we can discover whether or not the given relation is a function. In a function, to every member of the domain there corresponds one and only one member of the range. Hence, a relation is a function if, when a vertical line is drawn through each point of its graph, we find that to the abscissa of each point there corresponds one and only one ordinate. Note that the graph of relation A (Fig. 4) confirms that this relation is a function. However, the graph of relation B (Fig. 5) shows that relation B is not a function.

Relation A

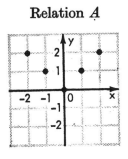

Fig. 4

Relation B

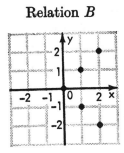

Fig. 5

MAPPING TEST FOR A FUNCTION

The mapping diagram at the right (Fig. 6) pictures a pairing that defines the function:

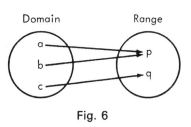

$$f = \{(a, p), (b, p), (c, q)\}$$

Observe that each member of the domain is mapped into one and only one member of the range.

Fig. 6

THE INVERSE OF A RELATION

Let us begin with the relation $A = \{(1, 4), (2, 6), (3, 8)\}$. By interchanging the two members of every ordered pair in relation A, we obtain relation $B = \{(4, 1), (6, 2), (8, 3)\}$. Relation B is called the *inverse* of relation A in accordance with the following definition:

The **inverse of a relation** is the relation that is formed when the members of each ordered pair of the given relation are interchanged.

From this definition, it can be seen that the domain of the original relation is the range of the inverse relation, and the range of the original relation is the domain of the inverse relation. For example, in relations A and B, which are inverse relations, the domain of relation A, $\{1, 2, 3\}$, is the range of relation B and the range of relation A, $\{4, 6, 8\}$, is the domain of relation B.

The inverse of a relation must always be a relation since interchanging the members of an ordered pair results in an ordered pair. For example, the inverse of $R = \{(1, 2), (3, 4), (3, 5)\}$ is the relation $S = \{(2, 1), (4, 3), (5, 3)\}$.

The inverse of a function need not be a function. For example, the inverse of the function $F = \{(1, 2), (3, 4), (5, 6)\}$ is the function $G = \{(2, 1), (4, 3), (6, 5)\}$. However, the inverse of the function $H = \{(1, 2), (3, 2), (5, 4)\}$ is $K = \{(2, 1), (2, 3), (4, 5)\}$, which is a relation that is not a function since both 1 and 3 in the range correspond to 2 in the domain.

When the inverse of a function is also a function, it is called the **inverse function.** Hence, in the previous paragraph, function G is the inverse function of function F.

The inverse of a function F may be indicated by the symbol F^{-1}, which is read "the inverse of F" or "F inverse." For example, if $F = \{(1, 2), (3, 4), (5, 6)\}$ then $F^{-1} = \{(2, 1), (4, 3), (6, 5)\}$.

Let us find the inverse of the function defined by $y = 3x - 1$, with the domain and range being the set of real numbers.

How To Proceed	*Solution*
1. In the equation, interchange x and y.	1. $y = 3x - 1$ $x = 3y - 1$
2. Solve for y in terms of x.	2. $x + 1 = 3y$ $\dfrac{x+1}{3} = y$

Therefore, the function defined by $y = \dfrac{x+1}{3}$ is the inverse function of the given function and it also has as its domain and range the set of real numbers.

There are cases where a given function does not have an inverse function. However, by restricting the domain of a function, it is often possible to find an inverse function. Such is the case in the function defined by $y = x^2$ whose domain is the set of real numbers and whose range is the set of non-negative real numbers. When we interchange x and y, we obtain $x = y^2$ or $y = \pm \sqrt{x}$, which does not define a function because with every positive number x there are associated *two* real values of y. For example, when $x = 4$, $y = 2$ and when $x = 4$, $y = -2$. However, if we limit the domain of x to the set of non-negative real numbers, we can confine the range of the inverse relation to the same set of numbers. Thus, when the function defined by $y = x^2$ has the set of non-negative numbers as its domain and range, then the inverse of this function is a function which may be defined by $x = y^2$, with its domain and range also being the set of non-negative real numbers.

~~~~~~~~~~~ *MODEL PROBLEMS* ~~~~~~~~~~~

1. If $A = \{(1, 3), (2, 5), (4, 9), (x, 6)\}$ is to be a function, list the integers which x may *not* represent.

   Solution: If A is to be a function, no first component of its ordered pairs may have more than one second component. Therefore, x may not be the numbers 1, 2, 4 because these numbers are already first components of ordered pairs in A.

   Answer: 1, 2, 4

2. If x and y are real numbers and the domain of the relation $\{(x, y) \,|\, y = 2x + 1\}$ is $\{x \,|\, 0 \leq x \leq 5\}$, find the range of the relation.

   Solution: The relation is defined for all members of the domain.
   When $x = 0$, $2x + 1 = 2(0) + 1 = 1 = y$
   When $x = 5$, $2x + 1 = 2(5) + 1 = 11 = y$
   Since the given relation is linear, when $0 \leq x \leq 5$ then $1 \leq y \leq 11$.
   Hence, the range is the set of real numbers between and including 1 and 11.

   Answer: $\{y \,|\, 1 \leq y \leq 11\}$

3. If x and y are real numbers, determine the restriction that must be placed on the domain of the function defined by $y = \dfrac{1}{x-3}$.

Solution: The domain is the set of all real numbers for which the equation has meaning. When $x = 3$, $y = \dfrac{1}{3-3} = \dfrac{1}{0}$, which is meaningless. Hence, x cannot be equal to 3.

Answer: $x \neq 3$

Exercises

In 1–4, (*a*) give the domain of the relation, (*b*) give the range of the relation, and (*c*) state whether or not the relation is a function.

1. $\{(1, 3), (3, 7), (5, 11), (7, 15)\}$ **2.** $\{(3, 4), (4, 3), (3, -4), (4, -3)\}$

3. $\{(2, 5), (4, 6), (-2, 5), (3, 4)\}$ **4.** $\{(0, 1), (0, 3), (0, -1), (0, -2)\}$

In 5–8, state whether or not the relation pictured by the mapping is a function. If it is not a function, tell why.

5. 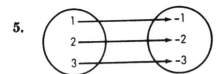   **6.**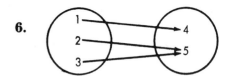

7. 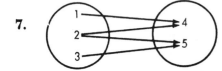   **8.**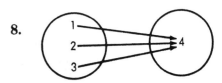

In 9–12, write the set of all ordered pairs pictured by the mapping shown.

9. 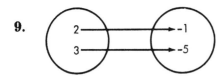   **10.**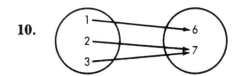

11. 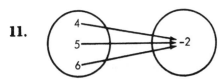   **12.**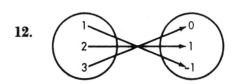

In 13–18, graph the relation. From the graph, determine whether or not the relation is a function.

13. $\{(x, y) \mid y = x\}$

14. $\{(x, y) \mid y = 2x - 3\}$

15. $\{(x, y) \mid y > 2x\}$

16. $\{(x, y) \mid y \leq 2x\}$

17. $\{(x, y) \mid x > 4\}$

18. $\{(x, y) \mid y = 4\}$

In 19–22, x and y are members of the set of real numbers. In each exercise:

a. Graph the relation described by the rule.

b. State whether the relation is a function.

c. State the range of the relation.

19. $\{(x, y) \mid y = 3x - 4\}$ when the domain is $\{x \mid -3 \leq x \leq 4\}$.

20. $\{(x, y) \mid y = 3\}$ when the domain of x is $\{x \mid x \geq 0\}$.

21. $\{(x, y) \mid y > x\}$ when the domain of x is $\{x \mid x \geq 5\}$.

22. $\{(x, y) \mid y < x + 4\}$ when the domain of x is $\{x \mid 0 \leq x \leq 2\}$.

In 23–31, each figure is the graph of a relation. The first and second members of each ordered pair of the relation are members of the set of real numbers. In each exercise:

a. State the domain of the relation.

b. State the range of the relation.

c. Determine whether or not the relation is a function by using the vertical line test.

23.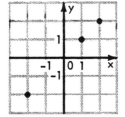

24.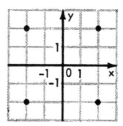

25.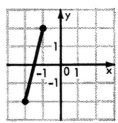

26.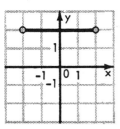

27.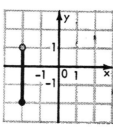

28.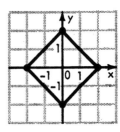

29.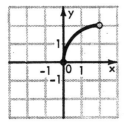

30.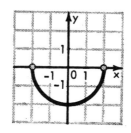

31.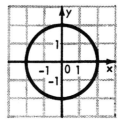

In 32–35:

a. Write the inverse of the relation.

b. State whether or not the inverse of the relation is a function.

c. Give the domain and range of the original relation.

d. Give the domain and range of the inverse of the original relation.

32. $\{(2, 5), (3, 8), (4, 11), (5, 14)\}$

33. $\{(1, 9), (-2, 3), (4, 9), (5, 7)\}$

34. $\{(-2, 5), (-4, 1), (-3, 7), (-5, 1)\}$

35. $\{(5, -3), (6, -2), (5, -8), (2, -7)\}$

In 36 and 37, list the integers that x may *not* represent if the given relation is to be a function.

36. $\{(3, 8), (x, 4), (1, 0), (0, -4)\}$

37. $\{(4, -1), (5, -1), (x, -2), (3, -2)\}$

In 38 and 39, list the integers that x may *not* be in order that the inverse of the given relation be a function.

38. $\{(1, 3), (2, 6), (3, 9), (4, x)\}$

39. $\{(3, 2), (3, -1), (5, x), (5, 3)\}$

In 40 and 41, write the set that is F^{-1} if the given set F is:

40. $\{(5, 2), (6, -1), (8, 0), (12, 9)\}$ **41.** $\{(0, 1), (5, 6), (9, -2), (8, 3)\}$

In 42 and 43, the replacement set is $\{3, 4, 5, 6, 7, 8\}$.

a. Determine whether or not the relation is a function.

b. Determine whether or not the inverse of the relation is a function.

42. $\{(x, y) \mid y = 2x\}$ **43.** $\{(x, y) \mid y > 2x\}$

44. State the range of the relation $\{(x, y) \mid x = 4\}$ and state whether or not the relation is a function.

In 45–47, write in the form of $y = mx + b$ the equation that defines the inverse of the function defined by the given equation.

45. $x = 5y - 2$ **46.** $y = 3x + 5$ **47.** $y = 6x$

In 48–51, if x and y are real numbers, determine the restriction, if any, that must be placed on the domain of the function.

48. $\{(x, y) \mid y = x - 5\}$ **49.** $\left\{(x, y) \mid y = \dfrac{1}{x - 5}\right\}$ **50.** $\{(x, y) \mid xy = 8\}$

51. $\{(x, y) \mid y = 4x\}$ when x represents the side of a square and y represents the perimeter of the square.

52. If F is a function in which no two ordered pairs have the same second member, will F^{-1} also be a function? Use an example to illustrate your answer. Explain your answer generally.

12. Understanding the Function Notation

We have learned that a letter such as f may be used to represent a function. When x represents an element of the domain of definition of a function f, we shall designate the corresponding element of the range which the function pairs with x—that is, assigns to x—as $f(x)$, read "f of x" or "f at x." Be careful to note that $f(x)$ does *not* mean "f times x." The number $f(x)$ is called the ***value of f at x***.

This function notation can be used to describe the pairings of the numbers which are the members of the ordered pairs of a function. For example, the function f which is described by the word statement "with each real number x, pair the real number $x + 2$" can be written:

$$f(x) = x + 2 \text{ for each real number } x, \text{ or } f = \{(x, y) \,|\, y = x + 2\}$$

Then, $f(1)$ means the number which the function f pairs with 1. To find $f(1)$, we find the number that $x + 2$ represents when x is replaced by 1. Thus, if $f(x) = x + 2$, then $f(1) = 1 + 2 = 3$; that is, f assigns to 1 the number 3. Likewise:

To find $f(0)$, we replace x by 0 and obtain $f(0) = 0 + 2$ or $f(0) = 2$.
To find $f(-3)$, we replace x by -3 and obtain $f(-3) = -3 + 2$ or $f(-3) = -1$.
To find $f(x^2)$, we replace x by x^2 and obtain $f(x^2) = x^2 + 2$.

Both $\{(x, y) \,|\, y = x + 2\}$ and $\{(x, f(x)) \,|\, f(x) = x + 2\}$ represent the same function. Since both y and $f(x)$ represent the number which the function assigns to x, we may write $y = f(x)$.

There are times when it is advantageous to use the function notation. If we begin with $y = x + 2$ and replace x with 1, then $y = 1 + 2$, or $y = 3$. However, if we begin with $f(x) = x + 2$ and replace x with 1, then $f(1) = 1 + 2$, or $f(1) = 3$. Compare the two results $y = 3$ and $f(1) = 3$. In $f(1) = 3$, we know that x was replaced with 1; but in $y = 3$, we have no way of knowing what the replacement was.

〜〜〜〜〜〜〜 *MODEL PROBLEMS* 〜〜〜〜〜〜〜

1. If the function f is defined by $f(x) = 2x + 5$ for each real number x, find:
       a. $f(-4)$ *b.* $f(|-2|)$ *c.* $f(3b)$ *d.* $f(x + 2)$

   Solution: It is given that $f(x) = 2x + 5$. In each part of the question, x is a placeholder to be replaced by the expression within parentheses.
   a. $f(-4) = 2(-4) + 5 = -8 + 5 = -3$ $f(-4) = -3$ *Ans.*
   b. Since $|-2| = 2$, $f(|-2|) = 2(2) + 5 = 9$ $f(|-2|) = 9$ *Ans.*
   c. $f(3b) = 2(3b) + 5 = 6b + 5$ $f(3b) = 6b + 5$ *Ans.*
   d. $f(x + 2) = 2(x + 2) + 5 = 2x + 9$ $f(x + 2) = 2x + 9$ *Ans.*

2. If f is the function defined by $f(x) = \dfrac{5x - 6}{2}$ for each real number x, find the solution set for $f(x) > 2x$.

Solution: To find the solution set of $f(x) > 2x$, we must find the number(s) that can replace x in the open phrase $\dfrac{5x - 6}{2}$ and give a result that is greater than twice that number x. That is, we must find the solution set of the open sentence $\dfrac{5x - 6}{2} > 2x$.

$$\frac{5x - 6}{2} > 2x$$

$$5x - 6 > 4x$$
$$5x - 4x > 6$$
$$x > 6 \qquad \textit{Answer: } \{x \mid x > 6\}$$

Exercises

In 1–6, the domain of the variable is the set of real numbers. For the function defined, compute the value of the function at $-2, -1, 0, 1, 2, |-3|, \frac{1}{2}, \frac{2}{3}$.

1. $f(x) = 6x$ **2.** $g(y) = 4y - 7$

3. $h(a) = -a + 3$ **4.** $g(x) = x^2 - 3$

5. $h(s) = -s^2 + 2$ **6.** $h(a) = a^3 - a^2 - a$

7. For f defined by $f(x) = x^2 - 3x + 5$, for each real number x, find:

   a. $f(6)$ *b.* $f(\frac{2}{3})$ *c.* $f(0)$ *d.* $f(|-3|)$
   e. $3[f(2)]$ *f.* $f(2a)$ *g.* $f(c + 4)$ *h.* $f(3b - 1)$

8. If f is the function defined by $\dfrac{3x - 1}{4}$ for each real number x, find the solution set of each of the following sentences:

   a. $f(x) = 2$ *b.* $f(x) = 0$ *c.* $f(x) = -7$
   d. $f(x) = x$ *e.* $f(x) = 2x + 6$ *f.* $f(x) = \frac{1}{2}x + 3$
   g. $f(x) \geq 2$ *h.* $f(x) < \frac{1}{2}x + 9$ *i.* $f(x) \geq \frac{1}{2}x - \frac{7}{2}$

In 9 and 10, find the range of f if:

9. $f(x) = 2x - 4$ and the domain of x is $-3 < x < 3$.

10. $f(x) = x^2 + 3x - 2$ and the domain of x is $-4 < x < 2$.

In 11–13, find the value(s) of x for which $f(x)$ is meaningless or undefined.

11. $f(x) = \dfrac{3}{x + 6}$ **12.** $f(x) = \dfrac{5}{3x - 2}$ **13.** $f(x) = \dfrac{1}{x} + \dfrac{1}{x - 1}$

13. Composition of Functions

We are familiar with operations in mathematics, such as addition and multiplication, and have used these operations to write polynomial functions. For example, $f(x) = 2x$ and $g(x) = x + 4$ are polynomial functions. When we add two polynomials, we are forming a new polynomial function called the **sum function;** and when we multiply two polynomials, we are forming a new function called the **product function.**

$$f(x) = 2x \qquad\qquad f(x) = 2x$$

$$g(x) = x + 4 \qquad\qquad g(x) = x + 4$$

$$f(x) + g(x) = 2x + (x + 4) \qquad f(x) \cdot g(x) = 2x(x + 4)$$

$$(f + g)(x) = 3x + 4 \qquad\qquad (f \cdot g)(x) = 2x^2 + 8x$$

The domain of $(f + g)(x)$ and $(f \cdot g)(x)$ is the intersection of the domains of f and of g—here, the set of real numbers.

Composition of functions is a binary operation in which the application of one function follows that of another. In other words, the function rule for one function is applied to the result of applying the function rule for another function.

COMPOSITION OF FUNCTIONS: METHOD 1

The composition of $f(x)$ following $g(x)$ can be written in the form $f(g(x))$. To evaluate this function, we treat the innermost function first. If $g(x) = x + 4$, then, for any real number x, $g(x)$ is the real number that is the sum of x and 4. If $f(x) = 2x$, then, for any real number x, $f(x)$ is the real number that is twice x. For example, to find $f(g(3))$:

1. First find $g(3)$. $\qquad\qquad\qquad\qquad g(3) = 3 + 4 = 7$
2. Then, using this result, 7, for x, find $f(7)$. $\quad f(g(3)) = f(7) = 2(7) = 14$

Similarly, to find $f\left(g\left(\frac{1}{2}\right)\right)$, we first apply the rule for g to find $g\left(\frac{1}{2}\right)$ and then apply the rule for f to that number:

$$f\left(g\left(\tfrac{1}{2}\right)\right) = f\left(\tfrac{1}{2} + 4\right) = f\left(4\tfrac{1}{2}\right) = 2\left(4\tfrac{1}{2}\right) = 9$$

In general, for any number x under the given functions: $\qquad f(g(x))$

1. Function g adds 4 to each value in its domain. $\qquad = f(x + 4)$
2. Function f doubles each value in its domain. $\qquad = 2(x + 4)$

$$= 2x + 8$$

Therefore, the function $f(g(x)) = 2x + 8$ states the rule for the composition where g is applied first, or where f *follows* g.

COMPOSITION OF FUNCTIONS: METHOD 2

The expression $(f \circ g)(x)$ is read as f *composition* g of x, or as f *following* g of x. The symbol $\circ$, indicating composition, is an open circle, placed in a raised position between f and g. As in method 1, g must be applied first. Let us examine the procedure for the same function, expressed in a different format.

$$x \xrightarrow{g} x + 4 \qquad x \xrightarrow{f} 2x$$

	Find $(f \circ g)(3)$.	Find $(f \circ g)\left(\frac{1}{2}\right)$.
1. Using g first, find the element in the range of g that corresponds to the given number from the domain of g.	$3 \xrightarrow{g} 7$	$\frac{1}{2} \xrightarrow{g} 4\frac{1}{2}$
2. Use the element from the range of g found in step 1 as an element of the domain of f, and apply the rule for f to this number.	$3 \xrightarrow{g} 7 \xrightarrow{f} 14$ $f \circ g$	$\frac{1}{2} \xrightarrow{g} 4\frac{1}{2} \xrightarrow{f} 9$ $f \circ g$
3. Write the composition in terms of the given number from the domain of g and the final number from the range of f.	$(f \circ g)(3) = 14$	$(f \circ g)\left(\frac{1}{2}\right) = 9$

This same procedure can be applied to the general case to find the rule for the single function that represents $(f \circ g)(x)$.

$$x \xrightarrow{g} (x + 4) \xrightarrow{f} 2(x + 4) = 2x + 8$$
$$f \circ g$$

Thus, $(f \circ g)(x) = 2x + 8$ states the rule for the composition where g is applied first, or where f *follows* g.

These examples illustrate the fact that there are two equivalent forms of the composition f *following* g, namely:

$$\mathbf{f(g(x)) = (f \circ g)(x)}$$

Note. For the composition f *following* g to be meaningful, each element in the range of g must be an element in the domain of f.

Exercises

In 1–8, using $f(x) = x + 5$ and $\dot{g}(x) = 4x$, evaluate each composition.

1. $f(g(2))$ **2.** $g(f(2))$ **3.** $f(g(-1))$ **4.** $g(f(-1))$

5. $f(g(0))$ **6.** $g(f(0))$ **7.** $g\left(f\left(\frac{1}{2}\right)\right)$ **8.** $f\left(g\left(\frac{1}{2}\right)\right)$

In 9–16, using $f(x) = 3x$ and $g(x) = x - 2$, evaluate each composition.

9. $(f \circ g)(4)$ **10.** $(g \circ f)(4)$ **11.** $(f \circ g)(-2)$ **12.** $(g \circ f)(-2)$

13. $(g \circ f)(0)$ **14.** $(f \circ g)(0)$ **15.** $(g \circ f)\left(\frac{2}{3}\right)$ **16.** $(f \circ g)\left(\frac{2}{3}\right)$

In 17–24, using $h(x) = x^2$ and $p(x) = 2x - 3$, evaluate each composition.

17. $(h \circ p)(2)$ **18.** $(p \circ h)(2)$ **19.** $(p \circ h)(1)$ **20.** $(h \circ p)(1)$

21. $(p \circ h)(-3)$ **22.** $(h \circ p)(-3)$ **23.** $(h \circ p)(1.5)$ **24.** $(p \circ h)(1.5)$

25. Let $f(x) = x + 6$ and $g(x) = 3x$.   *a.* Find the rule of the function $(f \circ g)(x)$.
   b. Find the rule of the function $(g \circ f)(x)$.   *c.* Does $(f \circ g)(x) = (g \circ f)(x)$?

26. Let $r(x) = x - 8$ and $t(x) = x^2$.   *a.* Find the rule of the function $(r \circ t)(x)$.
   b. Find the rule of the function $(t \circ r)(x)$.   *c.* Does $(r \circ t)(x) = (t \circ r)(x)$?

27. Let $d(x) = 2x + 3$ and $c(x) = x - 3$.
   a. Evaluate $(d \circ c)(2)$.   *b.* Find the rule of the function $(d \circ c)(x)$.
   c. Use the rule from part *b* to find the value of $(d \circ c)(2)$.
   d. Do the answers from parts *a* and *c* agree?

~~~~~~~~ *CALCULATOR APPLICATIONS*~~~~~~~~

A graphing calculator can be used to make a table of solutions for an equation in two varibles and to draw the graph of a linear inequality.

~~~~~~~~ *MODEL PROBLEMS*~~~~~~~~

1. *a.* Display a table of ordered pairs in the solution set of $3x + y = 14$.

   b. Solve $14 - 3x = 20$.

Solution:

a. First solve the equation for y:

$$3x + y = 14$$
$$y = 14 - 3x$$

Enter: $\boxed{\text{Y=}}$ 14 $\boxed{-}$ 3 $\boxed{\text{X,T,θ,}n}$ $\boxed{\text{ENTER}}$

Choose a starting value for x and some positive value by which to change (or increment) the values of x. Enter these values as TblMin and ΔTbl in the $\boxed{\text{TBLSET}}$ menu. Highlight Auto for both Indpnt and Depend in this menu.

Enter: $\boxed{\text{2nd}}$ $\boxed{\text{TBLSET}}$ $\boxed{(-)}$ 5 $\boxed{\text{ENTER}}$ .5 $\boxed{\text{ENTER}}$ $\boxed{▼}$ $\boxed{\text{ENTER}}$ $\boxed{\text{2nd}}$ $\boxed{\text{TABLE}}$

Display:

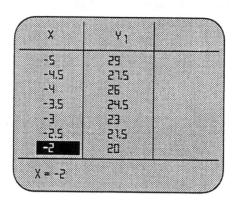

b. Examine the Y_1 column in the table to find 20; then read the corresponding x-value.

Answer: $x = -2$

2. Draw the graph of $y > 2x - 1$.

Solution: Use the standard WINDOW. Select the Shade(command from the DRAW menu. The left parenthesis must be followed by a lower boundary and an upper boundary. The calculator will shade the area between the two. If the inequality is "$y <$" or "$y \leq$," use the Ymin window value as the lower boundary and the equation of the line that is the plane divider as the upper boundary. If the inequality is "$y >$" or "$y \geq$," use the equation as the lower boundary and the Ymax window value as the upper boundary.

Be sure to clear the Y= list before using the Shade(command.

Enter: 2nd DRAW 7

2 X,T,θ,*n* − 1 , 10) ENTER

Display:

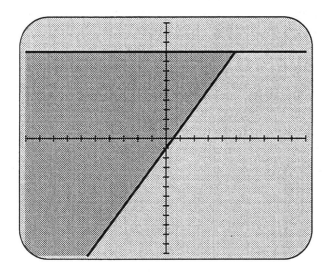

─────── *KEEP IN MIND* ───────
On the TI-83, it is also possible to specify optional shading patterns and shading resolutions. The default values shade the region completely in a vertical pattern. Consult the calculator manual for details.

CHAPTER VI

Transformation Geometry

1. Line Reflections and Line Symmetry

TRANSFORMATIONS

In a classroom, the seats are arranged in four rows with five seats in each row. On the first day of school, the 20 students in the class chose seats at random as they entered the classroom. The teacher made the diagram at the left below, showing where the students sat on that first day, but soon assigned new seats, as indicated in the diagram at the right below.

Seat						Seat				
5	Jeannie	Lori	Irina	Marc		5	Raj	Irina	Raul	Gemma
4	Raj	Peg	Huy	Cody		4	Rory	Peg	Sue	Aesha
3	Gemma	Aesha	Ben	Rory		3	Igor	Ben	Amy	Marc
2	Igor	Hima	Beth	Ada		2	Hima	Cody	Ada	Jeannie
1	Amy	Serji	Sue	Raul		1	Beth	Serji	Lori	Huy
	1	2	3	4			1	2	3	4
		Row						Row		

Nearly all of the students' seats were changed. For example, Amy, who had been seated in row 1, seat 1, was assigned to row 3, seat 3, but Sergi remained in his original place: row 2, seat 1.

The two charts that show the old and the new seating arrangements can be regarded as describing a transformation. On the basis of the seats to which the students were assigned, each seat in the classroom is associated with another seat or with itself. By extending this idea to points in a coordinate plane, we form the definition of a transformation of the plane.

A *transformation of the plane* is a one-to-one correspondence between the points in a plane such that each point is associated with itself or with some other point in the plane. A transformation demonstrates a change in position or a fixed position for each point in the plane.

An inifinite number of transformations can take place in a plane. In this chapter, we will study only a few special transformations, each of which follows a definite pattern or rule. Many transformations in the coordinate plane can be described by rules involving the variables x and y. Rules for the most common transformations will also be presented here.

LINE REFLECTIONS

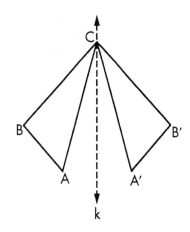

In the figure, $\triangle ABC \cong A'B'C'$. One triangle will "fit exactly" on top of the other if this page is folded along line k, the *line of reflection.* Thus, points A and A' correspond to each other, and points B and B' correspond to each other. Point C corresponds to itself and is called a *fixed point* because C is on the line of reflection.

The term *image* is used to describe the relationship between these points, as we may think of one point as being the mirror image of its corresponding point.

In Symbols	*In Words*
$A \rightarrow A'$, and $A' \rightarrow A$.	The image of A is A', and the image of A' is A.
$B \rightarrow B'$, and $B' \rightarrow B$.	The image of B is B', and the image of B' is B.
$C \rightarrow C$.	The image of C is C.

Many transformations in the plane can be described by rules involving the relationship between the coordinates of the image and those of the preimage. These rules are often discoverd by inductive reasoning.

REFLECTIONS IN THE Y-AXIS

The vertices of $\triangle ABC$ are $A(1, 6)$, $B(4, 3)$, and $C(2, -1)$. These vertices are reflected in the y-axis; and their images, when connected, form $\triangle A'B'C'$. We observe:

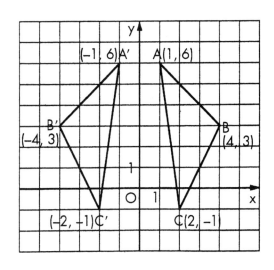

$$A(1, 6) \rightarrow A'(-1, 6)$$

$$B(4, 3) \rightarrow B'(-4, 3)$$

$$C(2, -1) \rightarrow C'(-2, -1)$$

From these examples, we form a general rule.

- Under a *reflection in the y-axis:*

$$P(x, y) \rightarrow P'(-x, y) \qquad \text{or} \qquad r_{y\text{-axis}}(x, y) = (-x, y)$$

REFLECTIONS IN THE X-AXIS

The vertices of quadrilateral *ABCD* are $A(1, 3)$, $B(2, 1)$, $C(7, 1)$, and $D(5, 5)$. These vertices are reflected in the *x*-axis; and their images, when connected, form quadrilateral $A'B'C'D'$. We observe:

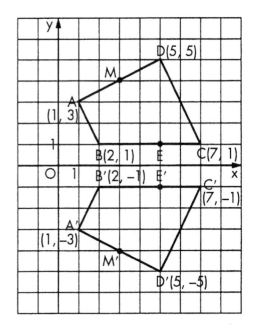

$$A(1, 3) \rightarrow A'(1, -3)$$

$$B(2, 1) \rightarrow B'(2, -1)$$

$$C(7, 1) \rightarrow C'(7, -1)$$

$$D(5, 5) \rightarrow D'(5, -5)$$

From these examples, we form a general rule.

- Under a *reflection in the x-axis:*

$$P(x, y) \rightarrow P'(x, -y) \qquad \text{or} \qquad r_{x\text{-axis}}(x, y) = (x, -y)$$

PROPERTIES UNDER A LINE REFLECTION

The figures shown above illustrate some properties preserved under a line reflection. The following examples refer to the quadrilateral reflected in the *x*-axis.

1. **Distance is preserved;** that is, each segment and its image are equal in length. For example, $\overline{BC} \rightarrow \overline{B'C'}$ and $BC = B'C' = 5$. Similarly, $\overline{AB} \rightarrow \overline{A'B'}$ and $AB = A'B' = \sqrt{5}$.

2. **Angle measure is preserved;** that is, each angle and its image are equal in measure. For example, $\angle DAB \rightarrow D'A'B'$. Since the slope of $\overline{DA}$ is $\frac{1}{2}$ and the slope of $\overline{AB}$ is $-\frac{2}{1}$, $\angle DAB$ is a right angle. Since the slope of $\overline{D'A'}$ is $-\frac{1}{2}$

and the slope of $\overline{A'B'}$ is $\frac{2}{1}$, $\angle D'A'B'$ is also a right angle. Thus, $m\angle DAB = m\angle D'A'B'$.

3. **Parallelism is preserved;** that is, if two lines are parallel, then their images will be parallel lines. Since the slope of $\overline{AB}$ is -2 and the slope of $\overline{DC}$ is -2, then $AB \parallel DC$. Examine their images: $\overline{AB} \rightarrow \overline{A'B'}$, and $\overline{DC} \rightarrow \overline{D'C'}$. Since the slope of $\overline{A'B'}$ is $+2$ and the slope of $\overline{D'C'}$ is $+2$, then $\overline{A'B'} \parallel \overline{D'C'}$.

4. **Collinearity is preserved;** that is, if three or more points lie on a straight line, their images will also lie on a straight line. For example, B, E, and C are collinear. Their images, B', E', and C', are also collinear.

5. **A midpoint is preserved;** that is, given three points such that one is the midpoint of the segment joining the other two, their images will be related in the same way. For example, the midpoint of the segment joining $A(1, 3)$ and $D(5, 5)$ is $M(3, 4)$. Find images A', D', and M'. Note that the midpoint of the segment joining $A'(1, -3)$ and $D'(5, -5)$ is $M'(3, -4)$.

The properties observed for this specific example will be true for any figure and its image under a line reflection.

OTHER LINE REFLECTIONS

In the same way that geometric figures may be reflected in the x-axis or in the y-axis, figures may be reflected in any line in the coordinate plane. For example, let us consider a reflection in the line whose equation is $y = x$.

The endpoints of $\overline{AB}$ are $A(3, 0)$ and $B(4, 3)$. These endpoints are reflected in the line whose equation is $y = x$, and $\overline{A'B'}$ is formed. We observe:

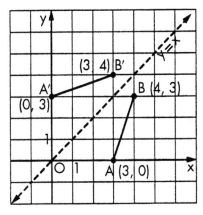

$$A(3, 0) \rightarrow A'(0, 3)$$

$$B(4, 3) \rightarrow B'(3, 4)$$

For these examples, we form a general rule.

- **Under a *reflection in the line y = x*:**

$$P(x, y) \rightarrow P'(y, x) \qquad \text{or} \qquad r_{y=x}(x, y) = (y, x)$$

~~~~~~~~~~~~~~~~~~~~~~~~~ *MODEL PROBLEM* ~~~~~~~~~~~~~~~~~~~~~~~~~

Given $\triangle ABC$, whose vertices are $A(1, 3)$, $B(-2, 0)$, and $C(4, -3)$.

    a. On one set of axes, draw $\triangle ABC$ and its image, $\triangle A'B'C'$, under a reflection in the y-axis.

    b. Find the coordinates of all points on the sides of $\triangle ABC$ that remain fixed under the given line reflection.

Solution:

    a. In step 1, draw and label $\triangle ABC$. In step 2, find the images of the vertices of $\triangle ABC$ by using the rule for a reflection in the y-axis, $P(x, y) \rightarrow P'(-x, y)$.

$$A(1, 3) \rightarrow A'(-1, 3)$$

$$B(-2, 0) \rightarrow B'(2, 0)$$

$$C(4, -3) \rightarrow C'(-4, -3)$$

Draw and label $\triangle A'B'C'$.

        Step 1   *Step 2*

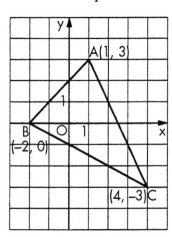
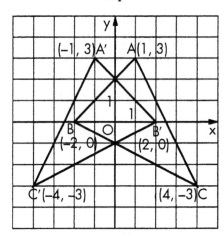

    b. Under a reflection in the y-axis, points on the y-axis remain fixed. The sides of $\triangle ABC$ intersect the y-axis at $(0, 2)$ and $(0, -1)$. As seen in the graph, only these two points are common to both triangles.

Answer: *a.* See the graph labeled *Step 2.*
           b. $(0, 2)$ and $(0, -1)$

~~~~~~

Exercises

1. What is the image of (x, y) under a reflection in the x-axis?
2. What is the image of (x, y) under a reflection in the y-axis?
3. What is the image of (x, y) under a reflection in the line $y = x$?

In 4–7, find the image of each point under a reflection in the x-axis.

4. (5, 7)   **5.** (6, −2)   **6.** (−1, −4)   **7.** (3, 0)

In 8–11, find the image of each point under a reflection in the y-axis.

8. (5, 7)   **9.** (−4, 10)   **10.** (0, 6)   **11.** (−1, −6)

In 12–15, find the image of each point under a reflection in the line $y = x$.

12. (5, 7)   **13.** (−3, 8)   **14.** (0, −2)   **15.** (6, 6)

In 16–21, the vertices of $\triangle ABC$ are given. In each case:

a. Find the coordinates of the images of the vertices, namely, A', B', and C', under the given reflection.
b. On one set of axes, draw $\triangle ABC$ and $\triangle A'B'C'$.
c. Find the coordinates of all points on the sides of $\triangle ABC$ that remain fixed under the given reflection.

In 16–18, the vertices of $\triangle ABC$ are $A(3, 0)$, $B(3, 6)$, and $C(0, 6)$. $\triangle ABC$ is reflected in:

16. the x-axis   **17.** the y-axis
18. the line $y = x$

In 19–21, the vertices of $\triangle ABC$ are $A(1, 4)$, $B(3, 0)$, and $C(-3, -4)$. $\triangle ABC$ is reflected in:

19. the x-axis   **20.** the y-axis
21. the line $y = x$

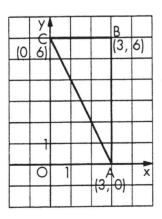

Ex. 16–18

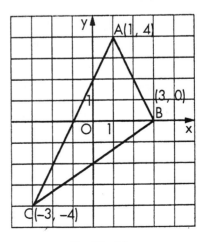

Ex. 19–21

22. *a.* Using the rule $(x, y) \to (x, -y)$, find the images of $C(1, 4)$, $A(5, 1)$, and $T(4, 5)$, namely, C', A', and T'.

  b. On one set of axes, draw $\triangle CAT$ and $\triangle C'A'T'$.

  c. Find the lengths of $\overline{CA}$ and $\overline{C'A'}$.

  d. Is distance preserved under the given transformation?

23. *a.* Using the rule $(x, y) \to (-x, y)$, find the images of $D(2, 3)$, $O(0,0)$, and $G(3, -2)$, namely, D', O', and G'.

  b. On one set of axes, draw $\triangle DOG$ and $\triangle D'O'G'$.

  c. Find the measure of $\angle DOG$ and $\angle D'O'G'$. (*Hint.* Look at the slopes.)

  d. Is angle measure preserved under the given transformation?

24. *a.* Using the rule $(x, y) \to (y, x)$, find the images of $B(7, 0)$, $I(7, 4)$, $R(4, 3)$, and $D(4, 1)$, namely, B', I', R', and D'.

  b. On one set of axes, draw quadrilateral *BIRD* and quadrilateral *B'I'R'D'*.

  c. Explain why $\overline{BI} \parallel \overline{DR}$, and why $\overline{B'I'} \parallel \overline{D'R'}$.

  d. Is parallelism preserved under the given transformation?

In 25–31, the image of $\triangle ABC$ under a line reflection is $\triangle A'B'C'$. In each case:

a. Using the given coordinates, draw $\triangle ABC$ and $\triangle A'B'C'$ on one set of axes.

b. Find an equation of the line of reflection.

25. $\triangle ABC$: $A(2, 4)$, $B(2, 1)$, and $C(-1, 1)$.
      $\triangle A'B'C'$: $A'(4, 4)$, $B'(4, 1)$, and $C'(7, 1)$.

26. $\triangle ABC$: $A(1, 3)$, $B(2, 5)$, and $C(5, 3)$.
      $\triangle A'B'C'$: $A'(1, 1)$, $B'(2, -1)$, and $C'(5, 1)$.

27. $\triangle ABC$: $A(1, 4)$, $B(2, 1)$, and $C(4, 2)$.
      $\triangle A'B'C'$: $A'(-5, 4)$, $B'(-6, 1)$, and $C'(-8, 2)$.

28. $\triangle ABC$: $A(4, 2)$, $B(6, 2)$, and $C(2, -1)$.
      $\triangle A'B'C'$: $A'(2, 4)$, $B'(2, 6)$, and $C'(-1, 2)$.

29. $\triangle ABC$: $A(-1, 0)$, $B(0, 2)$, and $C(4, 1)$.
      $\triangle A'B'C'$: $A'(-1, 8)$, $B'(0, 6)$, and $C'(4, 7)$.

30. $\triangle ABC$: $A(0, 1)$, $B(3, 1)$, and $C(3, 4)$.
      $\triangle A'B'C'$: $A'(-1, 2)$, $B'(-1, 5)$, and $C'(2, 5)$.

31. $\triangle ABC$: $A(2, -1)$, $B(4, 2)$, and $C(-1, 2)$.
      $\triangle A'B'C'$: $A'(1, -2)$, $B'(-2, -4)$, and $C'(-2, 1)$.

LINE SYMMETRY

In nature, in art, and in industry, we find many forms that contain a line of reflection.

If, in each of the figures below, every point in the figure moves to its image through a reflection in the given line, the figure will appear to be unchanged. This line of reflection is called an ***axis of symmetry,*** and each figure is said to have ***line symmetry.***

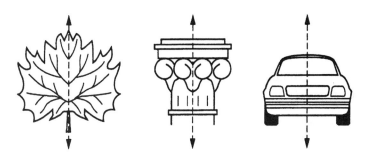

Line symmetry occurs in a figure when the figure is its own image under a reflection in a line. Such a line is called an ***axis of symmetry.***

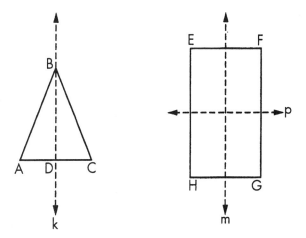

Triangle ABC is isosceles, with $AB = CB$. The axis of symmetry for $\triangle ABC$ is line k, drawn to contain altitude $\overline{BD}$ from vertex angle B to base $\overline{AC}$.

It is possible for a figure to have more than one axis of symmetry, or reflection line. Rectangle $EFGH$ has two axes of symmetry, namely, line m and line p.

Lines of reflection, or lines of symmetry, may be found for letters and for words.

～～～～～～ *MODEL PROBLEM* ～～～～～

On your paper, copy $\overline{AB}$ and line _m_. Then sketch as carefully as possible the image of $\overline{AB}$ under a reflection in line _m_. Let $A \to A'$, $B \to B'$, and label the image $\overline{A'B'}$.

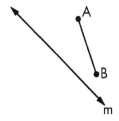

Solution: Use a ruler.

    Step 1. Hold the ruler perpendicular to line _m_ and touching point _A_. Measure the distance from _A_ to line _m_. Image _A'_ is on the other side of line _m_, the same distance away from _m_ as is point _A_. Mark _A'_.

    Step 2. Follow a similar plan to locate _B'_.

    Step 3. Draw $\overline{A'B'}$ by connecting _A'_ to _B'_.

        Step 1   *Step 2*   *Step 3*

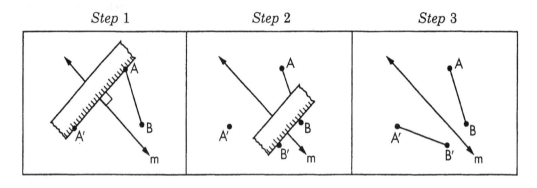

Exercises

In 1–8:   *a.* For each of the following words that has line symmetry, copy the word and draw the line of symmetry. *b.* For each word that does *not* have line symmetry, print the letters of the word in a vertical column. Which of these words now has (have) line symmetry?

  1. DAD   **2.** MOM   **3.** BOB   **4.** DEED
  5. RADAR   **6.** AVA   **7.** AXIOM   **8.** YOUTH

  9. On your paper, print all the capital letters of the alphabet that have line symmetry, and draw one or more lines of reflection for each of these letters.

In 10–16, the reflection of $\triangle ABC$ in line k is $\triangle DEC$.

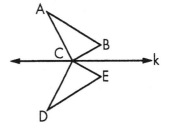

Ex. 10-16

10. What is the image of point A under the line reflection?

11. $r_k(B) = ?$ 12. $r_k(C) = ?$ 13. $r_k(D) = ?$

14. What is the preimage of point B under the line reflection?

15. $r_k(\angle ABC) = ?$ 16. $r_k(\overline{DE}) = ?$

In 17–30:   *a.* Copy the given figure or sketch the geometric figure named. *b.* Tell the number of lines of symmetry each figure has, if any, and sketch the lines on your drawing.

Ex. 17   Ex. 18   Ex. 19   Ex. 20   Ex. 21

22. parabola
25. parallelogram
28. line segment

23. square
26. rhombus
29. circle

24. rectangle
27. trapezoid
30. equilateral triangle

In 31–35:   *a.* Copy the given figure and line m.   *b.* Sketch, as carefully as possible, the image of the given figure under a reflection in line m. Label image points with prime marks, as in $A \to A'$.

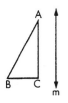 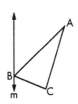 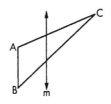 

Ex. 31   Ex. 32   Ex. 33   Ex. 34   Ex. 35

2. **Point Reflections and Point Symmetry**

POINT REFLECTIONS

Triangle *ABC* is reflected through point *P,* and its image, $\triangle A'B'C'$, is formed. To find this image under a reflection through point *P,* the following steps are taken:

Step 1. From each vertex of $\triangle ABC$, a segment is drawn through point *P* to its image in such a way that the distance from the vertex to point *P* is equal to the distance from point *P* to the image. Here, $\overline{AA'}$, $\overline{BB'}$, and $\overline{CC'}$ pass through point *P* so that $AP = PA'$, $BP = PB'$, and $CP = PC'$.

Step 2. Images A', B', and C' are connected to form $\triangle A'B'C'$, which is the reflection of $\triangle ABC$ through point *P*.

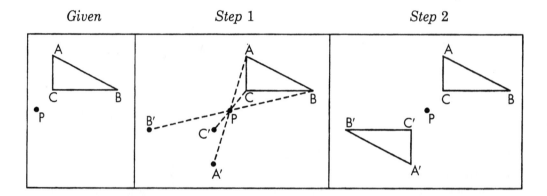

Given   *Step 1*   *Step 2*

For this point reflection, we can state that the image of *A* is A', and the image of A' is *A*, by writing $A \rightarrow A'$ and $A' \rightarrow A$.

A reflection in a point *P* is indicated in symbols as R_P Thus, to show specifically that the images are found under a reflection in point *P,* we write:

In Symbols   *In Words*

$R_P(A) = A'$.   Under a reflection in point *P,* the image of *A* is A'.

$R_P(B) = B'$.   Under a reflection in point *P,* the image of *B* is B'.

$R_P(C) = C'$.   Under a reflection in point *P,* the image of *C* is C'.

Under the point reflection just described, point P is the midpoint of each of segments $\overline{AA'}$, $\overline{BB'}$, and $\overline{CC'}$, leading to our definition. We notice, however, that one point remains fixed in the plane, namely, point P itself.

A *reflection in a point P* is a transformation of the plane such that:

1. The image of the fixed point P is P.

2. For all other points, if the image of K is K', then P is the midpoint of $\overline{KK'}$.

POINT SYMMETRY

The diagrams above show an advertising logo, a flag, a pinwheel, and a playing card. In each of these diagrams, every point in the figure moves to its image, which is also a point in the figure, through a point of reflection located in the "center" of the figure. This *point of reflection* is also called a *point of symmetry*, and each design is said to have *point symmetry.*

None of the figures pictured above has line symmetry. There is no way to fold any of these pictures over a line so that all points coincide; that is, there is no way to find a "mirror image."

We can think of point symmetry, however, as "turning the picture around" or "moving the picture through a half-turn." Try it. Turn the book upside down. Do the pictures look the same? They should if they have point symmetry.

Point symmetry occurs in a figure when the figure is its own image under a reflection in a point.

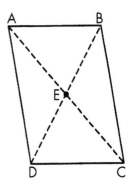

Quadrilateral $ABCD$ is a parallelogram whose diagonals $\overline{AC}$ and $\overline{BD}$ intersect at point E. This point E is a point of symmetry for $\square ABCD$. Thus, $A \to C$, $B \to D$, $C \to A$, $D \to B$, and so forth. While $\square ABCD$ has point symmetry, it does *not* have line symmetry.

It is possible, however, for a figure to have both line symmetry and point symmetry. Quadrilateral *EFGH* is a square containing four lines of reflection that intersect at *P,* a point of symmetry.

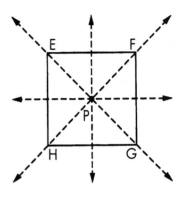

Points of symmetry may also be found for letters and for words.

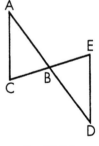

Exercises

1. On your paper, print all the capital letters of the alphabet that have point symmetry.

In 2–9, for each of the following words that has point symmetry, locate the reflection point in the word.

2. SIS   **3.** WOW   **4.** NOON   **5.** ZOO

6. OX   **7.** SWIMS   **8.** un   **9.** pod

In 10–16, the reflection of $\triangle ABC$ through point *B* is $\triangle DBE$.

10. What is the image of *A* under the point reflection?

11. $R_B(C) = ?$ **12.** $R_B(D) = ?$

13. $R_B(B) = ?$

14. What is the preimage of *C* under the point reflection?

15. $R_B(\angle CAB) = ?$ **16.** $R_B(\overline{CA}) = ?$

Ex. 10-16

In 17–25: *a.* Copy the given figure, or sketch the geometric figure named. *b.* For each figure that has a point of symmetry, locate the point on your drawing.

17. kite   **18.** star   **19.** rectangle   **20.** parallelogram
                                     21. rhombus   **22.** trapezoid
                                     23. circle   **24.** line segment
                                     25. equilateral triangle

In 26–29:   *a.* Copy the given figure.   *b.* Sketch as carefully as possible the image of the given figure under a reflection in point *A*. Label image points with prime marks, as in $B \rightarrow B'$.

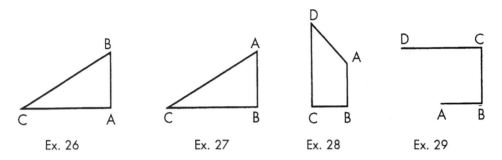

Ex. 26   Ex. 27   Ex. 28   Ex. 29

3. Rotations, Translations, and Dilations

POINT REFLECTIONS AND ROTATIONS

Although it is possible to use any point on the coordinate plane as a point of reflection or a point of rotation, the most commonly used point is the origin.

1. *Point reflection in the origin, or half-turn, or rotation of 180°*

   Here, the vertices of $\triangle ABC$ are $A(1, 2)$, $B(5, 5)$, and $C(5, 2)$. These vertices are reflected in the origin (point O). Their images are connected to form $\triangle A'B'C'$.

   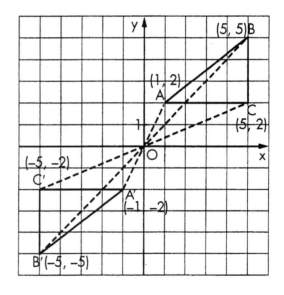

   $A(1, 2) \rightarrow A'(-1, -2)$

   $B(5, 5) \rightarrow B'(-5, -5)$

   $C(5, 2) \rightarrow C'(-5, -2)$

This example illustrates a general rule.

- **Under a reflection in point *O*, the origin:**

$$P(x, y) \rightarrow P'(-x, -y)$$

$$\text{or } R_O(x, y) = (-x, -y)$$

$$\text{or } R_{180°}(x, y) = (-x, -y)$$

2. *Rotation of 90°, counterclockwise about the origin*

Here, the vertices of $\triangle ABC$ are $A(1, 3)$, $B(5, 1)$, and $C(1, 1)$. These vertices are rotated counterclockwise about the origin (point O) through an angle of 90°, and their images are connected to form $\triangle A'B'C'$.

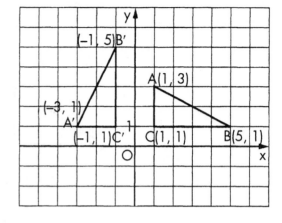

$$A(1, 3) \to A'(-3, 1)$$

$$B(5, 1) \to B'(-1, 5)$$

$$C(1, 1) \to C'(-1, 1)$$

The coordinates illustrate another general rule.

- **Under a rotation of 90° counterclockwise about point *O*, the origin:**

$$P(x, y) \to P'(-y, x) \qquad \text{or} \qquad R_{90°}(x, y) = (-y, x)$$

Note. If no point is mentioned, it is assumed that the rotation is taken about the origin. Thus, $R_{50°}$ means a 50° rotation about O, the origin.

The *properties preserved under a point reflection* and *under a rotation* include all five properties listed previously for a line reflection, namely, distance, angle measure, parallelism, collinearity, and midpoint.

TRANSLATIONS

A rule for a translation is easily stated in coordinate geometry. In the following figure, the segment $\overline{AB}$ is translated onto its image, $\overline{A'B'}$, by moving each point 3 units to the right and 2 units down. Thus, by counting, we form the rule for the translation:

$$P(x, y) \to P'(x + 3, y - 2)$$

$$\text{or}$$

$$T_{3, -2}(x, y) = (x + 3, y - 2)$$

- **Under a translation of *a* units horizontally and *b* units vertically:**

$$P(x, y) \to P'(x + a, y + b)$$

$$\text{or}$$

$$T_{a, b}(x, y) = (x + a, y + b)$$

The *properties preserved under a translation* include all five properties listed previously for a line reflection, namely, distance, angle measure, parallelism, collinearity, and midpoint.

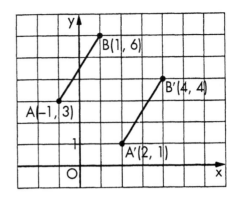

～～～～～ *MODEL PROBLEMS* ～～～～～

1. Given $\triangle ABC$, whose vertices are $A(1, 3)$, $B(-2, 0)$, $C(4, -3)$.

   a. On one set of axes, draw $\triangle ABC$ and its image, $\triangle A'B'C'$, under a point reflection in the origin.

   b. Is there another transformation that will result in the same image, $\triangle A'B'C'$?

Solution:

   a. In step 1, draw and label $\triangle ABC$.
   In step 2, find the images of the vertices of $\triangle ABC$ by using the rule $P(x, y) \rightarrow P'(-x, -y)$.
   Then draw and label $\triangle A'B'C'$.

   $A(1, 3) \rightarrow A'(-1, -3)$
   $B(-2, 0) \rightarrow B'(2, 0)$
   $C(4, -3) \rightarrow C'(-4, 3)$

Step 1

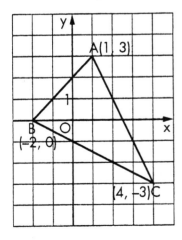

Step 2

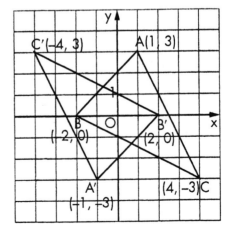

b. Under a counterclockwise rotation of 180°, the coordinates of △*ABC* will transform according to the rule $P(x, y) \rightarrow P'(-x, -y)$, and therefore result in the same image of △*A'B'C'* as shown in step 2.

Answer: a. See the graph labeled *Step 2.*   *b.* $R_{180°}$

2. A translation maps $A(2, 1)$ onto $A'(10, -4)$. Find the coordinates of B', the image of $B(8, 7)$, under the same translation.

Solution:

Use the coordinates of *A* and *A'* to discover the number of units shifted horizontally and vertically.

$$A(2, 1) \rightarrow A'(10, -4)$$
$$(x, y) \rightarrow (x + 8, y - 5)$$

Substitute the coordinates of *B* in the rule determined for the translation.

$$B(8, 7) \rightarrow B'(16, 2)$$

Answer: $B'(16, 2)$

3. The vertices of parallelogram *ABCD* are $A(1, 1)$, $B(3, 5)$, $C(9, 5)$, and $D(7, 1)$.

   a. Find the coordinates of the point of symmetry for □*ABCD*.
   b. Find the image of ∠*CAD* under this point reflection.

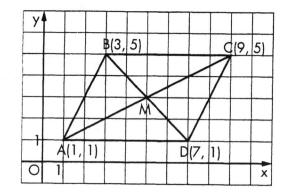

Solution:

a. Since the diagonals of a parallelogram bisect each other, the point of symmetry is the intersection point of the two diagonals, or the midpoint of either diagonal.

For $\overline{AC}$, the coordinates of midpoint *M* are:

$$\left(\frac{1+9}{2}, \frac{1+5}{2}\right) = \left(\frac{10}{2}, \frac{6}{2}\right) = (5, 3)$$

For $\overline{BD}$, the coordinates of midpoint *M* are:

$$\left(\frac{3+7}{2}, \frac{5+1}{2}\right) = \left(\frac{10}{2}, \frac{6}{2}\right) = (5, 3)$$

b. Under a reflection in point M: $C \rightarrow A$, $A \rightarrow C$, and $D \rightarrow B$. Therefore, $R_M(\angle CAD) = \angle ACB$.

Answer: a. (5, 3)   b. $\angle ACB$

~~~~~~

Exercises

In 1–5, find the image of each point under a reflection in the x-axis.

1. (3, 10)   **2.** (6, −4)   **3.** (−1, 8)   **4.** (3, 0)   **5.** (−2.7, 0)

In 6–10, find the image of each point under a reflection in the y-axis.

6. (3, 10)   **7.** (6, −4)   **8.** (−1, 8)   **9.** (0, 6)   **10.** (−8, −1.3)

In 11–15, find the image of each point under a reflection in the line $y = x$.

11. (3, 10)   **12.** (6, −4)   **13.** (−1, 8)   **14.** (−1.5, −7)   **15.** (8, 8)

In 16–20, find the image of each point under a reflection in the origin.

16. (3, 10)   **17.** (6, −4)   **18.** (−1, 8)   **19.** (0, 0)   **20.** (−10, −5)

In 21–25, find the image of point (−3, 2) under each given translation.

21. $T_{1,\,2}$ **22.** $T_{3,\,-6}$ **23.** $T_{-4,\,0}$ **24.** $T_{0,\,-7}$ **25.** $T_{3,\,-2}$

In 26–31, in each case find the rule for the translation so that the image of A is A'.

26. $A(3, 8) \rightarrow A'(4, 6)$ **27.** $A(1, 0) \rightarrow A'(0, 1)$ **28.** $A(2, 5) \rightarrow A'(-1, 1)$

29. $A(-1, 2) \rightarrow A'(-2, -3)$ **30.** $A(0 -3) \rightarrow A'(-7, -3)$ **31.** $A(4, -7), \rightarrow A'(4, -2)$

32. A translation maps $B(0, 2)$ onto $B'(5, 0)$. Find the coordinates of C', the image of $C(-3, 1)$, under the same translation.

33. A translation maps the origin to point (−1, 7). What is the image of (3, −7) under the same translation?

34. Translation $T_{-4,\,6}$ maps point $A(7, -4)$ to point B, and a second translation maps point B to the origin. What is the rule for the second translation?

DILATIONS

When a photograph is enlarged or reduced, a change, or transformation, takes place in its size. There is a constant ratio of the distance between points in the original photograph compared to the distances between their images in the enlargement or reduction. This type of transformation is called a *dilation*.

Given	*Step 1*	*Step 2*

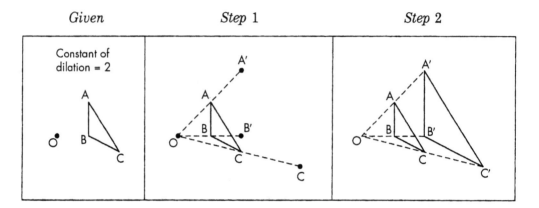

In the diagram, $\triangle ABC$ is to be dilated so that the *center of dilation* is point O and the *constant of dilation* is 2.

Step 1 From O, the center of dilation, rays are drawn to pass through each of the vertices of $\triangle ABC$. Then, the constant of dilation, 2, is used to locate the images, A', B', and C', on these rays so that $OA' = 2 \cdot OA$, $OB' = 2 \cdot OB$, and OC', $= 2 \cdot OC$.

Step 2 The images, A', B', and C', are connected to form $\triangle A'B'C'$.

This dilation, with a constant factor of 2, is written in symbols as D_2.

Although any point may be chosen as the center of dilation, we usually limit dilations to those where point O, the origin, is the center of dilation.

A *dilation* of k, where k is a positive number called the *constant of dilation*, is a transformation of the plane such that:

1. The image of point O, the center of dilation, is O.

2. For all other points, the image of P is P', where $\overrightarrow{OP}$ and $\overrightarrow{OP'}$ name the same ray and $OP' = k \cdot OP$.

A rule for a dilation is stated in coordinate geometry as follows:

- **Under a dilation of k (a positive number) whose center of dilation is the origin:**

$$P(x, y) \rightarrow P'(kx, ky) \qquad \text{or} \qquad D_k(x, y) = (kx, ky)$$

Exercises

In 1–5, use the rule $(x, y) \rightarrow (4x, 4y)$ to find the image of each given point.

1. $(3, 5)$ **2.** $(-3, 2)$ **3.** $(7, 0)$ **4.** $(-4, 9)$ **5.** $(0.125, 0.5)$

In 6–10, find the image of each given point under a dilation of 5.

6. $(12, 20)$ **7.** $(-9, -7)$ **8.** $(0, -8)$ **9.** $\left(\dfrac{3}{10}, -\dfrac{1}{5}\right)$ **10.** $(\sqrt{2}, 0)$

11. *a.* On graph paper, draw $\triangle ACE$ with vertices $A(2, 3)$, $C(2, -1)$, and $E(-1, -1)$.
  b. Using the same axes, graph $\triangle A'C'E'$ such that $D_3(\triangle ACE) = \triangle A'C'E'$.
  c. Using EA and $E'A'$, demonstrate that distance is *not* preserved under the dilation.
  d. Show that $\triangle ACE \sim \triangle A'C'E'$.

12. The vertices of quadrilateral $TRAP$ are $T(-2, 1)$, $R(0, 4)$, $A(5, 4)$, and $P(3, 1)$. Under a dilation of 2, $TRAP$ maps to $T'R'A'P'$.

  a. On one set of axes, draw and label quadrilaterals $TRAP$ and $T'R'A'P'$.
  b. True or False: Since $D_2 (TRAP) = T'R'A'P'$, the area of quadrilateral $T'R'A'P'$ is twice the area of quadrilateral $TRAP$.
  c. Explain your answer to part *b*, providing work to support your reasoning.

4. Compositions of Transformations

The combination of two transformations is called a ***composition of transformations*** when the first transformation produces an image, and the second transformation is performed on that image.

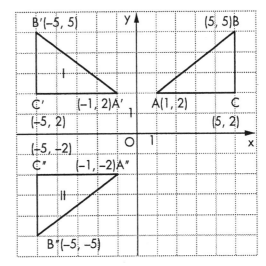

We start again with $\triangle ABC$, whose vertices are $A(1, 2)$, $B(5, 5)$, and $C(5, 2)$. For the composition of transformations, we will consider two line reflections. First, by reflecting $\triangle ABC$ in the y-axis, we form $\triangle A'B'C'$, or $\triangle\text{I}$. Then, by reflecting $\triangle A'B'C'$ in the x-axis, we form $\triangle A''B''C''$, or $\triangle\text{II}$. We observe how the vertices are related:

$A(1, 2) \rightarrow A'(-1, 2) \rightarrow A''(-1, -2)$

$B(5, 5) \rightarrow B'(-5, 5) \rightarrow B''(-5, -5)$

$C(5, 2) \rightarrow C'(-5, 2) \rightarrow C''(-5, -2)$

Now let us compare the original triangle, $\triangle ABC$, and the final image, $\triangle A''B''C''$, with the image of $\triangle ABC$ under a reflection in the origin. We observe:

- **The composition of a line reflection in the y-axis, followed by a line reflection in the x-axis, is equivalent to a single transformation, namely, a reflection through point O, the origin.**

We reflected the triangle first in the y-axis and then in the x-axis. If we had reflected the triangle first in the x-axis and then in the y-axis, would this composition also be equivalent to a reflection through point O, the origin? The answer is yes. However, not all compositions will act in the same way. In general, compositions of transformations are *not* commutative.

~~~~~~~~~~~ *MODEL PROBLEMS* ~~~~~~~~~~~

1. The coordinates of the endpoints of $\overline{RS}$ are $R(3, -2)$ and $S(5, -1)$.
    a. Graph the line segment $\overline{RS}$.
    b. Graph on the same set of axes the image $\overline{R'S'}$, a reflection of $\overline{RS}$ in the x-axis.
    c. Graph on the same set of axes $\overline{R''S''}$, a reflection of $\overline{R'S'}$ in the origin.
    d. Name the single transformation by which $\overline{RS} \rightarrow \overline{R''S''}$.

Solution:

a, b. $R(3, -2) \rightarrow R'(3, 2)$
        $S(5, -1) \rightarrow S'(5, 1)$

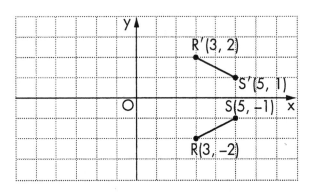

c. $R'(3, 2) \rightarrow R''(-3, -2)$
     $S'(5, 1) \rightarrow S''(-5, -1)$

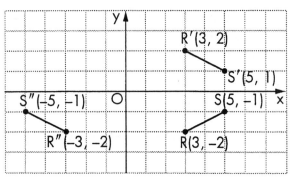

d. $\overline{RS} \rightarrow \overline{R''S''}$ is the same as a reflection of $\overline{RS}$ in the y-axis.
   $R(3, -2) \rightarrow R''(-3, -2)$
   $S(5, -1) \rightarrow S''(-5, -1)$

2. The coordinates of $\triangle ABC$ are $A(1, 3)$, $B(1, 1)$, and $C(5, 1)$. Is the image $\triangle A''B''C''$ the same if it is reflected in the origin, then rotated 90° clockwise about the origin as the image is if it is rotated 90° clockwise about the origin, then reflected 90° in the origin? Explain your answer.

   Solution: Under a reflection in the origin, then rotated 90° clockwise about the origin:

$A(1, 3) \rightarrow A'(-1, -3)$	$A'(-1, -3) \rightarrow A''(-3, 1)$
$B(1, 1) \rightarrow B'(-1, -1)$	$B'(-1, -1) \rightarrow B''(-1, 1)$
$C(5, 1) \rightarrow C'(-5, -1)$	$C'(-5, -1) \rightarrow C''(-1, 5)$

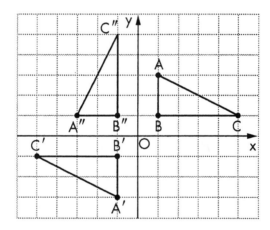

   Rotated 90° clockwise about the origin, then under a reflection in the origin:

$A(1, 3) \rightarrow A'(3, -1)$	$A'(3, -1) \rightarrow A''(-3, 1)$
$B(1, 1) \rightarrow B'(1, -1)$	$B'(1, -1) \rightarrow B''(-1, 1)$
$C(5, 1) \rightarrow C'(1, -5)$	$C'(1, -5) \rightarrow C''(-1, 5)$

   Answer: In each case $\triangle A''B''C''$ is the same. This composition of transformations is commutative.

Exercises

1. On graph paper draw $\triangle ABC$ whose vertices are $A(-5, 2)$, $B(-3, 6)$, and $C(0, 0)$.
   - *a.* Using the same set of axes, graph $\triangle A'B'C'$, the image of $\triangle ABC$ under a reflection in the origin.
   - *b.* Using the same set of axes, graph $\triangle A''B''C''$, a reflection of $\triangle A'B'C'$ in the y-axis.
   - *c.* Name the single transformation by which $\triangle ABC \rightarrow \triangle A''B''C''$.

   In 2 and 3, the vertices of $\triangle LMN$ are $L(1, 3)$, $M(1, 1)$, and $N(5, 1)$.

2. *a.* Draw $\triangle LMN$ and its image, $\triangle L'M'N'$, under a point reflection through the origin.
   - *b.* Using the same graph, reflect $\triangle L'M'N'$ in the x-axis to form its image, $\triangle L''M''N''$.
   - *c.* What single transformation is equivalent to the composition of a point reflection through the origin followed by a reflection in the x-axis?

3. *a.* Draw $\triangle LMN$ and its image $\triangle L'M'N'$ under a reflection in the x-axis.
   - *b.* Using the same graph, reflect $\triangle L'M'N'$ in the y-axis to form its image, $\triangle L''M''N''$.
   - *c.* What single transformation is equivalent to the composition of a reflection in the x-axis followed by a reflection in the y-axis?

4. Given $\triangle DEF$ and $\triangle D'E''F''$ as shown on the grid:

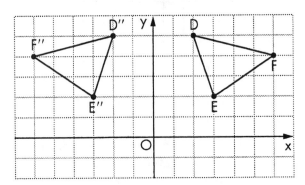

   - *a.* Describe one pair of transformations that will result in $\triangle DEF \rightarrow \triangle D''E''F''$.
   - *b.* Name a single transformation that will give the same result as the pair described in *a*.

CALCULATOR APPLICATIONS

The commands on the $\boxed{\text{DRAW}}$ menu can be used to draw triangles, rectangles, and other polygons. When the command 2:Line(is selected, the calculator draws a line segment between the coordinates (x_1, y_1) and (x_2, y_2).

Before you use the $\boxed{\text{DRAW}}$ instructions, clear existing drawings with ClrDraw and change the $\boxed{\text{WINDOW}}$ if necessary.

MODEL PROBLEM

Draw $\triangle ABC$ and its image $\triangle A'B'C'$. Identify the transformation.

$A(1, 3) \rightarrow A'(-3, 1)$
$B(5, 1) \rightarrow B'(-1, 5)$
$C(1, 1) \rightarrow C'(-1, 1)$

Solution: First draw $\overline{AB}$, $\overline{AC}$, and $\overline{CB}$; then draw $\overline{A'B'}$, $\overline{A'C'}$, and $\overline{C'B'}$.

Enter: $\boxed{\text{2nd}}$ $\boxed{\text{DRAW}}$ 2 1$\boxed{}$, 3$\boxed{}$, 5$\boxed{}$, 1$\boxed{}$) $\boxed{\text{ALPHA}}$ $\boxed{:}$

$\boxed{\text{2nd}}$ $\boxed{\text{DRAW}}$ 2 1$\boxed{}$, 3$\boxed{}$, 1$\boxed{}$, 1$\boxed{}$) $\boxed{\text{ALPHA}}$ $\boxed{:}$

$\boxed{\text{2nd}}$ $\boxed{\text{DRAW}}$ 2 1$\boxed{}$, 1$\boxed{}$, 5$\boxed{}$, 1$\boxed{}$) $\boxed{\text{ALPHA}}$ $\boxed{:}$

$\boxed{\text{2nd}}$ $\boxed{\text{DRAW}}$ 2 $\boxed{(-)}$ 3$\boxed{}$, 1$\boxed{}$, $\boxed{(-)}$1$\boxed{}$, 5$\boxed{}$) $\boxed{\text{ALPHA}}$ $\boxed{:}$

$\boxed{\text{2nd}}$ $\boxed{\text{DRAW}}$ 2 $\boxed{(-)}$ 3$\boxed{}$, 1$\boxed{}$, $\boxed{(-)}$1$\boxed{}$, 1$\boxed{}$) $\boxed{\text{ALPHA}}$ $\boxed{:}$

$\boxed{\text{2nd}}$ $\boxed{\text{DRAW}}$ 2 $\boxed{(-)}$ 1$\boxed{}$, 1$\boxed{}$, $\boxed{(-)}$1$\boxed{}$, 5$\boxed{}$) $\boxed{\text{ENTER}}$

Display:

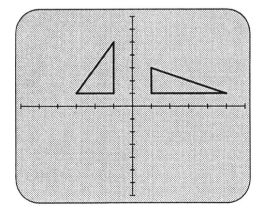

Note. The triangles are congruent, although they do not appear so in the calculator window.

Answer: Triangle ABC has been rotated 90° counterclockwise about the origin.

CHAPTER VII

Systems of Linear Open Sentences in Two or More Variables

1. The Algebra of Sets

We have performed operations on numbers. Now we will perform operations on sets.

UNION OF SETS

The **union** of two sets, A and B, is the set containing the elements present in both set A and set B.

The union of set A and set B is represented by the symbol $A \cup B$, read "A union B" or "A cup B." Thus, if $A = \{1, 2, 3, 4, 5\}$ and $B = \{3, 4, 5, 6, 7\}$, then $A \cup B = \{1, 2, 3, 4, 5, 6, 7\}$. The shaded region of the Venn diagram (Fig. 1) pictures $A \cup B$.

$A \cup B$ is shaded.

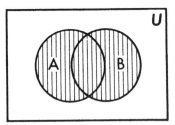

Fig. 1

If $A = \{$natural numbers between 10 and 20$\}$ and $B = \{$natural numbers between 15 and 25$\}$, then $A \cup B = \{$natural numbers between 10 and 25$\}$.

Superior College has only two teams, a baseball team and a basketball team. Hence, if $A = \{$baseball players in Superior College$\}$ and $B = \{$basketball players in Superior College$\}$, then $A \cup B = \{$athletes who play for Superior College$\}$.

Notice in each illustration that every element of $A \cup B$ is a member of at least one of the two given sets A and B.

When two open sentences are joined by the word *or*, the result is called a **disjunction** of two open sentences. For example, $x = 2$ or $x = -2$ is a disjunction of two open sentences. When we solve the disjunction $x = 2$ or $x = -2$, we are trying to find the set of numbers that belong to the solution set of at least one of the sentences. Therefore, the solution set of the disjunction is the union of the solution set of $x = 2$ and the solution set of $x = -2$.

We have learned that the solution set of $|x| = 2$ is $\{x \mid x = 2$ or $x = -2\}$. The sentence $x = 2$ or $x = -2$ is a disjunction. Therefore, the solution set of $|x| = 2$

256

is the union of the solution set of $x = 2$ and $x = -2$. Thus:

$$\{x : |x| = 2\} = \{x | x = 2\} \cup \{x | x = -2\}$$
$$= \{2\} \cup \{-2\}$$
$$= \{2, -2\}$$

The graph of the solution set $\{2, -2\}$ is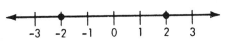

INTERSECTION OF SETS

The **intersection** of two sets, A and B, is the set containing those elements of set A that are also elements of set B.

The intersection of set A and set B is represented by the symbol $A \cap B$, read "A intersection B" or "A cap B." Thus, if $A = \{1, 2, 3, 4, 5\}$ and $B = \{2, 4, 6, 8, 10\}$, then $A \cap B = \{2, 4\}$. The shaded region of the Venn diagram (Fig. 2) pictures $A \cap B$.

$A \cap B$ is shaded.

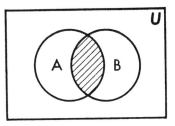

Fig. 2

If $A = \{$students in a school who study French$\}$ and $B = \{$students in the school who study German$\}$, then $A \cap B = \{$students in the school who study both French and German$\}$.

If $A = \{x | 2 < x\}$ and $B = \{x | x < 5\}$ then $A \cap B = \{x | 2 < x < 5\}$.

When two open sentences are joined by the word *and*, the result is called a **conjunction** of two open sentences. For example, $-2 < x - 1$ *and* $x - 1 < 2$ is a conjunction of two open sentences. When we solve the conjunction $-2 < x - 1$ and $x - 1 < 2$, we are trying to find the set of numbers that belong to the solution sets of both sentences. Therefore, the solution set of the conjunction is the intersection of the solution set of $-2 < x - 1$ and the solution set of $x - 1 < 2$.

We have learned that the solution of $|x - 1| < 2$ is $\{x | -2 < x - 1 < 2\}$. The sentence $-2 < x - 1 < 2$ is a conjunction which may be written

$$-2 < x - 1 \text{ and } x - 1 < 2$$

Therefore, the solution set of $|x - 1| < 2$ is the intersection of the solution set of $-2 < x - 1$ and the solution set of $x - 1 < 2$. Thus:

$$\{x : |x - 1| < 2\} = \{x | -2 < x - 1\} \cap \{x | x - 1 < 2\}$$
$$= \{x | -1 < x\} \cap \{x | x < 3\}$$
$$= \{x | -1 < x < 3\}$$

The graph of the solution set $\{x | -1 < x < 3\}$ is the "overlap" of the graphs of $\{x | -1 < x\}$ and $\{x | x < 3\}$ as shown:

graph of $\{x \mid -1 < x\}$

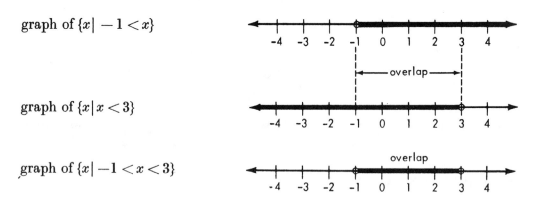

graph of $\{x \mid x < 3\}$

graph of $\{x \mid -1 < x < 3\}$

When two sets have no members in common they are called ***disjoint sets.*** The intersection of two disjoint sets is, therefore, the empty set. Thus, if $A = \{1, 3, 5\}$ and $B = \{2, 4, 6\}$, then A and B are disjoint sets. Hence, $A \cap B = \varnothing$. Therefore, if A and B are disjoint sets, the set $A \cap B$ may be pictured as shown in Fig. 3 or Fig. 4.

$A \cap B$ $A \cap B$

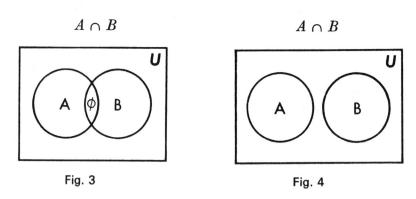

Fig. 3   Fig. 4

THE COMPLEMENT OF A SET

If A is a subset of a given universal set U, the ***complement*** of set A with respect to set U is a set whose elements are those elements of U that are not in A. The complement of set A is represented by A'. Thus, if $U = \{1, 2, 3, 4, 5, 6, 7, 8, 9\}$ and $A = \{2, 4, 6, 8\}$, then $A' = \{1, 3, 5, 7, 9\}$. The shaded region in the Venn diagram (Fig. 5) contains the elements of A'.

A' is shaded.

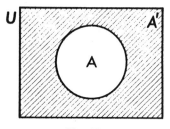

Fig. 5

If $U = \{$natural numbers$\}$ and $A = \{$natural numbers divisible by 5$\}$, then $A' = \{$natural numbers not divisible by 5$\}$.

If $U = \{$animals$\}$ and $A = \{$lions$\}$, then $A' = \{$animals that are not lions$\}$.

~~~~~~~~~~~~~~~~ **MODEL PROBLEMS** ~~~~~~~~~~~~~~

1. If $U = \{1, 2, 3, 4, \ldots, 15\}$ and A, B, and C are sets whose elements are those numbers shown in the Venn diagram (Fig. 6), write the following sets:

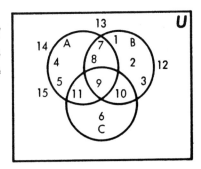

Fig. 6

a. A, B, and C b. $A \cap B$

c. $B \cup C$ d. $(B \cup C)'$

e. $(A \cap B) \cap C$ f. $A \cup (B \cup C)$

Solution:

a. $A = \{4, 5, 7, 8, 9, 11\}$ $B = \{1, 2, 3, 7, 8, 9, 10\}$ $C = \{6, 9, 10, 11\}$

b. $A \cap B = \{7, 8, 9\}$

c. $B \cup C = \{1, 2, 3, 6, 7, 8, 9, 10, 11\}$

d. $(B \cup C)'$ means the complement of $B \cup C$.
Since $B \cup C = \{1, 2, 3, 6, 7, 8, 9, 10, 11\}$, $(B \cup C)' = \{4, 5, 12, 13, 14, 15\}$.

e. $(A \cap B) \cap C$ means the intersection of the set $A \cap B$ and the set C.
Since $A \cap B = \{7, 8, 9\}$ and $C = \{6, 9, 10, 11\}$, then $(A \cap B) \cap C = \{9\}$.

f. $A \cup (B \cup C)$ means the union of set A and the set $B \cup C$.
Since $A = \{4, 5, 7, 8, 9, 11\}$ and $B \cup C = \{1, 2, 3, 6, 7, 8, 9, 10, 11\}$,
$A \cup (B \cup C) = \{1, 2, 3, 4, 5, 6, 7, 8, 9, 10, 11\}$.

2. A, B, and C are sets which are subsets of universal set U. If A, B, and C have non-empty intersections, use a Venn diagram to illustrate that $(A \cap B) \cap C = A \cap (B \cap C)$.

Solution:

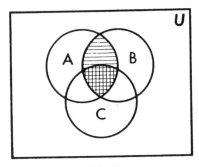

Fig. 7

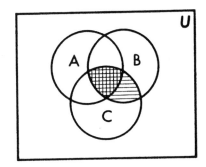

Fig. 8

Figs. 7 and 8 are Venn diagrams that picture sets A, B, and C. In Fig. 7, the horizontal hatched region pictures $A \cap B$ and the crosshatched region pictures $(A \cap B) \cap C$. In Fig. 8, the horizontal hatched region pictures

$B \cap C$ and the crosshatched region pictures $A \cap (B \cap C)$. Observe that the crosshatched region in Fig. 7 is exactly the same as the crosshatched region in Fig. 8. Therefore, $(A \cap B) \cap C = A \cap (B \cap C)$.

Exercises

In 1–13, using the roster notation, write the set that is indicated when $U = \{1, 2, 3, 4, \ldots, 12\}$, $A = \{1, 5, 7, 9\}$, $B = \{4, 5, 6, 7, 8\}$, and $C = \{3, 4, 11, 12\}$.

  1. $A \cap B$ **2.** $B \cap C$ **3.** $A \cap C$ **4.** $A \cup B$ **5.** $B \cup C$

  6. $A \cap (B \cap C)$ **7.** $(A \cup B) \cup C$ **8.** $(A \cap B) \cup C$ **9.** $A \cup (B \cap C)$

10. A' **11.** B' **12.** $(A \cap B)'$ **13.** $(A \cup C)'$

In 14–24, $U = \{$natural numbers less than 20$\}$, $A = \{$odd natural numbers less than 20 that are divisible by 3$\}$, $B = \{$odd natural numbers less than 20 that are divisible by 5$\}$, and $C = \{$even natural numbers less than 20 that are divisible by 5$\}$. Use the roster notation to designate the indicated set.

14. A **15.** B **16.** C **17.** $A \cap B$ **18.** $B \cap C$ **19.** $A \cap (B \cap C)$

20. $(A \cup B) \cup C$ **21.** A' **22.** B' **23.** $(A \cup B)'$ **24.** $(A \cap C)'$

25. Find the number of elements in A' when:

    a. the universal set has 20 elements and A has 5 elements.

    b. the universal set has y elements and A has x elements.

26. If $U = \{$triangles$\}$, $A = \{$isosceles triangles$\}$, $B = \{$right triangles$\}$, $C = \{$scalene triangles$\}$, picture with a Venn diagram:

    a. $A \cap B$ *b.* $A \cap C$ *c.* $A \cup B$ *d.* $A \cup C$

27. Draw a Venn diagram to show $A \cap B$ if:

    a. A and B are sets which have only some elements in common.

    b. A and B are sets which have no elements in common.

    c. Set B is a proper subset of set A.

28. Draw a Venn diagram to show $A \cup B$ if:

    a. A and B are sets which have only some elements in common.

    b. A and B are sets which have no elements in common.

    c. Set B is a proper subset of set A.

In 29–34, U is the universal set, and R, S, and T are proper subsets of U, each of which has some but not all of its elements in common with each of the other two sets. In a Venn diagram like the one shown at the right, shade the region that represents the indicated set.

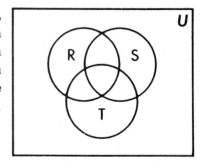

29. $(R \cap S) \cap T$ **30.** $(R \cup S) \cap T$

31. $R \cup (S \cup T)$ **32.** $(R \cap S) \cup T$

33. $(R \cap S) \cup (S \cap T)$ **34.** $(R \cap S) \cap (S \cap T)$

35. If universal set U is the set of all natural numbers, $A = \{2, 4, 6, 8, 10, 12, 15, 18\}$, $B = \{3, 6, 9, 12, 15, 18, 24\}$, and $C = \{5, 10, 15, 20\}$, draw a Venn diagram showing the relation of U, A, B, and C. Write the elements of sets A, B, and C in the proper regions in the diagram.

36. In a college, the languages which the romance language department teaches are French, Spanish, and Italian. There are 85 students who study French, 60 who study Spanish, and 25 who study Italian. Among those students, there are 18 who study both French and Spanish, 10 who study both French and Italian, 8 who study both Spanish and Italian, and 5 who study all three languages, French, Spanish, and 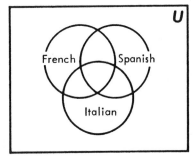 Italian. Use a Venn diagram like the one shown to help you find the number of students in the romance language department.

37. A is a subset of the universal set U, and $\varnothing$ is the empty set. Represent each of the following sets in a simpler way. Venn diagrams may help you discover the answers.

    a. $A \cap A$ *b.* $A \cup A$ *c.* $A \cap \varnothing$ *d.* $A \cup \varnothing$ *e.* $(A')'$

In 38–42, write for the given expression a simpler expression which does not contain parentheses.

38. $B \cup (A \cap B)$ **39.** $A \cap (A \cup B)$ **40.** $(A \cap B) \cap (A \cup B)$

41. $A \cup (A' \cap B)$ **42.** $A \cap (A' \cup B)$

43. If set A is equal to set B, then what must be true of the set $A \cup B$ and the set $A \cap B$?

In 44–48, A, B, and C are subsets of universal set U. Make a Venn diagram to illustrate that the statement is true.

44. $(A \cup B) \cup C = A \cup (B \cup C)$

45. $(A \cap B)' = A' \cup B'$

46. $(A \cup B)' = A' \cap B'$

47. $A \cap (B \cup C) = (A \cap B) \cup (A \cap C)$

48. $A \cup (B \cap C) = (A \cup B) \cap (A \cup C)$

In 49–54, state whether the word *or* should replace the question mark or the word *and* should replace the question mark to make the statement true.

49. $|x| = 4$ is equivalent to $x = -4 \; ? \; x = 4$.

50. $|y| > 2$ is equivalent to $-2 > y \; ? \; y > 2$.

51. $|t| \leq 6$ is equivalent to $-6 \leq t \; ? \; t \leq 6$.

52. $|x - 3| = 5$ is equivalent to $x - 3 = -5 \; ? \; x - 3 = 5$.

53. $|r - 4| < 8$ is equivalent to $-8 < r - 4$? $r - 4 < 8$.

54. $|s + 2| \geq 3$ is equivalent to $-3 \geq s + 2$? $s + 2 \geq 3$.

In 55–60, write a conjunction or a disjunction of open sentences which are equivalent to the given open sentence.

55. $|m| = 7$ **56.** $|t| > 1$ **57.** $|x| \leq 9$

58. $|x - 4| = 2$ **59.** $|2y - 1| < 5$ **60.** $|8 - 2x| \geq 10$

In 61–63, find and graph the solution set of the open sentence.

61. $|x - 5| = 8$ **62.** $|x + 2| \leq 6$ **63.** $|3y - 5| > 2$

2. Graphic Solution of a System of Linear Equations in Two Variables

We see in a coordinate plane (Fig. 1) the line L_1, which is the graph of the infinite number of ordered pairs in the solution set of the equation $y - x = 3$. In the same coordinate plane with the same set of axes, we see the line L_2, which is the graph of the infinite number of ordered pairs in the solution set of the equation $2x + y = 6$. The one point that these straight lines have in common is their point of intersection $P(1, 4)$. Hence, the ordered pair $(1, 4)$ which is associated with P is the common solution of the pair of equations $y - x = 3$ and $2x + y = 6$. Thus, the solution set of this pair is $\{(1, 4)\}$.

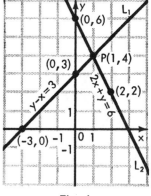

Fig. 1

If we represent the solution set of $y - x = 3$ by $A = \{(x, y) \mid y - x = 3\}$, and if we represent the solution set of $2x + y = 6$ by $B = \{(x, y) \mid 2x + y = 6\}$, then the set of ordered pairs common to A and B, that is, $A \cap B = \{(1, 4)\}$.

A pair of equations like $y - x = 3$ and $2x + y = 6$ is called a ***system of linear equations in two variables,*** or a ***set of simultaneous linear equations.***

A system of equations is really a conjunction of open sentences. Thus, the system $y - x = 3$ and $2x + y = 6$ is the conjunction

$$y - x = 3 \quad \text{and} \quad 2x + y = 6$$

The solution of this conjunction is the set of all ordered pairs of numbers that satisfy *both* equations of the system. Each of these ordered pairs is called a ***solution*** of the system. The set of all solutions of the system is called the ***solution set*** of the system. Hence, the solution set of the system $y - x = 3$ and $2x + y = 6$ is $\{(1, 4)\}$.

Procedure. To solve a pair of linear equations graphically:
1. Graph each equation, using the same set of axes.
2. Find the common solution, which is the ordered number pair associated with the point of intersection of the two graphs.
3. Check the resulting solution by verifying that the ordered pair satisfies both given equations.

KEEP IN MIND

The solution set of a system of equations is the intersection of the solution sets of the individual equations.

MODEL PROBLEM

Solve the system graphically and check: $2x + y = 6$
$$y - x = 3$$

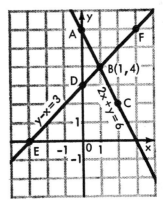

Solution:
1. Graph each equation of the system.

Graph $2x + y = 6$,	Graph $y - x = 3$,
or $y = -2x + 6$.	or $y = x + 3$.
(x, y)	(x, y)
$A(0, 6)$	$D(0, 3)$
$B(1, 4)$	$E(-3, 0)$
$C(2, 2)$	$F(3, 6)$

2. Find the common solution.
   The common solution is $(1, 4)$, the coordinates of B, the point of intersection.
3. *Check:* Substitute 1 for x and 4 for y in each given equation.

$2x + y = 6$	$y - x = 3$
$2(1) + 4 \stackrel{?}{=} 6$	$4 - 1 \stackrel{?}{=} 3$
$6 = 6$ (true)	$3 = 3$ (true)

Answer: The common solution is $(1, 4)$. The solution set is $\{(1, 4)\}$.

CONSISTENT, DEPENDENT, AND INCONSISTENT SYSTEMS OF EQUATIONS

When the graphs of two linear equations are drawn in a coordinate plane, they may be related to each other in only one of the following three ways:

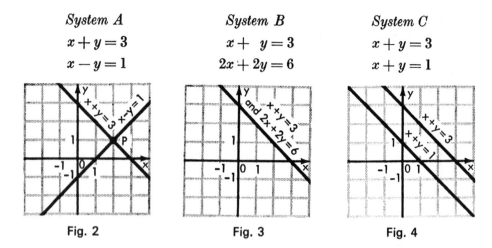

System A	System B	System C
$x + y = 3$	$x + y = 3$	$x + y = 3$
$x - y = 1$	$2x + 2y = 6$	$x + y = 1$

Fig. 2   Fig. 3   Fig. 4

In system A (Fig. 2), the graphs are lines that intersect in one and only one point. Since such lines have only one point of intersection, the solution set of such a system has one and only one ordered pair. A system of linear equations is a **consistent system** if one and only one ordered pair satisfies both of its equations. In a consistent system, if m_1 and m_2 are the slopes of the graphs of the equations, then $m_1 \neq m_2$.

In system B (Fig. 3), the graphs are lines that coincide. Since such lines have an infinite number of points in common, the solution set of such a system has an infinite number of ordered pairs. A system of linear equations is a **dependent system** if every ordered pair that satisfies one equation also satisfies the other equation. In a dependent system, if m_1 and m_2 are the slopes of the graphs of the equations, $m_1 = m_2$.

In system C (Fig. 4), the graphs are lines that are parallel. Since such lines have no points in common, the solution set of such a system has no ordered pairs; that is, it is the empty set $\varnothing$. A system of linear equations is an **inconsistent system** if no ordered pair satisfies both of its equations. In an inconsistent system, if m_1 and m_2 are the slopes of the graphs of the equations, $m_1 = m_2$.

PROPERTIES OF THE LINEAR SYSTEM
$$y = m_1 x + b_1 \text{ AND } y = m_2 x + b_2$$

Name of System	Relation Between Slopes	Relation Between Graphs	Number of Ordered Pairs of the Solution Set
Consistent	$m_1 \neq m_2$	intersecting	one
Dependent	$m_1 = m_2, b_1 = b_2$	coincident	infinite
Inconsistent	$m_1 = m_2, b_1 \neq b_2$	parallel	none

Exercises

In 1–12, solve the system of equations graphically. Check.

1. $x + y = 6$	**2.** $x + 2y = 15$	**3.** $x + 4y = 7$
$x - y = 2$	$y = 2x$	$x = 2y + 1$
4. $y - x = -2$	**5.** $5x + 2y = -5$	**6.** $3x - 4y = 6$
$x - 2y = 4$	$y - x = 1$	$y = 2x + 1$
7. $y = \frac{1}{3}x - 3$	**8.** $3x - y = 6$	**9.** $2x + 4y = 12$
$2x - y = 8$	$\frac{1}{3}y - x = 2$	$y = 3 - \frac{1}{2}x$
10. $x + y + 2 = 0$	**11.** $y + 2x + 6 = 0$	**12.** $5x + 3y + 9 = 0$
$x = y - 8$	$y = 2x$	$3x - 4y - 12 = 0$

In 13–20: (*a*) Determine whether the equations of the system are consistent, inconsistent, or dependent. (*b*) State whether the graphs of the equations of the system intersect, are parallel, or coincide. (*c*) State the number of ordered pairs that the solution set of the system contains.

13. $2x + y = 5$	**14.** $3x + y = 9$	**15.** $3x + y = 5$	**16.** $4x - 3y = 7$
$2x + y = 8$	$2x - y = 4$	$2y = 10 - 6x$	$8x - 6y = 7$
17. $x + y = 12$	**18.** $y = x$	**19.** $2x + 3y = 8$	**20.** $2x = 5y + 4$
$x - y = 2$	$3y - 3x = 4$	$3x + 2y = 10$	$4x - 10y = 8$

21. Write a system of two linear equations in two variables whose solution set contains (*a*) no members, (*b*) only one member, and (*c*) an infinite number of members.

3. Algebraic Solution of a System of Linear Equations in Two Variables Using Addition or Subtraction

Equivalent systems of equations are systems that have the same solution set. For example, the two systems shown below are equivalent systems because they have the same solution set, $\{(10, 3)\}$.

System S_1	*System S_2*
$x + 3y = 19$	$x = 10$
$x - 3y = 1$	$y = 3$

To solve a system of linear equations, such as S_1, whose solution set is not obvious, we make use of the properties of equality to transform the system into an equivalent system of equations, such as S_2, whose solution set is obvious.

～～～～～～ *MODEL PROBLEMS* ～～～～～～

1. Solve the system of equations and check: $x + 3y = 19$
$$x - 3y = 1$$

How To Proceed	*Solution*
1. The coefficients of the variable y have the same absolute value in both equations. Therefore, adding the members of equation (B) to the corresponding members of equation (A) will eliminate the variable y and will result in an equation which involves only the one variable x.	(A) $x + 3y = 19$ (B) $x - 3y = 1$ $\overline{2x = 20}$
2. Solve the resulting equation for the variable x.	$x = 10$
3. Replace x by its value 10 in any equation involving both variables, for example, equation (A).	(A) $x + 3y = 19$ $10 + 3y = 19$
4. Solve the resulting equation for the variable y.	$3y = 9$ $y = 3$

5. *Check:* Substitute 10 for x and 3 for y in each given equation.

$$x + 3y = 19 \qquad\qquad x - 3y = 1$$
$$10 + 3(3) \overset{?}{=} 19 \qquad\qquad 10 - 3(3) \overset{?}{=} 1$$
$$19 = 19 \text{ (true)} \qquad\qquad 1 = 1 \text{ (true)}$$

Answer: Since $x = 10$ and $y = 3$, the solution is (10, 3), or the solution set is $\{(10, 3)\}$.

2. Solve the system of equations and check: $3r = 4s + 17$
$$2r = -3s$$

How To Proceed	*Solution*
1. Transform each of the given equations (A) and (B) into equivalent equations (C) and (D) in which the terms containing the variables appear on one side and the constant appears on the other side.	(A) $3r = 4s + 17$ (B) $2r = -3s$ (C) $3r - 4s = 17$ (D) $2r + 3s = 0$
2. To eliminate r, multiply both members of equation (C) by 2 and multiply both members of equation (D) by 3 so that in the resulting equivalent equations (E) and (F) the absolute values of the coefficients of r are equal.	(E) $6r - 8s = 34$ (F) $6r + 9s = 0$ $\overline{}$
3. Subtract the members of equation (F) from the corresponding members of equation (E) to eliminate the variable r.	$-17s = 34$

<table>
<tr><td>4. Solve the resulting equation for the variable s.</td><td rowspan="6"></td><td>$s = -2$</td></tr>
</table>

4. Solve the resulting equation for the variable s.

$s = -2$

5. Replace s by its value -2 in any equation containing both variables, for example equation (A).

(A) $\quad 3r = 4s + 17$
$\quad\quad 3r = 4(-2) + 17$

6. Solve the resulting equation for the remaining variable, r.

$3r = 9$
$r = 3$

7. *Check:* Substitute -2 for s and 3 for r in both given equations to verify that the resulting sentences are true. This is left to the student.

Answer: Since $r = 3$ and $s = -2$, the solution is $(3, -2)$, or the solution set is $\{(3, -2)\}$.

Exercises

In 1–21, solve the system of equations by eliminating one of the variables, using the methods of addition or subtraction. Check.

1. $x + y = 9$
$x - y = 3$

2. $x + 2y = 8$
$x - 2y = 4$

3. $y + 3x = 8$
$y - 3x = 2$

4. $m + 2n = 14$
$3n + m = 18$

5. $2a + 3b = 12$
$5a + 3b = 14$

6. $3x + 2y = 9$
$x + y = 3$

7. $3a - b = 3$
$a + 3b = 11$

8. $2c - d = -1$
$c + 3d = 17$

9. $4x + 5y = -8$
$3y - 4x = -8$

10. $2x + y = -2$
$x + 3y = 9$

11. $3r + 7s = -4$
$2r + 5s = -3$

12. $4y - 6x = 15$
$6y - 4x = 10$

13. $6x + 10y = 7$
$15x - 4y = 3$

14. $m = 11 + n$
$3m = 3 - 2n$

15. $x = 2(8 + y)$
$3y = 2(1 + 2x)$

16. $5a = 4b$
$\frac{1}{2}a + 2b = 12$

17. $.04x + .05y = 44$
$x + y = 1000$

18. $.02x = .03y + 1$
$x + y = 800$

19. $\dfrac{3x + 8}{5} = \dfrac{3y - 1}{2}$

$\dfrac{x + y}{2} = 3 + \dfrac{x - y}{2}$

20. $\dfrac{a}{3} + \dfrac{a + b}{6} = 3$

$\dfrac{b}{3} - \dfrac{a - b}{2} = 6$

21. $\dfrac{x + y}{6} - \dfrac{x - y}{4} = 3$

$\dfrac{x + y}{4} + \dfrac{x - y}{2} = 8$

In 22–30, consider x and y as the variables of the system of equations. Solve the system by eliminating one of these variables, using the methods of addition or subtraction. Check. (*Note.* In 28–30, the equations of the system are not linear equations.)

22. $x + y = 7a$
$x - y = 3a$

23. $3x + 2y = 14b$
$4x - y = 15b$

24. $x + y = a$
$x - y = b$

25. $5x - 2y = c$
$3x - 2y = d$

26. $x + 2y = 4m$
$3x - y = 5m$

27. $2x + 3y = 13b$
$3x + 2y = 12b$

28. $\dfrac{1}{x} + \dfrac{1}{y} = 7$ **29.** $\dfrac{4}{x} + \dfrac{9}{y} = -1$ **30.** $\dfrac{1}{x} + \dfrac{1}{y} = 5a$

$\quad\; \dfrac{1}{x} - \dfrac{1}{y} = 1$ $\quad\; \dfrac{6}{x} - \dfrac{12}{y} = 7$ $\quad\; \dfrac{2}{x} + \dfrac{3}{y} = 13a$

In 31–36, consider x and y as the variables of the system of equations. Solve the system by eliminating one of these variables, using the methods of addition and subtraction. Check.

31. $\dfrac{x}{2} + \dfrac{y}{3} = \dfrac{a}{2}$ **32.** $\dfrac{x}{2} + y = b$ **33.** $x + \dfrac{y}{2} = \dfrac{c}{3}$

$\quad\; \dfrac{x}{4} - \dfrac{y}{3} = a$ $\quad\; x - \dfrac{y}{2} = \dfrac{b}{2}$ $\quad\; \dfrac{x}{3} - \dfrac{y}{4} = \dfrac{c}{4}$

34. $x + y = \dfrac{a}{3}$ **35.** $\dfrac{x}{6} + y = a + b$ **36.** $\dfrac{x}{2} + y = \dfrac{a}{2}$

$\quad\; x - y = \dfrac{b}{3}$ $\quad\; \dfrac{x}{3} - y = a - b$ $\quad\; -\dfrac{x}{2} + \dfrac{y}{2} = b$

4. Algebraic Solution of a System of Linear Equations in Two Variables Using Substitution

A second method, called the *substitution method*, can be used to solve a system of equations algebraically. This method depends upon transforming one of the equations of the system into an equivalent equation that contains only one variable.

∼∼∼∼∼∼∼∼ *MODEL PROBLEM* ∼∼∼∼∼∼∼∼

Solve the system by using the substitution method, and check: $2x - 5y = 3$
$\qquad\qquad\qquad\qquad\qquad\qquad\qquad\qquad\qquad\qquad 3y + \; x = 7$

How To Proceed	*Solution*
1. Transform one of the equations into an equivalent equation in which one of the variables can be expressed in terms of the other.	1. (A) $2x - 5y = 3$ (B) $3y + \; x = 7$ In equation (B), solve for x in terms of y. (C) $\qquad x = 7 - 3y$
2. In the other equation, substitute the resulting expression for the first variable, thus eliminating the first variable.	2. In equation (A), substitute $7 - 3y$ for x. $2(7 - 3y) - 5y = 3$

3. Solve the resulting equation for the second variable.

3. Solve for y.
$$14 - 6y - 5y = 3$$
$$-11y = -11$$
$$y = 1$$

4. Substitute the value of the variable obtained in step 3 in any equation involving both variables and solve the resulting equation for the remaining variable.

4. Substitute 1 for y in equation (C).
$$x = 7 - 3y$$
$$x = 7 - 3(1)$$
$$x = 7 - 3$$
$$x = 4$$

5. *Check:* Substitute 4 for x and 1 for y in each of the given equations to verify that the resulting sentences are true. This is left to the student.

Answer: Since $x = 4$ and $y = 1$, the solution is $(4, 1)$, or the solution set is $\{(4, 1)\}$.

Exercises

In 1–12, solve the system of equations by eliminating one of the variables, using the substitution method. Check.

1. $y = 3x$
$x + y = 8$

2. $x = 4y$
$3x + 8y = 5$

3. $x = -2y$
$4x + 2y = 6$

4. $y = 3x$
$x - y = 2$

5. $x = 2y + 1$
$x + y = -2$

6. $a - b = 3$
$2a - 3b = 1$

7. $y = 3x - 1$
$9x + 2y = 3$

8. $2x - y = 1$
$y - x = 2$

9. $5c + d + 2 = 0$
$c + 2d = 5$

10. $4x + y = 0$
$8x + \frac{1}{4}y = 7$

11. $2x = 3y$
$4x - 5y = 2$

12. $2x + 3y = 9$
$4x - 5y = 7$

In 13–16, consider x and y as the variables of the system of equations. Solve the system of equations, using the method of substitution. Check.

13. $x + y = m$
$x = y + n$

14. $6x - 4y = 2a$
$x = y - a$

15. $3x = y + a$
$5x = 3y + 7a$

16. $5x - 3y = -2d$
$2x + 5y = 24d$

In 17–20, consider x and y as the variables of the system of equations. Solve the system by eliminating one of these variables, using the method of substitution. Check.

17. $x + y = a$
$x = 2y$

18. $x = \dfrac{y}{2}$
$4x + y = b$

19. $x - y = a$
$x = b - y$

20. $x + y = c$
$x = d - 5y$

5. Solving Verbal Problems by Using Two Variables

Now we will learn how to solve verbal problems by translating the given relationships into a system of two equations involving two variables. Frequently, a problem can be solved more readily by using two variables than by using one variable.

Procedure. To solve verbal problems by using a system of two equations involving two variables:
1. **Use different variables to represent the different unknown quantities in the problem.**
2. **Translate the given relationships in the problem into a system of equations.**
3. **Solve the system of equations to determine the answer(s) to the problem.**
4. **Check the answer(s) in the original problem.**

NUMBER PROBLEMS

~~~~~~~~~~~~~~~~ *MODEL PROBLEM* ~~~~~~~~~~~~~~~~

The sum of two numbers is 56. Twice the smaller number exceeds one-half the larger number by 22. Find the numbers.

How To Proceed	*Solution*
1. Represent the two different unknown quantities.	1. Let $x =$ smaller number. Let $y =$ larger number.
2. Translate the given relationships in the problem into a system of equations.	2. *The sum of the two numbers is 56.* (A) $x + y = 56$ *Twice the smaller number is 22 more than one-half of the larger number.* (B) $2x = \frac{1}{2}y + 22$
3. Solve the system of equations: In (B), M_2. Then transpose. Copy equation (A). Eliminate the y-term by addition.	3. $4x = y + 44$ $4x - y = 44$ $x + y = 56$ $\overline{5x = 100}$ $x = 20$
Substitute 20 for x in equation (A).	$x + y = 56$ $20 + y = 56$ $y = 36$

Answer: The smaller number is 20; the larger number is 36.

~~~~~~

Exercises

In 1–5, solve the problem by using a system of two equations involving two variables.

1. The sum of two numbers is 50. If twice the larger is subtracted from 4 times the smaller, the result is 8. Find the numbers.

2. If 3 times the smaller of two numbers is increased by the larger, the result is 63. If 5 times the smaller is subtracted from twice the larger, the result is 16. Find the numbers.

3. The sum of two numbers is 90. If 20 is added to 3 times the smaller number, the result exceeds twice the larger number by 50. Find the numbers.

4. Twice the smaller of two numbers is one-half of the larger number. The larger is 10 more than 3 times the smaller. Find the numbers.

5. The sum of $\frac{1}{3}$ of the smaller of two numbers and $\frac{2}{5}$ of the larger is 34. Also, $\frac{1}{2}$ of the smaller number is equal to $\frac{1}{4}$ of the larger. Find the numbers.

COIN PROBLEMS

~~~~~~~~~~~~ *MODEL PROBLEM* ~~~~~~~~~~~~

A collection of coins consisting of nickels and quarters has a value of $4.50. The number of quarters is 4 less than twice the number of nickels. Find the number of coins of each kind in the collection.

Let $n =$ the number of nickels.
Let $q =$ the number of quarters.

	(of coins) Number	(¢) × Value of each =	(¢) Total value
Nickel	n	5	$5n$
Quarter	q	25	$25q$

The total value of the coins is 450 cents.

$$5n + 25q = 450$$

The number of quarters is 4 less than twice the number of nickels.

$$q = 2n - 4$$

(A) $5n + 25q = 450$

(B) $\qquad q = 2n - 4$

(C) $-2n + q = -4$

In (C), M$_{25}$. Then subtract (A).

$$-50n + 25q = -100$$
$$5n + 25q = \quad 450$$

$$\overline{\qquad -55n \qquad = -550}$$
$$n = 10$$

(B) $\qquad q = 2n - 4$
$$q = 2(10) - 4$$
$$q = 16$$

Answer: There are 10 nickels and 16 quarters.

Check in the given problem:

Is the total value $4.50?
Value of 16 quarters = $4.00
Value of 10 nickels   =   .50

Total value   = $4.50 (true)

Is 16, the number of quarters, 4 less than twice the number of nickels?
$16 \overset{?}{=} 2(10) - 4$
$16 = 16$ (true)

Exercises

In 1–5, solve the problem by using a system of two equations involving two variables.

1. Harry has a collection of coins consisting of dimes and quarters whose value is $17.60. The number of quarters exceeds twice the number of dimes by 8. Find the number of coins of each kind in the collection.
2. Mrs. Carey cashed a $600 check in her bank. She received $5 bills and $10 bills. The number of $10 bills was 10 less than 3 times the number of $5 bills. How many bills of each type did Mrs. Carey receive?
3. Ray has $7.60 in quarters and dimes. In all, he has 40 coins. How many coins of each kind does he have?
4. A class contributed $11.40 in dimes and quarters to a welfare fund. In all, there were 60 coins. How many coins of each kind were contributed?
5. In a collection box, there are dimes and quarters whose total value is $28. If there were as many quarters as there are dimes, and as many dimes as there are quarters, the total value would be $36.40. How many coins of each kind are in the collection box?

MIXTURE PROBLEMS

~~~~~~~~~~~~~~~~~ *MODEL PROBLEM* ~~~~~~~~~~~~~~~~~

A dealer has some candy worth $4.00 per pound and some worth $6.00 per pound. How many pounds of each kind should he use to make a mixture of 90 pounds that he can sell for $4.80 per pound?

Solution:
Let x = the number of pounds of $6.00 candy to be used.
Let y = the number of pounds of $4.00 candy to be used.

	(lb.) Number	×	($) Price per pound	=	($) Total value
$6.00 candy	x		6.00		$6x$
$4.00 candy	y		4.00		$4y$
Mixture	90		4.80		90(4.80)

The total number of pounds of candy is 90.

$$x + y = 90$$

The total value of the $6.00 candy and the $4.00 candy is 90($4.80).

$$6x + 4y = 90(4.80)$$

(A) $6x + 4y = 90(4.80)$
(B) $x + y = 90$
In (B), M_4: $4x + 4y = 360$
Subtract (A): $\underline{6x + 4y = 432}$
$-2x = -72$
$ x = 36$

(B) $x + y = 90$
$36 + y = 90$
$ y = 54$

Check in the given problem:

Is the total number of pounds 90?
$36 + 54 = 90$ (true)

Is 90 × $4.80 or $432 the total value?
Value of 36 lb. at $6 per lb. = $216
Value of 54 lb. at $4 per lb. = $\underline{$216}$
Total = $432

Total value of 90 lb. at $4.80 per lb. = $432
The values are the same.

Answer: He should use 36 pounds of the $6.00 and 54 pounds of the $4.00 candy.

Algebra II segment

Exercises

In 1–6, solve the problem by using two variables.

1. A dealer wishes to obtain 80 pounds of mixed cookies to sell for $5.00 per pound. If she mixes cookies worth $6.00 per pound with cookies worth $3.50 per pound, find the number of pounds of each kind she should use.
2. A dealer mixed coffee worth $5.10 per pound with coffee worth $3.30 per pound. How many pounds of each kind did he use to make a mixture of 120 pounds to sell at $4.50 per pound?
3. How many ounces of seed worth $1.05 per ounce must be mixed with 60 ounces of seed worth $.90 per ounce in order to produce a mixture to sell for $1.00 per ounce?
4. One bar of tin alloy is 25% pure tin and another bar is 10% pure tin. How many pounds of each alloy must be used to make 75 pounds of a new alloy which is 20% pure tin?
5. A chemist has a solution which is 18% pure salt and a second solution which is 45% pure salt. How many ounces of each solution should he use to make 24 ounces of a solution which is 36% pure salt?
6. A dairy has milk which contains 4% butterfat and cream which contains 40% butterfat. How many gallons of each should be used in order to produce 36 gallons of a mixture which contains 20% butterfat?

INVESTMENT PROBLEMS

MODEL PROBLEM

Mr. Curran invested a sum of money in 4% bonds and $3000 more than this amount in 6% bonds. His annual income from the 6% bonds exceeded the annual income from the 4% bonds by $240. Find the amount he invested in each type of bond.

Solution:

Let x = the number of dollars invested in the 4% bonds.
Let y = the number of dollars invested in the 6% bonds.

	($) Principal	×	Annual rate of interest	=	($) Annual income
4% bonds	x		.04		.04x
6% bonds	y		.06		.06y

The 6% investment is $3000 more than the 4% investment.

$$y = x + 3000$$

*The annual income from the 6% bonds is $240 more than
the annual income from the 4% bonds.*

$$.06y = .04x + 240$$

(A) $y = x + 3000$
(B) $.06y = .04x + 240$

In (B), M_{100}: $6y = 4x + 24,000$
Substitute $x + 3000$ for y.
$$6(x + 3000) = 4x + 24,000$$
$$6x + 18,000 = 4x + 24,000$$
$$2x = 6000$$
$$x = 3000$$

(A) $y = x + 3000$
$$y = 3000 + 3000$$
$$y = 6000$$

Check in the given problem:

Is the amount invested in 6% bonds
$3000 more than the amount in-
vested in 4% bonds?
$6000 = 3000 + 3000$ (true)

Does the annual income from the 6%
bonds exceed the annual income
from the 4% bonds by $240?
4% of $3000 = $120.00
6% of $6000 = $360.00
$360 exceeds $120 by $240 (true)

Answer: The amount invested in 4% bonds was $3000; the amount invested in
6% bonds was $6000.

Exercises

In 1–5, solve the problems by using two variables.

1. Mrs. Brand invested $7000, part at 8% and the rest at 5%. Her total annual income from these investments was $500. Find the amount she invested at each rate.
2. Mr. Trask invested a certain sum of money in bonds yielding $4\frac{1}{2}\%$ a year and twice as much in bonds yielding 7% a year. If his total annual income from these investments was $1850, how much did he invest in each type of bond?
3. Mr. Orsini invested $36,000, part at 5% and the rest at 7%. If his annual incomes from both investments were equal, find the amount he invested at each rate.
4. Mr. Dunn invested $35,000, part at 8% and the rest at 5%. His annual income from the 8% investment was $450 less than his annual income from the 5% investment. Find the amount he invested at each rate.
5. Ms. Walsh invested $16,000 at 7%. How much additional money must she invest at 4% so that her total annual income will be 5% of her entire investment?

BUSINESS PROBLEMS

~~~~~~~~~~~~~~ *MODEL PROBLEM* ~~~~~~~~~~~~~~

The owner of a men's clothing store bought 12 shirts and 6 hats for $225. A week later, at the same prices, he bought 9 shirts and 4 hats for $159. Find the price of a shirt and the price of a hat.

Solution:

Let s = the cost of a shirt in dollars. Let h = the cost of a hat in dollars.

$$12 \text{ shirts and 6 hats cost } \$225. \quad \text{(A)} \quad 12s + 6h = 225$$

$$9 \text{ shirts and 4 hats cost } \$159. \quad \text{(B)} \quad 9s + 4h = 159$$

1. In order to eliminate h, in (A), M_2; in (B), M_3.

$$24s + 12h = 450$$
$$27s + 12h = 477$$

2. Subtract.

$$-3s \quad\quad = -27$$
$$s = 9$$

3. In (A), substitute 9 for s.

$$\text{(A)} \quad 12s + 6h = 225$$
$$108 + 6h = 225$$
$$6h = 117$$
$$h = 19\tfrac{1}{2}$$

Answer: A shirt costs $9; a hat costs $19.50.

~~~~~~

Exercises

In 1–5, solve the problem by using two variables.

1. Mrs. Bond bought 3 cans of corn and 5 cans of tomatoes for $5.46. The following week, she bought 2 cans of corn and 3 cans of tomatoes for $3.33, paying the same prices. Find the cost of a can of corn and the cost of a can of tomatoes.

2. A baseball manager bought 4 bats and 9 balls for $135. On another day, he bought 3 bats and 1 dozen balls at the same prices and paid $138. How much did he pay for each bat and for each ball?

3. Eight roses and 9 carnations cost $10.05. At the same prices, one dozen roses and 5 carnations cost $11.25. Find the cost of a rose and the cost of a carnation.

4. One day 4 plumbers and 5 helpers earned $1750. At the same rate of pay, another group of 5 plumbers and 6 helpers earned $2150. How much does a plumber and how much does a helper earn each day?

5. The cost of a 17-minute phone call from one state to another is $1.25. The cost of a 24-minute call between the same two states is $1.60. The charge in each case is based upon a fixed charge for the first 10 minutes and an extra charge for each additional minute beyond 10. Find the charge for the first 10 minutes and the charge for each extra minute beyond 10.

PROBLEMS INVOLVING FRACTIONS

~~~~~~~~~~~~~~ *MODEL PROBLEM* ~~~~~~~~~~~~~~

If 1 is added to the numerator of a fraction and 3 is added to the denominator of the fraction, the value of the resulting fraction is $\frac{3}{5}$. If the numerator of the original fraction is decreased by 3, and the denominator of the original fraction is doubled, the value of the resulting fraction is $\frac{1}{7}$. Find the original fraction.

Solution:
Let $n =$ the numerator of the fraction.
Let $d =$ the denominator of the fraction.

(A) $\dfrac{n+1}{d+3} = \dfrac{3}{5}$

(B) $\dfrac{n-3}{2d} = \dfrac{1}{7}$

In (A), $M_{5(d+3)}$: $5(n+1) = 3(d+3)$
$\qquad\qquad\qquad 5n + 5 = 3d + 9$

(C) $\qquad 5n - 3d = 4$

In (B), M_{14d}: $7(n-3) = 2d$

(D) $\qquad 7n - 21 = 2d$

(E) $\qquad 7n - 2d = 21$

In (C), M_2: $10n - 6d = 8$
In (E), M_3: $\underline{21n - 6d = 63}$
$\qquad\qquad -11n \qquad\;\; = -55$
$\qquad\qquad\qquad\quad n = 5$

(D) $\qquad 7n - 21 = 2d$
Let $n = 5$: $7(5) - 21 = 2d$
$\qquad\qquad\qquad 14 = 2d$
$\qquad\qquad\qquad\; 7 = d$

Check in the given problem:
If 1 is added to the numerator of the original fraction and 3 is added to the denominator of the original fraction, is the value of the resulting fraction $\frac{3}{5}$?

$\dfrac{5+1}{7+3} = \dfrac{6}{10} = \dfrac{3}{5}$ (true)

If the numerator of the original fraction is decreased by 3 and the denominator of the original fraction is doubled, is the value of the resulting fraction $\frac{1}{7}$?

$\dfrac{5-3}{2(7)} = \dfrac{2}{14} = \dfrac{1}{7}$ (true)

Answer: The original fraction is $\frac{5}{7}$.

Exercises

In 1–5, solve the problem by using two variables.

1. If 1 is added to both the numerator and the denominator of a certain fraction, the value of the fraction becomes $\frac{1}{3}$. If 1 is added to the denominator of the original fraction, the value of the fraction becomes $\frac{1}{4}$. Find the original fraction.

2. The sum of the numerator and the denominator of a fraction is 20. If the numerator is increased by 3 and the denominator is decreased by 2, the value of the resulting fraction is $\frac{3}{4}$. Find the original fraction.

3. The denominator of a fraction exceeds its numerator by 6. If 24 is added to the numerator of the fraction and the denominator of the fraction is doubled, the value of the resulting fraction is 1. Find the original fraction.

4. If 3 is added to the numerator of a fraction and 2 is subtracted from the denominator of the fraction, the value of the resulting fraction is $\frac{3}{4}$. If 1 is added to the numerator of the reciprocal of the given fraction, the value of the resulting fraction is $\frac{5}{2}$. Find the original fraction.

5. The numerator and denominator of a fraction are in the ratio 3 : 2. If 3 is subtracted from the denominator of the reciprocal of the fraction, the ratio of the numerator of the resulting fraction to its denominator is 8 : 11. Find the original fraction.

DIGIT PROBLEMS

Preparing to Solve Digit Problems

In our decimal number system, every integer may be written by using only the ten symbols 0, 1, 2, 3, 4, 5, 6, 7, 8, 9, which are called **digits**.

When we write an integer, each place is given a value ten times the value of the place immediately at its right. For example, $685 = 6(100) + 8(10) + 5(1)$. Likewise, $79 = 7(10) + 9(1)$. In the number 79, we call 7 the *tens digit* and 9 the *units digit*. The value of the tens digit 7 is $7(10)$, or 70; the value of the units digit 9 is $9(1)$, or 9.

If we wish to represent a two-digit number whose tens digit is represented by t and whose units digit is represented by u, we write $t(10) + u(1)$, or $10t + u$. Notice that we may *not* represent the two-digit number by tu because "tu" means "t times u."

If we reverse the digits of the number 79, we obtain a new number, 97. Observe that $97 = 9(10) + 7(1)$. That is, the new number 97 can be represented by adding 10 times 9, the units digit of the original number 79, and 1 times 7, the tens digit of the original number 79. Likewise, if t represents the tens digit of a two-digit number, and u represents the units digit of the number, when the digits are reversed, the new number that is formed is represented by $10u + t$.

── KEEP IN MIND ──

If t represents the tens digit and u represents the units digit of a two-digit number:

$10t + u$ represents the original number.

$10u + t$ represents the original number with its digits reversed.

$t + u$ represents the sum of the digits of the original number.

Exercises

In 1–5, give the value of each digit in the number.

1. 47   **2.** 372   **3.** 5604   **4.** 706   **5.** 34027

In 6–8, represent the number which is described.

6. The number whose tens digit is 5 and whose units digit is 2.

7. The number whose units digit is 9 and whose tens digit is 5.

8. The number obtained by reversing the digits of 47.

Solving Digit Problems

~~~~~~~~~~ *MODEL PROBLEM* ~~~~~~~~~~

The sum of the digits of a two-digit number is 12. If the digits are reversed, the resulting number exceeds twice the original number by 15. Find the original number.

Solution:

Let $t =$ the tens digit of the number.

And $u =$ the units digit of the number.

Then $10t + u =$ the original number.

And $10u + t =$ the original number with the digits reversed.

The sum of the digits is 12.

(A) $t + u = 12$

The number with the digits reversed is 15 more than twice the original number.

(B) $10u + t = 2(10t + u) + 15$

Simplify (B): $10u + t = 20t + 2u + 15$

$$-19t + 8u = 15$$

In (A), M_8: $8t + 8u = 96$

$$-27t \qquad = -81$$
$$t = 3$$

(A) $t + u = 12$

In (A), let $t = 3$: $3 + u = 12$

$$u = 9$$
$$10t + u = 10(3) + 9 = 39$$

Check in the given problem:
Is the sum of the digits 12?
$3 + 9 = 12$ (true)

Does the number with the digits reversed, 93, exceed twice the original number, 39, by 15?
93 exceeds 2×39, or 78, by 15. (true)

Answer: The number is 39.

Exercises

1. The units digit of a two-digit number is 1 more than twice the tens digit. The sum of the digits is 10. Find the number.
2. The tens digit of a two-digit number is 1 less than 5 times the units digit. The sum of the digits is 11. Find the number.
3. The units digit of a two-digit number is 3 more than its tens digit. The sum of the digits is $\frac{1}{4}$ of the number. Find the number.
4. The units digit of a two-digit number is one more than twice the tens digit. If the digits are reversed, the resulting number is 27 more than the original number. Find the number.
5. The units digit of a two-digit number is 1 less than 5 times the tens digit. If the digits are reversed, the new number exceeds 3 times the original number by 5. Find the number.
6. The sum of the digits of a two-digit number is 12. If the digits are reversed, a new number is formed which is 12 less than twice the original number. Find the number.
7. The tens digit of a two-digit number exceeds the units digit by 2. If the digits are reversed, the resulting number is 4 times the sum of the digits. Find the original number.
8. The sum of the digits of a two-digit number is 13. If the digits are reversed, the new number is 27 less than the original number. Find the number.
9. The units digit of a two-digit number is one less than twice the tens digit. If the digits are reversed, the new number exceeds the original number by 27. Find the original number.
10. The sum of the digits of a two-digit number is 10. If 18 is added to the number, the sum is the number obtained by reversing the digits of the original number. Find the number.

11. The sum of the digits of a two-digit number is 9. The number is 7 times the sum of the digits. Find the number.
12. A two-digit number is 7 times the sum of the digits. The original number exceeds by 18 the number obtained by reversing the digits. Find the number.
13. The units digit of a two-digit number is 5 less than the tens digit. If the digits are reversed, a new number is formed which is $\frac{3}{8}$ of the original number. What is the original number?
14. The sum of the digits of a two-digit number is 12. If the digits are reversed, the new number exceeds $\frac{1}{3}$ of the original number by 20. Find the number.
15. A two-digit number is 9 more than the number obtained by reversing the digits. The number is also 6 more than 5 times the sum of the digits. Find the number.
16. The sum of the digits of a two-digit number is 6. If the number is divided by the sum of the digits, the quotient is 7. Find the number.
17. If a two-digit number is divided by the sum of the digits, the quotient is 4 and the remainder is 15. If the digits are reversed, the resulting number exceeds the original number by 18. Find the original number.
18. When a certain two-digit number is divided by the sum of its digits, the quotient is 4. If the digits are reversed, the resulting number is 18 more than the original number. Find the original number.
19. When a two-digit number is divided by the sum of the digits, the quotient is 5. The original number subtracted from the number obtained by reversing the digits gives a result of 9. Find the original number.
20. A two-digit number is equal to 3 times the sum of its digits. If the number obtained by reversing the digits is divided by the original number, the quotient is 2 and the remainder is 18. Find the original number.
21. In a two-digit number, the ratio of the tens digit to the units digit is 2 : 3. If the digits are reversed, the resulting number exceeds the original number by 18. Find the original number.

MOTION PROBLEMS INVOLVING WATER CURRENTS AND AIR CURRENTS

〰〰〰〰〰〰 *MODEL PROBLEM* 〰〰〰〰〰〰

A motorboat can travel 24 miles downstream in 2 hours. It requires 3 hours to make the return trip. Find the rate of the boat in still water and the rate of the stream.

Solution:
 Let r = rate of the boat in still water in mph.
 And c = rate of the stream in mph.

	(mph) Rate	(hr.) × Time	(mi.) = Distance
Downstream	$r + c$	2	$2(r + c)$
Upstream	$r - c$	3	$3(r - c)$

The distance downstream is 24 miles.

(A) $2(r + c) = 24$

The distance upstream is 24 miles.

(B) $3(r - c) = 24$

In (A), D_2: $r + c = 12$
In (B), D_3: $r - c = 8$
$$\overline{\quad 2r \quad\; = 20}$$
$$r = 10$$

$$r + c = 12$$
Let $r = 10$: $10 + c = 12$
$$c = 2$$

Check in the given problem:
Can the boat travel 24 miles downstream in 2 hours?
Rate downstream $= 10 + 2 = 12$ mph.
Distance downstream $= 2(12) = 24$ miles. (true)

Can the boat make the return trip in 3 hours?
Rate upstream $= 10 - 2 = 8$ mph.
Distance upstream $= 3(8) = 24$ miles. (true)

Answer: The rate of the boat in still water is 10 mph; the rate of the stream is 2 mph.

~~~~~~

Exercises

In 1-5, solve the problem by using two variables.
1. If a boat is rowed down a river a distance of 16 miles in 2 hours and it is rowed upstream the same distance in 8 hours, find the rate of rowing in still water and the rate of the stream.
2. A man rows 18 miles downstream in 2 hours. He finds that it takes 6 hours to row back. Find the rate of rowing in still water and the rate of the stream.
3. It takes a motorboat 50 minutes to travel upstream a distance of 10 miles, and 30 minutes to travel the same distance downstream. Find the rate of the boat in still water and the rate of the current.
4. A plane left an airport and flew with the wind for 3 hours, covering 1200 miles. It then returned over the same route to the airport against the wind in 4 hours. Find the rate of the plane in still air and the speed of the wind.

5. A motorboat which is driven at full speed (at its maximum speed in still water) upstream moves at the rate of 14 mph. When the boat is driven at half speed downstream, it moves at 10 mph. Find the boat's maximum speed in still water and the rate of the current.

6. Algebraic Solution of a System of Three First-Degree Equations in Three Variables

An equation such as $2x + 3y + 2z = 2$ is an example of a first-degree equation in three variables. Since the values $x = 2, y = -1, z = \frac{1}{2}$ satisfy this equation, the ordered triple $(2, -1, \frac{1}{2})$ is a solution of the equation. Note that $2(2) + 3(-1) + 2(\frac{1}{2}) = 2$ is a true sentence.

In space, the graph of the equation $2x + 3y + 2z = 2$ is a plane. In this course, we will not study the graphs of equations in space. However, we will learn how to solve a system of three first-degree equations in three variables algebraically. We will do this by making use of the properties of equality to transform the given system into an equivalent system whose solution is quite obvious.

〜〜〜〜〜〜〜〜〜 *MODEL PROBLEM* 〜〜〜〜〜〜〜〜〜

Solve the system of equations and check.

$$3x + y + 2z = 6$$
$$x + y + 4z = 3$$
$$2x + 3y + 2z = 2$$

How To Proceed

1. Eliminate one variable using any combination of the given equations. Thus, to eliminate z from equations (A) and (B), in equation (A) multiply by 2; then from the result subtract the members of equation (B).

2. Eliminate the same variable using a different combination of the given equations. Thus to eliminate z from equations (A) and (C), subtract the members of (C) from the members of (A).

Solution

(A) $3x + y + 2z = 6$
(B) $x + y + 4z = 3$
(C) $2x + 3y + 2z = 2$

In (A), M_2: $6x + 2y + 4z = 12$
Subtract (B): $\underline{x + y + 4z = 3}$
(D) $5x + y = 9$

(A) $3x + y + 2z = 6$
Subtract (C): $\underline{2x + 3y + 2z = 2}$
(E) $x - 2y = 4$

3. Solve the resulting equations (D) and (E) in two variables, x and y, to find the values of those variables.

In (D), M_2: $\quad 10x + 2y = 18$

(E) $\qquad x - 2y = 4$

$$\overline{ 11x = 22}$$

$$x = 2$$

In (D), let $x = 2$.

$$5(2) + y = 9$$
$$10 + y = 9$$
$$y = -1$$

4. Substitute the values of the variables found in step 3 in any convenient equation involving the third variable. Thus, in equation (A), replace x by 2 and y by -1 to find the value of z.

(A) $\qquad 3x + y + 2z = 6$

$$3(2) - 1 + 2z = 6$$
$$2z = 1$$
$$z = \tfrac{1}{2}$$

Check: Substitute 2 for x, -1 for y, and $\tfrac{1}{2}$ for z in all the given equations to verify that the resulting sentences are true. This is left to the student.

Answer: Since $x = 2$, $y = -1$, and $z = \tfrac{1}{2}$, the solution is $(2, -1, \tfrac{1}{2})$, or the solution set is $\{(2, -1, \tfrac{1}{2})\}$.

Note. A variable may be eliminated by using a combination of all three equations. Thus, z may be eliminated by using the combination (A) + (C) − (B), and x may be eliminated by using the combination (B) + (C) − (A).

∿∿∿

Exercises

In 1–15, solve the system of equations and check. (*Note.* In 13–15, the equations of the system are not first-degree equations.)

1. $x + 3y + 2z = 13$
$ x - 2y + 3z = 6$
$ 2x + 2y - z = 3$

2. $a + 3b + 2c = 6$
$ 3a - 6b - 2c = 9$
$ 2a - 3b - 4c = 9$

3. $r + 3s + t = 3$
$ r + 6s - t = 2$
$ 4r + 9s - 2t = 1$

4. $3x - 2y - 3z = -1$
$ 6x + y + 2z = 7$
$ 9x + 3y + 4z = 9$

5. $x - 2y + 20z = 1$
$ 3x + y - 4z = 2$
$ 2x + y - 8z = 3$

6. $x + 2y - z = 5$
$ 2x + z = -1$
$ 3x - 4y = 2z + 7$

7. $2x + y + 3z = -2$
$ 5x = 5 - 2y$
$ 2y + 3z = -13$

8. $3x + 2y = 5$
$ 4x = 3z + 7$
$ 6y - 6z = -5$

9. $3a + 2c = 2$
$ 4b + c = 6$
$ 2a + 3b = 10$

10. $2x + 3y = 2$
$ 8x - 4z = 3$
$ 3y - 8z = -1$

11. $2x + y = 43$
$ x + 3z = 47$
$ y - z = 14$

12. $4x + y = 5b$
$ 3x + 2z = 0$
$ 2y - 3z = 3b$

13. $\dfrac{2}{x} + \dfrac{1}{y} - \dfrac{1}{z} = 4$

$\dfrac{3}{x} - \dfrac{1}{y} + \dfrac{2}{z} = 15$

$\dfrac{1}{x} - \dfrac{3}{y} + \dfrac{4}{z} = 13$

14. $\dfrac{1}{x} - \dfrac{2}{y} - \dfrac{2}{z} = 1$

$\dfrac{3}{x} + \dfrac{4}{y} + \dfrac{6}{z} = -9$

$\dfrac{1}{x} + \dfrac{2}{y} + \dfrac{2}{z} = -5$

15. $\dfrac{2}{x} + \dfrac{3}{y} = 2$

$\dfrac{1}{x} - \dfrac{1}{z} = \dfrac{3}{10}$

$\dfrac{12}{y} - \dfrac{5}{z} = 3$

In 16–18, solve the problem using three variables.

16. The perimeter of a triangle is 24 in. The sum of the lengths of the first two sides of the triangle exceeds the length of its third side by 4 in. Twice the length of the first side increased by the length of the second side is equal to twice the length of the third side. Find the lengths of the sides of the triangle.

17. Harry has a bank in which there are nickels, dimes, and quarters. In all, there are 34 coins, whose value is $3.00. The number of nickels is 6 more than the number of dimes and the number of quarters put together. Find the number of coins of each kind that are in the bank.

18. The sum of the digits of a three-digit number is 14. The units digit is equal to the sum of the hundreds digit and the tens digit. The number with the digits reversed exceeds the original number by 297. Find the original number.

7. Graphing the Solution Set of a System of Linear Inequalities in Two Variables

The solution set of a system of linear inequalities consists of all ordered pairs which are common solutions of all the inequalities of the system. Therefore, the solution set of the system is the intersection of the solution sets of all the inequalities of the system.

Procedure. To graph the solution set of a system of inequalities:
1. **For each inequality, graph the related plane divider and shade the region which is the graph of the solution set of the inequality.**
2. **Determine the graph of the solution set of the given system by finding the region which is common to (is the intersection of) the graphs of the inequalities that were made in step 1.**

~~~~~~~~~~~~~ *MODEL PROBLEMS* ~~~~~~~~~~~~~

1. Graph the solution set of the system in the coordinate plane:

$$x + y \geq 4$$
$$y \leq 2x - 5$$

Solution:

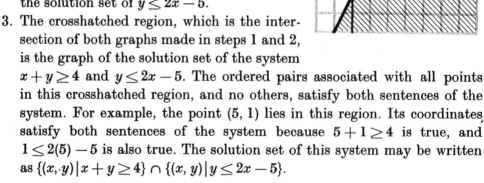

1. Transform the inequality $x + y \geq 4$ into the equivalent inequality $y \geq -x + 4$. Graph the inequality $y \geq -x + 4$ by first graphing the plane divider $y = -x + 4$. (In the figure, see the solid line $\overleftrightarrow{RS}$.) The line $y = -x + 4$ and the half-plane above this line are the graph of the solution set of $x + y \geq 4$.

2. Using the same set of axes, graph $y \leq 2x - 5$ by first graphing the plane divider $y = 2x - 5$. (In the figure, see the solid line $\overleftrightarrow{LM}$.) The line $y = 2x - 5$ and the half-plane below this line together form the graph of the solution set of $y \leq 2x - 5$.

3. The crosshatched region, which is the intersection of both graphs made in steps 1 and 2, is the graph of the solution set of the system
$x + y \geq 4$ and $y \leq 2x - 5$. The ordered pairs associated with all points in this crosshatched region, and no others, satisfy both sentences of the system. For example, the point (5, 1) lies in this region. Its coordinates satisfy both sentences of the system because $5 + 1 \geq 4$ is true, and $1 \leq 2(5) - 5$ is also true. The solution set of this system may be written as $\{(x, y) \mid x + y \geq 4\} \cap \{(x, y) \mid y \leq 2x - 5\}$.

2. Graph the solution set of the system in the coordinate plane:

$$3 \leq x \leq 5$$
$$|y| \geq 3$$

Solution:

1. The compound sentence $3 \leq x \leq 5$ is equivalent to $3 \leq x$ and $x \leq 5$, or $x \geq 3$ and $x \leq 5$ since $3 \leq x$ may be written as $x \geq 3$. Draw the graph of $x = 3$, which is $\overleftrightarrow{AC}$, and the graph of $x = 5$, which is $\overleftrightarrow{BD}$. The graph of $3 \leq x \leq 5$ is the rectangular region which is between $\overleftrightarrow{AC}$ and $\overleftrightarrow{BD}$ and includes these lines, the region hatched with horizontal lines.

2. The inequality $|y| \geq 3$ is equivalent to $y \geq 3$ or $y \leq -3$. Draw the graph of $y \geq 3$ and draw the graph of $y \leq -3$. The graph of $|y| \geq 3$ is those regions above the line $y = 3$, $\overleftrightarrow{AB}$, and below the line $y = -3$, $\overleftrightarrow{CD}$, which are hatched with vertical lines.

3. The graph of the solution set of the given system consists of all points in the open region that is crosshatched together with the points on the boundary lines of these regions. An example of such a point is $(4, 6)$ because $3 \leq 4 \leq 5$ is a true sentence, and $|6| \geq 3$ is also a true sentence. The solution set may be written as follows:

$$\{(x, y) \,|\, 3 \leq x \leq 5\} \cap \{(x, y) : |y| \geq 3\}$$

Exercises

In 1–18, graph the solution set of the system in a coordinate plane. Check one representative point in the given system.

1. $y \geq 2x$
   $y \geq x + 4$

2. $y \geq 3x - 1$
   $y < 3 - x$

3. $x + y > 4$
   $x - y < 6$

4. $2x + 3y < 6$
   $y \geq 2$

5. $x - y = 0$
   $x \geq 3$

6. $2x - y - 2 > 0$
   $x + y - 2 \leq 0$

7. $1 < x < 4$

8. $-1 \leq y \leq 3$

9. $-2 < x \leq 3$

10. $x < y < x + 3$

11. $x - 3 \leq y \leq x + 2$

12. $3 < y - x < 6$

13. $x + 2y \geq 4$
    $|x| \geq 2$

14. $|x| \geq 1$
    $|y| \leq 3$

15. $-2 < x < 3$
    $y > 2x$

16. $-3 < x < 2$
    $|y| \geq 2$

17. $x \leq 2$
    $y \geq 2$
    $x + y \geq 2$

18. $y \geq x$
    $x + y - 3 \geq 0$
    $x - 4 \leq 0$

‿‿‿‿‿‿‿ *CALCULATOR APPLICATIONS* ‿‿‿‿‿‿‿

A graphing calculator allows several equations to be graphed on the same screen. The solution to a system of linear equations is found at the point where the lines intersect. Different methods can be used to determine the coordinates (x, y) of the solution.

‿‿‿‿‿‿‿‿‿ *MODEL PROBLEMS* ‿‿‿‿‿‿‿‿‿

1. Solve the system of equations.

$$2x + y = 11 \qquad x + 3y = 18$$

Solution: First solve each equation for y.

$$y = -2x + 11 \qquad y = -\frac{1}{3}x + 6$$

Graph both equations in the standard ⌗WINDOW⌗.

Enter: ⌗Y=⌗ ⌗(−)⌗ 2 ⌗X,T,θ,*n*⌗ ⌗+⌗ 11 ⌗ENTER⌗

⌗(⌗ ⌗(−)⌗ 1 ⌗÷⌗ 3 ⌗)⌗ ⌗X,T,θ,*n*⌗ ⌗+⌗ 6 ⌗GRAPH⌗

Display:

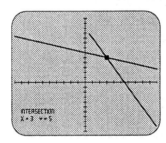

Note. Since an intersection point is contained in the portion of the graph shown, no ⌗WINDOW⌗ adjustment is needed.

Method 1 To determine the coordinates of the intersection point, use the INTERSECT feature.

Enter: ⌗2nd⌗ ⌗CALC⌗ 5 ⌗ENTER⌗

Use the arrow keys to move the cursor as close as possible to the intersection point. Then press ⌗ENTER⌗ ⌗ENTER⌗. The solution is $x = 3, y = 5$.

Method 2 The solution can also be found using the ⌗TABLE⌗ feature. Press ⌗2nd⌗ ⌗TABLE⌗. The table display shows coordinates of both lines. Examine the coordinates, using the arrow keys to scroll up or down as necessary. The row where $y_1 = y_2$ gives the solution.

X	Y₁	Y₂
0	11	6
1	9	5.6667
2	7	5.3333
3	5	5
4	3	4.6667
5	1	4.3333
6	−1	4

X=3

Answer: The solution is (3, 5).

The method for graphing a system of inequalities uses the Shade(command introduced in Chapter V. It shades above the first function entered and below the second function entered. Be sure to clear the $\boxed{\text{Y=}}$ list and clear any drawings that may be on the screen.

2. Graph the system of inequalities in the standard $\boxed{\text{WINDOW}}$.

$$y < x \qquad y \geq 3x + 2$$

Solution: The $\geq$ symbol in $y \geq 3x + 2$ indicates points on or above the line $3x + 2$. Similarly, the $<$ symbol in $y < x$ indicates points below the line $y = x$. Therefore, the boundary $y = 3x + 2$ will be entered *first* in the Shade(command and the boundary $y = x$ will be entered second.

Enter: $\boxed{\text{2nd}}$ $\boxed{\text{DRAW}}$ 7 3 $\boxed{\text{X,T,θ,n}}$ $\boxed{+}$ 2 $\boxed{,}$ $\boxed{\text{X,T,θ,n}}$ $\boxed{)}$ $\boxed{\text{ENTER}}$

Display:

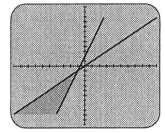

The shaded area includes points that satisfy both inequalities $y < x$ and $y \geq 3x + 2$.

3. Graph the system of inequalities in the standard $\boxed{\text{WINDOW}}$.

$$y \geq 2x \qquad y \geq x + 4$$

Solution: In this case, it is not obvious which function should be used as the upper boundary and which as the lower. However, the calculator can shade the region above each function using a different pattern, and then the region with both patterns can be identified. The command used is Shade(lower, upper, Xleft, Xright, pattern, resolution) where Xleft is −10, Xright is 10, 1 indicates vertical shading, 2 indicates horizontal shading, and 2 indicates every second line drawn.

Enter: $\boxed{\text{2nd}}$ $\boxed{\text{DRAW}}$ 7 2 $\boxed{\text{X,T,θ,n}}$ $\boxed{,}$ 10 $\boxed{,}$ $\boxed{(-)}$ 10 $\boxed{,}$ 10 $\boxed{,}$ 1 $\boxed{,}$ 2
$\boxed{)}$ $\boxed{\text{ALPHA}}$ $\boxed{:}$ $\boxed{\text{2nd}}$ $\boxed{\text{DRAW}}$ 7 $\boxed{\text{X,T,θ,n}}$ $\boxed{+}$ 4 $\boxed{,}$ 10 $\boxed{,}$ $\boxed{(-)}$ 10
$\boxed{,}$ 10 $\boxed{,}$ 2 $\boxed{,}$ 2 $\boxed{)}$ $\boxed{\text{ENTER}}$

Display:

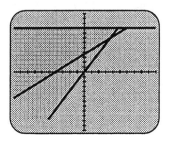

The area with both vertical and horizontal shading is the set of points common to both inequalities $y \geq 2x$ and $y \geq x + 4$.

CHAPTER VIII

Real Numbers and Radicals

1. Understanding More About Real Numbers

We have learned that:

A *rational number* is a real number that can be expressed in the form $\dfrac{a}{b}$ where a and b are integers and $b \neq 0$. Recall that any rational number may be represented by one and only one point on a number line. Examples of rational numbers are:

$$\tfrac{2}{3} \qquad 5 \qquad 0 \qquad -\tfrac{5}{7} \qquad -2 \qquad -5 \qquad -4.3$$

An *irrational number* is a real number than cannot be expressed in the form $\dfrac{a}{b}$ where a and b are integers and $b \neq 0$. It can be proved that any irrational number may be represented by one and only one point on a number line. Examples of irrational numbers are:

$$\frac{\sqrt{3}}{2} \qquad \frac{1}{\sqrt{2}} \qquad \sqrt{17} \qquad \sqrt[3]{4} \qquad -2\sqrt{3}$$

Also, the number π is an irrational number.

The set of real numbers may be considered to be the union of the set of rational numbers and the set of irrational numbers.

Hence, a real number is either a rational number or an irrational number.

A *perfect square* is the square of a rational number. For example, 36 is a perfect square since $6^2 = 36$. Also, $\tfrac{36}{49}$ is a perfect square since $(\tfrac{6}{7})^2 = \tfrac{36}{49}$.

EXPRESSING RATIONAL NUMBERS AS DECIMALS

To express a rational number as a decimal, we simply perform the indicated division. For example,

$$\frac{1}{2} = 2\overline{\smash{\big)}1.00000}^{\textstyle.50000} \qquad \frac{3}{4} = 4\overline{\smash{\big)}3.00000}^{\textstyle.75000} \qquad \frac{1}{16} = 16\overline{\smash{\big)}1.00000}^{\textstyle.06250}$$

In each of these examples, the division terminates. Decimals which result from such divisions (for example, .5, .75, and .0625) are called **terminating decimals.**

However, not all rational numbers can be expressed as terminating decimals. For example,

$$\frac{1}{3} = 3\overline{\smash{\big)}1.0000}^{\textstyle.3333...} \qquad \frac{2}{11} = 11\overline{\smash{\big)}2.0000}^{\textstyle.1818...} \qquad \frac{1}{6} = 6\overline{\smash{\big)}1.0000}^{\textstyle.1666...}$$

Decimals that keep repeating endlessly, such as .3333..., .1818..., and .1666..., are known as **non-terminating repeating decimals.**

A non-terminating repeating decimal may be written in an abbreviated form by placing a bar ($^-$) over the group of digits that is to be continually repeated. For example,

$$.3333... = .\overline{3} \qquad .1818... = .\overline{18} \qquad .1666... = .1\overline{6}$$

The previous examples, which we have studied, illustrate the truth of the following statement, which we will assume:

Every rational number can be expressed as either a terminating decimal or a non-terminating repeating decimal.

Note. Every terminating decimal can be treated as a repeating decimal. For example, $.35 = .34999... = .34\overline{9}$ and $.7 = .6999... = .6\overline{9}$.

EXPRESSING DECIMALS AS RATIONAL NUMBERS
IN THE FORM $\frac{a}{b}$ WHERE a AND b ARE INTEGERS

We have learned how to express a terminating decimal in the form $\frac{a}{b}$ where a and b are integers. For example,

$$.7 = \frac{7}{10} \qquad .49 = \frac{49}{100} \qquad .672 = \frac{672}{1000} \qquad .0513 = \frac{513}{10,000}$$

Study the following model problems to learn how to express a non-terminating repeating decimal in the form $\frac{a}{b}$ where a and b are integers.

~~~~~~~~~~~~~ *MODEL PROBLEMS* ~~~~~~~~~~~~~

1. Change .727272... to the form $\frac{a}{b}$ where a and b are integers.

Solution:

$$\text{Let } N = \;\;\;.727272\ldots$$
$$\text{Then } 100N = 72.727272\ldots$$
$$\text{Subtract: } N = \;\;\;.727272\ldots$$
$$\text{Hence, } 99N = 72$$
$$N = \frac{72}{99} \text{ or } \frac{8}{11}$$

Answer: $.727272\ldots = \dfrac{8}{11}$

2. Change .45555... to the form $\frac{a}{b}$ where a and b are integers.

Solution:

$$\text{Let } N = \;\;\;.45555\ldots$$
$$\text{Then } 10N = 4.55555\ldots$$
$$\text{Subtract: } N = \;\;\;.45555\ldots$$
$$\text{Hence, } 9N = 4.1$$
$$N = \frac{4.1}{9} \text{ or } \frac{41}{90}$$

Answer: $.45555\ldots = \dfrac{41}{90}$

Check: To verify the answer, divide the numerator by the denominator. See whether the quotient is the given decimal.

~~~~~~

The previous examples illustrate the truth of the following statement, which we will assume:

Every terminating and non-terminating repeating decimal can be expressed as a rational number in the form $\dfrac{a}{b}$ where a and b are integers.

EXPRESSING IRRATIONAL NUMBERS AS DECIMALS

There are decimals which are non-terminating and non-repeating. Examples of such decimals are .04004000400004... and −.27227222722227....

Since these non-terminating decimals are non-repeating, they cannot represent rational numbers. Hence, such decimals are irrational numbers.

It can be proved that $\sqrt{3}$ is a non-terminating and non-repeating decimal. Hence, $\sqrt{3}$ is an irrational number. The approximate value of $\sqrt{3}$ to six decimal places is 1.732051. Approximations of irrational numbers are rational numbers.

In general, if n is positive and not a perfect square, $\sqrt{n}$ is an irrational number.

PRINCIPLES OF OPERATIONS ON REAL NUMBERS

When we perform operations on real numbers, we will make use of the following principles, whose truth we will assume without proof:

Principle 1. The sum or difference of an irrational number and a rational number is an irrational number.

Thus, $3 + \sqrt{2}$ and $\sqrt{5} - 3$ are irrational numbers.

Principle 2. The product of an irrational number and a nonzero rational number is an irrational number.

Thus, $5\sqrt{3}$ and 15π are irrational numbers.

Principle 3. The quotient obtained by dividing a nonzero rational number by an irrational number is an irrational number.

Thus, $\dfrac{3}{\sqrt{2}}$ and $\dfrac{10}{\pi}$ are irrational numbers.

Principle 4. The quotient obtained by dividing an irrational number by a nonzero rational number is an irrational number.

Thus, $\dfrac{\sqrt{10}}{3}$ and $\dfrac{\pi}{20}$ are irrational numbers.

Principle 5. If zero is multiplied by any real number, the product is zero.
Thus, $0\sqrt{15} = 0$. Also, $0(7 - \sqrt{8}) = 0$.

Principle 6. If zero is divided by any nonzero real number, the quotient is zero.

Thus, $\dfrac{0}{\sqrt{20}} = 0$. Also, $\dfrac{0}{5 + \sqrt{2}} = 0$.

Principle 7. Division of a real number by zero is impossible.

Thus, $\sqrt{17} \div 0$ is impossible. $\dfrac{\sqrt{17}}{0}$ is meaningless.

Exercises

In 2–12, complete the table as illustrated in exercise 1. (A number may be in more than one category.)

	Real Number	Positive Integer	Negative Integer	Rational Number	Irrational Number
1.	5	yes	no	yes	no
2.	$\sqrt{5}$				
3.	-3.5				
4.	$3\sqrt{5}$				
5.	$-7 - \sqrt{25}$				
6.	$\dfrac{2}{\sqrt{3}}$				
7.	$4 + \sqrt{2}$				
8.	$-\frac{1}{2}\sqrt{3}$				
9.	$-2\sqrt{9}$				
10.	.171717...				
11.	.171171117...				
12.	$0(-5\sqrt{10})$				

In 13–24, change the decimal to the form $\dfrac{a}{b}$ where a and b are integers.

13. .333...   **14.** .777...   **15.** .272727...
16. .3666...   **17.** .13555...   **18.** .0454545...
19. 1.888...   **20.** 25.222...   **21.** 3.0141414...
22. .125125125...   **23.** 125.125125...   **24.** 10.123123123...

25. Which of the following is undefined or meaningless?

$a.$ $0(-\sqrt{5})$ $b.$ $0(2 - \sqrt{8})$ $c.$ $\dfrac{2\sqrt{5} + \sqrt{3}}{0}$ $d.$ $\dfrac{2 + \sqrt{9}}{0}$ $e.$ $2 + \dfrac{\sqrt{9}}{0}$

2. Understanding Roots and Radicals

Now we will learn how to work with real numbers such as $\sqrt{3}$ and $\sqrt[3]{9}$ which have a **radical sign,** $\sqrt{}$ These numbers occur frequently in the solutions of equations of the second degree or higher, and in the solutions to problems using the theorem of Pythagoras.

A **square root** of a number is one of its two equal factors.

Thus, since $5 \times 5 = 25$, then 5 is a square root of 25. Also, since $(-5)(-5) = 25$, then -5 is also a square root of 25. Hence, 25 has two square roots, which may be written ± 5. In fact, *every positive number has two square roots.* For example, .49 has two square roots, $\pm .7$, and $\frac{16}{25}$ has two square roots, $\pm(\frac{4}{5})$.

The ***principal square root*** of a positive number a, symbolized $\sqrt{a}$, is its positive square root. Hence,

$$\sqrt{25} = 5 \qquad \sqrt{.49} = .7 \qquad \sqrt{\frac{16}{25}} = \frac{4}{5}$$

Since the square root of a number is one of its two equal factors, it follows that $\sqrt{a}\sqrt{a} = a$. Thus, $\sqrt{5}\sqrt{5} = 5$, $\sqrt{\frac{2}{3}}\sqrt{\frac{2}{3}} = \frac{2}{3}$, and $\sqrt{3.47}\sqrt{3.47} = 3.47$.

In general, if $a > 0$, there is a number $b > 0$ such that $b^2 = a$. The number b is called the principal square root of a and is symbolized $\sqrt{a}$. Hence, $(\sqrt{a})^2 = a$.

To indicate that the negative square root of a positive number a is to be found, we place a minus sign in front of the radical. Hence, $-\sqrt{a}$ is the negative square root of a. For example,

$$-\sqrt{25} = -5 \qquad -\sqrt{.49} = -.7 \qquad -\sqrt{\frac{16}{25}} = -\frac{4}{5}$$

Note. If $b^2 = 25$, then $b = \pm\sqrt{25}$. In general, if $b^2 = a$, then $b = \pm\sqrt{a}$.

A ***cube root*** of a number is one of its three equal factors. For example, since $2 \times 2 \times 2 = 8$, then 2 is a cube root of 8.

Every nonzero real number has three cube roots, one of which is a real number called the ***principal cube root*** of the number. (The other two roots, to be discussed later, are members of the set of *complex numbers.*)

The ***principal cube root*** of a real number a, symbolized $\sqrt[3]{a}$, is the cube root of a which is a real number. Hence,

$$\sqrt[3]{27} = 3 \qquad \sqrt[3]{-27} = -3 \qquad \sqrt[3]{\frac{8}{27}} = \frac{2}{3} \qquad \sqrt[3]{-\frac{8}{27}} = -\frac{2}{3} \qquad \sqrt[3]{0} = 0$$

In general, if a is a real number, there is a real number b such that $b^3 = a$. The number b is called the principal cube root of a and is symbolized $\sqrt[3]{a}$. Hence, $(\sqrt[3]{a})^3 = a$.

In general, the nth root of a number is one of its n equal factors, where n is a natural number. The principal nth root of a real number a is symbolized by $\sqrt[n]{a}$. The number, n, which indicates the root to be taken is called the ***index;*** the number a is called the ***radicand;*** the symbol, $\sqrt[n]{a}$, is called a ***radical.***

For example, in the case of the radical $\sqrt[3]{64}$, the index is 3 and the radicand is 64. When no index is written, as in the case of $\sqrt{25}$, the index is understood to be 2, and the radical is a square root.

Note that when $\sqrt[n]{a}$ is a real number, $(\sqrt[n]{a})^n = a$.

MEANINGS OF THE SYMBOL $\sqrt[n]{a}$, CALLED THE PRINCIPAL nTH ROOT OF a

1. If n is an odd positive integer, then $\sqrt[n]{a}$ means the one real number b such that $b^n = a$. If a is positive, then b is positive; if a is negative, then b is negative; if a is zero, then b is zero. For example:

    (1) $\sqrt[3]{64} = 4$, since $(4)^3 = 64$ (2) $\sqrt[5]{-1} = -1$, since $(-1)^5 = -1$

    (3) $\sqrt[7]{0} = 0$, since $(0)^7 = 0$

2. If n is an even positive integer and a is a non-negative number, then $\sqrt[n]{a}$ means the real non-negative number b such that $b^n = a$. If a is positive, then b is positive; if a is zero, then b is zero. For example:

    (1) $\sqrt{81} = 9$, since $(9)^2 = 81$ (3) $\sqrt[4]{16} = 2$, since $(2)^4 = 16$

    (2) $\sqrt{0} = 0$, since $0^2 = 0$ (4) $\sqrt[6]{0} = 0$, since $(0)^6 = 0$

3. If n is an even positive integer and a is a negative number, then $\sqrt[n]{a}$ is not defined in the set of real numbers. For example, $\sqrt{-25}$ is not defined in the set of real numbers because there is no real number whose square is -25. These roots will be studied in the next chapter, The System of Complex Numbers.

    Note. We can simplify $\sqrt[n]{a^n}$ as follows: (1) If n is an odd positive integer, $\sqrt[n]{a^n} = a$. Thus, $\sqrt[3]{5^3} = 5$ and $\sqrt[3]{(-5)^3} = -5$. (2) If n is an even positive integer, $\sqrt[n]{a^n} = |a|$ because radicals of even index designate non-negative numbers. Thus, $\sqrt[4]{3^4} = |3| = 3$, and $\sqrt[4]{(-3)^4} = |-3| = 3$.

~~~~~~~~~ **MODEL PROBLEMS** ~~~~~~~~~

1. Find the value of   *a.* $(\sqrt{11})^2$ *b.* $(\sqrt[3]{17})^3$ *c.* $(\sqrt[5]{2})^5$

    Solution:

    a. Since $(\sqrt{a})^2 = a$, then $(\sqrt{11})^2 = 11$.   11 *Ans.*
    b. Since $(\sqrt[3]{a})^3 = a$, then $(\sqrt[3]{17})^3 = 17$.   17 *Ans.*
    c. Since $(\sqrt[5]{a})^5 = a$, then $(\sqrt[5]{2})^5 = 2$.   2 *Ans.*

2. Solve for x: $x^2 = 36$

    Solution: If $x^2 = a$, then $x = \pm \sqrt{a}$ when a is a positive number.

$x^2 = 36$	*Check:* $x^2 = 36$	$x^2 = 36$
$x = \pm\sqrt{36}$	$(+6)^2 \overset{?}{=} 36$	$(-6)^2 \overset{?}{=} 36$
$x = \pm 6$ *Ans.*	$36 = 36$ (true)	$36 = 36$ (true)

Exercises

In 2–10, complete the table as illustrated in exercise 1.

	Radical	Index	Radicand	Principal Root
1.	$\sqrt[3]{8}$	3	8	2
2.	$\sqrt[5]{32}$	5	?	?
3.	$\sqrt{144}$	2	?	?
4.	?	2	49	?
5.	?	2	?	12
6.	?	?	125	5
7.	?	3	1000	?
8.	$\sqrt{\frac{100}{81}}$	?	?	?
9.	?	4	?	10
10.	?	3	a^3	?

In 11–26, find the indicated root. (The variables represent positive numbers.)

11. $\sqrt{36}$ **12.** $\sqrt{144}$ **13.** $-\sqrt{16}$ **14.** $-\sqrt{81}$

15. $\sqrt{4x^2}$ **16.** $\sqrt{9y^6}$ **17.** $-\sqrt{49c^2}$ **18.** $-\sqrt{25x^4}$

19. $\sqrt[3]{64}$ **20.** $-\sqrt[3]{64}$ **21.** $\sqrt[3]{-64}$ **22.** $-\sqrt[3]{-64}$

23. $\sqrt{\dfrac{81}{25}}$ **24.** $-\sqrt{\dfrac{x^2}{64}}$ **25.** $\sqrt[3]{-\dfrac{1}{125}}$ **26.** $-\sqrt[3]{\dfrac{x^6}{8}}$

In 27–39, find the value of the expression.

27. $\sqrt{(7)^2}$ **28.** $\sqrt{(\frac{1}{3})^2}$ **29.** $\sqrt{(.3)^2}$ **30.** $\sqrt[3]{(2)^3}$

31. $\sqrt[3]{(-\frac{3}{4})^3}$ **32.** $(\sqrt{8})^2$ **33.** $(\sqrt{13})^2$ **34.** $(\sqrt[3]{27})^3$

35. $(\sqrt[3]{2})^3$ **36.** $\sqrt{(-4)^2}$ **37.** $\sqrt{(-3)^2}$ **38.** $\sqrt[4]{(5)^4}$

39. $\sqrt[4]{(-2)^4}$ **40.** $\sqrt[6]{(-1)^6}$ **41.** $(\sqrt{35})(\sqrt{35})$ **42.** $\sqrt{81}-\sqrt{49}$

43. $(\sqrt{15})^2 + (\sqrt{8})(\sqrt{8})$ **44.** $\sqrt[3]{(-7)^3} + (\sqrt[3]{15})^3$

In 45–52, solve for the variable.

45. $x^2 = 9$ **46.** $y^2 = 100$ **47.** $m^2 = \dfrac{9}{25}$ **48.** $c^2 = .64$

49. $a^2 - 4 = 0$ **50.** $x^2 - 49 = 0$ **51.** $3x^2 = 75$ **52.** $2y^2 - 72 = 0$

Algebra II

3. Square Roots and Cube Roots That Are Irrational Numbers

Recall that $\sqrt{36}$ is a rational number because $\sqrt{36} = \frac{6}{1}$. Also, $\sqrt{0}$, or 0, is a rational number.

In general:

1. If n is a non-negative integer that is a perfect square, then $\sqrt{n}$ is a rational number.
2. If n is a non-negative integer that is not a perfect square, then $\sqrt{n}$ is an irrational number.

Hence, $\sqrt{25}$ and $\sqrt{144}$ are rational numbers because 25 and 144 are perfect squares. However, $\sqrt{3}$ and $\sqrt{19}$ are irrational numbers because 3 and 19 are not perfect squares.

Similarly, $\sqrt[3]{8}$ and $\sqrt[3]{125}$ are rational numbers because 8 and 125 are perfect cubes. However, $\sqrt[3]{4}$ and $\sqrt[3]{25}$ are irrational numbers because 4 and 25 are not perfect cubes.

We know that $\sqrt{2}$ is a number which is between 1 and 2 because $(1)^2 = 1$ and $(2)^2 = 4$. It can be proved that $\sqrt{2}$ cannot be expressed in the form $\dfrac{a}{b}$ where a and b are integers. Therefore, $\sqrt{2}$ is an irrational number.

In general:

If the square root of an integer is between two consecutive integers, then the root is an irrational number.

Thus, $\sqrt{50}$ is an irrational number because $\sqrt{50}$ lies between 7 and 8, since $(7)^2 = 49$ and $(8)^2 = 64$. Also, $\sqrt[3]{32}$ is an irrational number because $\sqrt[3]{32}$ lies between 3 and 4 since $(3)^3 = 27$ and $(4)^3 = 64$.

Exercises

In 1–10, state whether the radical is a rational or an irrational number.

1. $\sqrt{11}$ 2. $\sqrt{81}$ 3. $-\sqrt{125}$ 4. $\sqrt{169}$ 5. $-\sqrt{400}$
6. $\sqrt[3]{1}$ 7. $\sqrt[3]{36}$ 8. $\sqrt[3]{-64}$ 9. $-\sqrt[3]{8}$ 10. $-\sqrt[3]{100}$

In 11–20, name two consecutive integers between which the given number lies.

11. $\sqrt{5}$ 12. $\sqrt{12}$ 13. $\sqrt{37}$ 14. $-\sqrt{20}$ 15. $-\sqrt{95}$
16. $\sqrt[3]{2}$ 17. $\sqrt[3]{16}$ 18. $\sqrt[3]{70}$ 19. $-\sqrt[3]{25}$ 20. $-\sqrt[3]{75}$

4. Finding Roots Using a Calculator

In our daily lives, we often use *approximate rational values* for irrational numbers. For example, if we have determined mathematically that a piece of wood $\sqrt{2}$ meters long is needed for a project, we would probably ask the salesperson at the lumberyard for a piece of wood slightly longer than 1.4 meters. Although $\sqrt{2} \neq 1.4$ and $\sqrt{2} \neq 1.414$, the rational numbers are close enough to $\sqrt{2}$ so that we can state $\sqrt{2} \approx 1.414$; that is, the square root of 2 is approximately equal to 1.414.

The calculator is the usual method of finding approximate values of $\sqrt{k}$ when k is not a perfect square. Every scientific and graphing calculator has a square root key, $\boxed{\sqrt{x}}$ or $\boxed{\sqrt{}}$. Depending on the type and model of calculator being used, one of the methods shown below will find an approximate value for an irrational number such as $\sqrt{15}$:

Enter:

    Method 1   15 $\boxed{\sqrt{x}}$

    Method 2 $\boxed{\sqrt{x}}$ 15 $\boxed{=}$

    Method 3 $\boxed{\text{2nd}}$ $\boxed{\sqrt{}}$ 15 $\boxed{)}$ $\boxed{\text{ENTER}}$

Display: $\boxed{3.872983346}$

Many scientific calculators have a key, $\boxed{\sqrt[x]{y}}$, that will find a rational approximation for any root of a number. The order in which the index and the radicand of the root are entered differs, however, and you must try different sequences of keys to find what is right for your calculator. In some cases, as shown in Methods 3 and 4 below, the $\boxed{\sqrt[x]{y}}$ key is accessed by first pressing the $\boxed{\text{2nd}}$, $\boxed{\text{SHIFT}}$, or $\boxed{\text{INV}}$ key. For example, to find an approximate value for $\sqrt[4]{24}$, most scientific calculators will use one of the following methods:

Enter:

    Method 1 24 $\boxed{\sqrt[x]{y}}$ 4 $\boxed{=}$

    Method 2   4 $\boxed{\sqrt[x]{y}}$ 24 $\boxed{=}$

    Method 3 24 $\boxed{\text{2nd}}$ $\boxed{\sqrt[x]{y}}$ 4 $\boxed{=}$

    Method 4   4 $\boxed{\text{2nd}}$ $\boxed{\sqrt[x]{y}}$ 24 $\boxed{=}$

Display: $\boxed{2.21336383}$

On some graphing calculators, the cube root and the nth root functions are in the ⌜MATH⌝ menu.

Enter: 4 ⌜MATH⌝ 5 24 ⌜ENTER⌝

The display is as above.

In Chapter XII, we will learn how nth roots can be expressed using exponents.

Exercises

In 1–20, find the square root of the number.

1. 324	**2.** 2,035	**3.** 1,296	**4.** 4,225	**5.** 784
6. 5,184	**7.** 90.25	**8.** 9,801	**9.** 289	**10.** 11,025
11. 15,376	**12.** 56.25	**13.** 441	**14.** 161.29	**15.** 1.1025
16. 16,900	**17.** 9.61	**18.** 17,689	**19.** 161,604	**20.** 667.489

In 21–40, find, to the nearest tenth, the square root of the number.

21. 12	**22.** 19	**23.** 37	**24.** 58	**25.** 60
26. 75	**27.** 79	**28.** 150	**29.** 200	**30.** 416
31. 84	**32.** 90	**33.** 18.25	**34.** 73.61	**35.** 205.78
36. 95	**37.** 108	**38.** 8.5	**39.** 61.7	**40.** 4.052

In 41–45, find, to the nearest thousandth, the indicated root of the number.

41. $\sqrt[3]{7}$	**42.** $\sqrt[3]{-52}$	**43.** $\sqrt[4]{100}$	**44.** $\sqrt[5]{268}$	**45.** $\sqrt[6]{600}$

5. The Distance and Midpoint Formulas

In Chapter V, Section 6, we defined the distance between two points as the length of the line segment that has these points as its endpoints. The distance between two points on the same vertical line is the absolute value of the difference of their y-coordinates (ordinates). Similarly, the distance between two points on the same horizontal line is the absolute value of the difference of their x-coordinates (abscissas). These results can be used to find the length of a line segment that is neither horizontal nor vertical.

To find the distance d from $A(2, 3)$ to $B(5, 7)$, two points that have different abscissas and different ordinates, first draw a horizontal line through A and a vertical line through B. These two lines intersect at point $C(5, 3)$. The distance from $A(2, 3)$ to $C(5, 3)$ is the absolute value of the difference of these abscissas:

$$AC = |5 - 2| = 3$$

Also, the distance from $B(5, 7)$ to $C(5, 3)$ is the absolute value of the difference of their ordinates:

$$BC = |7 - 3| = 4$$

Since $\triangle ABC$ is a right triangle whose hypotenuse is $\overline{AB}$, we can use the Pythagorean Theorem to find AB.

$$
\begin{aligned}
c^2 &= a^2 + b^2 \\
(AB)^2 &= (BC)^2 + (AC)^2 \\
&= 4^2 + 3^2 \\
&= 16 + 9 \\
&= 25 \\
AB &= 5
\end{aligned}
$$

Repeat the steps used to find AB to derive a formula for the distance d between any two points $R(x_1, y_1)$ and $S(x_2, y_2)$. It is possible to use the Pythagorean Theorem to derive the distance formula only because the axes that were chosen for the coordinate system are perpendicular to each other, and the scales on the axes are the same.

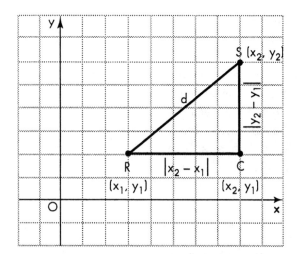

1. Form right triangle *RCS* by drawing a vertical line through *S* and a horizontal line through *R* with the two lines intersecting at *C*.

2. The coordinates of *C* are (x_2, y_1).

3. Therefore, $RC = |x_2 - x_1|$ and $CS = |y_2 - y_1|$. Let $RS = d$.

4. $(RS)^2 = \quad (RC)^2 \quad + \quad (CS)^2$

   $d^2 = |x_2 - x_1|^2 + |y_2 - y_1|^2$

   $\quad = (x_2 - x_1)^2 + (y_2 - y_1)^2$

5. Take the positive square root of each member of the equation:

   $$d = \sqrt{(x_2 - x_1)^2 + (y_2 - y_1)^2}$$

We can now state the **distance formula** in general terms:

- The distance *d* between any two points (x_1, y_1) and (x_2, y_2) in the coordinate plane is given by the formula

$$\boldsymbol{d = \sqrt{(x_2 - x_1)^2 + (y_2 - y_1)^2}}$$

~~~~~~~~~~~~ *MODEL PROBLEMS* ~~~~~~~~~~~~

1. Find the distance between $A(-4, -2)$ and $B(8, 3)$.

   Solution: Use the distance formula.

   Let $A = (x_1, y_1)$.   Then $x_1 = -4$, $y_1 = -2$.

   Let $B = (x_2, y_2)$.   Then $x_2 = 8$, $y_2 = 3$.

$$d = \sqrt{(x_2 - x_1)^2 + (y_2 - y_1)^2}$$

$$AB = \sqrt{[8 - (-4)]^2 + [3 - (-2)]^2}$$

$$= \sqrt{(8 + 4)^2 + (3 + 2)^2}$$

$$= \sqrt{12^2 + 5^2}$$

$$= \sqrt{144 + 25}$$

$$= \sqrt{169}$$

$$= 13$$

Answer: $AB = 13$

2. Express the length of $\overline{CD}$ in radical form if the coordinates of C are $(-4, 2)$ and the coordinates of D are $(-3, 5)$.

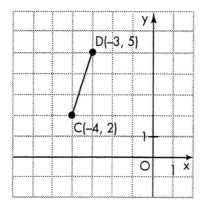

Solution: Let $C = (x_1, y_1)$. Then $x_1 = -4$, $y_1 = 2$.

Let $D = (x_2, y_2)$. Then $x_2 = -3$, $y_2 = 5$.

$$CD = \sqrt{(x_2 - x_1)^2 + (y_2 - y_1)^2}$$

$$= \sqrt{[-3 - (-4)]^2 + [5 - 2]^2}$$

$$= \sqrt{(-3 + 4)^2 + (5 - 2)^2}$$

$$= \sqrt{(1)^2 + (3)^2} = \sqrt{1 + 9} = \sqrt{10}$$

Answer: $CD = \sqrt{10}$

Note. The only acceptable answer for this question is $\sqrt{10}$. Recall that $\sqrt{10}$ is an irrational number whose exact value is shown by the radical. Any solution found by using a calculator will be a *rational approximation* of $\sqrt{10}$.

FINDING THE MIDPOINT OF A LINE SEGMENT

The *midpoint* of a line segment is the point that divides it into two segments with equal lengths. The general formula for finding a midpoint is:

• For any two points (x_1, y_1) and (x_2, y_2) in the coordinate plane, the midpoint of the line segment joining these points has coordinates

$$\left(\frac{x_1 + x_2}{2}, \frac{y_1 + y_2}{2}\right)$$

Notice that the formula for finding the midpoint of a line segment is in the form of an ordered pair. Each coordinate of the midpoint is equal to half the sum of the corresponding endpoints. That is,

the x-value of the midpoint is the average of the x-values of the endpoints
the y-value of the midpoint is the average of the y-values of the endpoints

-------------------------- **MODEL PROBLEM** --------------------------

Find the midpoint of the line segment if the coordinates of the endpoints are $(5, 2)$ and $(-1, -4)$.

Solution: It makes no difference which point is chosen as (x_1, y_1) and which is chosen as (x_2, y_2). Let (x_1, y_1) represent $(-1, -4)$ and (x_2, y_2) represent $(5, 2)$.

$$\text{midpoint} = \left(\frac{x_1 + x_2}{2}, \frac{y_1 + y_2}{2}\right)$$

$$= \left(\frac{-1 + 5}{2}, \frac{-4 + 2}{2}\right)$$

$$= (2, -1)$$

Answer: The point $(2, -1)$ is halfway between the points $(5, 2)$ and $(-1, -4)$.

Exercises

Throughout this section, answers that are not whole numbers should be left in radical form.

In 1–12, find the distance between each pair of points.
1. (2, 7) and (12, 7)
2. (–3, 4) and (5, 4)
3. (–1, –2) and (–1, 4)
4. (0, 0) and (4, 3)
5. (–6, –8) and (0, 0)
6. (1, 4) and (4, 8)
7. (4, 2) and (–2, 10)
8. (6, 4) and (3, 6)
9. (–5, 0) and (–9, 6)
10. (0, 0) and (–2, 5)
11. (0, c) and (b, 0)
12. (0, 0) and (a, b)

In 13–16, find in each case the length of the line segment joining the points and the coordinates of the midpoint.
13. (5, 2) and (8, 6)
14. (–5, 1) and (7, 6)
15. (0, 5) and (–3, 3)
16. (–4, –5) and (1, –2)

In 17–22, find in each case the lengths of the sides of a triangle whose vertices are:
17. (0, 0), (8, 0), (4, 3)
18. (1, 5), (5, 5), (5, 1)
19. (3, 6), (–1, 3), (5, –5)
20. (6, –3), (0, 4), (8, –1)
21. (–1, 7), (0, 0), (8, 4)
22. (–4, 2), (–1, 6), (5, 4)

23. Find the length of the shortest side of the triangle whose vertices are R(–2, –1), S(1, 3), and T(1, 10).

In 24–27, show that each triangle with the given vertices is isosceles.
24. (2, 3), (5, 7), (1, 4)
25. (1, 0), (5, 0), (3, 4)
26. (7, –1), (2, –2), (3, 3)
27. (4, –7), (–3, –4), (7, 0)

6. Simplifying a Radical Whose Radicand Is a Product

Since $\sqrt{4 \cdot 25} = \sqrt{100} = 10$ and $\sqrt{4} \cdot \sqrt{25} = 2 \cdot 5 = 10$, then $\sqrt{4 \cdot 25} = \sqrt{4} \cdot \sqrt{25}$. This example illustrates the following property of radicals:

Property. The square root of a product of non-negative numbers is equal to the product of the square roots of the numbers.

In general, if a and b are non-negative numbers and n is a natural number, it can be proved that

$$\sqrt[n]{a \cdot b} = \sqrt[n]{a} \cdot \sqrt[n]{b}$$

Hence, $\sqrt{a \cdot b} = \sqrt{a} \cdot \sqrt{b}$. Also, $\sqrt[3]{a \cdot b} = \sqrt[3]{a} \cdot \sqrt[3]{b}$.

These rules permit us to transform a radical into an equivalent radical, as is illustrated in the following example:

$$\sqrt{12} = \sqrt{4 \cdot 3} = \sqrt{4}\sqrt{3} = 2\sqrt{3}$$

Observe that we expressed 12 as a product of 4, the largest perfect square factor of 12, and 3. Then we expressed $\sqrt{4}$ as 2. Also,

$$\sqrt[3]{24} = \sqrt[3]{8 \cdot 3} = \sqrt[3]{8}\sqrt[3]{3} = 2\sqrt[3]{3}$$

Observe that we expressed 24 as the product of 8, the largest perfect cube factor of 24, and 3. Then we expressed $\sqrt[3]{8}$ as 2.

When we expressed $\sqrt{12}$ as $2\sqrt{3}$, we simplified $\sqrt{12}$. When we expressed $\sqrt[3]{24}$ as $2\sqrt[3]{3}$, we simplified $\sqrt[3]{24}$.

A radical is considered to be in the **simplest form** when:

1. The radicand does not have a factor whose indicated root may be taken exactly. Thus, $\sqrt{80}$ and $\sqrt[3]{24}$ are *not* in simplest form. However, $4\sqrt{5}$ and $2\sqrt[3]{3}$ are in simplest form.

2. The radicand is not a fraction. Thus $\sqrt{\dfrac{5}{9}}$ is not in simplest form. Later we will learn how to transform $\sqrt{\dfrac{5}{9}}$ to $\frac{1}{3}\sqrt{5}$, which is in simplest form.

Note. In radicals whose index is an even number, if the radicand involves variables, assume that these variables represent positive numbers only.

~~~~~~~~~~ **MODEL PROBLEMS** ~~~~~~~~~~

In 1–6, simplify the radical.

1. $\sqrt{80}$ **2.** $\sqrt[3]{48}$ **3.** $3\sqrt{75}$ **4.** $\frac{1}{2}\sqrt{48}$ **5.** $-4\sqrt[3]{54}$ **6.** $\sqrt{8c^3d^6}$

Solution:

	1. $\sqrt{80}$	**2.** $\sqrt[3]{48}$
How To Proceed	*Solution*	*Solution*
1. Factor the radicand, using the largest perfect power as one of the factors.	1. $\sqrt{80} = \sqrt{16 \cdot 5}$ (16 is the largest perfect square factor of 80.)	1. $\sqrt[3]{48} = \sqrt[3]{8 \cdot 6}$ (8 is the largest perfect cube factor of 48.)

2. Express the root of the product as the product of the roots of the factors, using the same index.

2. $\sqrt{80} = \sqrt{16} \cdot \sqrt{5}$

2. $\sqrt[3]{48} = \sqrt[3]{8} \cdot \sqrt[3]{6}$

3. Simplify the radical having the perfect power.

3. $\sqrt{80} = 4\sqrt{5}$ *Ans.*

3. $\sqrt[3]{48} = 2\sqrt[3]{6}$ *Ans.*

3. $3\sqrt{75} = 3\sqrt{25 \cdot 3} = 3\sqrt{25} \cdot \sqrt{3} = 3 \cdot 5\sqrt{3} = 15\sqrt{3}$ *Ans.*

4. $\frac{1}{2}\sqrt{48} = \frac{1}{2}\sqrt{16 \cdot 3} = \frac{1}{2}\sqrt{16} \cdot \sqrt{3} = \frac{1}{2} \cdot 4\sqrt{3} = 2\sqrt{3}$ *Ans.*

5. $-4\sqrt[3]{54} = -4\sqrt[3]{27 \cdot 2} = -4\sqrt[3]{27} \cdot \sqrt[3]{2} = -4 \cdot 3\sqrt[3]{2} = -12\sqrt[3]{2}$ *Ans.*

6. $\sqrt{8c^3d^6} = \sqrt{4c^2d^6 \cdot 2c} = \sqrt{4c^2d^6} \cdot \sqrt{2c} = 2cd^3\sqrt{2c}$ *Ans.*

Exercises

In 1–40, simplify the radical.

1. $\sqrt{45}$
2. $\sqrt{40}$
3. $-\sqrt{28}$
4. $\sqrt{98}$
5. $\sqrt{108}$
6. $-\sqrt{200}$
7. $\sqrt{162}$
8. $\sqrt{300}$
9. $-\sqrt{500}$
10. $\sqrt[3]{24}$
11. $\sqrt[3]{250}$
12. $-\sqrt[3]{40}$
13. $5\sqrt{8}$
14. $4\sqrt{28}$
15. $-7\sqrt{20}$
16. $3\sqrt{80}$
17. $2\sqrt{45}$
18. $-4\sqrt{98}$
19. $\frac{1}{3}\sqrt{27}$
20. $\frac{1}{2}\sqrt{48}$
21. $-\frac{1}{3}\sqrt{50}$
22. $\frac{3}{4}\sqrt{80}$
23. $\frac{3}{8}\sqrt{80}$
24. $-\frac{4}{5}\sqrt{150}$
25. $\sqrt{x^3}$
26. $\sqrt{x^2y}$
27. $5\sqrt{rs^4}$
28. $\sqrt{r^2s^3}$
29. $\sqrt{5x^3y}$
30. $3\sqrt{3x^2y^5}$
31. $\sqrt{25y}$
32. $\sqrt{4r^3s^2}$
33. $5\sqrt{9x^5y^3}$
34. $\sqrt{40s^2}$
35. $\sqrt{12r^4s^3}$
36. $6\sqrt{8x^3y^5}$
37. $\sqrt[3]{54}$
38. $3\sqrt[3]{5x^3}$
39. $\sqrt[3]{16y^7}$
40. $-\sqrt[3]{8y^4}$

In 41–44, express the radical in simplest form and then use a calculator to find its value, correct to the nearest tenth.

41. $\sqrt{300}$
42. $-\frac{2}{5}\sqrt{175}$
43. $-4 + \sqrt{200}$
44. $-5 - \sqrt{500}$

45. *a.* Does $\sqrt{36 + 64} = \sqrt{36} + \sqrt{64}$?

    b. Is finding a square root always distributive over addition?

    c. When does $\sqrt{x + y} = \sqrt{x} + \sqrt{y}$?

    d. Does $\sqrt{100 - 64} = \sqrt{100} - \sqrt{64}$?

    e. Is finding a square root always distributive over subtraction?

    f. When does $\sqrt{x - y} = \sqrt{x} - \sqrt{y}$?

46. *a.* Use the properties of the set of real numbers to prove that $\sqrt{a \cdot b} = \sqrt{a} \cdot \sqrt{b}$ where a and b are non-negative. (*Hint:* Show $(\sqrt{a} \cdot \sqrt{b})^2 = ab$.)

    b. Is finding a square root of the product of two non-negative numbers always distributive over multiplication?

47. *a.* Use the properties of the set of real numbers to prove that $\sqrt[n]{a \cdot b} = \sqrt[n]{a} \cdot \sqrt[n]{b}$ where a and b are non-negative.

    b. Is finding the nth root of the product of two non-negative numbers always distributive over multiplication?

7. Simplifying a Radical Whose Radicand Is a Fraction

Since $\sqrt{\dfrac{4}{25}} = \dfrac{2}{5}$ and $\dfrac{\sqrt{4}}{\sqrt{25}} = \dfrac{2}{5}$, then $\sqrt{\dfrac{4}{25}} = \dfrac{\sqrt{4}}{\sqrt{25}}$. This example illustrates the following property of radicals:

Property. The square root of a quotient of non-negative numbers is equal to the quotient of the square roots of the numbers.

In general, if a and b are non-negative numbers, $b \neq 0$, and n is a positive integer, it can be proved that

$$\sqrt[n]{\frac{a}{b}} = \frac{\sqrt[n]{a}}{\sqrt[n]{b}}$$

Hence, $\sqrt{\dfrac{a}{b}} = \dfrac{\sqrt{a}}{\sqrt{b}}$ and $\sqrt[3]{\dfrac{a}{b}} = \dfrac{\sqrt[3]{a}}{\sqrt[3]{b}}$. Thus, $\sqrt{\dfrac{4}{9}} = \dfrac{\sqrt{4}}{\sqrt{9}} = \dfrac{2}{3}$ and $\sqrt[3]{\dfrac{125}{27}} = \dfrac{\sqrt[3]{125}}{\sqrt[3]{27}} = \dfrac{5}{3}$.

〜〜〜〜〜〜〜 *MODEL PROBLEMS* 〜〜〜〜〜〜〜

In 1–6, simplify the radical

1. $\sqrt{\dfrac{2}{3}}$ **2.** $\sqrt[3]{\dfrac{1}{2}}$ **3.** $\sqrt{\dfrac{7}{12}}$ **4.** $4\sqrt[3]{\dfrac{1}{4}}$

Solution:

$$1. \sqrt{\frac{2}{3}} \qquad 2. \sqrt[3]{\frac{1}{2}}$$

How To Proceed	*Solution*	*Solution*
1. Transform the radicand into an equivalent fraction whose denominator is the least perfect power.	1. $\sqrt{\frac{2}{3}} = \sqrt{\frac{2}{3} \cdot \frac{3}{3}}$ $= \sqrt{\frac{6}{9}}$	1. $\sqrt[3]{\frac{1}{2}} = \sqrt[3]{\frac{1}{2} \cdot \frac{4}{4}}$ $= \sqrt[3]{\frac{4}{8}}$
2. Express the root of the quotient as the quotient of the root of the numerator divided by the root of the denominator, using the same index.	2. $= \frac{\sqrt{6}}{\sqrt{9}}$	2. $= \frac{\sqrt[3]{4}}{\sqrt[3]{8}}$
3. Find the root of the denominator.	3. $= \frac{\sqrt{6}}{3}$ *Ans.*	3. $= \frac{\sqrt[3]{4}}{2}$ *Ans.*

3. $\sqrt{\frac{7}{12}} = \sqrt{\frac{7}{12} \cdot \frac{3}{3}} = \sqrt{\frac{21}{36}} = \frac{\sqrt{21}}{\sqrt{36}} = \frac{\sqrt{21}}{6}$ or $\frac{1}{6}\sqrt{21}$ *Ans.*

4. $4\sqrt[3]{\frac{1}{4}} = 4\sqrt[3]{\frac{1}{4} \cdot \frac{2}{2}} = 4\sqrt[3]{\frac{2}{8}} = 4\frac{\sqrt[3]{2}}{\sqrt[3]{8}} = 4\frac{\sqrt[3]{2}}{2} = 2\sqrt[3]{2}$ *Ans.*

Exercises

In 1–24, simplify the radical.

1. $\sqrt{\frac{3}{4}}$ 2. $8\sqrt{\frac{7}{16}}$ 3. $\sqrt{\frac{12}{25}}$ 4. $2\sqrt{\frac{75}{64}}$

5. $\sqrt{\frac{1}{2}}$ 6. $\sqrt{\frac{1}{3}}$ 7. $10\sqrt{\frac{1}{5}}$ 8. $3\sqrt{\frac{1}{6}}$

9. $\sqrt{\frac{2}{3}}$ 10. $5\sqrt{\frac{3}{5}}$ 11. $6\sqrt{\frac{11}{2}}$ 12. $\sqrt{\frac{8}{7}}$

13. $8\sqrt{\frac{9}{2}}$ 14. $9\sqrt{\frac{4}{3}}$ 15. $\sqrt{\frac{7}{18}}$ 16. $\sqrt{\frac{9}{32}}$

17. $5\sqrt{\frac{49}{20}}$ 18. $\sqrt[3]{\frac{1}{2}}$ 19. $\sqrt[3]{\frac{4}{9}}$ 20. $9\sqrt[3]{\frac{8}{3}}$

21. $\sqrt{\frac{a}{b^2}}$ 22. $\sqrt{\frac{x^2}{y}}$ 23. $\sqrt{\frac{s}{6}}$ 24. $\sqrt{\frac{2s}{g}}$

In 25–28, find the value of the radical, correct to the nearest tenth.

25. $\sqrt{\frac{5}{4}}$ **26.** $\sqrt{\frac{1}{3}}$ **27.** $20\sqrt{\frac{1}{5}}$ **28.** $6\sqrt{\frac{9}{2}}$

8. Adding or Subtracting Radicals That Have the Same Index

ADDING OR SUBTRACTING LIKE RADICALS

Like radicals are radicals that have the *same index* and the *same radicand*. Thus, $5\sqrt{3}$ and $2\sqrt{3}$ are like radicals, as are $8\sqrt[3]{5}$ and $3\sqrt[3]{5}$. However, $3\sqrt{18}$ and $5\sqrt{2}$ are not like radicals; also, $5\sqrt{3}$ and $2\sqrt[3]{3}$ are not like radicals.

To add or subtract like radicals, use the distributive property as follows:

$$5\sqrt{3} + 2\sqrt{3} = (5+2)\sqrt{3} = 7\sqrt{3}$$
$$8\sqrt[3]{5} - 3\sqrt[3]{5} = (8-3)\sqrt[3]{5} = 5\sqrt[3]{5}$$

Procedure. To add (or subtract) like radicals:
1. Add (or subtract) the coefficients of the radicals.
2. Multiply the sum (or difference) obtained by the common radical.

ADDING OR SUBTRACTING UNLIKE RADICALS

The sum of the unlike radicals $\sqrt{7}$ and $\sqrt{2}$ is indicated as $\sqrt{7} + \sqrt{2}$; the difference is indicated as $\sqrt{7} - \sqrt{2}$. Neither of these can be represented as a single term.

However, when it is possible to transform the unlike radicals into equivalent radicals all of which are like radicals, the resulting like radicals can be added or subtracted. For example,

$$\sqrt{12} + \sqrt{27} = \sqrt{4 \cdot 3} + \sqrt{9 \cdot 3}$$
$$= \sqrt{4}\sqrt{3} + \sqrt{9}\sqrt{3}$$
$$= 2\sqrt{3} + 3\sqrt{3} = (2+3)\sqrt{3} = 5\sqrt{3}$$

〰〰〰〰〰〰 *MODEL PROBLEMS* 〰〰〰〰〰〰

In 1 and 2, combine the radicals.

     1. $5\sqrt{18} - \sqrt{72}$ **2.** $\sqrt{\frac{1}{3}} + \sqrt{75} - \sqrt{2}$

How To Proceed	Solution	Solution
1. Simplify each radical.	1. $5\sqrt{18} - \sqrt{72}$ $= 5\sqrt{9} \cdot \sqrt{2} - \sqrt{36} \cdot \sqrt{2}$ $= 15\sqrt{2} - 6\sqrt{2}$	1. $\sqrt{\frac{1}{3}} + \sqrt{75} - \sqrt{2}$ $= \sqrt{\frac{1}{3} \cdot \frac{3}{3}} + \sqrt{25} \cdot \sqrt{3} - \sqrt{2}$ $= \frac{\sqrt{3}}{3} + 5\sqrt{3} - \sqrt{2}$
2. Combine like radicals by using the distributive property.	2. $= (15-6)\sqrt{2}$ $= 9\sqrt{2}$ *Ans.*	2. $= (\frac{1}{3}+5)\sqrt{3} - \sqrt{2}$ $= 5\frac{1}{3}\sqrt{3} - \sqrt{2}$ *Ans.*

Exercises

In 1–19, combine the radicals.

1. $5\sqrt{2} + 6\sqrt{2}$
2. $4\sqrt{3} - 4\sqrt{3}$
3. $\sqrt{45} + \sqrt{20}$
4. $\sqrt{27} + \sqrt{12}$
5. $\sqrt{50} - \sqrt{8}$
6. $\sqrt{45} - \sqrt{80}$
7. $7\sqrt{28} - 4\sqrt{63}$
8. $3\sqrt[3]{16} + 5\sqrt[3]{54}$
9. $\frac{1}{2}\sqrt{20} + \sqrt{45}$
10. $4\sqrt{18} - \frac{3}{4}\sqrt{32}$
11. $\sqrt{32} + \sqrt{\frac{1}{2}}$
12. $9\sqrt{\frac{1}{3}} - 4\sqrt{\frac{1}{12}}$
13. $\sqrt{16x} + \sqrt{25x}$
14. $4\sqrt{49c} - \frac{1}{2}\sqrt{4c}$
15. $9\sqrt{c^3} - c\sqrt{4c}$
16. $\sqrt{32} - 5\sqrt{8} + 2\sqrt{50}$
17. $\sqrt{27x^2} + 2\sqrt{75x^2} - 2x\sqrt{12}$
18. $5\sqrt{6} - 4\sqrt{\frac{3}{2}} + 9\sqrt{\frac{2}{3}}$
19. $\sqrt[3]{24} + 5\sqrt[3]{3} - 6\sqrt[3]{\frac{1}{9}}$

SUMS INVOLVING IRRATIONAL NUMBERS

The number $5 + \sqrt{5}$, which represents the sum of the rational number 5 and the irrational number $\sqrt{5}$, is an irrational number. This example illustrates the following statement, whose truth we will assume:

The sum of a rational number and an irrational number is an irrational number. (See principle 1, page 293.)

The sum $4\sqrt{3} + 2\sqrt{3} = 6\sqrt{3}$ is an irrational number, whereas the sum $(4 + \sqrt{3}) + (2 - \sqrt{3}) = 6$ is a rational number.

Therefore, we say:

The sum of two irrational numbers may be either rational or irrational.

Exercises

In 1–4, state whether the sum is a rational number or an irrational number.

1. $(5 + \sqrt{2}) + (\sqrt{4} + \sqrt{9})$

2. $(\sqrt{7} + 2\sqrt{3}) + (\sqrt{7} - 2\sqrt{3})$

3. $(6 + 2\sqrt{8}) + (-2 - 2\sqrt{8})$

4. $(9 + 2\pi) + (25 + 8\pi)$

9. Multiplying Radicals That Have the Same Index

MULTIPLYING MONOMIALS CONTAINING RADICALS

We have learned that $\sqrt[n]{a \cdot b} = \sqrt[n]{a} \cdot \sqrt[n]{b}$, when a and b are non-negative and n is a positive integer. By using the symmetric property of equality, we can interchange the members of the equation and obtain the following rule:

In general, if a and b are non-negative numbers and n is a positive integer,

$$\sqrt[n]{a} \cdot \sqrt[n]{b} = \sqrt[n]{a \cdot b}$$

Therefore, we say:

The product of two radicals with the same index is equivalent to the root of the product of their radicands, using the same index.

For example, $\sqrt{5} \cdot \sqrt{7} = \sqrt{35}$ and $\sqrt[3]{2} \cdot \sqrt[3]{5} = \sqrt[3]{10}$. To multiply $7\sqrt{3}$ by $5\sqrt{2}$, we use the commutative and associative properties of multiplication as follows:

$$(7\sqrt{3})(5\sqrt{2}) = (7)(5)(\sqrt{3})(\sqrt{2}) = (5 \cdot 7)(\sqrt{3 \cdot 2}) = 35\sqrt{6}$$

~~~~~~~~~~ *MODEL PROBLEMS* ~~~~~~~~~~

1. Multiply: $10\sqrt{2} \cdot \frac{1}{2}\sqrt{10}$

How To Proceed	*Solution*
1. Multiply the coefficients; multiply the radicands.	1. $10\sqrt{2} \cdot \frac{1}{2}\sqrt{10}$ $= (10 \cdot \frac{1}{2})(\sqrt{2} \cdot \sqrt{10})$ $= 5\sqrt{20}$
2. Simplify the resulting radical.	2. $= 5\sqrt{4 \cdot 5}$ $= 5\sqrt{4} \cdot \sqrt{5}$ $= 10\sqrt{5}$ *Ans.*

In 2–4, multiply the radicals.

2. $(3\sqrt{2})^2$ **3.** $\frac{2}{3}\sqrt[3]{9} \cdot 6\sqrt[3]{6}$ **4.** $\sqrt{8x^3} \cdot \sqrt{3x^2y}$

Solution:

2. $(3\sqrt{2})^2 = 3\sqrt{2} \cdot 3\sqrt{2} = (3 \cdot 3)\sqrt{2} \cdot \sqrt{2} = 9(2) = 18$ *Ans.*

3. $\frac{2}{3}\sqrt[3]{9} \cdot 6\sqrt[3]{6} = (\frac{2}{3} \cdot 6)\sqrt[3]{9} \cdot \sqrt[3]{6} = 4\sqrt[3]{54} = 4\sqrt[3]{27} \cdot \sqrt[3]{2} = 4 \cdot 3\sqrt[3]{2} =$

$$12\sqrt[3]{2} \quad Ans.$$

4. $\sqrt{8x^3}\sqrt{3x^2y} = \sqrt{24x^5y} = \sqrt{4x^4 \cdot 6xy} = \sqrt{4x^4}\sqrt{6xy} = 2x^2\sqrt{6xy}$ *Ans.*

MULTIPLYING POLYNOMIALS CONTAINING RADICALS

To find the product $\sqrt{2}(\sqrt{5} + \sqrt{7})$, we use the distributive property as follows:

$$\sqrt{2}(\sqrt{5} + \sqrt{7}) = \sqrt{2}\sqrt{5} + \sqrt{2}\sqrt{7} = \sqrt{10} + \sqrt{14}$$

To find the product $(3 + \sqrt{5})(4 - \sqrt{5})$, we use the distributive property of multiplication and arrange the solution in the same way that was previously used in finding the product of two binomials.

$$\begin{array}{r} 3 + \sqrt{5} \\ 4 - \sqrt{5} \\ \hline 12 + 4\sqrt{5} \\ -3\sqrt{5} - 5 \\ \hline 12 + \sqrt{5} - 5 = 7 + \sqrt{5} \end{array}$$

MODEL PROBLEMS

1. Multiply: $3\sqrt{3}(4\sqrt{6} - 3\sqrt{24})$

Solution:

$3\sqrt{3}(4\sqrt{6} - 3\sqrt{24})$
$= (3\sqrt{3})(4\sqrt{6}) - (3\sqrt{3})(3\sqrt{24})$
$= 12\sqrt{18} - 9\sqrt{72}$
$= 12\sqrt{9}\sqrt{2} - 9\sqrt{36}\sqrt{2}$
$= 36\sqrt{2} - 54\sqrt{2}$
$= -18\sqrt{2}$ *Ans.*

2. Multiply: $(5 + 2\sqrt{3})(2 - 3\sqrt{3})$

Solution:

$$\begin{array}{r} 5 + 2\sqrt{3} \\ 2 - 3\sqrt{3} \\ \hline 10 + 4\sqrt{3} \\ -15\sqrt{3} - 6\sqrt{9} \\ \hline 10 - 11\sqrt{3} - 18 \\ = -8 - 11\sqrt{3} \quad Ans. \end{array}$$

Exercises

In 1–36, multiply or raise to the power as indicated. Then simplify the result.

1. $\sqrt{8} \cdot \sqrt{2}$

2. $2\sqrt{50} \cdot 3\sqrt{2}$

3. $\sqrt{6} \cdot \sqrt{3}$

4. $\sqrt{40} \cdot \sqrt{5}$

5. $2\sqrt{6} \cdot \sqrt{2}$

6. $4\sqrt{18} \cdot 5\sqrt{3}$

7. $\frac{3}{5}\sqrt{24} \cdot 10\sqrt{3}$

8. $\frac{1}{2}\sqrt{12} \cdot 10\sqrt{2}$

9. $4\sqrt{\frac{1}{2}} \cdot 8\sqrt{18}$

10. $4\sqrt{\frac{1}{3}} \cdot 2\sqrt{96}$

11. $\sqrt{x} \cdot \sqrt{x^3}$

12. $\sqrt{3b}\sqrt{12b^4}$

13. $\sqrt{4x} \cdot \sqrt{xy}$

14. $\sqrt{4c^3d}\sqrt{6cd^2}$

15. $(\sqrt{5})^2$

16. $(\sqrt{x})^2$

17. $(5\sqrt{3})^2$

18. $(\frac{1}{2}\sqrt{3})^2$

19. $\sqrt[3]{2} \cdot \sqrt[3]{4}$

20. $(\sqrt[3]{5})^3$

21. $5\sqrt[3]{4} \cdot 2\sqrt[3]{2}$

22. $\sqrt[3]{4x^2} \cdot \sqrt[3]{6x^4}$

23. $(\sqrt{3x})^2$

24. $(5\sqrt{3y})^2$

25. $(\sqrt{x+1})^2$

26. $(\sqrt{3x-4})^2$

27. $2(\sqrt{5x-3})^2$

28. $\sqrt{2}(5\sqrt{2} + 3\sqrt{8} - 4\sqrt{32})$

29. $2\sqrt{3}(4\sqrt{5} - 3\sqrt{20} - \sqrt{45})$

30. $(4 + \sqrt{2})(3 + \sqrt{2})$

31. $(5 + \sqrt{7})^2$

32. $(2 + \sqrt{5})(8 - \sqrt{5})$

33. $(4 + \sqrt{3})(4 - \sqrt{3})$

34. $(5 + 6\sqrt{3})(5 - 6\sqrt{3})$

35. $(4\sqrt{3} + \sqrt{5})(4\sqrt{3} - \sqrt{5})$

36. $(\sqrt{x} + \sqrt{y})(\sqrt{x} + \sqrt{y})$

37. If $x = 3 + \sqrt{2}$, the value of x^2 is

    (1) 11 (2) $11 + 6\sqrt{2}$ (3) $9 + \sqrt{2}$ (4) 5

38. If $x = 3 - \sqrt{2}$, find the value of $x^2 - 6x + 7$.

PRODUCTS INVOLVING IRRATIONAL NUMBERS

The number $5\sqrt{3}$, which represents the product of the rational number 5 and the irrational number $\sqrt{3}$, is an irrational number.

This example illustrates the following statement, whose truth we will assume:

The product of a nonzero rational number and an irrational number is an irrational number. (See principle 2, page 293).

The product $(\sqrt{8})(\sqrt{2}) = 4$ is a rational number; the product $(\sqrt{5})(\sqrt{2}) = \sqrt{10}$ is an irrational number.

Therefore, we say:

The product of two irrational numbers may be either rational or irrational.

Exercises

In 1–4, state whether the product is a rational number or an irrational number.

1. $(\sqrt{9} - \sqrt{1})(2 + \sqrt{3})$

2. $(7 + \sqrt{5})(3 - \sqrt{5})$

3. $(3\sqrt{7} + 2\sqrt{7})(\sqrt{7})$

4. $(\sqrt{4} + \sqrt{16})\pi$

10. Dividing Radicals That Have the Same Index

We already know that

$$\sqrt[n]{\frac{a}{b}} = \frac{\sqrt[n]{a}}{\sqrt[n]{b}}$$

when a is non-negative, b is positive, and n is a positive integer. By using the symmetric property of equality, we can interchange the members of the equation and obtain the following rule:

In general if a is non-negative, b is positive, and n is a positive integer:

$$\frac{\sqrt[n]{a}}{\sqrt[n]{b}} = \sqrt[n]{\frac{a}{b}}$$

Therefore, we say:

The quotient of two radicals with the same index is equivalent to the root of the quotient of their radicands, using the same index.

For example, $\dfrac{\sqrt{36}}{\sqrt{4}} = \sqrt{\dfrac{36}{4}} = \sqrt{9} = 3$ and $\dfrac{\sqrt[3]{64}}{\sqrt[3]{8}} = \sqrt[3]{\dfrac{64}{8}} = \sqrt[3]{8} = 2.$

To divide $8\sqrt{15}$ by $2\sqrt{3}$, we use the property of fractions, $\dfrac{ac}{bd} = \dfrac{a}{b} \cdot \dfrac{c}{d}$. See how the division is performed:

$$\frac{8\sqrt{15}}{2\sqrt{3}} = \frac{8}{2} \cdot \frac{\sqrt{15}}{\sqrt{3}}$$

$$= \frac{8}{2}\sqrt{\frac{15}{3}}$$

$$= 4\sqrt{5}$$

~~~~~~~~~~ *MODEL PROBLEMS* ~~~~~~~~~~

1. Divide: $50\sqrt{24} \div 5\sqrt{2}$ **2.** Divide: $8\sqrt[3]{20c^4} \div 2\sqrt[3]{5c}$

How To Proceed	*Solution*	*Solution*
1. Divide the coefficients; divide the radicands.	1. $\quad 50\sqrt{24} \div 5\sqrt{2}$ $= \left(\dfrac{50}{5}\right)\left(\dfrac{\sqrt{24}}{\sqrt{2}}\right)$ $= 10\sqrt{12}$	1. $\quad 8\sqrt[3]{20c^4} \div 2\sqrt[3]{5c}$ $= \left(\dfrac{8}{2}\right)\left(\dfrac{\sqrt[3]{20c^4}}{\sqrt[3]{5c}}\right)$ $= 4\sqrt[3]{4c^3}$
2. Simplify the resulting radical.	2. $= 10\sqrt{4 \cdot 3}$ $= 10\sqrt{4} \cdot \sqrt{3}$ $= 20\sqrt{3} \quad Ans.$	2. $= 4\sqrt[3]{c^3 \cdot 4}$ $= 4\sqrt[3]{c^3} \cdot \sqrt[3]{4}$ $= 4c\sqrt[3]{4} \quad Ans.$

Exercises

In 1–17, divide. Then simplify the quotient.

1. $\sqrt{50} \div \sqrt{2}$ 　　　　 **2.** $6\sqrt{45} \div 3\sqrt{5}$ 　　　　 **3.** $\sqrt{24} \div \sqrt{3}$

4. $8\sqrt{60} \div 4\sqrt{5}$ 　　　　 **5.** $10\sqrt{3} \div 5\sqrt{3}$ 　　　　 **6.** $4\sqrt{7} \div 8\sqrt{7}$

7. $\dfrac{15\sqrt{150}}{5\sqrt{2}}$ 　　 **8.** $\dfrac{6\sqrt{27}}{12\sqrt{3}}$ 　　 **9.** $\dfrac{2\sqrt{24}}{4\sqrt{3}}$ 　　 **10.** $\dfrac{18\sqrt{72}}{6\sqrt{6}}$

11. $\dfrac{\sqrt[3]{16}}{\sqrt[3]{2}}$ 　　 **12.** $\dfrac{4\sqrt[3]{15}}{2\sqrt[3]{3}}$ 　　 **13.** $\dfrac{\sqrt{12d^5}}{\sqrt{3d}}$ 　　 **14.** $\dfrac{\sqrt{75x^3y}}{\sqrt{3xy}}$

15. $\dfrac{\sqrt{32} + \sqrt{50}}{\sqrt{2}}$ 　　 **16.** $\dfrac{\sqrt{15} - \sqrt{180}}{\sqrt{3}}$ 　　 **17.** $\dfrac{10\sqrt{72} + 15\sqrt{18}}{\sqrt{3}}$

In 18–21, simplify the expression. Then approximate the result to the nearest tenth.

18. $\dfrac{\sqrt{27} + \sqrt{15}}{\sqrt{3}}$ 　　 **19.** $\dfrac{9 + \sqrt{18}}{3}$ 　　 **20.** $\dfrac{4 - \sqrt{8}}{2}$ 　　 **21.** $\dfrac{-5 - \sqrt{50}}{5}$

QUOTIENTS INVOLVING IRRATIONAL NUMBERS

The number $\dfrac{\sqrt{3}}{5}$, which represents the quotient of the irrational number $\sqrt{3}$ and the rational number 5, is an irrational number. Also, the number $\dfrac{5}{\sqrt{3}}$, which represents the quotient of the rational number 5 and the irrational number $\sqrt{3}$, is an irrational number.

These examples illustrate the following statement, whose truth we will assume:

A quotient involving a nonzero rational number and an irrational number is an irrational number. (See principle 3, page 293.)

The quotient $(\sqrt{8}) \div (\sqrt{2}) = \sqrt{4} = 2$ is a rational number, whereas the quotient $(\sqrt{15}) \div (\sqrt{5}) = \sqrt{3}$ is an irrational number.

Therefore, we say:

The quotient of two irrational numbers may be either rational or irrational.

Exercises

In 1–4, state whether the quotient is a rational number or an irrational number.

1. $\sqrt{32} \div \sqrt{8}$ **2.** $6 \div \sqrt[3]{7}$ **3.** $(\sqrt{2} + 8) \div 2$ **4.** $\sqrt{48} \div \sqrt{2}$

11. Rationalizing an Irrational Monomial Radical Denominator

To find the approximate value of $\dfrac{1}{\sqrt{3}}$, we can use 1.732 as the approximate value of $\sqrt{3}$ and then divide 1 by 1.732. The result obtained by this inconvenient computation is $\dfrac{1}{\sqrt{3}} \approx .577$.

To simplify the computation, we *rationalize the denominator* of the fraction $\dfrac{1}{\sqrt{3}}$. That is, we transform the fraction $\dfrac{1}{\sqrt{3}}$, which has an irrational denominator, into an equivalent fraction which has a rational denominator. We multiply

$\dfrac{1}{\sqrt{3}}$ by 1 in the form of $\dfrac{\sqrt{3}}{\sqrt{3}}$ and obtain:

$$\frac{1}{\sqrt{3}}=\frac{1}{\sqrt{3}}\cdot\frac{\sqrt{3}}{\sqrt{3}}=\frac{\sqrt{3}}{3}\approx\frac{1.732}{3}\approx .577$$

MODEL PROBLEMS

In 1 and 2, rationalize the denominator.　**1.** $\dfrac{15}{\sqrt{20}}$ 　　**2.** $\dfrac{5\sqrt[3]{3}}{7\sqrt[3]{2}}$

How To Proceed

1. Multiply the given fraction by 1 represented as a fraction whose numerator and denominator are both the least radical needed to make the denominator of the resulting fraction a rational number.

2. Simplify the resulting fraction.

Solution

1. $\dfrac{15}{\sqrt{20}}$

$=\dfrac{15}{\sqrt{20}}\cdot\dfrac{\sqrt{5}}{\sqrt{5}}$

(Use $\sqrt{5}$ rather than $\sqrt{20}$.)

$=\dfrac{15\sqrt{5}}{\sqrt{100}}$

2. $=\dfrac{15\sqrt{5}}{10}$

$=\dfrac{3\sqrt{5}}{2}$ *Ans.*

Solution

1. $\dfrac{5\sqrt[3]{3}}{7\sqrt[3]{2}}$

$=\dfrac{5\sqrt[3]{3}}{7\sqrt[3]{2}}\cdot\dfrac{\sqrt[3]{4}}{\sqrt[3]{4}}$

$=\dfrac{5\sqrt[3]{12}}{7\sqrt[3]{8}}$

2. $=\dfrac{5\sqrt[3]{12}}{7\cdot 2}$

$=\dfrac{5\sqrt[3]{12}}{14}$ *Ans.*

Exercises

In 1–25, rationalize the denominator.

1. $\dfrac{5}{\sqrt{2}}$ 　**2.** $\dfrac{7}{\sqrt{3}}$ 　**3.** $\dfrac{3}{\sqrt{5}}$ 　**4.** $\dfrac{2}{\sqrt{7}}$ 　**5.** $\dfrac{9}{\sqrt{11}}$

6. $\dfrac{4}{\sqrt{2}}$ 　**7.** $\dfrac{15}{\sqrt{5}}$ 　**8.** $\dfrac{14}{\sqrt{7}}$ 　**9.** $\dfrac{12}{\sqrt{6}}$ 　**10.** $\dfrac{12}{\sqrt{18}}$

11. $\dfrac{4}{\sqrt{12}}$ 12. $\dfrac{25}{\sqrt{50}}$ 13. $\dfrac{15}{\sqrt{20}}$ 14. $\dfrac{6}{2\sqrt{3}}$ 15. $\dfrac{40}{2\sqrt{20}}$

16. $\dfrac{18}{4\sqrt{12}}$ 17. $\dfrac{14}{3\sqrt{8}}$ 18. $\dfrac{\sqrt{3}}{\sqrt{5}}$ 19. $\dfrac{10\sqrt{2}}{\sqrt{5}}$ 20. $\dfrac{6\sqrt{3}}{\sqrt{8}}$

21. $\dfrac{15\sqrt{2}}{\sqrt{12}}$ 22. $\dfrac{6}{\sqrt[3]{4}}$ 23. $\dfrac{10}{\sqrt[3]{25}}$ 24. $\dfrac{8\sqrt[3]{4}}{\sqrt[3]{2}}$ 25. $\dfrac{7\sqrt[3]{18}}{3\sqrt[3]{9}}$

In 26–30, transform the fraction into an equivalent fraction that does not have a radical in the denominator.

26. $\dfrac{1}{\sqrt{x}}$ 27. $\dfrac{\sqrt{c}}{\sqrt{d}}$ 28. $\dfrac{9}{\sqrt{3a}}$ 29. $\dfrac{cd}{\sqrt{d}}$ 30. $\dfrac{15x}{\sqrt{5x}}$

In 31–34, rationalize the denominator and simplify the resulting fraction.

31. $\dfrac{\sqrt{18}+\sqrt{8}}{\sqrt{2}}$ 32. $\dfrac{\sqrt{5}-1}{\sqrt{5}}$ 33. $\dfrac{2\sqrt{3}-8}{\sqrt{2}}$ 34. $\dfrac{\sqrt{3}-\sqrt{6}}{\sqrt{3}}$

In 35–39, approximate the value of the fraction to the nearest tenth.

35. $\dfrac{15}{\sqrt{3}}$ 36. $\dfrac{4}{\sqrt{8}}$ 37. $\dfrac{6}{\sqrt{6}}$ 38. $\dfrac{9}{\sqrt{2}}$ 39. $\dfrac{\sqrt{3}-1}{\sqrt{3}}$

40. The fraction $\dfrac{\sqrt{3}+\sqrt{2}}{\sqrt{2}}$ is equivalent to (1) $\sqrt{3}$ (2) $\dfrac{\sqrt{3}+2}{2}$ (3) $\dfrac{\sqrt{6}+2}{2}$

12. Rationalizing a Binomial Radical Denominator

The expressions $2+\sqrt{3}$ and $2-\sqrt{3}$ are called ***conjugate binomial radicals.*** Either of the binomials is called the ***conjugate*** of the other. Additional examples of conjugate binomial radicals are:

$$\sqrt{2}+\sqrt{5} \text{ and } \sqrt{2}-\sqrt{5}$$

$$2\sqrt{7}+5\sqrt{11} \text{ and } 2\sqrt{7}-5\sqrt{11}$$

Observe that in each example one of the binomials is the indicated sum of two numbers and the other is the indicated difference of the same two numbers. The product $(5+\sqrt{3})(5-\sqrt{3})=25-3$, or 22, is a rational number. Also, the product $(\sqrt{5}+\sqrt{3})(\sqrt{5}-\sqrt{3})=5-3$, or 2, is a rational number.

In general, when a and b are positive rational numbers, the following products are rational numbers:

$$(a + \sqrt{b})(a - \sqrt{b}) = a^2 - b$$

$$(\sqrt{a} + \sqrt{b})(\sqrt{a} - \sqrt{b}) = a - b$$

We say:

The product of two conjugate binomial radicals is a rational number.

We make use of this fact in rationalizing a binomial radical denominator.

~~~~~~~~~~~~~ *MODEL PROBLEMS* ~~~~~~~~~~~~~

In 1 and 2, express each fraction as an equivalent fraction with a rational denominator.

1. $\dfrac{9}{3 - \sqrt{3}}$ 2. $\dfrac{3 - 5\sqrt{2}}{5 + 2\sqrt{2}}$

How To Proceed	*Solution*	*Solution*
1. Multiply the given fraction by 1 represented as a fraction whose numerator and denominator are both the conjugate of the denominator of the original fraction.	1. $\dfrac{9}{3 - \sqrt{3}}$ $= \dfrac{9}{3 - \sqrt{3}} \cdot \dfrac{3 + \sqrt{3}}{3 + \sqrt{3}}$ $= \dfrac{9(3 + \sqrt{3})}{9 - 3}$	1. $\dfrac{3 - 5\sqrt{2}}{5 + 2\sqrt{2}}$ $= \dfrac{3 - 5\sqrt{2}}{5 + 2\sqrt{2}} \cdot \dfrac{5 - 2\sqrt{2}}{5 - 2\sqrt{2}}$ $= \dfrac{15 - 31\sqrt{2} + 20}{25 - 8}$
2. Simplify the resulting fraction.	2. $= \dfrac{\overset{3}{\cancel{9}}(3 + \sqrt{3})}{\underset{2}{\cancel{6}}}$ $= \dfrac{3(3 + \sqrt{3})}{2}$ *Ans.*	2. $= \dfrac{35 - 31\sqrt{2}}{17}$ *Ans.*

~~~~~~

Exercises

In 1–12, rationalize the denominator of the fraction.

1. $\dfrac{9}{3-\sqrt{2}}$

2. $\dfrac{12}{3-\sqrt{5}}$

3. $\dfrac{18}{\sqrt{3}-3}$

4. $\dfrac{12}{\sqrt{5}-2}$

5. $\dfrac{22}{2\sqrt{3}+1}$

6. $\dfrac{44}{2\sqrt{5}-3}$

7. $\dfrac{\sqrt{2}+4}{\sqrt{2}-1}$

8. $\dfrac{6+\sqrt{3}}{4-\sqrt{3}}$

9. $\dfrac{2\sqrt{3}-1}{2\sqrt{3}+1}$

10. $\dfrac{4-5\sqrt{2}}{7+3\sqrt{2}}$

11. $\dfrac{\sqrt{5}-\sqrt{3}}{\sqrt{5}-\sqrt{3}}$

12. $\dfrac{2\sqrt{5}+3\sqrt{2}}{3\sqrt{5}-\sqrt{2}}$

In 13–15, express the fraction as an equivalent fraction with a rational denominator.

13. $\dfrac{5}{3+\sqrt{2}}$

14. $\dfrac{1}{\sqrt{7}-2}$

15. $\dfrac{3}{\sqrt{6}+2}$

16. The expression $\dfrac{2}{\sqrt{3}-1}$ is equivalent to (1) $\dfrac{2\sqrt{3}+1}{2}$ (2) $\sqrt{3}+1$
(3) $\dfrac{\sqrt{3}+1}{2}$ (4) $\sqrt{3}$

17. The fraction $\dfrac{6-\sqrt{2}}{1+\sqrt{2}}$ is equivalent to (1) $7\sqrt{2}-8$ (2) $\dfrac{8-7\sqrt{2}}{3}$
(3) $4-7\sqrt{2}$ (4) $8-7\sqrt{2}$

13. Solving Radical or Irrational Equations

A *radical equation* or *irrational equation* in one variable is an equation which has the variable in a radicand. For example, $\sqrt{x}=5$ and $\sqrt{4x-1}=1$ are radical equations.

To solve a radical equation in which the radical is the only term of one side of the equation, we square both members of the equation if the radical is a square root, cube both members if the radical is a cube root, etc., in order to transform the equation into one which does not contain radicals.

To illustrate, let us solve the radical equations $\sqrt{x+1}=5$ and $\sqrt{x+1}=-5$.

Solution

$$\sqrt{x+1}=5$$
$$(\sqrt{x+1})^2=(5)^2$$
$$x+1=25$$
$$x=24$$

Check

$$\sqrt{x+1}=5$$
$$\sqrt{24+1}\overset{?}{=}5$$
$$\sqrt{25}\overset{?}{=}5$$
$$5=5\ (\text{true})$$

Observe that {24} is the solution set of both the original equation $\sqrt{x+1}=5$ and the derived equation $x+1=25$. Therefore, these two equations are equivalent equations whose solution set is {24}.

Solution

$$\sqrt{x+1}=-5$$
$$(\sqrt{x+1})^2=(-5)^2$$
$$x+1=25$$
$$x=24$$

Check

$$\sqrt{x+1}=-5$$
$$\sqrt{24+1}\overset{?}{=}-5$$
$$\sqrt{25}\overset{?}{=}-5$$
$$5=-5\ (\text{not true})$$

Observe that {24} is the solution set of the derived equation $x+1=25$, but {24} is not the solution set of the original equation. The original equation $\sqrt{x+1}=-5$ and the derived equation $x+1=25$ are not equivalent equations. Therefore, 24 is an ***extraneous value*** and must be rejected.

We see that when we solve an equation by squaring or cubing both members, the "squared" or the "cubed" equation and the original equations may not be equivalent equations. We must be careful to check the roots of the "squared" or the "cubed" equation in the given equation to see that these roots also satisfy the given equation. If they do not, they are extraneous values and we reject them.

The following procedure applies to the solution of a radical equation having a single radical in which the variable appears in the radicand.

Procedure. To solve a radical equation containing only one radical:
1. **Isolate the radical on one side of the equation by transposing the remaining terms to the other side.**
2. **If the radical is a square root, square both sides; if the radical is a cube root, cube both sides; etc.**
3. **Solve the resulting equation.**
4. **Check to determine if the roots of the resulting equation are roots of the original equation. If not, reject any such root as an extraneous value.**

Special procedures for radical equations containing two radicals are indicated in model problems 4 and 5 that follow.

~~~~~~~~~~~~~ *MODEL PROBLEMS* ~~~~~~~~~~~~~

In 1 and 2, solve and check.

1. $\sqrt{2x+1} - 1 = 4$

Solution:

$$\sqrt{2x+1} - 1 = 4$$
$$\sqrt{2x+1} = 5 \quad \text{Isolating the radical.}$$
$$2x+1 = 25 \quad \text{Squaring both sides.}$$
$$2x = 24$$
$$x = 12$$

Check:

$$\sqrt{2x+1} - 1 = 4$$
$$5 - 1 \overset{?}{=} 4$$
$$4 = 4 \text{ (true)}$$

Answer: $x = 12$, or the solution set is {12}.

2. $\sqrt{2x+1} - 1 = -4$

Solution:

$$\sqrt{2x+1} - 1 = -4$$
$$\sqrt{2x+1} = -3$$
$$2x+1 = 9$$
$$2x = 8$$
$$x = 4$$

Check:

$$\sqrt{2x+1} - 1 = -4$$
$$3 - 1 \overset{?}{=} -4$$
$$2 = -4 \text{ (not true)}$$

Hence, 4 is an extraneous value and must be rejected.

Answer: The equation has no root, or the solution set is the empty set $\varnothing$.

3. Solve and check: $\sqrt{x^2+9} - x = 1$

Solution:

$$\sqrt{x^2+9} - x = 1$$
$$\sqrt{x^2+9} = x + 1$$
$$(\sqrt{x^2+9})^2 = (x+1)^2$$
$$x^2 + 9 = x^2 + 2x + 1$$
$$8 = 2x$$
$$4 = x$$

Check:

$$\sqrt{x^2+9} - 4 = 1$$
$$\sqrt{25} - 4 \overset{?}{=} 1$$
$$5 - 4 \overset{?}{=} 1$$
$$1 = 1 \text{ (true)}$$

Answer: $x = 4$, or the solution set is {4}.

In 4 and 5, solve and check: (Special Procedures)

4. $\sqrt{2x+1} = \dfrac{15}{\sqrt{2x+1}}$

Solution:

$$\sqrt{2x+1} = \dfrac{15}{\sqrt{2x+1}}$$

Multiply each side by $\sqrt{2x+1}$.

$$2x+1 = 15$$
$$x = 7 \quad Ans.$$

5. $\sqrt{3y-1} = 2\sqrt{8-2y}$

Solution:

$$\sqrt{3y-1} = 2\sqrt{8-2y}$$

Square both sides.

$$(\sqrt{3y-1})^2 = (2\sqrt{8-2y})^2$$
$$3y-1 = 4(8-2y)$$
$$3y-1 = 32-8y$$
$$11y = 33$$
$$y = 3 \quad Ans.$$

(The checks are left to the student.)

~~~~~~

Exercises

In 1–18, solve the equation and check.

1. $\sqrt{x} = 7$ **2.** $\sqrt[3]{a} = 2$ **3.** $\sqrt{5x} = 5$

4. $\sqrt[3]{2b} = 4$ **5.** $4\sqrt{x} = 8$ **6.** $5\sqrt{5a} = 20$

7. $\sqrt{x-5} = 3$ **8.** $\sqrt[3]{3a-1} = 2$ **9.** $\sqrt{x+3} = 5$

10. $7 - 3\sqrt{x} = 1$ **11.** $\sqrt{5a-1} - 3 = 0$ **12.** $4\sqrt{a+7} - 5 = 11$

13. $\sqrt{x^2+3} = x+1$ **14.** $\sqrt{x^2+27} - 3 = x$

15. $x = 4 + \sqrt{x^2-32}$ **16.** $\dfrac{3}{\sqrt{x-5}} = \sqrt{x-5}$

17. $\sqrt{5x+6} = \sqrt{9x-2}$ **18.** $\sqrt[3]{5x+1} - \sqrt[3]{2x-8} = 0$

19. Solve for A: $R = \sqrt{\dfrac{A}{\pi}}$. **20.** Solve for x: $7 - \sqrt{2x+1} = 4$.

21. What value of x satisfies the equation $\sqrt{x^2-7} = x-1$?

22. What value of x satisfies the equation $\sqrt{x^2-2x+4} = x$?

23. Solve the equation $\sqrt{x^2+27} = 2x$ for x.

24. In the equation $\sqrt{2x+3} - x = 0$, x is equal to (1) 3 only (2) -1 only (3) 3 and -1 (4) -3 and 1

25. The equation $x + \sqrt{x^2 + 3} = 3x$ has (1) both $+1$ and -1 as its roots (2) $+1$ as its only root (3) -1 as its only root (4) neither $+1$ nor -1 as its roots

26. Which equation has both 3 and 6 as roots? (1) $\sqrt{x - 2} = \dfrac{3}{x}$

(2) $\sqrt{x - 2} = \dfrac{x}{3}$ (3) $\sqrt{x - 2} = -x + 4$ (4) $\sqrt{x - 2} = x - 4$

27. Solve the equation $\sqrt[3]{x - 2} = \frac{1}{2}$ for the value of x.

In 28–33, show that the solution set is the empty set $\varnothing$.

28. $\sqrt{2x + 1} = -3$ **29.** $2\sqrt{x - 1} = -4$ **30.** $\sqrt{3x - 2} + 1 = 0$

31. $3 + \sqrt{5 - x} = 2$ **32.** $2 - \sqrt{3x} = 5$ **33.** $3 - 2\sqrt{x^2 - 1} = 5$

~~~~~~~~ *CALCULATOR APPLICATIONS* ~~~~~~~~

The TI-83 and other graphing calculators have commands for converting terminating or repeating decimals to fraction form and fractions to decimal form.

~~~~~~~~ *MODEL PROBLEMS* ~~~~~~~~

1. Express each rational number as a fraction.

   a. .3125   *b.* $0.1\overline{3}$

   Solution:

   a. The command that changes decimals to fractions is on the MATH menu. The decimal value must be entered before the command.

   Enter: .3125 MATH 1 ENTER

   Display: 5/ 16

   Answer: The fraction equivalent of .3125 is $\frac{5}{16}$.

   b. With repeating decimals, it is necessary to enter a sufficient number of places to obtain the correct conversion. Look what happens.

   Enter: .133333 MATH 1 ENTER

   Display: .133333

   The original decimal was returned. If the entered decimal cannot be simplified or the resulting denominator is more than three digits, the entered decimal is displayed. Try again.

   Enter: .1333333333333 MATH 1 ENTER

   Display: 2/ 15

   This time the calculator interpreted the entry as a repeating decimal, and the desired fraction was returned.

   Answer: The fraction equivalent of $.1\overline{3}$ is $\frac{2}{15}$.

2. Add $\frac{1}{7} + \frac{3}{11}$ and express the answer in fraction and decimal forms.

Solution:

Enter: 1 $\boxed{\div}$ 7 $\boxed{+}$ 3 $\boxed{\div}$ 11 $\boxed{\text{MATH}}$ 1 $\boxed{\text{ENTER}}$

Display: $\boxed{\text{32/77}}$

Enter: $\boxed{\text{2nd}}$ $\boxed{\text{ANS}}$ $\boxed{\text{MATH}}$ 2 $\boxed{\text{ENTER}}$

Display: $\boxed{\text{.2337662338}}$

The decimal form also could have been found by dividing the numerator of the fraction by the denominator.

Answer: The sum $\frac{1}{7} + \frac{3}{11}$ is equal to $\frac{32}{77}$ or .2338 (nearest thousandth).

CHAPTER IX

The System of Complex Numbers

1. Pure Imaginary Numbers

EXTENDING THE NUMBER SYSTEM TO INCLUDE THE SET OF PURE IMAGINARY NUMBERS

To solve the equation $x^2 = -1$, or an equivalent equation such as $x^2 + 1 = 0$, we must have a number whose square is -1. There can be no such number in the system of real numbers since the square of a real number must be positive or zero and cannot be negative. Hence, it becomes desirable to extend the number system so that the equation $x^2 = -1$ and in general $x^2 = a$, when a is negative, will have a solution.

On several previous occasions, we extended the number system because it proved inadequate to solve a given equation or perform a given operation.

1. *The set of counting numbers, or natural numbers*, is adequate to solve $x - 2 = 4$, $x - 4 = 2$, and $x + 2 = 4$; but not $x + 4 = 4$. Hence, we extended the number system by inventing the set of whole numbers, which includes 0. In the set of whole numbers, the root of $x + 4 = 4$ is 0.

2. *The set of whole numbers* is adequate to solve $x + 4 = 4$, but not $x + 4 = 2$. Hence, we extended the number system by inventing the set of integers, which includes negative numbers. In the set of integers, the root of $x + 4 = 2$ is -2.

3. *The set of integers* is adequate to solve $x + 4 = 2$, $\dfrac{x}{2} = 4$, and $2x = 4$; but not $4x = 2$. Hence, we extended the number system by inventing the set of rational numbers, which includes numbers that are the quotient of two integers, except 0 as a divisor. In the set of rational numbers, the root of $4x = 2$ is $\frac{1}{2}$.

4. *The set of rational numbers* is adequate to solve $4x = 2$, $x^2 = 4$, and $x^2 = \frac{1}{4}$; but not $x^2 = 2$. Hence, we extended the number system to include the set of irrational numbers, which includes real numbers that are not rational. In the set of irrational numbers, the roots of $x^2 = 2$ are $\sqrt{2}$ and $-\sqrt{2}$. With the creation of the set of irrational numbers, the system of real numbers becomes complete.

5. *The system of real numbers, which includes the set of irrational numbers*, is adequate to solve $x^2 = 2$, $x^2 = \frac{1}{2}$, $x^2 = 16$, and, in general, $x^2 = a$, where a is a non-negative real number. However, the set of real numbers is not adequate to

solve $x^2 = -1$, $x^2 = -2$, $x^2 = -\frac{1}{4}$, and, in general, $x^2 = a$, where a is a negative real number. The roots of $x^2 = -1$ are $+\sqrt{-1}$ and $-\sqrt{-1}$, which cannot be real numbers, because the square of a real number cannot be a negative number. These numbers, $\sqrt{-1}$ and $-\sqrt{-1}$, are numbers belonging to the set of *pure imaginary numbers*. Hence, mathematicians invented the system of complex numbers, which includes the set of pure imaginary numbers.

6. *The system of complex numbers, which includes the set of pure imaginary numbers*, is adequate to solve equations such as $x^2 = -1$, $x^2 = -2$, $x^2 = -6$, $x^2 = -\frac{1}{4}$, and, in general, $x^2 = a$, where a is a negative number.

PURE IMAGINARY NUMBERS

We will understand what complex numbers are if we consider first complex numbers such as $\sqrt{-9}$ and $\sqrt{-20}$, which are called pure imaginary numbers, according to the following definition:

A ***pure imaginary number*** is a number that can be expressed as the product of a real number and $\sqrt{-1}$. If b is a real number, then $b\sqrt{-1}$ is a pure imaginary number.

For example, $\sqrt{-9} = \sqrt{9}\sqrt{-1} = 3\sqrt{-1}$.

Also, $\sqrt{-20} = \sqrt{20}\sqrt{-1} = \sqrt{4}\sqrt{5}\sqrt{-1} = 2\sqrt{5}\sqrt{-1}$.

IMAGINARY UNIT *i* OR $\sqrt{-1}$

Since $\sqrt{-1}$ is a factor of every pure imaginary number, the symbol i is used to represent $\sqrt{-1}$. The imaginary number i is called the ***imaginary unit.***

Hence, $\sqrt{-9} = 3\sqrt{-1} = 3i$ and $\sqrt{-20} = 2\sqrt{5}\sqrt{-1} = 2i\sqrt{5}$.
In general, if b is a real number,

$$\sqrt{-b^2} = \sqrt{b^2}\sqrt{-1} = bi$$

Since pure imaginary numbers are not real numbers, they cannot be graphed using the real number line.

POWERS OF *i*

Since $i = \sqrt{-1}$, then $i^2 = -1$. Since $i^3 = i^2 \cdot i$, then $i^3 = -1 \cdot i$ or $-i$.
Since $i^4 = i^2 \cdot i^2$, then $i^4 = (-1)(-1) = 1$.

Since $i^5 = i^4 \cdot i$, then $i^5 = 1 \cdot i$ or i.

Continuing in this way to obtain successive powers of i, we find that the first four powers of i are all different, as the following display shows; but, thereafter, there is a repetition in cycles of four:

$$i^1 = i \qquad\qquad i^5 = i \qquad\qquad i^9 = i$$
$$i^2 = -1 \qquad\qquad i^6 = -1 \qquad\qquad i^{10} = -1$$
$$i^3 = -i \qquad\qquad i^7 = -i \qquad\qquad i^{11} = -i$$
$$i^4 = 1 \qquad\qquad i^8 = 1 \qquad\qquad i^{12} = 1$$

In general, if n is a natural number, then

$$i^{4n} = 1 \qquad i^{4n+1} = i \qquad i^{4n+2} = -1 \qquad i^{4n+3} = -i$$

Thus,

$$i^{20} = i^{4(5)} = 1; \ i^{25} = i^{4(6)+1} = i; \ i^{102} = i^{4(25)+2} = -1; \ i^{403} = i^{4(100)+3} = -i.$$

PROPERTIES OF *i*

Property 1. $i \cdot i = i^2 = -1$. $i^2 = -1$ because $\sqrt{-1}\sqrt{-1} = (\sqrt{-1})^2 = -1$. This property can be used to obtain higher powers of i such as i^4, i^6, etc.

Caution. When multiplying $\sqrt{-1}$ and $\sqrt{-1}$, do not apply the rule $\sqrt{a}\sqrt{b} = \sqrt{ab}$ since this rule applies to situations where a and b are not both negative; that is, $a \geq 0$ or $b \geq 0$. Keep in mind that $\sqrt{-1}\sqrt{-1} \neq \sqrt{1}$ or 1.

Property 2. $0i = 0 \cdot i = 0$. This property is the multiplicative property of zero.

Property 3. $i + 0 = i$. Here, 0 or $0i$ is the additive identity.

Property 4. $i + (-i) = 0$. Hence, i and $-i$ are additive inverses.

Property 5. $1i = 1 \cdot i = i$. Here, 1 is the multiplicative identity. It follows that $\dfrac{i}{i} = 1$.

Property 6. $i(-i) = 1$. Hence, i and $-i$ are multiplicative inverses, or reciprocals of each other. Proof: $i(-i) = -i^2 = -(-1) = 1$.

In the following model problems dealing with imaginary numbers, problems 1 and 2 involve addition and subtraction; problems 3–5 involve multiplication and division.

∿∿∿∿∿ *MODEL PROBLEMS* ∿∿∿∿∿

In 1 and 2, simplify in terms of i and combine.

	1. $\sqrt{-49} - 2\sqrt{-4}$	**2.** $3\sqrt{-2} + \sqrt{-8}$
How To Proceed	*Solution*	*Solution*
1. Express each number in terms of i.	$\sqrt{49}\sqrt{-1} - 2\sqrt{4}\sqrt{-1}$ $7i - 4i$	$3\sqrt{2}\sqrt{-1} + \sqrt{4}\sqrt{2}\sqrt{-1}$ $3i\sqrt{2} + 2i\sqrt{2}$
2. Factor the highest common factor.	$i(7-4)$	$i\sqrt{2}(3+2)$
3. Perform the indicated operations.	$3i$ *Ans.*	$5i\sqrt{2}$ *Ans.*

In 3–5, simplify in terms of i and perform the indicated operation.

	3. $\sqrt{-2} \cdot \sqrt{-8}$	**4.** $\dfrac{\sqrt{-20}}{\sqrt{-5}}$	**5.** $\dfrac{-\sqrt{-48}}{\sqrt{-6}}$
How To Proceed	*Solution*	*Solution*	*Solution*
1. Express each number in terms of i.	$i\sqrt{2} \cdot i\sqrt{8}$	$\dfrac{i\sqrt{20}}{i\sqrt{5}}$	$\dfrac{-i\sqrt{48}}{i\sqrt{6}}$
2. Perform the indicated operations, simplifying powers of i as needed.	$i^2\sqrt{16}$ $(-1)4$	$\dfrac{\sqrt{4}\sqrt{5}}{\sqrt{5}}$	$-\sqrt{8}$ $-\sqrt{4}\sqrt{2}$
	-4 *Ans.*	2 *Ans.*	$-2\sqrt{2}$ *Ans.*

The preceding model problems illustrate the following rules:

Rule 1. The product of two pure imaginary numbers is a real number.

Rule 2. The quotient of two pure imaginary numbers is a real number.

Exercises

In 1–32, express in terms of the unit i.

1. $\sqrt{-25}$	**2.** $\sqrt{-16}$	**3.** $\sqrt{-64}$	**4.** $\sqrt{-100}$
5. $8\sqrt{-4}$	**6.** $3\sqrt{-49}$	**7.** $2\sqrt{-36}$	**8.** $5\sqrt{-81}$
9. $\frac{1}{2}\sqrt{-16}$	**10.** $\frac{3}{4}\sqrt{-64}$	**11.** $\frac{2}{3}\sqrt{-81}$	**12.** $\frac{5}{6}\sqrt{-144}$
13. $\sqrt{-\frac{1}{9}}$	**14.** $\sqrt{-\frac{4}{25}}$	**15.** $8\sqrt{-\frac{9}{16}}$	**16.** $10\sqrt{-\frac{49}{25}}$
17. $\sqrt{-5}$	**18.** $\sqrt{-13}$	**19.** $2\sqrt{-7}$	**20.** $6\sqrt{-15}$
21. $\sqrt{-8}$	**22.** $\sqrt{-12}$	**23.** $-\sqrt{-75}$	**24.** $-\sqrt{-32}$
25. $4\sqrt{-12}$	**26.** $\frac{1}{3}\sqrt{-63}$	**27.** $-\frac{1}{2}\sqrt{-48}$	**28.** $-\frac{2}{5}\sqrt{-50}$
29. $\sqrt{-\frac{1}{2}}$	**30.** $\sqrt{-\frac{1}{7}}$	**31.** $-6\sqrt{-\frac{1}{3}}$	**32.** $-10\sqrt{-\frac{1}{5}}$

In 33–39, indicate whether the power of i is equal to 1, i, -1, or $-i$.

33. i^9 **34.** i^{40} **35.** i^{11} **36.** i^{102} **37.** i^{22} **38.** i^{100} **39.** i^{401}

In 40–42, express as a single term the sum of:

40. $5\sqrt{-1}$ and $2i$ **41.** $2\sqrt{-9}$ and $5i$ **42.** $\sqrt{-16}$ and $-i$

In 43–52, simplify in terms of i and combine.

43. $\sqrt{-16} + \sqrt{-49}$ **44.** $\sqrt{-25} - \sqrt{-144}$

45. $5\sqrt{-36} + 2\sqrt{-81}$ **46.** $8\sqrt{-100} - 4\sqrt{-64}$

47. $4\sqrt{-1} - \sqrt{-4}$ **48.** $10\sqrt{-\frac{1}{25}} + 3\sqrt{-9}$

49. $\frac{3}{4}\sqrt{-64} - \frac{2}{3}\sqrt{-81}$ **50.** $20\sqrt{-\frac{1}{100}} - 12\sqrt{-\frac{1}{36}}$

51. $\sqrt{-72} + \sqrt{-32}$ **52.** $5\sqrt{-27} - 3\sqrt{-12}$

In 53–56, explain the error and state the correct answer.

53. A student replaced $\sqrt{-36}$ by $6\sqrt{i}$.

54. A student found the product of $\sqrt{-4}$ and $\sqrt{-9}$ to be 6.

55. A student said that $\sqrt{-49} - \sqrt{-9} = -4i$.

56. A student found the quotient of $\dfrac{-\sqrt{64}}{\sqrt{-4}}$ to be 4.

In 57–60, state the greatest positive integral value of x for which the radical is imaginary.

57. $\sqrt{x-8}$ **58.** $\sqrt{x^2-8}$ **59.** $\sqrt{2x-8}$ **60.** $\sqrt{2x^2-8}$

In 61–64, state the least positive integral value of x for which the radical is imaginary.

61. $\sqrt{8-x}$ **62.** $\sqrt{8-x^2}$ **63.** $\sqrt{8-2x}$ **64.** $\sqrt{8-2x^2}$

65. State the domain of x if $\sqrt{16-x^2}$ is a real number.

66. State the domain of x if $\sqrt{x^2-25}$ is imaginary.

In 67–70, simplify in terms of i and multiply.

67. $\sqrt{-1} \cdot \sqrt{-4}$ **68.** $\sqrt{-3} \cdot \sqrt{-12}$

69. $\sqrt{-4} \cdot \sqrt{-9} \cdot \sqrt{-100}$ **70.** $(2\sqrt{-8})(5\sqrt{-18})$

In 71–75, simplify in terms of i and divide.

71. $\dfrac{\sqrt{-18}}{\sqrt{-2}}$ **72.** $\dfrac{\sqrt{-3}}{\sqrt{-75}}$ **73.** $\dfrac{8\sqrt{-1}}{2\sqrt{-4}}$ **74.** $\dfrac{-\sqrt{-25}}{-\sqrt{5}}$ **75.** $\dfrac{\sqrt{-5}}{-\sqrt{25}}$

In 76–78, if x is a real number, simplify the expression in terms of i.

76. $\sqrt{-x^2} + \sqrt{-9x^2}$ **77.** $10\sqrt{-x^2} - \sqrt{-49x^2}$ **78.** $x\sqrt{-25} + \sqrt{-100x^2}$

In 79–82, state the roots of the equation in terms of i.

79. $x^2 + 4 = 0$ **30.** $2x^2 + 4 = 0$ **81.** $\dfrac{x^2}{2} + 4 = 0$ **82.** $18 + 3x^2 = x^2$

In 83–86, show that if a and b are positive real numbers, $a \neq b$, then:

83. $\sqrt{-a^2} + \sqrt{-b^2}$ is a pure imaginary number.

84. $\sqrt{-a^2} - \sqrt{-b^2}$ is a pure imaginary number.

85. $(\sqrt{-a})(\sqrt{-b})$ is a real negative number.

86. $\dfrac{\sqrt{-a}}{\sqrt{-b}}$ is a real positive number.

2. Complex Numbers

Numbers such as $2 + 3i$, $\frac{1}{2} - i\sqrt{3}$, and $\sqrt{2} + \frac{1}{4}i$ are *complex numbers*. Note that each of these numbers is the sum of a real number and a pure imaginary number. By definition, a **complex number** is a number that may be expressed in the form of $a + bi$ where a and b are real numbers and $i = \sqrt{-1}$.

The set of real numbers and the set of imaginary numbers are subsets of the system of complex numbers, for the following reasons:

1. A real number is expressible as a complex number, $a + bi$, if $b = 0$. Thus, the real numbers -6 and $3\sqrt{2}$ are expressible as the complex numbers $-6 + 0i$ and $3\sqrt{2} + 0i$.
2. An imaginary number is expressible as a complex number, $a + bi$, if $b \neq 0$. If $a = 0$, then the imaginary number $a + bi$ becomes $0 + bi$, or bi, which is a pure imaginary number. Thus, $2 + 3i$, $\frac{1}{2} - i\sqrt{3}$, and $2 + \frac{1}{4}i$ are imaginary numbers, whereas $0 + 3i$, or $3i$, and $0 + \frac{1}{4}i$, or $\frac{1}{4}i$, are pure imaginary numbers.

The following chart shows how the set of complex numbers includes the sets of real numbers, imaginary numbers, and pure imaginary numbers:

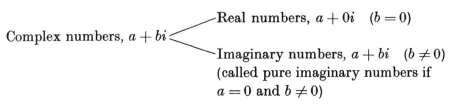

Complex numbers, $a + bi$
- Real numbers, $a + 0i$ $(b = 0)$
- Imaginary numbers, $a + bi$ $(b \neq 0)$
(called pure imaginary numbers if $a = 0$ and $b \neq 0$)

GRAPHING COMPLEX NUMBERS

The terms *real* and *imaginary* were first used to refer to number systems in 1637 by the French mathematician and philosopher René Descartes. The term *imaginary* indicates the fact that when the imaginary numbers were first discovered, they were considered to be curiosities of no real significance. It was Karl Friedrich Gauss, one of the greatest mathematicians of all time, who demonstrated their importance in the development of mathematical theory. In 1832, Gauss used the term *complex number*. In modern times, Charles Steinmetz, an American engineer, through his use of complex numbers in the analysis of electrical problems, proved beyond all doubt the enormous practicality of the complex number system in the physical world.

Gauss gave a physical meaning to complex numbers by representing them on a two-dimensional plane. The *real number line* is drawn horizontally, and the *pure imaginary number line* is drawn vertically, as shown at the right. Since $0i = 0$, it is natural that these number lines intersect at a point that represents 0 on the real number line and $0i$ on the imaginary number line. Therefore, this point of intersection represents the complex number $0 + 0i$.

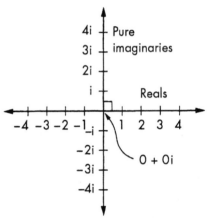

These two axes are used to form the **complex number plane.** The real number axis is the x-axis, and the pure imaginary number axis is called the yi-axis.

The complex number plane is similar to the rectangular coordinate system studied earlier. In the same way that we located point $(x, y) = (2, 5)$, we now locate the point for the complex number $x + yi = 2 + 5i$. In other words, as shown at the right, we use the rectangular grid to locate the point of intersection of the real component 2 and the imaginary component $5i$.

By studying other points on the complex number plane, we observe:

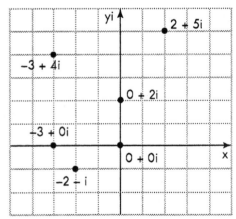

1. Complex numbers that are real numbers, for example, $-3 + 0i = -3$, are represented by a point on the real number axis.
2. Complex numbers that are pure imaginary numbers, for example, $0 + 2i = 2i$, are represented by a point on the pure imaginary axis.
3. Complex numbers $a + bi$ where $a \neq 0$ and $b \neq 0$, for example, $-3 + 4i$, $-2 - i$, $2 + 5i$, are represented by points on the complex plane that are not on either axis.

Recall that on the real number line the absolute value of a real number is the distance between the graph of the number and 0. The *absolute value of a complex number* $a + bi$, represented by $|a + bi|$, is $\sqrt{a^2 + b^2}$ by the Pythagorean Theorem.

In the diagram at the right, the absolute value of $3 + 2i$ is represented by $|3 + 2i| = \sqrt{3^2 + 2^2} = \sqrt{9 + 4} = \sqrt{13}$.

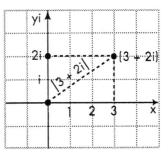

Exercises

In 1–10, graph each number on the complex plane and find its absolute value.

1. $5 + 2i$	2. $4i$	3. $-1 - 5i$	4. $7i$	5. $-3 + 6i$
6. -2	7. $-i$	8. $6 - 2i$	9. 0	10. $4 + 3i$

In 11 and 12, select the *numeral* preceding the expression that best answers the question: Which number is included in the shaded area?

11. (1) $-1.5 + 3.5i$
    (2) $1.5 - 3.5i$
    (3) 3.5
    (4) $4.5i$

12. (1) $4 + 0.5i$
    (2) $-4 - 0.5i$
    (3) $2.5i$
    (4) $2.4 - i$

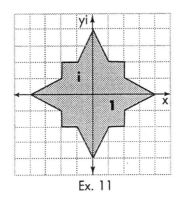

Ex. 11

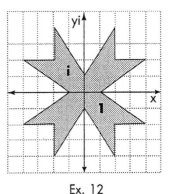

Ex. 12

3. Properties of the System of Complex Numbers

We have assumed that the set of complex numbers has two special properties, namely, that the set has a special element i and that every complex number can be written in the form of $a + bi$.

Thus, $\sqrt{-81} = 9i$, which can be written as $0 + 9i$.

In addition to the two special properties, *the set of complex numbers has all the properties of the set of real numbers.* Among these properties are the commutative, associative, and distributive properties, which the following illustrate:

1. $i(2i) = (2i)i$ Commutative property of multiplication
   $= 2(i \cdot i)$ Associative property of multiplication
   $= 2i^2 = 2(-1) = -2$ Properties of the special element i

2. $5i + 17 + 3i = 17 + 5i + 3i$ Commutative property of addition
   $= 17 + (5 + 3)i$ Distributive property
   $= 17 + 8i$ Expressing the complex number in the form $a + bi$

Note in the two illustrations that the special complex element i can be treated in exactly the same way as any real number. Keep in mind that i has its own special property which is $i \cdot i = i^2 = -1$.

IDENTITY AND INVERSE ELEMENTS IN THE SYSTEM OF COMPLEX NUMBERS

The *additive identity element* in the set of complex numbers is $0i$ or 0. The *multiplicative identity element* is $1 + 0i$, or 1.

Thus, $(a + bi) + 0 = a + bi$. Also, $1(a + bi) = a + bi$. In each case, a and b are real numbers.

Each complex number has one and only one additive inverse. The additive inverse of the general complex number $a + bi$ is $-(a + bi)$, which can be proved equal to $-a - bi$.

Thus, the additive inverse of $5 - 3i$ is $-(5 - 3i)$, or $-5 + 3i$.

Each complex number except $0i$ has one and only one multiplicative inverse. The multiplicative inverse of the general complex number $a + bi$ is its reciprocal $\dfrac{1}{a + bi}$. Thus, the multiplicative inverse of $5i$ is $\dfrac{1}{5i}$. If the denominator is rationalized, we obtain $\dfrac{1}{5i} \cdot \dfrac{i}{i} = \dfrac{i}{5i^2} = \dfrac{i}{-5} = -\dfrac{i}{5}$, or $-\dfrac{1}{5}i$.

EQUALITY OF TWO COMPLEX NUMBERS

Two complex numbers are equal if and only if their real parts are equal and, also, their imaginary parts are equal.

Hence, if $a + bi$ and $c + di$ are equal complex numbers, $a + bi = c + di$ if and only if $a = c$ and $b = d$. For example, $\frac{1}{2} + xi = y + 7i$ if and only if $x = 7$ and $y = \frac{1}{2}$.

COMPLEX NUMBER CONJUGATES

Complex numbers such as $4 + 5i$ and $4 - 5i$ are *conjugates* of each other. If we assume that the familiar properties of real numbers are true for complex numbers, then we obtain the following sum, difference, and product of these conjugates:

Sum: $(4 + 5i) + (4 - 5i) = 4 + 4 = 8$
Difference: $(4 + 5i) - (4 - 5i) = 5i + 5i = 10i$
Product: $(4 + 5i)(4 - 5i) = 16 - 25i^2 = 16 - 25(-1) = 16 + 25 = 41$

The general forms of complex number conjugates are $a + bi$ and $a - bi$.

General Rules for the Sum, Difference, and Product of Any Two Complex Number Conjugates

Sum: $(a + bi) + (a - bi) = 2a$

Difference: $(a + bi) - (a - bi) = 2bi$

Product: $(a + bi)(a - bi) = a^2 - b^2i^2 = a^2 - b^2(-1) = a^2 + b^2$
      Hence, $(a + bi)(a - bi) = a^2 + b^2$.

KEEP IN MIND

If $a + bi$ and $a - bi$ are complex number conjugates, then:

1. Their sum, $2a$, is a real number.
2. Their difference, $2bi$, $b \neq 0$, is a pure imaginary number.
3. Their product, $a^2 + b^2$, is a positive real number unless both a and b are 0.

~~~~~~~ *MODEL PROBLEMS* ~~~~~~~

Express in the form $a + bi$ and combine.

$$(6 + \sqrt{-50}) - (4 - \sqrt{-18})$$

How To Proceed	*Solution*
1. Express in the form $a + bi$.	$(6 + 5i\sqrt{2}) - (4 - 3i\sqrt{2})$
2. Combine the complex numbers as real numbers are combined, and express the result in the form $a + bi$.	$(6 - 4) + (5i\sqrt{2} + 3i\sqrt{2})$ $2 + 8i\sqrt{2}$ *Ans.*

In general, the sum and the difference of complex numbers $(a + bi)$ and $(c + di)$ are:

$$(a + bi) + (c + di) = (a + c) + (b + d)i$$

$$(a + bi) - (c + di) = (a - c) + (b - d)i$$

Note. Since a, b, c, and d are real numbers, then $(a + c)$, $(a - c)$, $(b + d)$, and $(b - d)$ are real numbers. Therefore, the sum or the difference of two complex numbers is a complex number.

Addition of complex numbers can be modeled on a coordinate plane. To find $(2 + 3i) + (5 + i)$, use these steps.

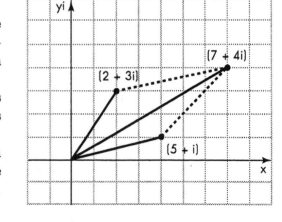

1. Graph $2 + 3i$ and $5 + i$ on the complex plane. Draw the line segment joining the origin and each point.
2. Draw a parallelogram that has the two segments drawn above as sides.
3. The diagonal of the parallelogram from the origin represents the sum of the two complex numbers.

$$(2 + 3i) + (5 + i) = 7 + 4i$$

~~~~~~~ *MODEL PROBLEMS* ~~~~~~~

In 1 and 2, find the product.

1. $(6 + \sqrt{-49})(3 - \sqrt{-16})$ **2.** $(4 + \sqrt{-27})(1 - \sqrt{-12})$

How To Proceed	*Solution*	*Solution*
1. Express each complex number in the form $a + bi$.	$(6 + 7i)(3 - 4i)$	$(4 + 3i\sqrt{3})(1 - 2i\sqrt{3})$

2. Multiply the complex numbers as real numbers are multiplied, expressing the result in the form $a + bi$.

$$18 - 3i - 28i^2$$
$$18 - 3i + 28$$
$$46 - 3i \quad Ans.$$

$$4 - 5i\sqrt{3} - 18i^2$$
$$4 - 5i\sqrt{3} + 18$$
$$22 - 5i\sqrt{3} \quad Ans.$$

In general, the product of complex numbers $(a + bi)$ and $(c + di)$ is:

$$(a + bi)(c + di) = ac + adi + bci + bdi^2$$
$$= (ac - bd) + (ad + bc)i$$

Note. Since a, b, c, and d are real numbers, then $(ac - bd)$ and $(ad + bc)$ are real numbers. Therefore, the product of two complex numbers is a complex number.

Exercises

In 1–3, find the sum and the difference of the pair of conjugates.

1. $(1 + i)$ and $(1 - i)$ **2.** $(5 + 3i)$ and $(5 - 3i)$ **3.** $(4 + i\sqrt{3})$ and $(4 - i\sqrt{3})$

In 4–6, find the product of the pair of conjugates.

4. $(2 + 5i)$ and $(2 - 5i)$ **5.** $(12 - 7i)$ and $(12 + 7i)$
6. $(3 + 2i\sqrt{5})$ and $(3 - 2i\sqrt{5})$

In 7–12, find real numbers x and y for which the equation is true.
7. $x + 4i = -5 + yi$ **8.** $2.8 + xi = y - 3.5i$
9. $2(5 - i) = x + yi$ **10.** $x + 6i + 3 = 12 + yi$
11. $2(x + 2i) = 4(8 - yi)$ **12.** $(x + 3i) + 7i = 8 + 2(8 + yi)$

In 13–19, find the conjugate and the reciprocal of the complex number.
13. i **14.** $3i$ **15.** $-8i$ **16.** $\frac{1}{4}i$ **17.** $2 + i$
18. $5 - \sqrt{-25}$ **19.** $-3 + \sqrt{-2}$

In 20–25, express the fraction in the complex number form $a + bi$.

20. $\dfrac{5}{i}$ **21.** $\dfrac{1}{3i}$ **22.** $\dfrac{7}{4 - i}$ **23.** $\dfrac{1}{3 + 2i}$ **24.** $\dfrac{2i}{4 - i}$ **25.** $\dfrac{10 - i}{10 + i}$

In 26–50, find the complex number $a + bi$ equal to the expression.
26. $(3 + 5i) + (5 - 2i)$ **27.** $(4 - 7i) + (15 + 2i)$
28. $(7 + i\sqrt{2}) + (1 - 2i\sqrt{2})$ **29.** $(11 + \sqrt{-1}) + (-3 - \sqrt{-16})$
30. $(9 - \sqrt{-5}) + (-4 - \sqrt{-20})$ **31.** $(20 + \sqrt{-28}) + (4 - \sqrt{-7})$

32. $(5 + 2i) + (7 - 4i) + (15 - 8i)$

33. $3i + 8i - 4i - 2i + (6 - 7i) + (3 + 2i)$

34. $(15 + 3i) - (-10 + 4i)$

35. $(3 + 15i) - (-4 + 7i)$

36. $(2.5 + i) - (1.7 - 2i)$

37. $(9 - \sqrt{-100}) - \sqrt{-81}$

38. $2\sqrt{-25} - (2 - \sqrt{-36})$

39. $(4\sqrt{5} + \sqrt{-50}) - 3\sqrt{20}$

40. $6i(5i)$

41. $6(5 + i)(5 - i)$

42. $6i(10 - 2i)$

43. $(3 + i)(2 - i)$

44. $(1 + 3i)(1 - 4i)$

45. $10(1 + \sqrt{-4})(2 + \sqrt{-9})$

46. $(1 + i)^2$

47. $(2 - i)^2$

48. $(3 + 7i)^2$

49. $(2 + \sqrt{-49})^2$

50. $(1 + i)(1 - i)(1 - i^2)(1 + i^2)$

In 51–59, if v, w, x, and y are real numbers, write the expression in the form of $a + bi$.

51. $(v + wi) + (v - wi)$ **52.** $(x + yi) - (x - yi)$ **53.** $(v + xi)(v - xi)$

54. $(x + yi)^2$ **55.** $(v - wi)^2$ **56.** $(x + yi)^2 + (x - yi)^2$

57. $(x + yi)^2 - (x - yi)^2$ **58.** $(v + wi) + (x + yi)$ **59.** $(x - yi) - (x - wi)$

In 60–68, if a, b, c, and d are real numbers, prove the statement.

60. $(a + bi) + (0 + 0i) = a + bi$ **61.** $(c + di) + (-c - di) = 0$

62. The sum of any complex number $a + bi$ and its conjugate is a real number.

63. The difference between any complex number $a + bi$ and its conjugate is a pure imaginary number.

64. The product of any complex number $c + di$ and its conjugate is a real positive number, if either c or d does not equal 0.

65. The multiplicative inverse of bi is $-\dfrac{i}{b}$, or $-\dfrac{1}{b}i$.

66. The multiplicative inverse of $a + bi$ equals $\dfrac{a - bi}{a^2 + b^2}$, or $\dfrac{a}{a^2 + b^2} - \dfrac{b}{a^2 + b^2}i$.

(*Hint:* Multiply both numerator and denominator by $a - bi$.)

67. The conjugate of the sum of two complex numbers $(a + bi)$ and $(c + di)$ is the sum of the conjugates of the numbers.

68. The conjugate of the product of two complex numbers $(a + bi)$ and $(c + di)$ is the product of the conjugate of the two numbers.

69. By showing that $(a + bi) + (c + di) = (c + di) + (a + bi)$, prove that the set of complex numbers is commutative under addition.

70. By showing that $(a + bi)(c + di) = (c + di)(a + bi)$, prove that the set of complex numbers is commutative under multiplication.

CALCULATOR APPLICATIONS

Any attempt to display an imaginary number on a calculator will result in an ERROR message.

$\sqrt{-4} \rightarrow$ *Enter:* 2nd $\boxed{\sqrt{}}$ (−) 4 ENTER

Display: ERR: nonREAL Ans

The TI-83 and other graphing calculators, however, have features for performing operations with complex numbers. If complex numbers are entered in $a + bi$ form, operations can be carried out in REAL mode or $a + bi$ mode. To use radicals, the calculator must be in $a + bi$ mode. The MATH CPX menu includes special commands for the complex conjugate, real and imaginary parts, and absolute value of complex numbers.

MODEL PROBLEMS

1. Find the absolute value of $7 + 9i$.

   Solution: With the calculator in REAL mode:

   Enter: MATH ▶ ▶ 5

   7 + 9 2nd i) ENTER

   Display: 11.40175425

   Note that the absolute value is given in decimal form. To express the absolute value in radical form, use the property that $\left(\sqrt{x}\right)^2 = x$.

   Enter: 2nd ANS ENTER

   Display: 130 ◄— This is $\left(\sqrt{a^2 + b^2}\right)^2$.

   Answer: The absolute value of $7 + 9i$ is $\sqrt{130}$, or approximately 11.4018.

2. Find the complex number equal to the expression $\left(13 - \sqrt{-4}\right) + \left(-5 + \sqrt{-9}\right)$.

   Solution:

   Method 1   The calculator is in REAL mode.
   First, express each number in $a + bi$ form.

   $$13 - \sqrt{-4} = 13 - 2i$$
   $$-5 + \sqrt{-9} = -5 + 3i$$

Enter: 13 ⎡−⎤ 2 ⎡2nd⎤ ⎡ *i* ⎤ ⎡ + ⎤ ⎡(−)⎤ 5 ⎡ + ⎤ 3 ⎡2nd⎤ ⎡ *i* ⎤ ⎡ENTER⎤

Display: ⎡ 8 + *i* ⎤

Method 2   Put the calculator into $a + bi$ mode.

Enter: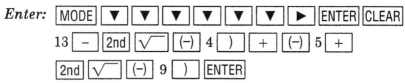

Display: ⎡ 8 + *i* ⎤

Answer: $\left(13 - \sqrt{-4}\right) + \left(-5 + \sqrt{-9}\right) = 8 + i$

3. Simplify $\dfrac{6 - 5i}{2 - i}$.

Solution: Note that on the calculator, it is not necessary to multiply numerator and denominator by the conjugate of $2 - i$. However, parentheses must be used for the numerator and the denominator.

Enter: ⎡ (⎤ 6 ⎡−⎤ 5 ⎡2nd⎤ ⎡ *i* ⎤ ⎡) ⎤ ⎡ ÷ ⎤ ⎡ (⎤ 2 ⎡−⎤ ⎡2nd⎤ ⎡ *i* ⎤ ⎡) ⎤ ⎡ENTER⎤

Display: ⎡ 3.4 − .8*i* ⎤

To convert to fraction form:

Enter: ⎡2nd⎤ ⎡ANS⎤ ⎡MATH⎤ 1 ⎡ENTER⎤

Display: ⎡ 17/5 − 4/5*i* ⎤

Had the approach been to multiply numerator and denominator by the conjugate, the number of keystrokes needed would have been much greater, increasing the risk of entry error. However, the final answer would have been the same. Try this method as an exercise.

4. Simplify i^{39}.

Solution:

Enter: ⎡2nd⎤ ⎡ *i* ⎤ ⎡ ∧ ⎤ 39 ⎡ENTER⎤

Display: ⎡ − 1E −13 − *i* ⎤

In this case, the calculator uses $-1\text{E}-13$ to represent 0.

Answer: $i^{39} = -i$

CHAPTER X

Quadratic Equations

1. Understanding Quadratic Equations in One Variable

STANDARD FORM OF A QUADRATIC EQUATION: $ax^2 + bx + c = 0, \; a \neq 0$

A *quadratic equation*, or a *second-degree equation* in x, is one that can be written in the *standard quadratic form*, $ax^2 + bx + c = 0$, in which a, b, and c are real numbers and $a \neq 0$.

Thus, $5x^2 + 7x + 6 = 0$ is in standard quadratic form; $a = 5$, $b = 7$, and $c = 6$.

TRANSFORMING A QUADRATIC EQUATION INTO STANDARD FORM

Note that if a quadratic equation is written in standard form, the expression $ax^2 + bx + c$ is "equated to 0." Hence, in order to transform a quadratic equation into standard form, it may be necessary to perform one or more of the following operations:

1. *Clear of fractions.* The equation $3x = \dfrac{8}{x} + 4$ is cleared of fractions by multiplying each side by x. By equating to 0, the result, $3x^2 = 8 + 4x$ is transformed into an equation in standard form, $3x^2 - 4x - 8 = 0$; $a = 3$, $b = -4$, and $c = -8$.

2. *Remove parentheses.* The parentheses in the equation $2(x^2 - 7) = 12x$ are removed by applying the distributive law. By equating to 0, the result, $2x^2 - 14 = 12x$, is transformed into an equation that is in standard form, $2x^2 - 12x - 14 = 0$; $a = 2$, $b = -12$, and $c = -14$.

3. *Remove radical signs.* The radical sign in the equation $x + 1 = \sqrt{3x + 7}$ is removed by squaring each side of the equation. By equating to 0, the result, $x^2 + 2x + 1 = 3x + 7$ is transformed into an equation in standard form, $x^2 - x - 6 = 0$; $a = 1$, $b = -1$, and $c = -6$.

4. *Collect like terms.* The like terms in the equation $x^2 + 2x + 1 = 3x + 3$ are

collected and combined by subtracting $3x + 3$ from each side. The result is $x^2 + 2x + 1 - 3x - 3 = 0$. By combining like terms, the result is transformed into an equation in standard form, $x^2 - x - 2 = 0$; $a = 1$, $b = -1$, and $c = -2$.

SETS OF NUMBERS INVOLVED IN QUADRATIC EQUATIONS

The real numbers a, b, and c in the standard quadratic form, $ax^2 + bx + c = 0$, could be either rational or irrational. However, *unless otherwise stated, a, b, and c are to be considered as rational numbers.*

KEEP IN MIND

In the standard quadratic form:
a is the coefficient of x^2, or the coefficient of the second-degree term.
b is the coefficient of x, or the coefficient of the first-degree term.
c is the constant term.

The roots of a quadratic equation need not be restricted to the set of real numbers but may be members of the set of complex numbers.

SOLVING A QUADRATIC EQUATION

A quadratic equation in one variable is solved when the replacements for the variable that satisfy the equation are found. The values of the variable that satisfy the equation are the roots of the equation and the elements of its solution set.

Thus, the equation $x^2 - x - 6 = 0$ is solved when its roots, 3 and -2, are found. The solution set of $x^2 - x - 6 = 0$ is $\{3, -2\}$.

CHECKING THE ROOTS OF A QUADRATIC EQUATION

The roots of a quadratic equation are checked by testing each of them in the *original* equation. There are two important reasons for this requirement:

1. A derived equation may be derived incorrectly.
2. A derived equation may be derived correctly, but its roots may not satisfy the original equation. Such roots, called **extraneous values**, may occur when both sides of an equation are squared, cubed, or raised to a power to remove the radical sign.

Thus, when the radical sign in $x + 1 = \sqrt{3x + 7}$ is removed by squaring, the derived equation, $x^2 - x - 6 = 0$, has two roots, 3 and -2. However, the value -2 is a root of the derived equation $x^2 - x - 6 = 0$, but is not a root of the original equation $x + 1 = \sqrt{3x + 7}$. Hence, -2 is an extraneous value that does not satisfy the original equation.

The quadratic equation $x^2 = 25$ or $x^2 - 25 = 0$ has two unequal roots, 5 and -5. However, the quadratic equation $x^2 - 10x + 25 = 0$, or $(x - 5)^2 = 0$, has two equal roots, 5 and 5.

Rule. A quadratic equation has two roots which may be equal or unequal.

Equal roots are referred to as **multiple roots.**

Exercises

In 1–12, transform the equation into standard quadratic form, $ax^2 + bx + c = 0$ with $a > 0$. Then state the values of a, b, and c in the resulting equation.

1. $3x = \dfrac{9}{x} + 26$ **2.** $2x + \dfrac{6}{x} = 13$ **3.** $5x - 8 = \dfrac{13}{x}$

4. $3(x^2 - 2) = 7x$ **5.** $x(6 - x) = 5$ **6.** $2(x - 3) = \dfrac{8}{x}$

7. $\sqrt{x^2 - 3x} = 2$ **8.** $\sqrt{8x + 9} = x + 2$ **9.** $x = \sqrt{2x^2 + 3x}$

10. $x^2 = 2x^2 + 7x$ **11.** $9x^2 - 4x^2 = 20$ **12.** $7x^2 + 4x = x^2 + 3x + 5$

In 13–20, by substituting the given values in parentheses, determine which of the values are roots of the given equation.

13. $x(x - 2) = 15$ (5 and -3)   **14.** $x - \dfrac{10}{x} = -3$ (2 and 5)

15. $\sqrt{x^2 - 3x} = 2$ (4 and -1)   **16.** $\sqrt{x^2 - 3x} = -2$ (4 and -1)

17. $\sqrt{x(5 - x)} = x - 2$ (4 and $\frac{1}{2}$)   **18.** $9x^2 - 4x^2 = 20$ (2 and -2)

19. $3(x^2 - x) = 12 - 3x$ (2 and -2)   **20.** $25(x - x^2) = 0$ (-1 and 0)

2. Solving Quadratic Equations by Factoring

Principle. The product of two real numbers is zero if and only if one of the factors is zero. That is, *for all real numbers a and b, ab = 0 if and only if a = 0 or b = 0.*

Thus, $(x - 4)(x + 5) = 0$ if and only if $x - 4 = 0$ or $x + 5 = 0$. Hence, $x = 4$ or $x = -5$. Also, $(x - 7)^2 = 0$, or $(x - 7)(x - 7) = 0$, if and only if $x - 7 = 0$, or $x = 7$. Hence, $(x - 7)^2 = 0$ has two equal roots, 7 and 7.

~~~~~~~~~~~~~ *MODEL PROBLEMS* ~~~~~~~~~~~~~

In 1 and 2, solve and check.

1. $x(x+4)=21$ **2.** $2x-1=\dfrac{10}{x}$

How To Proceed	*Solution*	*Solution*
1. Equate to 0.	$x^2+4x=21$	$2x^2-x=10$
	$x^2+4x-21=0$	$2x^2-x-10=0$
2. Factor ax^2+bx+c.	$(x-3)(x+7)=0$	$(2x-5)(x+2)=0$
3. Let each factor $=0$.	$x-3=0 \mid x+7=0$	$2x-5=0 \mid x+2=0$
4. Solve each resulting equation.	$x=3 \mid x=-7$	$x=2\frac{1}{2} \mid x=-2$
5. Check the roots in the original equation. (The check is left to the student.)	Check in the original equation, $x(x+4)=21$. $x=3$ or $x=-7$ *Ans.*	Check in the original equation, $2x-1=\dfrac{10}{x}$. $x=2\frac{1}{2}$ or $x=-2$ *Ans.*

3. Find the solution set of

$$2(x^2+1)=5x$$

Solution:
$$2(x^2+1)=5x$$
$$2x^2+2=5x$$
$$2x^2-5x+2=0$$
$$(2x-1)(x-2)=0$$
$$2x-1=0 \mid x-2=0$$
$$x=\tfrac{1}{2} \mid x=2$$

Answer: $\{\tfrac{1}{2}, 2\}$

4. List the members of the following set:

$$\{x \mid x^2-2x=15\}$$

Solution:
$$x^2-2x=15$$
$$x^2-2x-15=0$$
$$(x-5)(x+3)=0$$
$$x-5=0 \mid x+3=0$$
$$x=5 \mid x=-3$$

Answer: $\{5, -3\}$

5. Solve the equation $\dfrac{x^2+1}{5}=\dfrac{x+1}{6}$ for x.

Solution: If we multiply both members of the given equation by 30, we obtain:

$$6(x^2+1)=5(x+1)$$
$$6x^2+6=5x+5$$
$$6x^2-5x+1=0$$
$$(3x-1)(2x-1)=0$$
$$3x-1=0 \mid 2x-1=0$$
$$x=\tfrac{1}{3} \mid x=\tfrac{1}{2}$$

Answer: $x=\tfrac{1}{3}$, $x=\tfrac{1}{2}$

~~~~~~

Exercises

In 1–27, solve and check the equation.

1. $x^2 - 5x + 4 = 0$
2. $y^2 + 10 = 7y$
3. $c^2 + 6c + 8 = 0$
4. $r^2 + 8r = -15$
5. $x^2 - x - 6 = 0$
6. $y^2 = y + 12$
7. $x^2 - 36 = 0$
8. $25y^2 = 16$
9. $d^2 + 7d = -10$
10. $8m = m^2 + 15$
11. $3x^2 - 8x + 4 = 0$
12. $5x^2 + 4 = 12x$
13. $9x - 10 = 2x^2$
14. $2x^2 + x = 15$
15. $x + 4 = 3x^2$
16. $\frac{1}{3}x^2 + \frac{4}{3}x + 1 = 0$
17. $\frac{9}{4}x = \frac{1}{2}x^2 + 1$
18. $1 = \frac{1}{2}x^2 - \frac{7}{6}x$
19. $x = \dfrac{24}{x - 2}$
20. $\dfrac{c}{3} - 1 = \dfrac{6}{c}$
21. $\dfrac{2}{x - 1} = \dfrac{x}{x + 2}$
22. $\dfrac{4}{x - 1} = \dfrac{5}{2x - 2} + \dfrac{3x}{4}$
23. $\dfrac{200}{x} - 1 = \dfrac{200}{x + 10}$
24. $\sqrt{2x^2 + x} = \sqrt{x^2 + 6}$
25. $4\sqrt{25 - x^2} = 3x$
26. $\sqrt{3x + 10} = x + 2$
27. $\sqrt{x^2 - 16} + 7 = 2x$

In 28–35, find the solution set of the equation.

28. $x^2 - 2x - 8 = 0$
29. $2x^2 - x = 0$
30. $4x^2 - 1 = 0$
31. $5x^2 + 2 = 11x$
32. $10x^2 = 7x - 1$
33. $\sqrt{x^2 - 5} = 2$
34. $\sqrt{2x^2 - 2} = x - 1$
35. $\sqrt{2 - x^2} - 1 = 2x$

In 36–38, list the members of the set.

36. $\{x \mid x(x - 5) = 6\}$
37. $\{y \mid y^2 = 2(y + 4)\}$
38. $\left\{z \mid z - 5 = \dfrac{6}{z}\right\}$

39. If $(x - a)(x - b) = m$, then $(x - a) = m$ and $(x - b) = m$ when m is equal to either   (1) 1 or -1 (2) 1 or 0   (3) -1 or 0

3. Solving Quadratic Equations of the Form $ax^2 + bx = 0$

A quadratic equation lacking the constant term, c, is expressible in the form $ax^2 + bx = 0$. For example, $3x^2 + 5x = 0$ and $\dfrac{x^2}{3} - \dfrac{x}{2} = 0$ are quadratic equations lacking the constant term.

By factoring, $ax^2 + bx = 0$ becomes $x(ax + b) = 0$. Equating each factor to 0 yields $x = 0$ and $ax + b = 0$. Thus, $ax = -b$, and $x = -\dfrac{b}{a}$. The roots are 0 and $-\dfrac{b}{a}$.

Rule. A quadratic equation of the form $ax^2 + bx = 0$ has two roots, 0 and $-\dfrac{b}{a}$.

~~~~~~~~~~~~~~ **MODEL PROBLEMS** ~~~~~~~~~~~~~~

In 1 and 2, find the solution set by using the preceding rule.

1. $3(x^2 - x) = x^2$ **2.** $\dfrac{x^2}{3} + \dfrac{x}{9} = 0$

How To Proceed	*Solution*	*Solution*
1. Express in the form $ax^2 + bx = 0$.	$3x^2 - 3x = x^2$ $2x^2 - 3x = 0$	Multiply by 9. $3x^2 + x = 0$
2. Evaluate $-\dfrac{b}{a}$.	Since $a = 2$ and $b = -3$, $x = -\dfrac{-3}{2} = \dfrac{3}{2}$	Since $a = 3$ and $b = 1$, $x = -\tfrac{1}{3}$
3. State the roots, 0 and $-\dfrac{b}{a}$.	Roots are 0 and $\tfrac{3}{2}$. $\{0, \tfrac{3}{2}\}$ *Ans.*	Roots are 0 and $-\tfrac{1}{3}$. $\{0, -\tfrac{1}{3}\}$ *Ans.*

In 3 and 4, solve by factoring.

3. Solve for x and check:

$3x^2 = 9x$

Solution:

$$3x^2 = 9x$$
$$3x^2 - 9x = 0$$
$$3x(x - 3) = 0$$

$3x = 0 \mid x - 3 = 0$

$\quad x = 0 \mid \quad\quad x = 3$

Check: $3x^2 = 9x$

If $x = 0$, $3(0)^2 \overset{?}{=} 9(0)$

$\qquad\qquad 0 = 0$

If $x = 3$, $3(9) \overset{?}{=} 9(3)$

$\qquad\qquad 27 = 27$

Answer: $x = 0, x = 3$

4. Solve for x:

$$2x^2 + 3(x - 1) = 2x - 3$$

Solution:

$$2x^2 + 3(x - 1) = 2x - 3$$
$$2x^2 + 3x - 3 = 2x - 3$$
$$2x^2 + 3x - 2x = -3 + 3$$
$$2x^2 + x = 0$$
$$x(2x + 1) = 0$$

$x = 0 \mid 2x + 1 = 0$

$\qquad\quad \mid \qquad x = -\tfrac{1}{2}$

Answer: $x = 0, x = -\tfrac{1}{2}$

~~~~~~

Exercises

In 1–8, solve for x.

1. $x^2 - 4x = 0$ **2.** $x^2 + 3x = 0$ **3.** $2x^2 - 5x = 0$ **4.** $3x^2 - 2x = 0$

5. $x^2 = 5x$ **6.** $x^2 = -x$ **7.** $4(x^2 - 2x) = x$ **8.** $3x = 5x(x - 2)$

In 9–11, find the solution set.

9. $\dfrac{x^2}{2} - 3x = 0$　　　　**10.** $\dfrac{x^2}{3} + \dfrac{x}{2} = 0$　　　　**11.** $x^2 = \tfrac{1}{3}x$

In 12–17, solve for x.

12. $x^2 - 4x = 3x^2$　　　　**13.** $5x^2 + 2x = 3x^2 + 6x$　　**14.** $rx^2 + sx = 0$
15. $cx^2 = dx$　　　　　　　**16.** $\tfrac{1}{2}x^2 - 5ax = 0$　　　　**17.** $\tfrac{1}{2}mx^2 = \tfrac{1}{3}tx$

In 18–22, find the solution set.

18. $2x^2 - ax = 0$
19. $4x^2 = bx$
20. $3x^2 = \sqrt{3}x$
21. $3(x^2 + c) = 6x + 3c$
22. $5 + 2(ax^2 - 1) = x + 3$

4. Solving Quadratic Equations of the Form $ax^2 = k$

A quadratic equation lacking the first-degree term, bx, is expressible in the form $ax^2 + c = 0$, or in the form $ax^2 = k$, where $k = -c$.

Dividing each side by a, $ax^2 = k$ becomes $x^2 = \dfrac{k}{a}$. Taking the square root of each side yields $x = \pm\sqrt{\dfrac{k}{a}}$. The roots are $\pm\sqrt{\dfrac{k}{a}}$.

Rule. A quadratic equation of the form $ax^2 = k$ has two roots, $\pm\sqrt{\dfrac{k}{a}}$.

〰〰〰〰〰〰〰 *MODEL PROBLEMS* 〰〰〰〰〰〰〰

In 1 and 2, solve the equation by using the preceding rule.

1. $4x^2 - 20 = 80$　　　　　　**2.** $7x^2 = 4(x^2 + 6)$

How To Proceed	*Solution*	*Solution*
1. Express in the form $ax^2 = k$.	$4x^2 = 100$	$7x^2 = 4x^2 + 24$ $3x^2 = 24$
2. Evaluate $\dfrac{k}{a}$.	Since $a = 4$, and $k = 100$,	Since $a = 3$ and $k = 24$,
	$\dfrac{k}{a} = \dfrac{100}{4} = 25$	$\dfrac{k}{a} = \dfrac{24}{3} = 8$

3. State the roots, $\pm\sqrt{\dfrac{k}{a}}$, in simplest radical form.

| Roots are $\pm\sqrt{25}$, or ± 5. $\{5, -5\}$ *Ans.* | Roots are $\pm\sqrt{8}$, or $\pm 2\sqrt{2}$. $\{2\sqrt{2}, -2\sqrt{2}\}$ *Ans.* |

In 3 and 4, solve the equation by taking the square root of each side. Do *not* use the preceding rule.

3. Solve: $5x^2 - 8 = 3x^2$

Solution:
$$5x^2 - 8 = 3x^2$$
$$5x^2 - 3x^2 = 8$$
$$2x^2 = 8$$
$$x^2 = 4$$
$$x = \pm 2$$

Answer: $x = +2$, $x = -2$

4. Solve: $(2x - 3)^2 = 5$

Solution:
$$(2x - 3)^2 = 5$$
$$2x - 3 = \pm\sqrt{5}$$

$2x - 3 = \sqrt{5}$	$2x - 3 = -\sqrt{5}$
$2x = 3 + \sqrt{5}$	$2x = 3 - \sqrt{5}$
$x = \dfrac{3 + \sqrt{5}}{2}$	$x = \dfrac{3 - \sqrt{5}}{2}$

Answer: $x = \dfrac{3 + \sqrt{5}}{2}$, $x = \dfrac{3 - \sqrt{5}}{2}$

Exercises

In 1–8, solve for x.

1. $x^2 = 25$
2. $x^2 = 8$
3. $4x^2 = 16$
4. $3x^2 = 21$
5. $(x + 4)^2 = 36$
6. $(2x - 1)^2 = 81$
7. $(x - 3)^2 = 3$
8. $(3x + 2)^2 = 2$

In 9–17, find the solution set of the equation.

9. $\dfrac{x^2}{3} = 12$
10. $5 = \dfrac{x^2}{4}$
11. $x + 7 = \dfrac{1}{x - 7}$
12. $\dfrac{x - 3}{2} = \dfrac{8}{x + 3}$
13. $\dfrac{5 + x}{7} = \dfrac{1}{5 - x}$
14. $\dfrac{x - 1}{x + 1} + \dfrac{x + 1}{x - 1} = 3$
15. $9x^2 = a^2$
16. $4x^2 - r^2 = 0$
17. $mx^2 - r = 0$

18. Solve for x: $x^2 + a^2 = c^2$
19. Solve for r: $A = \pi r^2$
20. Solve for r: $S = 4\pi r^2$

21. Solve for t: $S = \frac{1}{2}at^2$

22. Solve for t: $t^2 - 9 = 0$

23. Solve for c: $4c^2 - 12 = 0$

24. Solve for s: $5s^2 = s^2 + 1$

25. Solve for c: $3c^2 = 5c^2 - 1$

26. Solve for s: $3s^2 + (1 - s^2) = 2$

27. Solve for s: $1 + (s^2 - 1) = 2s^2 - 9$

28. Solve for c: $1 + c^2 = 3(c^2 + 1) - 8$

29. Find the diagonal of a square whose side is 20.

30. Find the altitude of an equilateral triangle whose side is 10.

5. Solving Quadratic Equations by Completing the Square

Frequently, quadratic equations cannot be solved readily by factoring. If a quadratic equation has real coefficients, the method of *completing the square* can be used to find the roots, whether or not the method of factoring is applicable.

A ***perfect square trinomial*** is a trinomial which is the square of a binomial. Hence, a perfect square trinomial has two equal binomial factors.

Thus, the expression $x^2 + 10x + 25$ is a perfect square trinomial since $x^2 + 10x + 25 = (x + 5)^2$.

COMPLETING THE PERFECT SQUARE TRINOMIAL OR "COMPLETING THE SQUARE"

Suppose we are given $x^2 + 10x$, the first two terms of the perfect square trinomial $x^2 + 10x + 25$. To complete the perfect square trinomial, or simply to "complete the square," the last term, 25, must be obtained. This is done by squaring one-half of 10, the coefficient of x, in accordance with the following:

Rule. To obtain a perfect square trinomial, whose first two terms are $x^2 + px$, add $(\frac{1}{2}p)^2$, or $\frac{1}{4}p^2$, thus obtaining the perfect square trinomial, $x^2 + px + \dfrac{p^2}{4}$, or $\left(x + \dfrac{p}{2}\right)^2$.

Note that $\frac{1}{2}p$ is one-half the coefficient of x. The process of adding the term $\frac{1}{4}p^2$ is called "completing the square."

~~~~~~~~~~~~~~~~~~ **MODEL PROBLEMS** ~~~~~~~~~~~~~~~

In 1 and 2, complete the square and express the resulting perfect square trinomial as the square of a binomial.

	1. $6x + x^2$	**2.** $2x^2 - x^2 - \frac{3}{2}x$
How To Proceed	*Solution*	*Solution*
1. Express in the form $x^2 + px$.	$x^2 + 6x\ [p = 6]$	$x^2 - \frac{3}{2}x\ \left[p = -\frac{3}{2}\right]$
2. Evaluate $\left(\frac{p}{2}\right)^2$, or $\frac{p^2}{4}$.	Since $\dfrac{p}{2} = \dfrac{6}{2} = 3$, then $\dfrac{p^2}{4} = (3)^2 = 9$.	Since $\dfrac{p}{2} = \dfrac{1}{2}\left(-\dfrac{3}{2}\right) = -\dfrac{3}{4}$, then $\dfrac{p^2}{4} = \left(-\dfrac{3}{4}\right)^2 = \dfrac{9}{16}$.
3. Complete the square by adding $\dfrac{p^2}{4}$.	$x^2 + 6x + 9$	$x^2 - \dfrac{3}{2}x + \dfrac{9}{16}$
4. Express as $\left(x + \dfrac{p}{2}\right)^2$.	$(x + 3)^2$	$(x - \tfrac{3}{4})^2$

~~~~~~

Procedure. To solve a quadratic equation by completing the square:
1. Transform the equation so that all terms containing the variable are on one side of the equation with the constant term on the other side.
2. If the coefficient of x^2 is not one, divide both members of the equation by the coefficient of x^2 to obtain an equation of the form $x^2 + px = q$, where q is a constant.
3. Square one-half the coefficient of x and add the result to both members of of the equation obtained in step 2.
4. Express the left member as the square of a binomial and simplify the right member.
5. Find the square root of both members. Write $\pm$ before the square root of the right member.
6. Let the square root of the left member equal the positive $(+)$ square root of the right member and solve the resulting equation.
7. Let the square root of the left member equal the negative $(-)$ square root of the right member and solve the resulting equation.

~~~~~~~~~~~~ **MODEL PROBLEMS** ~~~~~~~~~~~~

1. Solve by completing the square and check: $2x^2 - 3x - 2 = 0$

Solution:

$$2x^2 - 3x - 2 = 0$$
$$2x^2 - 3x = 2$$
$$D_2: x^2 - \tfrac{3}{2}x = 1$$
$$x^2 - \tfrac{3}{2}x + (-\tfrac{3}{4})^2 = 1 + (-\tfrac{3}{4})^2$$
$$x^2 - \tfrac{3}{2}x + \tfrac{9}{16} = 1 + \tfrac{9}{16}$$
$$(x - \tfrac{3}{4})^2 = \tfrac{25}{16}$$
$$R_2: x - \tfrac{3}{4} = \pm\tfrac{5}{4}$$

$$x - \tfrac{3}{4} = +\tfrac{5}{4} \qquad x - \tfrac{3}{4} = -\tfrac{5}{4}$$
$$x = \tfrac{5}{4} + \tfrac{3}{4} \qquad x = -\tfrac{5}{4} + \tfrac{3}{4}$$
$$x = \tfrac{8}{4} = 2 \qquad x = -\tfrac{2}{4} = -\tfrac{1}{2}$$

Answer: $x = 2, x = -\tfrac{1}{2}$

Check: $\quad 2x^2 - 3x - 2 = 0$

If $x = 2$, $2(2)^2 - 3(2) - 2 \overset{?}{=} 0$
$$8 - 6 - 2 \overset{?}{=} 0$$
$$0 = 0$$

If $x = -\tfrac{1}{2}$, $2(-\tfrac{1}{2})^2 - 3(-\tfrac{1}{2}) - 2 \overset{?}{=} 0$
$$2(\tfrac{1}{4}) + \tfrac{3}{2} - 2 \overset{?}{=} 0$$
$$\tfrac{1}{2} + \tfrac{3}{2} - 2 \overset{?}{=} 0$$
$$0 = 0$$

Note. "R_2" means "Take the square root of each side of the equation."

2. By completing the square solve: $4x^2 + 4x = 3$.

Solution:

1. $4x^2 + 4x = 3$
2. $D_4: x^2 + x = \tfrac{3}{4}$
3. $x^2 + x + (\tfrac{1}{2})^2 = \tfrac{3}{4} + (\tfrac{1}{2})^2$
4. $x^2 + x + \tfrac{1}{4} = \tfrac{3}{4} + \tfrac{1}{4}$
5. $(x + \tfrac{1}{2})^2 = 1$

6. $R_2: x + \tfrac{1}{2} = \pm 1$
7. $x + \tfrac{1}{2} = 1 \quad \mid \quad x + \tfrac{1}{2} = -1$
8. $x = \tfrac{1}{2} \quad \mid \quad x = -\tfrac{3}{2}$

Answer: $x = \tfrac{1}{2}, x = -\tfrac{3}{2}$

Exercises

In 1–20, complete the square by replacing the question mark with a number and express the resulting trinomial as the square of a binomial.

1. $x^2 + 6x + (?)$ **2.** $d^2 - 4d + (?)$ **3.** $r^2 - 12r + (?)$ **4.** $y^2 + 16y + (?)$
5. $a^2 + 3a + (?)$ **6.** $x^2 + 5x + (?)$ **7.** $b^2 - b + (?)$ **8.** $m^2 - 7m + (?)$
9. $x^2 - \tfrac{1}{2}x + (?)$ **10.** $y^2 + \tfrac{1}{3}y + (?)$ **11.** $z^2 - \tfrac{2}{3}z + (?)$ **12.** $y^2 - \tfrac{4}{3}y + (?)$
13. $c^2 + 2c + (?)$ **14.** $t^2 - 4t + (?)$
15. $s^2 - s + (?)$ **16.** $c^2 + 3c + (?)$
17. $s^2 + \tfrac{1}{2}s + (?)$ **18.** $c^2 - \tfrac{1}{3}c + (?)$
19. $t^2 + \tfrac{2}{3}t + (?)$ **20.** $s^2 + \tfrac{3}{5}s + (?)$

In 21–35, solve the equation by completing the square. Write the solution set of the equation, expressing irrational roots in the simplest radical form.

21. $x^2 + 2x = 4$ **22.** $y^2 + 4y = 21$ **23.** $c^2 - 6c = 16$

24. $x^2 - 8x + 12 = 0$ **25.** $y^2 + 2y - 48 = 0$ **26.** $x^2 = 4x + 12$
27. $x^2 + 3x + 2 = 0$ **28.** $y^2 + 5y - 6 = 0$ **29.** $t^2 = 14 - 5t$
30. $3x^2 + 2x - 1 = 0$ **31.** $2c^2 - 3c - 2 = 0$ **32.** $5d^2 + 3d = 2$
33. $t^2 - 2t - 5 = 0$ **34.** $m^2 + 6m - 4 = 0$ **35.** $a^2 - 5a - 4 = 0$

In 36–38, using the method of completing the square, list the members of the set.

36. $\{y \mid y^2 = 6 - 5y\}$ **37.** $\{x \mid 2x^2 - 10x = 1\}$ **38.** $\{x \mid 3x(x - 1) = 1\}$

6. Solving Quadratic Equations by the Quadratic Formula

The following quadratic formula can be used to solve any quadratic equation expressed in standard quadratic form:

If a quadratic equation is of the form $ax^2 + bx + c = 0$, then

$$x = \frac{-b \pm \sqrt{b^2 - 4ac}}{2a}$$

Using r_1 and r_2 to represent the roots,

$$r_1 = \frac{-b + \sqrt{b^2 - 4ac}}{2a} \quad \text{and} \quad r_2 = \frac{-b - \sqrt{b^2 - 4ac}}{2a}$$

DERIVATION OF THE QUADRATIC FORMULA BY COMPLETING THE SQUARE

If $ax^2 + bx + c = 0$, the formula for the roots of this equation in terms of a, b, and c may be derived as follows:

1. $ax^2 + bx + c = 0$

2. $x^2 + \frac{b}{a}x + \frac{c}{a} = 0$ In equation 1, divide by a.

3. $x^2 + \frac{b}{a}x = -\frac{c}{a}$ In equation 2, subtract $\frac{c}{a}$ from each side.

To complete the square, take $\frac{1}{2}$ of the coefficient of x and square the result.

$$\frac{1}{2}\left(\frac{b}{a}\right) = \frac{b}{2a} \; ; \; \left(\frac{b}{2a}\right)^2 = \frac{b^2}{4a^2}$$

4. $x^2 + \dfrac{b}{a}x + \dfrac{b^2}{4a^2} = \dfrac{b^2}{4a^2} - \dfrac{c}{a}$ In equation 3, add $\dfrac{b^2}{4a^2}$ to each side.

5. $\left(x + \dfrac{b}{2a}\right)^2 = \dfrac{b^2 - 4ac}{4a^2}$ Combine the fractions.

6. $x + \dfrac{b}{2a} = \pm\dfrac{\sqrt{b^2 - 4ac}}{2a}$ In equation 5, take the square root of each side.

7. $x = -\dfrac{b}{2a} \pm \dfrac{\sqrt{b^2 - 4ac}}{2a}$ In equation 6, subtract $\dfrac{b}{2a}$ from each side.

8. $x = \dfrac{-b \pm \sqrt{b^2 - 4ac}}{2a}$ Combine the fractions.

MODEL PROBLEMS

1. Find the solution set of $2x^2 + 2 = 5x$.

How To Proceed

1. Equate to 0.

2. Determine a, b, and c by comparing with $ax^2 + bx + c = 0$.

3. Substitute the values of a, b, and c in the quadratic formula.

4. Compute the values of x by evaluating the result obtained in step 3.

Solution

$$2x^2 + 2 = 5x$$
$$2x^2 - 5x + 2 = 0$$
$$a = 2,\ b = -5,\ c = 2$$

$$x = \frac{-b \pm \sqrt{b^2 - 4ac}}{2a}$$

$$x = \frac{-(-5) \pm \sqrt{(-5)^2 - 4(2)(2)}}{2(2)}$$

$$x = \frac{5 \pm \sqrt{25 - 16}}{4}$$

$$x = \frac{5 \pm \sqrt{9}}{4} = \frac{5 \pm 3}{4}$$

$$x = \frac{5 + 3}{4} \qquad x = \frac{5 - 3}{4}$$

$$x = \tfrac{8}{4} = 2 \qquad x = \tfrac{2}{4} = \tfrac{1}{2}$$

Answer: $\{2, \tfrac{1}{2}\}$

2. Solve the equation $3x^2 - 3x = 2$ for values of x to the nearest tenth.

Solution:

Equate to 0: $3x^2 - 3x = 2$ becomes $3x^2 - 3x - 2 = 0$

Compare with $ax^2 + bx + c = 0$: $a = 3$, $b = -3$, $c = -2$

Quadratic formula is $x = \dfrac{-b \pm \sqrt{b^2 - 4ac}}{2a}$.

Substitute: $x = \dfrac{-(-3) \pm \sqrt{(-3)^2 - 4(3)(-2)}}{2(3)}$

$$x = \frac{3 \pm \sqrt{9 + 24}}{6}$$

$$x = \frac{3 \pm \sqrt{33}}{6} = \frac{3 \pm 5.74}{6}$$

$$\begin{array}{r} 5.\,7\ \ 4 \\ \sqrt{33.00\,00} \\ 25 \\ \hline \end{array}$$

$$\begin{array}{r|r} 107 & 800 \\ & 749 \\ \hline 1144 & 5100 \\ & 4576 \\ \hline & 524 \end{array}$$

$x = \dfrac{3 + 5.74}{6}$　　　　　　$x = \dfrac{3 - 5.74}{6}$

$x = \dfrac{8.74}{6} = 1.45+$　　　　$x = \dfrac{-2.74}{6} = -.45+$

$x = 1.5$, to the nearest tenth　　　$x = -.5$, to the nearest tenth

Answer: 1.5 or $-.5$

Note. In finding the square root, carry it out to one decimal place more than the number of decimal places required in the answer.

Exercises

In 1–9, using the quadratic formula, find the solution set of the equation. Express irrational roots in simplest radical form.

1. $x^2 + 3x - 40 = 0$　　　**2.** $x^2 + 16 = 8x$　　　**3.** $3x^2 + 5x = -2$

4. $2(x^2 - 3) = x$　　　**5.** $c^2 - 10c + 15 = 0$　　　**6.** $2 + 4r = 5r^2$

7. $2x - \dfrac{4}{x} = 5$　　　**8.** $1 + \dfrac{2}{x^2} = \dfrac{7}{2x}$　　　**9.** $\dfrac{3}{x-2} - \dfrac{1}{x-1} = 2$

In 10–24, find, to the nearest tenth, the roots of the equation.

10. $x^2 - 3x - 3 = 0$　　　**11.** $y^2 + 4y - 2 = 0$　　　**12.** $2c^2 + 7c + 1 = 0$

13. $3d^2 = 5d + 4$　　　**14.** $3 = 2x^2 + 4x$　　　**15.** $2x^2 - 5x + 1 = 0$

16. $x^2 + 2x - 4 = 0$　　　**17.** $3x^2 - 2x - 6 = 0$　　　**18.** $2x^2 - 10x = 9$

19. $2x^2 - 8x + 1 = 0$　　　**20.** $2x^2 - 2x = 3$　　　**21.** $3x^2 + 5x - 1 = 0$

22. $x^2 - 20x - 10 = 0$　　　**23.** $x^2 + 9x - 12 = 0$　　　**24.** $x^2 + 4x - 16 = 0$

25. Given the equation $t^2 - 3t + 1 = 0$. Find, to the nearest tenth, the values of t that satisfy this equation.

26. Given the equation $s^2 + s - 1 = 0$. Find, to the nearest tenth, the positive value of s that satisfies this equation.

27. Given the equation $2t^2 + 5t = 8$. Find, to the nearest tenth, the values of t that satisfy this equation.

28. Solve the equation $3s^2 - 2 = 3s$ for the negative value of s correct to the nearest hundredth.

7. Determining the Nature of the Roots of a Quadratic Equation

DISCRIMINANT OF A QUADRATIC EQUATION, $b^2 - 4ac$

We shall see that the value of $b^2 - 4ac$, the radicand, (that is, the expression under the radical sign) in the quadratic formula, can be used to determine the nature of the roots of any quadratic equation of the form $ax^2 + bx + c = 0$. Hence, $b^2 - 4ac$ is called the ***discriminant*** of the quadratic equation. If a, b, and c are rational numbers and $a \neq 0$, the discriminant reveals whether the roots are (1) real or imaginary, (2) equal or unequal, (3) rational or irrational, according to the following table:

USING THE DISCRIMINANT, $b^2 - 4ac$, TO DETERMINE THE NATURE
OF THE ROOTS OF A QUADRATIC EQUATION WHEN a, b, AND
c ARE RATIONAL NUMBERS

Case	*Value of the Discriminant, $b^2 - 4ac$*	*Nature of the Roots*
(1)	zero or positive	real
	negative	imaginary
(2)	zero	equal
	nonzero	unequal
(3)	a perfect square	rational
	positive and not a perfect square	irrational

It can be seen from the table that the nature of the roots can be determined using one of the following rules, given a, b, and c are rational numbers, $a \neq 0$:

Rule 1. If $b^2 - 4ac$ is positive and not a perfect square, the roots are real, irrational, and unequal. This rule would apply if $b^2 - 4ac$ were 7.

Rule 2. If $b^2 - 4ac$ is a perfect square and not equal to 0, the roots are real, rational, and unequal. This rule would apply if $b^2 - 4ac$ were 9.

Rule 3. If $b^2 - 4ac = 0$, the roots are real, rational, and equal.

Rule 4. If $b^2 - 4ac$ is negative, the roots are imaginary. This rule would apply if $b^2 - 4ac$ were -4.

If the roots are imaginary, then these roots are conjugate complex numbers. (Recall that conjugate complex numbers are expressible in the form $a + bi$ and $a - bi$.) Conjugate complex numbers cannot be equal since equal complex numbers must have equal real parts and also equal imaginary parts. Furthermore, conjugate complex numbers can be neither rational nor irrational numbers since such numbers are real numbers.

～～～～～～～ MODEL PROBLEMS ～～～～～～～

In 1 and 2, state the nature of the roots of the equation.

1. $x^2 + 12 = 8x$ **2.** $4 = x(2 - x)$

How To Proceed	*Solution*	*Solution*
1. Equate to 0.	$x^2 - 8x + 12 = 0$	$4 = 2x - x^2$ $x^2 - 2x + 4 = 0$
2. Determine values of a, b, and c.	$a = 1, b = -8, c = 12$	$a = 1, b = -2, c = 4$
3. Evaluate $b^2 - 4ac$.	$b^2 - 4ac$ $(-8)^2 - 4(1)(12)$ $64 - 48 = 16$	$b^2 - 4ac$ $(-2)^2 - 4(1)(4)$ $4 - 16 = -12$
4. State the nature of the roots.	Since $b^2 - 4ac$ is a positive perfect square, the roots are real, rational, and unequal.	Since $b^2 - 4ac$ is negative, the roots are imaginary. (The roots are conjugate complex numbers.)

3. Find the value of k for which the equation $x^2 - 6x + k = 0$ has equal roots.

Solution: From $x^2 - 6x + k = 0$, $a = 1$, $b = -6$, $c = k$.

If the equation has equal roots,

$$b^2 - 4ac = 0$$
$$(-6)^2 - 4(1)(k) = 0$$
$$36 - 4k = 0$$
$$36 = 4k$$
$$9 = k \quad Answer: k = 9$$

Exercises

In each of the following exercises, assume that the coefficients of the quadratic equations are rational numbers, unless there is a statement to the contrary.

In 1–6, find the discriminant.

1. $x^2 - 5x + 6 = 0$ **2.** $c^2 + 2c - 10 = 0$ **3.** $r^2 = 7r + 5$
4. $8 = 2x^2 - 3x$ **5.** $4x^2 + 1 = 4x$ **6.** $2z^2 + 5 = 2z$

In 7–21, state the nature of the roots of a quadratic equation if its discriminant is:

7. 49 **8.** 0 **9.** 16 **10.** −4 **11.** 17
12. 100 **13.** −49 **14.** 23 **15.** −64 **16.** 121
17. 54 **18.** −13 **19.** 86 **20.** −25 **21.** 225

In 22–30, determine the nature of the roots of the equation without solving the equation.

22. $x^2 - 7x + 10 = 0$ **23.** $2x^2 - 5x - 4 = 0$ **24.** $y^2 + 3y + 4 = 0$
25. $3y^2 - 7y = 4$ **26.** $9c^2 = 6c$ **27.** $4d^2 = 3d - 5$
28. $5y^2 + 2 = 11y$ **29.** $2d - 10 = d^2$ **30.** $25x^2 + 4 = 0$

In 31–36, determine the values of k that will make the roots of the equation equal.

31. $y^2 - 10y + k = 0$ **32.** $ky^2 - 12y + 9 = 0$ **33** $x^2 - kx + 16 = 0$
34. $x^2 + 12x = k$ **35.** $y^2 + 2kx + 16 = 0$ **36.** $kx^2 - 12x + k + 5 = 0$

In 37–39, determine the values of k for which the roots of the equation will be imaginary.

37. $x^2 + 8x + k = 0$ **38.** $kx^2 - 10x + 5 = 0$ **39.** $2x^2 - 12x + 8 = k$

40. If the discriminant of a quadratic equation with real coefficients is 0, the roots of the equation are (1) real and equal (2) real and unequal (3) imaginary

41. The roots of the equation $2x^2 - 3x + 3 = 0$ are (1) real, equal, and rational (2) real, unequal, and irrational (3) real, unequal, and rational (4) imaginary

42. The roots of the equation $3x^2 + kx - 4 = 0$ are real, rational, and unequal if k is equal to (1) 1 (2) 2 (3) 0 (4) 6

43. The roots of the equation $ax^2 + bx + c = 0$ are real, rational, and unequal. What integer between 13 and 20 may be the value of the discriminant of the equation?

44. In the equation $s^2 + s + 1 = 0$, the values of s are (1) real and equal (2) imaginary (3) real and unequal

45. In the equation $6c^2 - 5c + 1 = 0$, the values of c are (1) imaginary (2) real and equal (3) real and rational

8. Expressing a Quadratic Equation in Terms of Its Roots: Sum and Product of the Roots

If 3 and 5 are the roots of a quadratic equation, then the equation in factored form is $(x-3)(x-5)=0$. Multiplying the two binomials to remove parentheses, we obtain the quadratic equation in standard quadratic form, $x^2-8x+15=0$.

In general, if r_1 and r_2 represent the roots of a quadratic equation, then the equation in factored form is $(x-r_1)(x-r_2)=0$. Multiplying the two binomials to remove parentheses, as shown on the right, we obtain the quadratic equation in standard quadratic form:

$$x-r_1$$
$$x-r_2$$
$$\overline{x^2-r_1x}$$
$$-r_2x+r_1r_2$$
$$\overline{x^2-(r_1+r_2)x+r_1r_2}$$

$$(1)\quad x^2-(r_1+r_2)x+r_1r_2=0$$

RELATING THE ROOTS OF A QUADRATIC EQUATION TO *a, b,* AND *c* OF THE STANDARD QUADRATIC FORM

If each side of the standard quadratic form, $ax^2+bx+c=0$, is divided by a, we obtain the equation

$$(2)\quad x^2+\frac{b}{a}x+\frac{c}{a}=0$$

Note in equations (1) and (2) that x^2 is the second-degree term of each. Hence, equating the coefficients of the first-degree term, we find that $-(r_1+r_2)=\frac{b}{a}$. Therefore,

$$r_1+r_2=-\frac{b}{a}$$

Also, by equating the constant terms of (1) and (2), we find that

$$r_1r_2=\frac{c}{a}$$

Rules for the Sum and Product of the Roots of a Quadratic Equation

If a quadratic equation is in standard form, $ax^2+bx+c=0$, then:

Rule 1. The product of the roots, $r_1r_2=\frac{c}{a}$.

Rule 2. The sum of the roots, $r_1+r_2=-\frac{b}{a}$.

~~~~~~~~~~~~~~~ **MODEL PROBLEMS** ~~~~~~~~~~~~~~~

1. Find the sum and the product of the roots of $2x^2 - 8x + 3 = 0$.

Solution:

$a = 2, \; b = -8, \; c = 3$

$$r_1 + r_2 = \frac{-b}{a} = \frac{-(-8)}{2} = \frac{8}{2} = 4 \quad Ans. \qquad\qquad r_1 r_2 = \frac{c}{a} = \frac{3}{2} \text{ or } 1\tfrac{1}{2} \quad Ans.$$

2. If one root of the equation $x^2 - 4x + c = 0$ is 1, find the other root.

Solution: $\quad x^2 - 4x + c = 0 \qquad\qquad a = 1, \quad b = -4, \quad r_1 = 1$

$$\begin{aligned} r_1 + r_2 &= 4 \\ r_1 &= 1 \end{aligned} \qquad\qquad r_1 + r_2 = \frac{-b}{a} = 4$$

Therefore, $\qquad\qquad \overline{ r_2 = 3} \quad Ans.$

CHECKING THE ROOTS OF A QUADRATIC EQUATION BY USING THE SUM AND PRODUCT OF THE ROOTS

~~~~~~~~~~~~~~~ **MODEL PROBLEM** ~~~~~~~~~~~~~~~

Is $\{3 + \sqrt{2}, 3 - \sqrt{2}\}$ the solution set of $x^2 - 6x + 7 = 0$?

Solution: If the solution set is $\{3 + \sqrt{2}, 3 - \sqrt{2}\}$, then $r_1 = 3 + \sqrt{2}$ and $r_2 = 3 - \sqrt{2}$. Since the equation in standard form is $x^2 - 6x + 7 = 0$, then $a = 1$, $b = -6$, and $c = 7$.

(a) Is the product, $r_1 r_2 = \dfrac{c}{a}$?

$$(3 + \sqrt{2})(3 - \sqrt{2}) \overset{?}{=} \frac{7}{1}$$

$$7 = 7$$

(b) Is the sum, $r_1 + r_2 = -\dfrac{b}{a}$?

$$(3 + \sqrt{2}) + (3 - \sqrt{2}) \overset{?}{=} -\frac{-6}{1}$$

$$6 = 6$$

Answer: $\{3 + \sqrt{2}, 3 - \sqrt{2}\}$ is the solution set of $x^2 - 6x + 7 = 0$.

When checking roots that have been expressed to the nearest tenth or other decimal, the sum and the product method of checking can be used. The approximations obtained for the sum and the product of the roots will, if the roots are correct, closely approximate $-\dfrac{b}{a}$ and $\dfrac{c}{a}$, respectively.

Exercises

In 1–9, find the sum and the product of the roots of the equation.

1. $x^2 + 6x + 2 = 0$ 2. $y^2 + 4y - 5 = 0$ 3. $2c^2 + 4c - 7 = 0$
4. $3d^2 + 7d + 3 = 0$ 5. $4x^2 - 3x = 9$ 6. $2r^2 + 7 = 10r$
7. $2y^2 = 4y + 1$ 8. $6 = 3y^2 + 7y$ 9. $8y = 2 - y^2$

In 10–17, if the number in parentheses is one root of the equation, find the other root.

10. $x^2 - 5x + c = 0$ (2)   11. $2y^2 + 4y + e = 0$ (−3)
12. $x^2 + dx + 6 = 0$ (2)   13. $2c^2 + 9c + 4 = 0$ $(-\tfrac{1}{2})$
14. $r^2 + 4r + d = 0$ (1)   15. $x^2 + cx - 5 = 0$ (1)
16. $3x^2 + ex - 4 = 0$ $(\tfrac{1}{3})$ 17. $6x^2 + 1 = 5x$ $(\tfrac{1}{2})$

In 18–23, by using the rules for the sum and product of the roots of a quadratic equation, determine whether the statement is *true* or *false*.

18. The roots of $2x^2 - 5x + 2 = 0$ are $x = 2$ and $x = \tfrac{1}{2}$.
19. The roots of $4c^2 - 7c + 3 = 0$ are $c = -\tfrac{3}{4}$ and $c = -1$.
20. The roots of $6x^2 + 13x - 5 = 0$ are $x = \tfrac{1}{3}$ and $x = -2\tfrac{1}{2}$.
21. The roots of $x^2 - 4x + 2 = 0$ are $x = 2 + \sqrt{2}$ and $x = 2 - \sqrt{2}$.
22. The roots of $6s^2 - 5s + 1 = 0$ are $s = \tfrac{1}{2}$ and $s = \tfrac{1}{3}$.
23. The roots of $4c^2 - 5c + 1 = 0$ are $c = \tfrac{1}{4}$ and $c = 1$.

In 24–28, find the value of k in the equation $2x^2 - 3x + k = 0$ if one of the roots of the equation is the given number.

24. 5   25. −5   26. $\tfrac{1}{2}$ 27. 2.1   28. $-1\tfrac{1}{4}$

29. If 2 is a root of $x^2 - 5x + n = 0$, find n.
30. Find c so that $\tfrac{1}{2}$ will be a root of $4x^2 + c = 4x$.
31. If $\tfrac{2}{3}$ is a root of $3x^2 - cx + 2 = 0$, find c.
32. Find k so that $-\tfrac{1}{5}$ will be a root of $5x^2 + kx = 4$.
33. Find the value of c so that one root of $x^2 - 3x + c = 0$ will be twice the other.
34. In the quadratic equation $x^2 - 5x + 2 = 0$, the sum of the roots exceeds the product of the roots by (1) −3   (2) 3   (3) 7
35. The sum of the roots of the equation $x^2 - px + p = 0$ is (1) always (2) sometimes   (3) never   equal to their product.
36. If the roots of the equation $x^2 + 4x + q = 0$ are equal, what is the value of q?
37. If 0 is one root of the equation $x^2 + px + q = 0$, what must be the value of q?
38. If the roots of the equation $x^2 + px + q = 0$ are numerically equal but opposite in sign, what is the value of p?

39. In the equation $t^2 - 5t + c = 0$, one value of t is 3.5. Find the remaining value of t.

9. Forming a Quadratic Equation Whose Roots Are Given

~~~~~~~~~~ *MODEL PROBLEMS* ~~~~~~~~~~

In 1 and 2, write a quadratic equation with integral coefficients whose solution set is the given set.

1. $\{2, -\frac{1}{2}\}$ **2.** $\{3\sqrt{2}, -3\sqrt{2}\}$

Method 1:

How To Proceed	Solution	Solution
1. Substitute for r_1 and r_2 in the equation $(x - r_1)(x - r_2) = 0$.	Since $r_1 = 2$ and $r_2 = -\frac{1}{2}$, $(x - 2)(x + \frac{1}{2}) = 0$	Since $r_1 = 3\sqrt{2}$ and $r_2 = -3\sqrt{2}$, $(x - 3\sqrt{2})(x + 3\sqrt{2}) = 0$
2. Multiply the binomials.	$x^2 - \frac{3}{2}x - 1 = 0$	
3. Clear of fractions to obtain integral coefficients.	$2x^2 - 3x - 2 = 0$ *Ans.*	$x^2 - 18 = 0$ *Ans.*

Method 2:

How To Proceed	Solution	Solution
1. Find $(r_1 + r_2)$ and $r_1 r_2$.	Since $r_1 = 2$ and $r_2 = -\frac{1}{2}$, $r_1 + r_2 = \frac{3}{2}$ and $r_1 r_2 = -1$.	Since $r_1 = 3\sqrt{2}$ and $r_2 = -3\sqrt{2}$, $r_1 + r_2 = 0$ and $r_1 r_2 = -18$.
2. Substitute in the equation $x^2 - (r_1 + r_2)x + r_1 r_2 = 0$.	$x^2 - \frac{3}{2}x - 1 = 0$	$x^2 - 0x - 18 = 0$ $x^2 - 18 = 0$ *Ans.*
3. Clear of fractions to obtain integral coefficients.	$2x^2 - 3x - 2 = 0$ *Ans.*	

Exercises

In 1–19, form a quadratic equation, with integral coefficients, whose roots are:

1. 5, 2   **2.** 3, 7   **3.** $-1, -3$ **4.** $-4, -5$

5. 5, -9 **6.** $-3, 7$ **7.** $\frac{1}{2}, \frac{1}{5}$ **8.** $\frac{2}{3}, 3$

9. $-\frac{1}{2}, \frac{3}{4}$ **10.** $-\frac{3}{4}, -\frac{1}{3}$ **11.** 2.5, 1.2   **12.** $-1.4, -.5$

13. $-1.5, .2$ **14.** $.2, -.5$ **15.** $\sqrt{2}, -\sqrt{2}$ **16.** $\sqrt{5}, -\sqrt{5}$

17. $2 + \sqrt{5}, 2 - \sqrt{5}$ **18.** $3 - \sqrt{7}, 3 + \sqrt{7}$ **19.** $-2 + \sqrt{3}, -2 - \sqrt{3}$

In 20–31, write a quadratic equation, with integral coefficients, the sum and the product of whose roots are the given numbers.

20. 9 and 4   **21.** 7 and 1   **22.** -5 and -2 **23.** -2 and -4

24. -3 and -1 **25.** 1 and -3 **26.** -6 and 8   **27.** $\frac{1}{3}$ and $\frac{1}{9}$

28. $-\frac{5}{6}$ and $\frac{1}{6}$ **29.** $-\frac{3}{4}$ and $\frac{3}{8}$ **30.** $-\frac{5}{3}$ and -2 **31.** $-\frac{2}{5}$ and $\frac{1}{25}$

In 32–40, write a quadratic equation in the form $x^2 + px + q = 0$, whose roots are:

32. 5, 3   **33.** $-5, -3$ **34.** $-5, 3$

35. $\sqrt{5}, -\sqrt{5}$ **36.** $\sqrt{3}, -\sqrt{3}$ **37.** $5 + \sqrt{3}, 5 - \sqrt{3}$

38. $\sqrt{3} + 5, \sqrt{3} - 5$ **39.** $3 + \sqrt{5}, 3 - \sqrt{5}$ **40.** $\sqrt{5} + 3, \sqrt{5} - 3$

10. Solving a Linear-Quadratic System of Equations Consisting of a First-Degree Equation and a Second-Degree Equation

We have learned how to solve a system of two linear equations in two variables. Now we will extend the ideas we have learned in a linear-linear system to a linear-quadratic system and also to a quadratic-quadratic system. In the linear-quadratic system, one of the equations is a first-degree equation, while the other is a second-degree equation. In the quadratic-quadratic system, both equations are second-degree equations. All equations to be treated involve two variables.

The following model problem shows how to solve a linear-quadratic system algebraically using the method of substitution:

〰〰〰〰〰 *MODEL PROBLEM* 〰〰〰〰〰

Solve and check: $\begin{cases} y + 3 = 2x \\ 3x^2 = 4 + xy \end{cases}$

How To Proceed	*Solution*

1. In the first-degree equation, solve for one variable in terms of the other.

Since $y + 3 = 2x$, then
$$y = 2x - 3$$

2. In the second-degree equation, substitute the resulting expression to eliminate one variable.

In $3x^2 = 4 + xy$, substitute $2x - 3$ for y to obtain $3x^2 = 4 + x(2x - 3)$.

3. Solve the resulting quadratic equation to obtain two values of the remaining variable.

$$3x^2 = 4 + 2x^2 - 3x$$
$$3x^2 - 2x^2 + 3x - 4 = 0$$
$$x^2 + 3x - 4 = 0$$
$$(x + 4)(x - 1) = 0$$

$x + 4 = 0$	$x - 1 = 0$
$x = -4$	$x = 1$

4. Substitute each of the resulting two values in the equation found in (1) to obtain the corresponding values of the variable that had been eliminated.

In $y = 2x - 3$, substitute -4 and 1 for x.

If $x = -4$,	If $x = 1$,
$y = 2(-4) - 3$	$y = 2(1) - 3$
$= -11$	$= -1$

5. Group the *two* common ordered number pairs as your answer.

Ans.

x	-4	1
y	-11	-1

or

x	y
-4	-11
1	-1

6. Check in the original equations. (Check each of the ordered number pairs in each of the original equations, a total of four checks.)

Check for $x = 1$ and $y = -1$:

$y + 3 = 2x$	$3x^2 = 4 + xy$
$(-1) + 3 \overset{?}{=} 2(1)$	$3(1)^2 \overset{?}{=} 4 + (1)(-1)$
$2 = 2$	$3 = 3$

Check for $x = -4$ and $y = -11$:

$y + 3 = 2x$	$3x^2 = 4 + xy$
$(-11) + 3 \overset{?}{=} 2(-4)$	$3(-4)^2 \overset{?}{=} 4 +$
$-8 = -8$	$(-4)(-11)$
	$48 = 48$

Exercises

In 1–30, solve the system of equations; group your answers and check one set.

1. $x^2 + y^2 = 18$
$x = y$

2. $4x^2 + y^2 = 40$
$2x = 3y$

3. $x^2 + y^2 = 41$
$x + y = 9$

4. $x^2 + 3y^2 = 7$
$x - y = 1$

5. $x^2 + y = 5$
$y = x + 5$

6. $x^2 + 3 = y$
$y = 5 - x$

7. $3x^2 + y^2 = 12$
   $4x + y = 1$

8. $xy = 8$
   $y = x + 2$

9. $x^2 - xy = 10$
   $x + y = 8$

10. $y^2 = 2xy + 3$
    $y = x - 2$

11. $x - 2y = 0$
    $5xy - x^2 = 54$

12. $2x^2 - y^2 = 31$
    $2x - y = 9$

13. $2x^2 - y^2 - 2 - 0$
    $2x - y - 2 = 0$

14. $x^2 - 2y + 10$
    $3x - y = 9$

15. $x^2 - 3y^2 = 13$
    $1 + 2y = x$

16. $x^2 + y^2 = 10$
    $2x - y = 5$

17. $x^2 - 3y \stackrel{.}{=} 7$
    $2x - y = 4$

18. $x^2 = 2y + 6$
    $5x - y = 15$

19. $x^2 - 3y^2 = 6$
    $x + 2y = -1$

20. $2x^2 = 2 - y^2$
    $2x - y = -2$

21. $x^2 + y^2 - 10y = 24$
    $y = x - 2$

22. $x^2 - 2y = 11$
    $x = y + 4$

23. $x^2 + y^2 = 25$
    $y = 2x + 5$

24. $y^2 + 4x = 21$
    $y - 2x + 3 = 0$

25. $3x^2 - xy = 3$
    $6x - y = 10$

26. $y^2 + xy = 6$
    $3y = x + 2$

27. $x^2 - xy - x = 18$
    $x + y = -1$

28. $x^2 - xy + y^2 = 12$
    $x - y = 2$

29. $x^2 - xy + y = 7$
    $2y + x = 5$

30. $3xy + y^2 = 4$
    $x + y = 3$

In 31–33, find the solution set of the system and show that, in the ordered pairs (x, y) of the solution set, x and y are not both real numbers.

31. $xy = 8$
    $y = -x$

32. $y = x^2 - 5$
    $y = -6$

33. $x^2 + y^2 = 4$
    $2y = 5$

11. Solving a Quadratic-Quadratic System of Equations Consisting of Two Second-Degree Equations

Recall that in solving a system of linear equations, the methods of addition, subtraction, and substitution are used. These methods are also applied to the solution of many systems of second-degree equations. In the first model problem, the method of subtraction is used; in the second model problem, the method of substitution is applied.

~~~~~~~~~~ *MODEL PROBLEMS* ~~~~~~~~~~

1. Solve and check: $\begin{cases} 2x^2 + 3y^2 = 17 \\ x^2 + y^2 = 7 \end{cases}$

How To Proceed

1. For the given system, use the method of subtraction to eliminate one variable.

Solution

1 Since $x^2 + y^2 = 7$, then

$$3x^2 + 3y^2 = 21$$

Subtract: $\underline{2x^2 + 3y^2 = 17}$

$$x^2 \qquad\ = 4$$

2. Solve the resulting equation to obtain two values of the remaining variable.

3. Substitute each of the resulting values in a given equation to obtain corresponding values of the other variable.

4. Group the *four* common ordered number pairs as your answer.

5. Check in the original equations. (Check each of the common ordered number pairs in each of the original equations, a total of eight checks.)

---

2. $x^2 = 4$. Thus, $x = \pm 2$.

3. In $x^2 + y^2 = 7$, substitute ± 2 for x. If $x = 2$, then $2^2 + y^2 = 7$. Transposing, $y^2 = 3$; $y = \pm\sqrt{3}$. If $x = -2$, then $(-2)^2 + y^2 = 7$. Transposing, $y^2 = 3$; $y = \pm\sqrt{3}$.

4.

x	2	2	-2	-2
y	$\sqrt{3}$	$-\sqrt{3}$	$\sqrt{3}$	$-\sqrt{3}$

or $\{(2, \sqrt{3}), (2, -\sqrt{3}), (-2, \sqrt{3}), (-2, -\sqrt{3})\}$ *Ans.*

5. The following check is for the first ordered pair, $(2, \sqrt{3})$. The check of the remaining pairs is left to the student.

$$2x^2 + 3y^2 = 17 \qquad\qquad x^2 + y^2 = 7$$
$$2(2^2) + 3(\sqrt{3})^2 \overset{?}{=} 17 \qquad 2^2 + (\sqrt{3})^2 \overset{?}{=} 7$$
$$8 + 9 \overset{?}{=} 17 \qquad\qquad 4 + 3 \overset{?}{=} 7$$
$$17 = 17 \qquad\qquad 7 = 7$$

---

2. Solve and check: $\begin{cases} x^2 - y^2 = 5 \\ xy = 6 \end{cases}$

Solution:

1. Since $xy = 6$, then $y = \dfrac{6}{x}$.

2. Substitute $\dfrac{6}{x}$ for y in $x^2 - y^2 = 5$ and solve.

$$x^2 - \left(\frac{6}{x}\right)^2 = 5$$

$$x^2 - \frac{36}{x^2} = 5$$

$\text{M}_{x^2}:$ $x^4 - 36 = 5x^2$

Equate to 0: $x^4 - 5x^2 - 36 = 0$

Factor: $(x^2 - 9)(x^2 + 4) = 0$

$x^2 - 9 = 0$	$x^2 + 4 = 0$
$x^2 = 9$	$x^2 = -4$
$x = \pm 3$	$x = \pm\sqrt{-4} = \pm 2i$

3. In $xy = 6$, substitute 3, -3, $2i$, and $-2i$ for x to obtain corresponding values for y: If $x = 3$, then $y = 2$; if $x = -3$, then $y = -2$; if $x = 2i$, then $y = -3i$; if $x = -2i$, then $y = 3i$.

4.

x	3	-3	$2i$	$-2i$
y	2	-2	$-3i$	$3i$

or $\{(3, 2), (-3, -2), (2i, -3i), (-2i, 3i)\}$ *Ans.*

5. Check in the original equations. (The check of each of the four ordered number pairs in each of the original equations, a total of eight checks, is left to the student.)

Exercises

In 1–12, solve the system of equations; group your answers and check one set.

1. $x^2 + y^2 = 5$
   $x^2 - y^2 = 3$

2. $x^2 + y^2 = 8$
   $x^2 - y^2 = 2$

3. $x^2 + 2y^2 = 27$
   $x^2 - 2y^2 = 23$

4. $3x^2 + 2y^2 = 19$
   $3x^2 + y^2 = 17$

5. $4x^2 + 2y^2 = 19$
   $8x^2 + 3y^2 = 29$

6. $2x^2 + 3y^2 = 47$
   $3x^2 - 2y^2 = 38$

7. $x^2 + y^2 = 10$
   $xy = 3$

8. $x^2 + 2y^2 = 18$
   $xy = 6$

9. $x^2 + 5y^2 = 25$
   $x^2 + y^2 = 9$

10. $2x^2 + 3y^2 = 9.$
    $x^2 - y^2 = 2$

11. $x^2 + y^2 = 25$
    $3x^2 - 2y^2 = 30$

12. $2x^2 + 3y^2 = 23$
    $3x^2 - y^2 = 7$

In 13–16, solve the system of two equations in which one of the equations is $x^2 + y^2 = 25$ and the remaining equation is:

13. $x^2 - 4y = 4$ 14. $y^2 = 5x + 1$ 15. $xy = 12$ 16. $xy = -12$

12. Number Problems Involving Quadratic Equations

～～～～～～ *MODEL PROBLEM* ～～～～～～

Find three positive consecutive odd integers such that twice the square of the first is one less than the product of the second and the third.

Solution: Let $x = $ the first consecutive odd integer.
   Then $x + 2 = $ the second consecutive odd integer,
      and $x + 4 = $ the third consecutive odd integer.
Since twice the square of the first is one less than the product of the other two, then

$$2x^2 = (x+2)(x+4) - 1$$
$$2x^2 = x^2 + 6x + 8 - 1$$
$$x^2 - 6x - 7 = 0$$
$$(x-7)(x+1) = 0$$

$$x - 7 = 0 \quad | \quad x + 1 = 0$$
$$x = 7 \quad | \quad x = -1$$

The root -1 is rejected since the problem calls for positive integers. Since $x = 7$, then $x + 2 = 9$ and $x + 4 = 11$.

Answer: 7, 9, 11

Exercises

1. If 4 is subtracted from 9 times a certain number, the result is equal to 2 times the square of the number. Find the number.
2. The larger of two integers exceeds the smaller by 3. The sum of the squares of the two integers is 89. Find the integers.
3. Find three consecutive positive odd integers such that the square of the first exceeds 6 times the sum of the second and third by 9.
4. The sum of the reciprocals of two consecutive integers is $\frac{5}{6}$. Find the integers.
5. The sum of two positive integers is 10. The difference of their reciprocals is $\frac{3}{8}$. Find both positive integers.
6. Two positive integers are in the ratio 1 : 2. If their product is added to their sum, the result is 5. Find the integers.

13. Area Problems Involving Quadratic Equations

In this and succeeding sections, we will consider the solution of quadratic equations in verbal problems involving formulas and rules having relationships of the form $XY = Z$. In section 13, the formula of this type to be studied is the area formula, $LW = A$. In section 14, the motion formula, $D = RT$, is applied to the solution of verbal problems. Sections 15 and 16 involve the application of the cost formula, $NP = C$, to "business problems" and the work formula, $RT = W$, to problems where people work together.

Note that all of these formulas are relationships of the general form $XY = Z$. The number of such formulas in mathematics and science is a large one and only a few of these are listed in the following table:

TABLE OF FORMULAS AND RULES HAVING RELATIONSHIPS OF THE FORM $XY = Z$

Formula	Rule
$LW = A$	Length of a rectangle $\times$ Width of the rectangle $=$ Area of the rectangle
$RT = D$	Rate of speed $\times$ Time of travel $=$ Distance traveled
$NP = C$	Number of units $\times$ Price of each $=$ Cost of the units
$RT = W$	Rate of work $\times$ Time of work $=$ Work accomplished
$PR = I$	Principal $\times$ Rate of interest $=$ Interest earned
$BR = P$	Base $\times$ Rate $=$ Percentage
$SN = D$	Scale of map $\times$ Number of units in a map length $=$ actual Distance
$FD = N$	Value of a Fraction $\times$ Denominator $=$ Numerator
$PV = k$	Pressure of a confined gas $\times$ Volume of the gas $=$ Constant (Boyle's Law)
$WD = M$	Weight $\times$ Distance from fulcrum $=$ Moment of the weight

∿∿∿∿∿ *MODEL PROBLEMS* ∿∿∿∿∿

1. The perimeter of a rectangle is 16 inches and its area is 15 square inches. Find the dimensions of the rectangle.

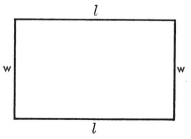

Solution: Let $l =$ length of the rectangle in inches.

       $w =$ width of the rectangle in inches.

Then $2l + 2w =$ perimeter of the rectangle in inches.

       $lw =$ area of the rectangle in square inches.

Since the area is 15 square inches, then $lw = 15$.

Since the perimeter is 16 inches, then $2l + 2w = 16$

$$\mathbf{D_2}: l + w = 8$$
$$w = 8 - l$$

In equation $lw = 15$, substitute $(8 - l)$ for w.

$$l(8 - l) = 15$$
$$\text{Equate to 0: } 8l - l^2 - 15 = 0$$
$$l^2 - 8l + 15 = 0$$

$$\text{Factor: } (l - 5)(l - 3) = 0$$

$$
\begin{array}{c|c}
l - 5 = 0 & l - 3 = 0 \\
l = 5 & l = 3 \\
w = 3 & w = 5
\end{array}
$$

Answer: The length is 5 inches and the width is 3 inches.
    Or, the length is 3 inches and the width is 5 inches.

2. A rectangular lawn is 40 feet by 50 feet. How wide a uniform strip has been mowed around the edge if $\frac{3}{10}$ of the lawn has not yet been mowed?

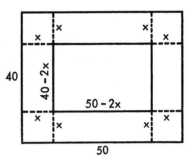

Solution: Let $x =$ width of the uniform strip in feet.

Then $50 - 2x =$ one side of rectangle not mowed in feet.

And $40 - 2x =$ other side of rectangle not mowed in feet.

Note that the inner rectangle is the area that has not yet been mowed. Since the area of the inner rectangle is $\frac{3}{10}$ of the area of the lawn,

$$(50 - 2x)(40 - 2x) = \tfrac{3}{10}(50 \times 40)$$
$$2000 - 180x + 4x^2 = 600$$
$$\text{Equate to 0: } 4x^2 - 180x + 1400 = 0$$
$$x^2 - 45x + 350 = 0$$
$$\text{Factor: } (x - 10)(x - 35) = 0$$
$$x = 10 \quad | \quad x = 35 \quad \text{Impossible}$$

Answer: Strip is 10 feet wide.

Exercises

In 1–4, find the dimensions of a rectangle whose width in inches is represented by w, whose length in inches is represented by $w + 3$, and whose area in square inches is:

1. 10 **2.** 40 **3.** 70 **4.** 6.75

In 5–8, find the dimensions of a rectangle whose area is 72 square feet and whose perimeter in feet is:

5. 34 **6.** 36 **7.** 54 **8.** 76

In 9–12, find the dimensions of a rectangle whose perimeter in yards is 32 and whose area in square yards is:

9. 15 **10.** 63 **11.** 60 **12.** 63.75

13. The area of a rectangle is 600 square inches. Find the dimensions of the rectangle if they are in the ratio of 2 : 3.

14. A rectangular lot is 50 feet wide and 60 feet long. If both the width and the length are increased by the same amount, the area is increased by 1200 square feet. Find the amount by which both the width and the length are increased.

15. A rectangular lawn is 60 feet by 80 feet. How wide a uniform strip must be cut around the edge when mowing the grass in order that half of the grass be cut?

16. A picture 9 inches by 12 inches is surrounded by a frame of uniform width. If the area of the frame exceeds the area of the picture by 124 square inches, find the width of the frame.

In 17–20, the width of the rectangle is 2 inches more than the length. Find its length to the nearest tenth of an inch if the area in square inches is:

17. 5   **18.** 12   **19.** 20   **20.** 32

In 21–24, one side of a square is increased 2 feet and the adjacent side is decreased 3 feet. Find the side of the square in feet if the resulting rectangle has an area in square feet of:

21. 24   **22.** 66   **23.** 126   **24.** 29.75

25. A floor can be covered with 1728 small square tiles. If tiles 2 inches longer on each side are used, the floor can be covered with 432 tiles. Find the length of éach side of the smaller tile.

26. The altitude of a triangle exceeds the base by 4 inches. If the area of the triangle is 30 square inches, find the base and altitude of the triangle.

27. A rectangular piece of cardboard is twice as long as it is wide. From each of its four corners a square piece 2 inches on a side is cut out. The flaps are then turned up to form an uncovered box.

   a. If the length of the shorter side of the original piece of cardboard is represented by x, express the volume V of the box in terms of x.

   b. If $V = 320$ cubic inches, find x.

14. Motion Problems Involving Quadratic Equations

〰〰〰〰〰〰 *MODEL PROBLEM* 〰〰〰〰〰〰

A train made a trip of 600 miles between two cities. If it had averaged 10 miles per hour more, it would have required 2 hours less to make the trip. Find the rate of speed of the train.

Method 1: Apply $D = RT$.

Solution: Let $r =$ rate of speed of the train in miles per hour (mph).

	(mph) Rate	$\times$	(hr.) Time	$=$	(mi.) Distance
Slow trip	r		t		600
Fast trip	$(r+10)$		$(t-2)$		600

For each trip, the distance was 600 miles.

$$(r+10)(t-2) = 600$$
$$rt + 10t - 2r - 20 = 600$$
$$600 + 10t - 2r - 20 = 600 \quad \text{(since } rt = 600\text{)}$$
$$10t - 2r - 20 = 0$$

Since $rt = 600$, then $\dfrac{600}{r}$ may be substituted for t.

$$\text{Equate to 0: } 10\left(\frac{600}{r}\right) - 2r - 20 = 0$$

$$\begin{aligned}
\text{M}_r\colon \quad & 6000 - 2r^2 - 20r = 0 \\
& 2r^2 + 20r - 6000 = 0 \\
\text{D}_2\colon \quad & r^2 + 10r - 3000 = 0 \\
\text{Factor:} \quad & (r - 50)(r + 60) = 0
\end{aligned}$$

$$
\begin{array}{c|c}
r - 50 = 0 & r + 60 = 0 \\
r = 50 & r = -60 \quad \text{Reject}
\end{array}
$$

Answer: The rate of speed of the train was 50 miles per hour.

Method 2: Apply $T = \dfrac{D}{R}$

Solution: If r and $r + 10$ represent the rates of the train in mph, the following is obtained:

Time for slow trip is 2 hours more than time for fast trip.

$$\text{M}_{r(r+10)}\colon \quad \frac{600}{r} = \frac{600}{r+10} + 2$$

Exercises

1. Two cars made the same trip of 200 miles. One traveled 15 miles an hour faster than the other and took 3 hours less to make the trip. Find the rate of each car.

2. A salesman made a trip of 90 miles to see a customer and returned home. On his return trip, he traveled 4 miles per hour faster than on his trip out and made it in 15 minutes less time. Find his rate each way.

3. Two men, *A* and *B*, traveled 96 miles and 100 miles respectively. *A*'s average rate was 8 miles an hour less than *B*'s, and *A*'s trip took one-half hour more than *B*'s. Find the average rate of each.

4. A woman makes a trip of 135 miles. By traveling 5 miles an hour faster, she can arrive 18 minutes earlier. Find both rates.

5. Two trains made the same run of 300 miles. One traveled 10 miles an hour faster than the other and used $1\frac{1}{2}$ hours less time. Find the rate at which each train traveled.

6. A train traveling at 8 miles an hour less than its usual rate arrives at its destination 5 hours late. The destination was 800 miles from the starting point. What was the usual rate of the train?

7. Sam drove 140 miles in $\frac{1}{2}$ hour less time than it took Tom to drive 176 miles. Sam's rate was 4 miles per hour less than Tom's rate. Find the rate of each.

8. On a 75-mile trip, Miss Jones' average rate for the first 15 miles was 10 miles per hour less than her average rate for the remainder of the trip. Her time for the entire trip was two hours. Find her average rate for the first 15 miles.

15. Business Problems Involving Quadratic Equations

~~~~~~~~~~ *MODEL PROBLEM* ~~~~~~~~~~

Some girls rented a cottage for $80 a day. Just before they moved in, another girl joined them. This reduced by $4 a day the amount of rent each of the original group of girls had expected to pay. How many girls were in the original group?

Method 1: Apply *NS = T.*

Solution: Let *n* = number of girls in the original group.
And *r* = daily rent each girl was to pay in dollars.

	(of girls) Number ×	($) Share of rent =	($) Total rent
Original group	*n*	*r*	*nr*
New group	(*n* + 1)	(*r* − 4)	(*n* + 1)(*r* − 4)

Since the rent is $80, then $nr = 80$ and $(n + 1)(r - 4) = 80$.

$$(n + 1)(r - 4) = 80$$
$$nr + r - 4n - 4 = 80$$
$$80 + r - 4n - 4 = 80$$
$$r = 4n + 4$$

Substitute $4n + 4$ for r in the equation $nr = 80$.

$$n(4n + 4) = 80$$

Equate to 0: $\quad 4n^2 + 4n - 80 = 0$

$$n^2 + n - 20 = 0$$

Factor: $\quad (n + 5)(n - 4) = 0$

$n + 5 = 0$	$n - 4 = 0$
$n = -5$	$n = 4$
Reject	

Answer: There were four girls in the original group.

Method 2: Apply $S = \dfrac{T}{N}$.

Solution: If n and $n + 1$ represent respectively the number of girls in the original group and in the new group, the following is obtained:

The new rent for each girl was $4 less than the old rent.

$$\mathbf{M}_{n(n+1)}: \quad \frac{80}{n + 1} = \frac{80}{n} - 4$$

The remainder of the solution is the same as the previous solution.

~~~~~~

Exercises

1. A group of boys agreed to buy a used car for $900. Just before they paid for the car, 2 more boys joined the group. This reduced the amount that each boy had to pay by $75. How many boys were in the original group?

2. A man buys a certain number of shares of stock for $528. Had he bought the stock when each share was $2 less, he could have purchased 2 more shares for the same amount of money. How many shares did he buy?

3. A teenager mowed lawns a certain number of days to earn $300. If he had received $3 less per day, he would have had to work 5 days longer to earn the same amount. How many days did he work?

4. A dealer bought a number of birds for $440. After 5 birds had died, he sold the rest of the birds at a profit of $2 each, thereby making $60 on the whole transaction. How many birds did he buy?

5. At a certain school, the total receipts for a junior dance were $150. If 5 more couples had attended the dance, the price per couple could have been reduced one dollar without causing any change in the total receipts. What was the price per couple?

6. A worker received $320 for a certain number of days' work. If she had received $8 less per day, it would have taken her 2 days longer to earn the same amount. How many days did she work?

16. Work Problems Involving Quadratic Equations

〜〜〜〜〜〜〜〜〜〜 *MODEL PROBLEM* 〜〜〜〜〜〜〜〜〜

It takes Thomas 3 hours less time to paint a fence than it takes Larry. If the two boys work together, they can paint the fence in 2 hours. How many hours would each boy, working alone, need to paint the fence?

Solution: Let $x =$ number of hours Larry needs to do the job alone.
  Then $x - 3 =$ number of hours Thomas needs to do the job alone.

	(part of job per hr.) Rate of work $\times$	(hr.) Time of work $=$	(part of job) Work done
Larry	$\dfrac{1}{x}$	2	$\dfrac{2}{x}$
Thomas	$\dfrac{1}{x-3}$	2	$\dfrac{2}{x-3}$

If the job is finished, the sum of the fractional parts done
by each boy must equal 1.

$$\text{Hence, } \frac{2}{x} + \frac{2}{x-3} = 1$$

Multiply each side by the L.C.D., which is $x(x - 3)$.

$$2(x - 3) + 2x = x(x - 3)$$
$$2x - 6 + 2x = x^2 - 3x$$

Equate to 0: $x^2 - 7x + 6 = 0$

Factor: $(x-1)(x-6) = 0$

$$x = 1 \qquad \mid \qquad x = 6$$

Reject

Answer: Larry requires 6 hours; Thomas requires 3 hours.

Exercises

1. It takes Bill 6 hours longer than John to plow a certain field. Together they can plow it in 4 hours. How long would it take each man alone to plow the field?
2. A company contracted to make a certain number of planes. Its two factories can make the planes in 12 days. Working alone, one factory requires 7 days longer than the other. Find the time in which each factory alone can fulfill the contract.
3. Sally and Wilma working together can complete a job in $1\frac{1}{3}$ hours. If it takes Sally working alone 2 hours less time than Wilma to do the job, find the time required by each woman working alone.
4. It takes Tom three hours longer to do a certain job than it takes his brother Bill. They worked together for three hours; then Tom left and Bill finished the job in one hour. How many hours would it have taken Bill to do the job alone?

17. Right Triangle Problems Involving Quadratic Equations

MODEL PROBLEM

A and B start from the same point and travel along roads that are at right angles to each other. A travels 3 miles an hour faster than B. At the end of 2 hours, they are 30 miles apart. Find their rates.

Solution: Let r = rate of B in mph.

Then $r + 3$ = rate of A in mph.

	(mph) Rate	×	(hr.) Time	=	(mi.) Distance
A	$r + 3$		2		$2r + 6$
B	r		2		$2r$

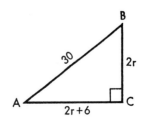

Apply the theorem of Pythagoras: $\overset{a^2}{(\text{leg})^2} + \overset{b^2}{(\text{leg})^2} = \overset{c^2}{(\text{hypotenuse})^2}$

$$(2r)^2 + (2r+6)^2 = (30)^2$$
$$4r^2 + 4r^2 + 24r + 36 = 900$$

Equate to 0: $\qquad 8r^2 + 24r - 864 = 0$
$$r^2 + 3r - 108 = 0$$

Factor: $\qquad (r-9)(r+12) = 0$

$r - 9 = 0$	$r + 12 = 0$
$r = 9$	$r = -12$
$r + 3 = 12$	Reject

Answer: A's rate is 12 mph; B's rate is 9 mph.

Exercises

1. One leg of a right triangle exceeds the other leg by 5 inches. The hypotenuse is 25 inches. Find the length of each leg of the triangle.
2. The hypotenuse of a right triangle is 2 inches longer than one leg and 4 inches longer than the other leg. Find the length of each side of the triangle.
3. The perimeter of a right triangle is 40 inches. If the hypotenuse is 17 inches, find the length of each leg.
4. Find the dimensions of a rectangle if its perimeter is 34 feet and its diagonal is 13 feet.
5. The hypotenuse of a right triangle exceeds the longer of the two legs by 2. If the perimeter of the triangle is 40, find the lengths of the three sides of the triangle.
6. Peter and William start from the point of intersection of two roads which are at right angles to each other. Peter travels along one road at the rate of 5 miles per hour and William travels along the other road at the rate of 12 miles per hour. In how many hours will they be 26 miles apart?
7. A ship sails due east from a certain point at a speed of 16 mph. Two hours later, a ship leaves the same point and sails due north at a speed of 20 mph. In how many hours after the first ship starts will the two ships be 100 miles apart?

In 8–10, A and B start from the same point and travel along straight roads that are at right angles to each other. Find their rates if:

8. A travels 7 miles per hour faster than B and, at the end of an hour, they are 13 miles apart.
9. A travels 7 miles an hour slower than B and, at the end of 2 hours, they are 34 miles apart.
10. The ratio of A's speed to B's speed is $3:4$ and, at the end of 5 hours, they are 125 miles apart.

—————— *CALCULATOR APPLICATIONS* ——————

Calculators are very useful for performing many of the operations involved in solving quadratic equations. Remember that when roots are irrational, the calculator will display approximate solutions.

——————— *MODEL PROBLEMS* ———————

1. Solve $x^2 - 10x + 13 = 0$ for values of x to the nearest hundredth.

   Solution: To solve the problem using a scientific calculator:

   Compare $x^2 - 10x + 13 = 0$ with $ax^2 + bx + c = 0$ to determine that $a = 1$, $b = -10$, and $c = 13$.

   Substitute these values in the quadratic formula:

   $$x = \frac{-b \pm \sqrt{b^2 - 4ac}}{2a} = \frac{-(-10) \pm \sqrt{(-10)^2 - 4(1)(13)}}{2(1)}$$

   Use the calculator to evaluate the radical part of the expression above and store the value in memory.

   Enter: $\boxed{(}$ $\boxed{10}$ $\boxed{+/-}$ $\boxed{)}$ $\boxed{x^2}$ $\boxed{-}$ $\boxed{4}$ $\boxed{\times}$ $\boxed{1}$ $\boxed{\times}$ $\boxed{1}$ $\boxed{3}$ $\boxed{=}$ $\boxed{\sqrt{x}}$ $\boxed{\text{STO} \blacktriangleright}$

   The value of 6.92820323 is stored.

   (Some calculators may use $\boxed{\text{M+}}$ key to enter into memory and $\boxed{\text{MR}}$ to recall.)

   Now find the two solutions.

   Enter: $\boxed{(}$ $\boxed{10}$ $\boxed{+/-}$ $\boxed{)}$ $\boxed{+}$ $\boxed{\text{RCL}}$ $\boxed{=}$ $\boxed{\div}$ $\boxed{2}$ $\boxed{=}$

   Display: $\boxed{8.464101615}$ The root is 8.46 to the nearest hundredth.

   Enter: $\boxed{(}$ $\boxed{10}$ $\boxed{+/-}$ $\boxed{)}$ $\boxed{-}$ $\boxed{\text{RCL}}$ $\boxed{=}$ $\boxed{\div}$ $\boxed{2}$ $\boxed{=}$

   Display: $\boxed{1.535898385}$ The root is 1.54 to the nearest hundredth.

   Answer: To the nearest hundredth, the solutions are $x = 8.46$ and $x = 1.54$.

The solution process can be carried out much more efficiently with a TI-83 or similar graphing calculator. One approach is described below, but you may develop others.

2. Solve $6x^2 + 3x = 4x + 2$

   Solution: First express the equation in the form $ax^2 + bx + c = 0$.

   $$6x^2 + 3x = 4x + 2$$
   $$6x^2 - x - 2 = 0$$

   So $a = 6$, $b = -1$, and $c = -2$.

Enter the formula for one solution as Y_1 and the formula for the other solution as Y_2.

Enter: Y= ((−) ALPHA B + 2nd √ ALPHA B x^2

─ 4 ALPHA A ALPHA C)) ÷ (2 ALPHA A

) ENTER

((−) ALPHA B ─ 2nd √ ALPHA B x^2

─ 4 ALPHA A ALPHA C)) ÷ (2 ALPHA A

) ENTER

Press 2nd QUIT to return to the home screen. Now store the values of a, b, and c.

Enter: 6 STO ▶ ALPHA A ALPHA : (−) 1 STO ▶ ALPHA B ALPHA

: (−) 2 STO ▶ ALPHA C ENTER

Display: | 6 → A: −1 → B: −2 → C | The last value stored is displayed on
| −2 | the right side.

Now the solutions can be found. First, instruct the calculator to compute Y_1.

Enter: VARS ▶ ENTER ENTER ENTER

Display: .6666666667

To show the solution as a fraction:

Enter: MATH 1 ENTER

Display: 2/3

To find the second solution, use Y_2.

Enter: VARS ▶ ENTER 2 ENTER

Display: −.5

Enter: MATH 1 ENTER

Display: −1/2

Answer: The solutions are $x = \frac{2}{3}$ and $x = -\frac{1}{2}$.

A whole group of quadratic equations can be solved by changing the values stored and then using the formulas entered in Y_1 and Y_2.

CHAPTER XI

Graphs of Quadratic Functions, Relations, Equations, and Inequalities

1. The Parabola as a Graph

PARABOLAS OF THE FORM $y = ax^2 + bx + c$

A quadratic function is a function whose values are given by a quadratic polynomial of the form $ax^2 + bx + c$, where a, b, and c are real numbers, $a \neq 0$.

A quadratic function may be defined as:
1. the equation $y = ax^2 + bx + c$, $a \neq 0$
2. the set $\{(x, y) \mid y = ax^2 + bx + c, \ a \neq 0\}$
3. $f(x) = ax^2 + bx + c$, $a \neq 0$, using function notation

If, in the quadratic equation $y = ax^2 + bx + c$, a value is assigned to x, there is one and only one corresponding value of y. The set of ordered pairs (x, y) that satisfy $y = ax^2 + bx + c$ defines a function. If the ordered pairs of the solution set of $y = ax^2 + bx + c$ are graphed, the resulting points can be joined to form a curve called a *parabola*. This is illustrated in the graphing of the equation $y = x^2 - 4x + 3$, which follows.

In the discussions to follow, we will use the expression "*the parabola $y = ax^2 + bx + c$*" as the simplified form of the expression "the parabola that is the graph of $y = ax^2 + bx + c$, where a, b, and c are real numbers, $a \neq 0$."

We begin by graphing the parabola $y = x^2 - 4x + 3$ for values of x from -1 to 5 inclusive. We will then consider several important features and properties of the parabola.

〰〰〰〰〰〰〰 *MODEL PROBLEM* 〰〰〰〰〰〰〰

Draw the graph of $y = x^2 - 4x + 3$ for values of x from -1 to 5 inclusive.

How To Proceed	*Solution*

TABLE OF ORDERED PAIRS

1. Obtain a *table of ordered pairs* by substituting a set of consecutive integral values for x and finding the corresponding values of y.

x	$x^2 - 4x + 3 = y$	
-1	$1 + 4 + 3$	8
0	$0 + 0 + 3$	3
1	$1 - 4 + 3$	0
2	$4 - 8 + 3$	-1
3	$9 - 12 + 3$	0
4	$16 - 16 + 3$	3
5	$25 - 20 + 3$	8

2. Plot the points whose coordinates represent the ordered pairs (x, y) in the table. (Note each of the points in Fig. 1.)

3. Join the points with a smooth curve. The resulting curve, the graph of $y = x^2 - 4x + 3$, is a *parabola*.

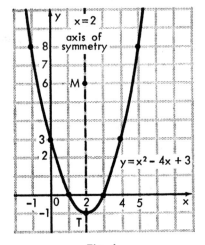

Fig. 1

A study of the table of values and the graph of $y = x^2 - 4x + 3$ in Fig. 1 leads to the following observations and conclusions:

1. As x increases from -1 to 2, y decreases from 8 to -1; then as x continues to increase from 2 to 5, y increases from -1 to 8.

2. The point $T(2, -1)$, where the values of y stop decreasing and begin to increase, is called the **turning point**, or **vertex**, of the parabola. Note that T is the lowest point of the graph. The point T is called the **minimum point** because y or its equivalent expression $x^2 - 4x + 3$ has its minimum or smallest value, -1, at T.

3. The **axis of symmetry** $(\overleftrightarrow{TM})$ of the parabola is the line $x = 2$. Note that the axis of symmetry passes through the turning point T and is parallel to the y-axis. If the parabola were folded over on the axis of symmetry, its left half would coincide with its right half.

MAXIMUM AND MINIMUM POINTS OF A PARABOLA $y = ax^2 + bx + c$

We shall now graph the parabola, $y = -x^2 + 4x - 3$ and consider several of its important features and properties. Note in the table of ordered pairs that the set of consecutive integral values substituted for x is the same as that used in the graphing of the parabola $y = x^2 - 4x + 3$.

TABLE OF
ORDERED PAIRS

x	$-x^2 + 4x - 3 = y$	
-1	$-1 - 4 - 3$	-8
0	$-0 + 0 - 3$	-3
1	$-1 + 4 - 3$	0
2	$-4 + 8 - 3$	1
3	$-9 + 12 - 3$	0
4	$-16 + 16 - 3$	-3
5	$-25 + 20 - 3$	-8

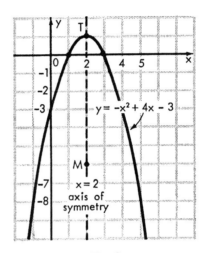

Fig. 2

Compare the expressions $x^2 - 4x + 3$ and $-x^2 + 4x - 3$ and note that either one is the negative of the other. Hence, for any assigned value of x, the resulting values of the expressions are negatives of each other. Verify this by examining the tables of ordered pairs that were used to graph $y = x^2 - 4x + 3$ and $y = -x^2 + 4x - 3$.

A study of the table of values and the graph of $y = -x^2 + 4x - 3$ in Fig. 2 leads to the following observations and conclusions:

1. As x increases from -1 to 2, y increases from -8 to 1; then as x continues to increase from 2 to 5, y decreases from 1 to -8.

2. The point $T(2, 1)$, where the values of y stop increasing and begin to decrease, is the *turning point*, or *vertex*, of the parabola. Note that T is the highest point of the graph. The point T is called the **maximum point** because y or its equivalent expression $-x^2 + 4x - 3$ has its maximum or greatest value, $+1$, at T.

3. The *axis of symmetry* $(\overleftrightarrow{TM})$ of the parabola is the line $x = 2$. Here again, the axis of symmetry passes through the turning point T and is parallel to the y-axis.

An examination of the two parabola graphs in Figs. 1 and 2 shows that:

The parabola $y = x^2 - 4x + 3$, in which the coefficient of x^2 is $+1$, has a *minimum turning point* and *opens upward*.

The parabola $y = -x^2 + 4x - 3$, in which the coefficient of x^2 is -1, has a *maximum turning point* and *opens downward*.

The two parabolas that we have considered in Figs. 1 and 2 illustrate the following:

Rule 1. If $y = ax^2 + bx + c$ and a is positive, $a > 0$, the graph of the equation is a parabola that *opens upward and has a minimum turning point* (Fig. 3).

Rule 2. If $y = ax^2 + bx + c$ and a is negative, $a < 0$, the graph of the equation is a parabola that *opens downward and has a maximum turning point* (Fig. 4).

Graph of $y = ax^2 + bx + c$
$a > 0$

Graph of $y = ax^2 + bx + c$
$a < 0$

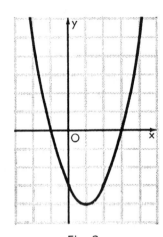 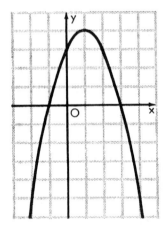

Fig. 3   Fig. 4

THE MEANING OF THE EXPRESSION
"QUADRATIC FUNCTION $ax^2 + bx + c$"

Examine Figs. 3 and 4 again. Note that in the case of any parabola whose equation is of the form $y = ax^2 + bx + c$, for each value of x there is one and only one value of y. If the vertical test of a function is applied by drawing vertical lines crossing either parabola, the vertical lines cannot pass through more than one point. Hence, the set of all ordered pairs that satisfy $y = ax^2 + bx + c$ is a *relation that is a function*.

Keep in mind that the expression "*quadratic function* $ax^2 + bx + c$" is a simplified form of the expression "quadratic function that is the set of ordered pairs satisfying $y = ax^2 + bx + c$ where a, b, and c are real numbers, $a \neq 0$."

AXIS OF SYMMETRY OF A PARABOLA $y = ax^2 + bx + c$

Recall that the *axis of symmetry of a parabola* is the line over which one-half of the parabola may be folded so that both halves can be made to coincide. The equation of the axis of symmetry can be found by applying the following rule:

Rule. The equation $x = -\dfrac{b}{2a}$ is the equation of the axis of symmetry of the parabola $y = ax^2 + bx + c$.

A justification of the above rule for the equation of the axis of symmetry can be found on pages 394 and 395.

Thus, if $y = x^2 - 4x + 3$, $a = 1$ and $b = -4$. Using $x = -\dfrac{b}{2a}$ to find the equation of the axis of symmetry of the parabola $y = x^2 - 4x + 3$, then $x = -\dfrac{-4}{2}$, or $x = 2$. In the case of the parabola $y = -x^2 + 4x - 3$, the equation of the axis of symmetry is $x = -\dfrac{4}{-2}$ or $x = 2$. By examining Fig. 5, verify that the line $x = 2$ is the axis of symmetry of each of the parabolas.

Another kind of symmetry may be seen in Fig. 5. Note that either parabola may be obtained by revolving the other parabola 180° about the x-axis. For this reason, any point on one parabola is a "reflection" through the x-axis of a point on the other parabola. Hence, the x-axis is the axis of symmetry of the combined figure of both parabolas, in the sense that the x-axis is the line over which one parabola can be folded to make it coincide with the other. Think of the x-axis serving as a mirror through which a point on one parabola is reflected to a corresponding "image" point on the other parabola. Thus, the turning point of one parabola $(2, 1)$ is the *mirror image* (or simply *image*) of the turning point of the other parabola $(2, -1)$.

In parabolas that have a vertical axis of symmetry, the equation of the axis of symmetry is useful in determining the x-values that should be chosen in a table of values needed to graph the parabola. We can now understand why the integral values from -1 to 5 were

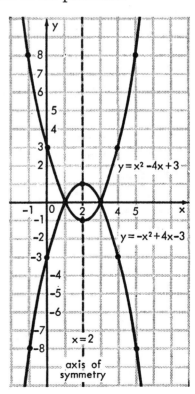

Fig. 5

assigned to x in the graphing of both $y = x^2 - 4x + 3$ and $y = -x^2 + 4x - 3$. The middle value, 2, of the set $\{-1, 0, 1, 2, 3, 4, 5\}$ is the value of x obtainable from the equation of the axis of symmetry. Had more values been needed, the set of consecutive integers from -2 to 6 or from -3 to 7 could be used, keeping 2 as the middle value of the set.

If a problem requires the graphing of a parabola in the form $y = ax^2 + bx + c$, and no set of values is assigned to x, it is desirable to obtain the equation of the axis of symmetry. The value of x thus obtained should be made the middle value of a set consisting of a total of at least 7 consecutive integers.

FINDING THE COORDINATES OF THE TURNING POINT OF THE PARABOLA $y = ax^2 + bx + c$

The turning point of a parabola is one of the points of its axis of symmetry. Hence, the abscissa of the turning point of $y = ax^2 + bx + c$ is $x = -\dfrac{b}{2a}$. The ordinate of the turning point can be determined by substituting $-\dfrac{b}{2a}$ for x in the equation $y = ax^2 + bx + c$.

〜〜〜〜〜〜〜〜〜 *MODEL PROBLEMS* 〜〜〜〜〜〜〜〜〜

In 1 and 2, find the equation of the axis of symmetry and the coordinates of the turning point of the parabola.

1. $y = x^2 - 8x + 15$　　　　**2.** $y = -3x^2 + 12x + 7$

How To Proceed	*Solution*	*Solution*
1. State the values of a and b.	$a = 1, b = -8$	$a = -3, b = 12$
2. Find the equation of the axis of symmetry, using $x = -\dfrac{b}{2a}$.	$x = -\dfrac{b}{2a}$ $= -\dfrac{-8}{2} = 4$	$x = -\dfrac{b}{2a}$ $= -\dfrac{12}{-6} = 2$
3. Find the y-coordinate of the turning point by substituting $-\dfrac{b}{2a}$ for x in the equation $y = ax^2 + bx + c$.	$y = x^2 - 8x + 15$ $= 4^2 - 8(4) + 15$ $= 16 - 32 + 15 = -1$ Axis of symmetry, $x = 4$; coordinates of turning point, $(4, -1)$.　*Ans.*	$y = -3x^2 + 12x + 7$ $= -3(2^2) + 12(2) + 7$ $= -12 + 24 + 7 = 19$ Axis of symmetry, $x = 2$; coordinates of turning point, $(2, 19)$.　*Ans.*

PARABOLAS OF THE FORM $x = ay^2 + by + c$

The equation $x = ay^2 + by + c$ is obtained by interchanging x and y in the equation $y = ax^2 + bx + c$. When equations of the form $x = ay^2 + by + c$ are graphed, as in Figs. 6 and 7, the graphs are parabolas having an axis of symmetry parallel to the x-axis. The equation of the axis of symmetry of the parabolas of the form $x = ay^2 + by + c$, is $y = -\dfrac{b}{2a}$. The equation $y = -\dfrac{b}{2a}$ is obtained simply by substituting y for x in the equation $x = -\dfrac{b}{2a}$, which we found to be the axis of symmetry of the previous vertical parabolas.

Thus, using $y = -\dfrac{b}{2a}$, we can show that the line $y = 1$ is the axis of symmetry of both the parabola $x = y^2 - 2y$ (Fig. 6) and the parabola $x = -y^2 + 2y$ (Fig. 7).

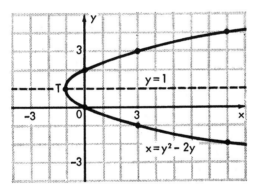

Fig. 6

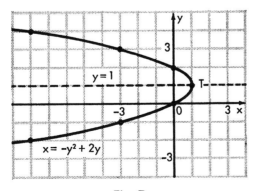

Fig. 7

THE RELATION $\{(x, y)\,|\,x = ay^2 + by + c\}$ IS NOT A FUNCTION

Note in Figs. 6 and 7 that the parabolas do not meet the vertical line test for a function, since vertical lines may pass through more than one point of each of these parabolas. In general, parabolas of the form $x = ay^2 + by + c$ do not meet the vertical line test of a function. Hence, the set of ordered pairs that satisfy $x = ay^2 + by + c$ is a *relation that is not a function.*

Figs. 6 and 7 illustrate the fact that parabolas of the form $x = ay^2 + by + c$ do not have a maximum or a minimum point. Instead, the turning point or vertex, T, may be a leftmost point, (Fig. 6) or a rightmost point, (Fig. 7).

Summary of the Properties of a Parabola $y = ax^2 + bx + c$

1. If a is positive, $a > 0$, the parabola opens upward and has a minimum turning point.
   If a is negative, $a < 0$, the parabola opens downward and has a maximum turning point.

2. The equation of the axis of symmetry is $x = -\dfrac{b}{2a}$.

   The x-coordinate of the turning point is also equal to $-\dfrac{b}{2a}$.

3. When $a > 0$: From left to right, the values of y decrease until the minimum point is reached; then the values of y increase.
   When $a < 0$: From left to right, the values of y increase until the maximum point is reached; then the values of y decrease.

4. The set of all ordered pairs that satisfy $y = ax^2 + bx + c$ is a relation that is a function.

Summary of the Properties of a Parabola $x = ay^2 + by + c$

1. If a is positive, $a > 0$, the parabola opens to the right and has a leftmost point.
   If a is negative, $a < 0$, the parabola opens to the left and has a rightmost point.

2. The equation of the axis of symmetry is $y = -\dfrac{b}{2a}$.

   The y-coordinate of the turning point is also equal to $-\dfrac{b}{2a}$.

3. The set of all ordered pairs that satisfy $x = ay^2 + by + c$ is a relation that is *not* a function.

Exercises

In 1–13, for the quadratic function: (*a*) draw a graph using the indicated values of x, (*b*) mark the turning point with the letter T and state whether it is a minimum or a maximum point, (*c*) state the coordinates of the turning point, and (*d*) draw the axis of symmetry and state its equation.

1. $y = x^2$ from $x = -5$ to $x = 5$ 2. $y = 3x^2$ from $x = -2$ to $x = 2$
3. $y = -2x^2$ from $x = -2$ to $x = 2$ 4. $y = x^2 - 9$ from $x = -4$ to $x = 4$
5. $y = -x^2 + 4$ from $x = -3$ to $x = 3$
6. $y = x^2 - 4x$ from $x = -2$ to $x = 6$
7. $y = x^2 - 6x + 8$ from $x = 0$ to $x = 6$
8. $y = x^2 - 4x + 3$ from $x = 0$ to $x = 4$
9. $y = -x^2 + 6x - 8$ from $x = 0$ to $x = 6$
10. $y = x^2 - 3x + 2$ from $x = -1$ to $x = 4$
11. $y = -x^2 + x + 2$ from $x = -2$ to $x = 3$
12. $y = 2x^2 - 5x + 2$ from $x = -1$ to $x = 4$
13. $y = -2x^2 + 7x - 3$ from $x = 0$ to $x = 4$

In 14–25, for the quadratic function: (*a*) prepare a table of values, (*b*) draw a graph, (*c*) mark the turning point with the letter T, (*d*) state the coordinates of the turning point, and (*e*) draw the axis of symmetry and state its equation.

14. $y = 2x^2$ 15. $y = -3x^2$ 16. $y = x^2 - 4$
17. $y = -x^2 + 9$ 18. $y = x^2 - 6x$ 19. $y = -x^2 + 4x$
20. $y = x^2 - 4x + 3$ 21. $y = x^2 - x - 2$ 22. $y = -x^2 + 3x + 2$
23. $x = 2y^2$ 24. $x = -y^2$ 25. $x = y^2 - 4y$

26. Which, if any, of the pairs of x and y values given in the table at the right are *not* roots of $x^2 - 3x = y$?

x	-1	2	3
y	4	2	0

27. The parabola whose equation is $y = ax^2$ passes through the point $(2, 3)$. Find the value of a.
28. The graph of the equation $y = x^2 + 3x + k$ passes through the point $(2, 0)$. Find k.
29. The graph of the equation $x^2 + bx + c = y$ (1) always   (2) sometimes   (3) never   passes through the origin.

In 30–35, give the equation of the axis of symmetry of the graph of the function.

30. $y = x^2 - 6x + 5$ 31. $y = x^2 + 8x + 7$ 32. $y = 3x^2 - 6x$
33. $y = x^2 - 16$ 34. $y = -x^2 - 2x + 3$ 35. $y = 2x^2 + 5x + 2$

In 36–41, find the turning point of the graph of the function.

36. $y = 4x^2$ 37. $y = x^2 - 2x - 8$ 38. $y = x^2 - 6x + 5$
39. $y = x^2 - 3x$ 40. $y = -x^2 - x + 6$ 41. $y = 2x^2 + 5x + 2$

2. Studying the Roles That *a*, *b*, and *c* Play in Changing Parabolas of the Form $y = ax^2 + bx + c$

Now we will study how changes in the values of a, b, and c in the equation $y = ax^2 + bx + c$ change parabolas having this form.

CHANGING THE WIDTH OF A PARABOLA BY CHANGING THE VALUE OF *a*

To simplify our discussion, we will study the effect of changing the value of a alone by considering the quadratic function defined by $y = ax^2$. If $y = ax^2$ and $a = 1$, we obtain $y = x^2$. The graph of $y = x^2$ is the parabola shown in Fig. 1.

The parabola $y = x^2$ may be obtained by using the following table of ordered pairs and joining the points representing these ordered pairs by a smooth curve (Fig. 1).

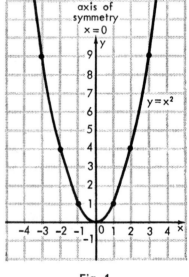

Fig. 1

TABLE OF
ORDERED PAIRS

If $x = 3$ or $x = -3$,
  then, $y = 9$ ⟶ $(-3, 9)$ $(3, 9)$
If $x = 2$ or $x = -2$,
  then, $y = 4$ ⟶ $(-2, 4)$ $(2, 4)$
If $x = 1$ or $x = -1$,
  then, $y = 1$ ⟶ $(-1, 1)$ $(1, 1)$
If $x = 0$, then $y = 0$ ⟶ $(0, 0)$

Note the following properties of the parabola $y = x^2$:

1. The origin, $(0, 0)$, is the vertex of the parabola and a minimum turning point.
2. The y-axis, $x = 0$, is the axis of symmetry of the parabola.
3. Since a is positive, the parabola faces upward.
4. Since $y = x^2$, then for all real values of x, the values of y must be non-negative. Hence, the parabola passes through the origin and is entirely within quadrants I and II; that is, the parabola is above the x-axis and tangent to it.

In Fig. 2, note the parabolas $y = x^2$, $y = 2x^2$, and $y = \frac{1}{2}x^2$, which face upward; also note the parabolas $y = -x^2$, $y = -2x^2$, and $y = -\frac{1}{2}x^2$, which face downward. Observe that all six parabolas have a turning point, or vertex, which is the origin. The origin serves as the minimum point for the parabolas that face upward and the maximum point for the parabolas that face downward. The y-axis is the axis of symmetry of each parabola. The x-axis is the axis of symmetry for the pair of parabolas $y = 2x^2$ and $y = -2x^2$, the pair $y = x^2$ and $y = -x^2$, and also the pair $y = \frac{1}{2}x^2$ and $y = -\frac{1}{2}x^2$. Note that in each of these pairs, the values of a are opposites, which means the absolute value of a is the same. In each of the pairs, the parabolas are congruent and, for this reason, have the same "width."

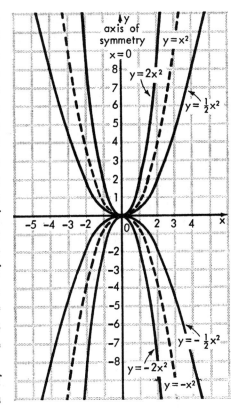

Fig. 2

Note in Fig. 2 that the narrowest pair of parabolas are those having the equations $y = 2x^2$ and $y = -2x^2$. Note further that when the absolute value of a is decreased, the parabolas widen. Thus, the parabolas $y = \frac{1}{2}x^2$ and $y = -\frac{1}{2}x^2$ are wider than the parabolas $y = x^2$ and $y = -x^2$; in turn, the parabolas $y = x^2$ and $y = -x^2$ are wider than the parabolas $y = 2x^2$ and $y = -2x^2$.

We may generalize the role played by a in the changing of the width of parabolas of the form $y = ax^2$ as follows:

Rule. In the quadratic function $y = ax^2$, increasing the absolute value of a, $|a|$, *narrows* the parabola; decreasing $|a|$ *widens* the parabola.

In a more general way, it may be shown that in the quadratic function $y = ax^2 + bx + c$, if the values of b and c are kept constant, increasing the absolute value of a, $|a|$, narrows the parabola; decreasing $|a|$ widens the parabola.

Thus, the parabola $y = 3x^2 + 4x - 5$ is narrower than the parabola $y = x^2 + 4x - 5$.

CHANGING THE VERTICAL POSITION OF A PARABOLA BY CHANGING THE VALUE OF c

In Fig. 3, note that the parabolas $y = \frac{1}{2}x^2 + 2$, $y = \frac{1}{2}x^2$, and $y = \frac{1}{2}x^2 - 2$ have the same axis of symmetry and are congruent. Note, however, that the parabolas differ in their y-intercept: The y-intercept is 2 in the case of $y = \frac{1}{2}x^2 + 2$; -2 in the case of $y = \frac{1}{2}x^2 - 2$; and 0 in the case of $y = \frac{1}{2}x^2$. Since the y-intercept of a graph is the ordinate of the point where the graph intersects the y-axis, $x = 0$. Hence, at this point, to find the y-intercept, simply substitute 0 for x. In the case of the general parabola, $y = ax^2 + bx + c$, if 0 is substituted for x, we see that $y = c$; that is, the y-intercept is c, the constant term.

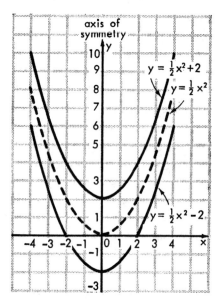

Fig. 3

Hence, increasing the constant term c results in an upward shift in the parabola; decreasing c results in a downward shift. Verify this in Fig. 3 by using transparent paper and making a trace of the parabola $y = \frac{1}{2}x^2$. Note that the trace can be made to coincide with each of the other parabolas, showing that the parabolas are congruent, having the same size and shape. By moving the trace vertically, we can obtain other parabolas such as $y = \frac{1}{2}x^2 + 4$, $y = \frac{1}{2}x^2 - 4$, $y = \frac{1}{2}x^2 + 4.5$, and $y = \frac{1}{2}x^2 - \frac{1}{4}$.

In general, if the equations of parabolas differ only in the constant term, as in the case of the parabola $y = ax^2 + bx + c$ and the parabola $y = ax^2 + bx + c'$, then:

1. The parabolas are congruent.

2. The parabolas have different y-intercepts: c in the case of $y = ax^2 + bx + c$, and c' in the case of $y = ax^2 + bx + c'$.

3. The parabolas have the same axis of symmetry. The reason that the axis of symmetry is the same for the parabolas is that a change in the value of c alone does not affect the equation of the axis of symmetry, $x = -\dfrac{b}{2a}$.

CHANGING THE AXIS OF SYMMETRY OF A PARABOLA

In Fig. 4, note that the parabolas $y = x^2 + 2x + 1$ and $y = x^2 - 2x + 1$ have the same y-intercept and are congruent. Note, however, that the parabolas differ in their axis of symmetry: The axis of symmetry is $x = 1$ in the case of $y = x^2 - 2x + 1$; it is $x = -1$ in the case of $y = x^2 + 2x + 1$. Note also in the two equations that the value of b has changed, while that of a is unchanged.

If a parabola is of the form $y = ax^2 + bx + c$, changing the value of b while keeping the value of a fixed will change the axis of symmetry. The reason is that the value of $-\dfrac{b}{2a}$ must change if there is a change in b and no change in a.

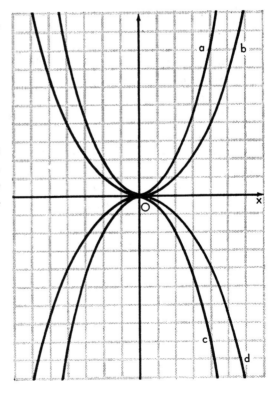

Fig. 4

Exercises

The parabolas shown in the figure at the right represent the graphs of $y = x^2$, $y = -x^2$, $y = \frac{1}{2}x^2$, and $y = -\frac{1}{2}x^2$. In 1–4, using the figure, state the equation represented by:

1. parabola a **2.** parabola b

3. parabola c **4.** parabola d

In 5–13, sketch the graphs of the given equations and explain how a change from one equation to another changes the graph represented by the equation.

5. $y = x^2$, $y = 2x^2$, $y = 3x^2$

6. $y = x^2$, $y = \frac{1}{2}x^2$, $y = \frac{1}{4}x^2$

7. $y = -x^2$, $y = -2x^2$, $y = -3x^2$

8. $y = -x^2$, $y = -\frac{1}{2}x^2$, $y = -\frac{1}{4}x^2$

9. $y = x^2$, $y = x^2 + 2$, $y = x^2 + 4$

10. $y = x^2$, $y = x^2 - 2$, $y = x^2 - 4$

11. $y = -x^2 - 3$, $y = -x^2$, $y = -x^2 + 3$

12. $y = x^2 + 2x$, $y = x^2 + 4x$, $y = x^2 - 2x$

13. $y = x^2 + 2x + 1$, $y = x^2 - 2x + 1$, $y = x^2 + 4x + 4$

14. By using the equation $x = -\dfrac{b}{2a}$, show that the axis of symmetry is the same for the parabolas $y = x^2 + 2x$, $y = 2x^2 + 4x$, and $y = -3x^2 - 6x$.

15. Graph $y = x^2$ and $y = 2x^2 - 3$ and show how changes in the coefficient of x^2 and the constant term affect the character of the graph.

16. Graph $y = x^2 + x + 3$ and $y = x^2 + x - 3$ and show how the change in the constant term affects the graph of $y = x^2 + x + 3$.

17. Graph $y = 2x^2 + 4x$ and $y = 2x^2 + 6x$ and show how the change in the coefficient of x affects the graph of $y = 2x^2 + 4x$.

18. Graph $y = 3x^2$ and $y = -3x^2$ and show how the change in the coefficient of x^2 affects the graph of $y = 3x^2$.

19. Graph $y = 3x^2$ and $y = \frac{1}{3}x^2$ and show how the change in the coefficient of x^2 affects the graph of $y = 3x^2$.

3. Using Parabolas To Solve Quadratic Equations

The quadratic equation, $x^2 = 4x + 5$, can be solved graphically, as in the model problem on page 395. This is done by applying the following rule:

Rule. If the parabola $y = ax^2 + bx + c$ intersects the x-axis, the x-intercepts are the real roots of the quadratic equation $ax^2 + bx + c = 0$.

This rule can be justified in the following manner, using the figure at the right:

If the parabola $y = ax^2 + bx + c$ intersects the x-axis at the points $P_1(x_1, 0)$ and $P_2(x_2, 0)$, then these points lie on both the parabola and the x-axis. Hence, the ordered pairs $(x_1, 0)$ and $(x_2, 0)$ satisfy the equations $y = ax^2 + bx + c$ and $y = 0$. If 0 is substituted for y, it follows that x_1 and x_2 are the real roots of $ax^2 + bx + c = 0$.

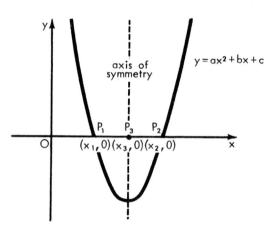

When the equation $ax^2 + bx + c = 0$ has real roots, the equation of the axis of symmetry of the parabola $y = ax^2 + bx + c$, $x = -\dfrac{b}{2a}$, can be justified as follows:

If x_1 and x_2 represent the roots of $ax^2 + bx + c = 0$, then $x_1 + x_2 = -\dfrac{b}{a}$. In the figure, $P_3(x_3, 0)$, a point on the axis of symmetry, is the midpoint of $\overline{P_1 P_2}$. Hence, x_3 is the average of x_1 and x_2. Therefore, $x_3 = \dfrac{1}{2}(x_1 + x_2) = \dfrac{1}{2}\left(-\dfrac{b}{a}\right) = -\dfrac{b}{2a}$.

Since all points on the axis of symmetry of the parabola, $y = ax^2 + bx + c$, have the same abscissa, the axis of symmetry is the set of all points whose abscissa, x, is $-\dfrac{b}{2a}$. Thus, the equation of the axis of symmetry is $x = -\dfrac{b}{2a}$.

~~~~~~~~~~ **MODEL PROBLEM** ~~~~~~~~~~

Find the solution set of $x^2 = 4x + 5$ graphically.

How To Proceed

1. Transform the given equation into the form $ax^2 + bx + c = 0$.

2. Graph the parabola $y = ax^2 + bx + c$.
3. Determine the real roots of the given equation by finding the x-intercepts. That is, find the x-coordinates of the points where the parabola intersects the x-axis.

Note. The parabola $y = x^2 - 4x - 5$ intersects the x-axis at the points A and B. At these points, $y = 0$. Hence, the x-coordinates of A and B are the real roots of the equation $x^2 - 4x - 5 = 0$ and are also the real roots of the equation $x^2 = 4x + 5$. The x-coordinates of A and B are -1 and 5.

Answer: The solution set is $\{-1, 5\}$.

Solution

$$x^2 = 4x + 5$$
$$x^2 - 4x - 5 = 0$$

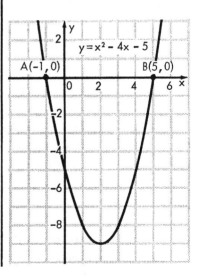

USING A PARABOLA $y = ax^2 + bx + c$ TO SOLVE QUADRATIC EQUATIONS OF THE FORM $ax^2 + bx + c = k$

Quadratic equations, such as $x^2 - 4x - 5 = 7$ and $x^2 - 4x - 5 = -5$, are of the form $ax^2 + bx + c = k$, where k is a constant. The real roots of such equations can be obtained graphically. In general, to solve graphically a quadratic equation of the form $ax^2 + bx + c = k$, use the following procedure:

Procedure:
1. Graph the parabola $y = ax^2 + bx + c$.
2. On the same set of axes, graph the line $y = k$.
3. Determine the real roots of the given equation by finding the point(s) of intersection of the two graphs.

Recall that $\{-1, 5\}$, the solution set of $x^2 - 4x - 5 = 0$, was found by noting points A and B, the intersection of the parabola $y = x^2 - 4x - 5 = 0$ and the x-axis, $y = 0$.

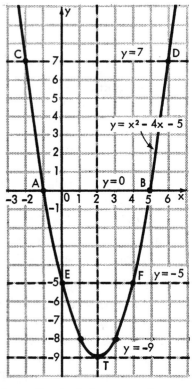

Using the figure at the right, let us now discover where the parabola $y = x^2 - 4x - 5$ intersects other lines of the form $y = k$, where k is a constant. Note that the parabola $y = x^2 - 4x - 5$ and the line $y = 7$ intersect at points C and D, whose x-coordinates are -2 and 6. It follows that $\{-2, 6\}$ is the solution set of $x^2 - 4x - 5 = 7$. Verify this by solving $x^2 - 4x - 5 = 7$ algebraically. Note further on the graph that 0 and 4, the roots of the equation $x^2 - 4x - 5 = -5$, are the x-coordinates of E and F, the points where the parabola $y = x^2 - 4x - 5$ intersects the line $y = -5$.

The parabola $y = x^2 - 4x - 5$ can be used to determine the values of k for which equations of the form $x^2 - 4x - 5 = k$ have no real roots. Note on the graph that when k is less than -9, the parabola does not intersect a line of the form $y = k$. Hence, equations such as $x^2 - 4x - 5 = -10$ and $x^2 - 4x - 5 = -20$ do not have real roots.

As you probably guessed, the equation $x^2 - 4x - 5 = -9$ has equal roots because the parabola is tangent to the line $y = -9$. Notice that -9 is the y-coordinate of the turning point of the parabola. The x-coordinate of the turning point, 2, is the value of each of the equal roots of $x^2 - 4x - 5 = -9$. Verify this algebraically.

Exercises

1. *a.* Draw the graph of $y = x^2 - 2x$ from $x = -3$ to $x = 5$ inclusive.
   b. From this graph, determine the roots of the following four equations:
   (1) $x^2 - 2x = 0$ (2) $x^2 - 2x = 3$ (3) $x^2 - 2x = 8$ (4) $x^2 - 2x = 5$

2. *a.* Draw the graph of $y = x^2 + 2x - 3$ from $x = -4$ to $x = 2$.
   b. From this graph, determine the roots of the following four equations:
   (1) $x^2 + 2x - 3 = 0$ (2) $x^2 + 2x - 3 = 5$
   (3) $x^2 + 2x - 3 = -4$ (4) $x^2 + 2x = 0$

In 3–8, solve the equation graphically.

3. $x^2 - 6x = 0$ **4.** $x^2 + 3x = 0$ **5.** $x^2 - 4 = 0$
6. $x^2 - 3x + 2 = 0$ **7.** $x^2 - x - 6 = 0$ **8.** $x^2 - 2x + 1 = 0$

9. *a.* Draw the graph of the equation $y = x^2 + 4x - 3$ from $x = -5$ to $x = 1$ inclusive.
   b. From the graph drawn in answer to *a*:
   1. Estimate, to the nearest tenth, the roots of the equation $x^2 + 4x - 3 = 0$.
   2. Write the equation of the axis of symmetry.
   3. Find a value of k for which the roots of the equation $x^2 + 4x - 3 = k$ are imaginary.

10. *a.* Draw the graph of $y = x^2 - 4x + 6$ from $x = -1$ to $x = 5$ inclusive.
   b. On the graph made in answer to *a*, indicate, by letters A and B, the points whose abscissas are the roots of the equation $x^2 - 4x + 6 = 8$.
   c. From the graph, determine the least value of k for which the roots of $x^2 - 4x + 6 = k$ are real.

11. *a.* Draw the graph of $y = x^2 - 2x - 4$ from $x = -2$ to $x = 4$ inclusive.
   b. From the graph, estimate, to the nearest tenth, the roots of $x^2 - 2x - 4 = 0$.
   c. On the same set of axes used in *a*, draw the graph $y = -8$.
   d. From the graphs made in answer to *a* and *c*, what conclusion can you draw about the roots of $x^2 - 2x - 4 = -8$?

In 12–17, solve the equation graphically; give the answers correct to the nearest tenth.

12. $x^2 + 2x - 5 = 0$ **13.** $x^2 - 4x + 1 = 0$ **14.** $x^2 - 2x - 2 = 0$
15. $x^2 - x - 3 = 0$ **16.** $x^2 + x - 1 = 0$ **17.** $x^2 - 3x + 1 = 0$

18. *a.* Draw the graph of the equation $y = -x^2 + 4x$, using all integral values of x from $x = -1$ to $x = 5$, inclusive.
   b. The value of K for which the roots of the equation $-x^2 + 4x = K$ are imaginary is   (1) $K = \pm 4$ (2) $K > 4$ (3) $K < 0$ (4) $0 < K < 4$

In 19–23, find the x-intercepts of the graph of the parabola defined by:

19. $y = x^2 - 6x + 8$ **20.** $y = x^2 - 5x$
21. $y = 16 - x^2$ **22.** $\{(x, y) | y = 2x^2 - 5x + 2\}$
23. $\{(x, y) | y = 3x^2\}$

4. Using Parabolas To Determine the Nature of the Roots of a Quadratic Equation

The nature of the roots of a quadratic equation of the form $ax^2 + bx + c = 0$ can be determined by an inspection of the position of parabola $y = ax^2 + bx + c$, relative to the x-axis. Note, in the figure, the positions of the three parabolas a, b, and c. Then examine the following table:

Parabola	Intersections with x-axis	Nature of Roots	Quadratic Equation
a: $y = x^2 - 4x - 5$	2	real, unequal	$x^2 - 4x - 5 = 0$
b: $y = x^2 - 4x + 4$	1	real, equal	$x^2 - 4x + 4 = 0$
c: $y = x^2 - 4x + 7$	0	not real	$x^2 - 4x + 7 = 0$

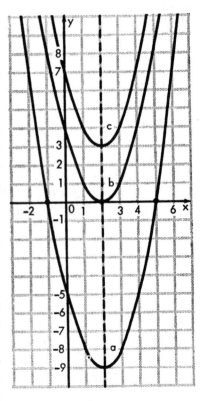

Recall that if the parabola $y = ax^2 + bx + c$ intersects the x-axis, the x-intercepts are the real roots of the quadratic equation $ax^2 + bx + c = 0$. Two intersections with the x-axis, as in the case of parabola a, mean that there are two roots, real and unequal. A single intersection, such as in the case of parabola b which is tangent to the x-axis, can only mean that the x-intercept is the value of each of the equal real roots. In the case of parabola c, there are no intersections with the x-axis. Hence, there are no real roots. Keep in mind that, in each case, the quadratic expression $ax^2 + bx + c$ is the same in both equations $y = ax^2 + bx + c$ and $ax^2 + bx + c = 0$.

Rule 1. If a parabola $y = ax^2 + bx + c$ intersects the x-axis in two points, then the quadratic equation $ax^2 + bx + c = 0$ has two real and unequal roots. Here, $b^2 - 4ac > 0$.

Rule 2. If a parabola $y = ax^2 + bx + c$ is tangent to the x-axis, then the quadratic equation $ax^2 + bx + c = 0$ has two real and equal roots. Here, $b^2 - 4ac = 0$.

Rule 3. If a parabola $y = ax^2 + bx + c$ does not intersect the x-axis, then the quadratic equation $ax^2 + bx + c = 0$ has no real roots. Here, $b^2 - 4ac < 0$.

Note. The roots of the equation $f(x) = 0$ where $f(x) = ax^2 + bx + c$ are sometimes called the **zeros** of the quadratic function.

COMPARING ALGEBRAIC AND GRAPHIC METHODS OF DETERMINING THE NATURE OF THE ROOTS

The three foregoing rules are used in the graphic methods of determining the nature of the roots of a quadratic equation. Recall that the algebraic method made use of the discriminant $b^2 - 4ac$. If $b^2 - 4ac > 0$, then the roots are real and unequal; if $b^2 - 4ac = 0$, then the roots are real and equal; if $b^2 - 4ac < 0$, then the roots are not real. The two methods are compared in the following table:

Parabola $y = ax^2 + bx + c$	*Discriminant* $b^2 - 4ac$	*Nature of Roots of* $ax^2 + bx + c = 0$
intersects x-axis (2 points)	positive	real and unequal
tangent to x-axis (1 point)	zero	real and equal
does not meet the x-axis	negative	not real

Exercises

In 1 and 2, indicate whether the discriminant of the equation $ax^2 + bx + c = 0$ is positive, zero, or negative if:

1. the parabola $y = ax^2 + bx + c$ is tangent to the x-axis.
2. the parabola $y = ax^2 + bx + c$ does not meet the x-axis.

In 3 and 4, indicate whether the roots of $ax^2 + bx + c = 0$ are real and unequal, real and equal, or not real if:

3. the parabola $y = ax^2 + bx + c$ intersects the x-axis in two points.
4. the parabola $y = ax^2 + bx + c$ is tangent to the x-axis.

In 5–10, without drawing the graph of the equation, determine the number of points that the graph of the equation and the x-axis have in common.

5. $y = x^2 - 2x - 3$ **6.** $y = x^2 - 2x + 3$ **7.** $y = x^2 - 2x + 1$
8. $y = x^2 + 2x + 1$ **9.** $y = x^2 + 2x$ **10.** $y = x^2 + 2x + 2$

In 11–15, indicate the number of points that the parabola $y = ax^2 + bx + c$ and the x-axis have in common if the roots of $ax^2 + bx + c = 0$ are:

11. 2 and -2 **12.** $2i$ and $-2i$
13. $3 + \sqrt{2}$ and $3 - \sqrt{2}$ **14.** $3 + \sqrt{-2}$ and $3 - \sqrt{-2}$
15. 12 and 12

5. Graphs of Second-Degree Equations: Conic Sections

When a second-degree equation is graphed, the resulting graph may be a *circle*, an *ellipse*, a *parabola*, or a *hyperbola*. Each of these curves is called a **conic section** since the curves may be obtained by cutting a right circular cone by a plane. In special cases, when a plane cuts a right circular cone, a *single point*, a *straight line*, or *two intersecting straight lines* may also be obtained. It can be shown that the graph of every second-degree equation in two variables must be either a conic section or one of the three special cases.

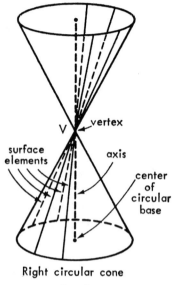

Right circular cone

Fig. 1

The right circular cone is the form of the familiar ice cream cone and also of many drinking cups. As shown in Fig. 1, the base is a circle. The *axis* of the cone is the line joining the vertex, *V*, to the center of the base. Any line joining the vertex to any point on the circle which is the base of the cone is an *element* of the cone. If the elements of the cone are extended, a second cone is obtained.

Figs. 2–8 illustrate the conic sections and also the special cases that result when a plane cuts one or both cones.

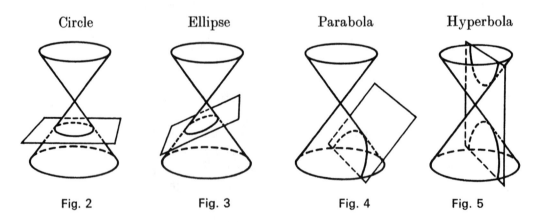

Circle	Ellipse	Parabola	Hyperbola
Fig. 2	Fig. 3	Fig. 4	Fig. 5

Fig. 2. A *circle* is formed when the cutting plane is parallel to the circular base of the cone.

Fig. 3. An *ellipse* is formed when the cutting plane cuts all the elements of the cone.

Point   Line   Double Line

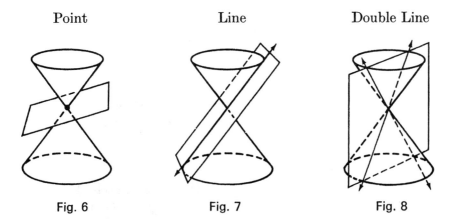

Fig. 6   Fig. 7   Fig. 8

Fig. 4. A *parabola* is formed when the cutting plane is parallel to one of the elements of the cone.

Fig. 5. A *hyperbola* is formed when the cutting plane is parallel to the axis of the cone and cuts both the cone and also its extension.

Figs. 6–8. By properly positioning the cutting plane, the special cases of intersections which are a *single point*, a *straight line*, or *two straight lines* are obtained. These three cases are considered as *limiting forms* of the conic sections.

In the discussions that follow, we will consider the graphs of second-degree equations associated with each of the conic sections.

SECOND-DEGREE EQUATIONS WHOSE GRAPH IS A POINT, A LINE, OR TWO LINES

A Point Graph

The solution set of $x^2 + y^2 = 0$ has one ordered pair, $(0, 0)$. Hence, the graph of this equation is a *point*. The point $(0, 0)$ is also the graph of $2x^2 + 3y^2 = 0$. The graph of $(x - 5)^2 + (y + 3)^2 = 0$ is the point $(5, -3)$.

A Single Straight-Line Graph

If the second-degree equation $(x - 5)^2 = 0$ is solved, then $x = 5$. The graph of $x = 5$ is a straight line parallel to the y-axis. The same line is the graph of equations equivalent to $(x - 5)^2 = 0$, such as $x^2 - 10x + 25 = 0$ and $x^2 = 10x - 25$. The line $y = -4$ is the graph of $(y + 4)^2 = 0$ or any of its equivalent transformations.

A Double Straight-Line Graph

If the second-degree equation $xy = 0$ is solved, then either $x = 0$ or $y = 0$. Hence, the graph of $xy = 0$ consists of the x-axis, which is the graph of $y = 0$ and the y-axis, which is the graph of $x = 0$. The lines $x = 5$ and $y = -2$ constitute the graph of the equation $(x - 5)(y + 2) = 0$ or any of its equivalent transformations.

Exercises

In 1–4, find the coordinates of the point that is the graph of the equation.

1. $x^2 + 2y^2 = 0$ **2.** $x^2 + (y - 3)^2 = 0$
3. $(x + 2)^2 + 3(y - 3)^2 = 0$ **4.** $(x - 1)^2 + y^2 = 0$

In 5–7, find the equation of the line that is the graph of the equation.

5. $(x - 3)^2 = 0$ **6.** $5(y + 3)^2 = 0$ **7.** $y^2 = 4(y - 1)$

In 8–16, state the equations of the two lines that form the graph of the equation.

8. $x(y - 2) = 0$ **9.** $(x - 3)(y + 2) = 0$ **10.** $(2x - 1)(4y + 1) = 0$
11. $(x - y)(x - 2y) = 0$ **12.** $x(x + 3y) = 0$ **13.** $x^2 = 5xy$
14. $xy = x + y - 1$ **15.** $2x^2 = 6xy - x^2$ **16.** $4x^2 - 9y^2 = 0$

6. The Circle as a Graph

THE CIRCLE WHOSE CENTER IS THE ORIGIN

The simplest second-degree equation whose graph is a circle is $x^2 + y^2 = r^2$, $r > 0$. This circle has its center at the origin and a radius r, as shown in the figure.

The equation $x^2 + y^2 = r^2$ is obtained by applying the theorem of Pythagoras in right triangle OPQ, where $P(x, y)$ is any point on the circle.

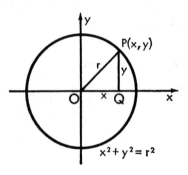

Thus, as shown in the following model problem, the graph of $x^2 + y^2 = 9$ is a circle whose center is the origin and whose radius is $\sqrt{9}$, or 3. To draw the circle using compasses, place the point of the compass at the origin and open the compass to 3 units. The following model problem shows how the circle can be graphed by means of a table of values:

〰〰〰〰〰〰 **MODEL PROBLEM** 〰〰〰〰〰〰

Using a table of values, graph the equation $x^2 + y^2 = 9$.

How To Proceed	*Solution*

How To Proceed

1. Solve the equation for y in terms of x.

2. Prepare a table of ordered pairs by assigning consecutive integral values for x and finding the corresponding values of y. (Use a table of square roots to obtain or to check values correct to the nearest tenth.)

3. Plot the points representing the ordered pairs in the table and join them with a smooth curve.

Solution

$$y^2 = 9 - x^2$$
$$y = \pm\sqrt{9 - x^2}$$

TABLE OF VALUES FOR
$$x^2 + y^2 = 9$$

x	0	± 1	± 2	± 3
y	± 3	± 2.8	± 2.2	0

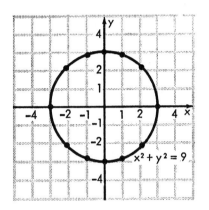

Note in the preceding graph of $x^2 + y^2 = 9$ that the x-coordinates of points on the circle are the values between -3 and 3 inclusive; that is, $-3 \le x \le 3$. These values constitute the domain of the relation. Similarly, the y-coordinates, whose values constitute the range of the relation, are restricted to $-3 \le y \le 3$.

The table of values for $x^2 + y^2 = 9$ shows ordered pairs that are real values of x and y. By using $y = \pm\sqrt{9 - x^2}$, the tabular values of y, to the nearest tenth, are obtained. If x equals 4, -4, or any value such that $|x| > 3$, the corresponding values of y would not be real but would be imaginary.

Note further that the circle graph does not pass the vertical line test of a function. Indeed, the equation $y = \pm\sqrt{9 - x^2}$ shows that each real value of x for $-3 < x < 3$ is not associated with a unique value of y. Hence, the relation, $\{(x, y) \mid x^2 + y^2 = 9\}$, is *not* a function.

The equation of a circle that has its center at the origin may be written as $ax^2 + by^2 = c$ where a, b, and c are all positive and $a = b$. If this equation is used, then $r = \sqrt{\dfrac{c}{a}}$.

Thus, the graph of $3x^2 + 3y^2 = 75$ is a circle whose center is at the origin. Since $c = 75$ and $a = 3$, then $\dfrac{c}{a} = 25$. Hence, the radius, $r = \sqrt{25} = 5$.

Exercises

In 1–6, state the center and radius of the graph of the equation.

1. $x^2 + y^2 = 121$ **2.** $x^2 + y^2 = 8$ **3.** $x^2 + y^2 = 75$

4. $x^2 = 400 - y^2$ **5.** $5x^2 + 5y^2 = 45$ **6.** $3x^2 + 3y^2 = 15$

In 7–15, draw the graph of the equation.

7. $x^2 + y^2 = 4$ **8.** $x^2 + y^2 = 49$ **9.** $x^2 + y^2 = 100$

10. $x^2 + y^2 = 12$ **11.** $x^2 + y^2 = 7$ **12.** $x^2 = 64 - y^2$

13. $y^2 = 15 - x^2$ **14.** $2x^2 + 2y^2 = 50$ **15.** $3x^2 + 3y^2 = 33$

In 16–21, write an equation of a circle whose center is at the origin and whose radius is:

16. 8   **17.** 13   **18.** $\frac{7}{2}$ **19.** $\frac{5}{3}$ **20.** $\sqrt{17}$ **21.** $\sqrt{20}$

In 22–26, write an equation of a circle whose center is at the origin and passes through the point:

22. $(4, 3)$ **23.** $(-5, 12)$ **24.** $(6, -8)$ **25.** $(0, -5)$ **26.** $(1, 2)$

27. Write an equation of a circle whose center is at the origin and whose radius is $\sqrt{3}$.

28. The locus of points whose distance from the origin is r is given by the equation   (1) $x^2 + y^2 = r^2$ (2) $y = r$ (3) $x = r$

In 29–32, show, by transforming each equation into the form $x^2 + y^2 = r^2$, that the graph of the equation is a circle.

29. $x^2 = 25 - y^2$ **30.** $y^2 = (2 - x)(2 + x)$

31. $(x - 2)^2 = 10 - y^2 - 4x$ **32.** $9 + x(y - x) = y^2 + xy$

THE CIRCLE WHOSE CENTER IS THE POINT (h, k)

In the figure at the right, $P(x, y)$ is any point of a circle whose center is $C(h, k)$ and whose radius is r.

The general equation of a circle is obtained by applying the distance formula to the distance between the points C and P, as follows:

$$d = \sqrt{(x_2 - x_1)^2 + (y_2 - y_1)^2}$$

Hence, $r = \sqrt{(x - h)^2 + (y - k)^2}$

Squaring, $r^2 = (x - h)^2 + (y - k)^2$.

The equation $(x - h)^2 + (y - k)^2 = r^2$ is called the *standard form* of the equation of a circle. If the equation is expanded and

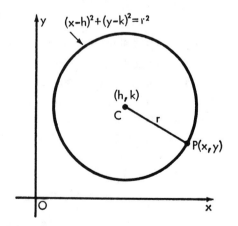

simplified, we obtain the general form of the equation of a circle, as follows:

$$x^2 - 2hx + h^2 + y^2 - 2ky + k^2 - r^2 = 0$$
$$x^2 - 2hx + y^2 - 2ky + h^2 + k^2 - r^2 = 0$$
$$x^2 + y^2 - 2hx - 2ky + h^2 + k^2 - r^2 = 0$$

The last equation is in the *general form* of the equation of a circle, $x^2 + y^2 + Dx + Ey + F = 0$ where $D = -2h$, $E = -2k$, and $F = h^2 + k^2 - r^2$.

KEEP IN MIND

The equation of a circle with center (h, k) and radius $r > 0$ is

$$(x - h)^2 + (y - k)^2 = r^2$$

∿∿∿∿∿∿ MODEL PROBLEMS ∿∿∿∿∿∿

1. State the standard form and the general form of the equation of the circle whose center is $(-4, 3)$ and whose radius is 10.

   Solution: Substitute -4 for h, 3 for k, and 10 for r in the standard form of the equation.
   $$(x - h)^2 + (y - k)^2 = r^2$$
   $$(x + 4)^2 + (y - 3)^2 = 100 \quad \textit{Ans.} \text{ (equation in standard form)}$$

   Expand and simplify.
   $$x^2 + 8x + 16 + y^2 - 6y + 9 = 100$$
   $$x^2 + y^2 + 8x - 6y - 75 = 0 \quad \textit{Ans.} \text{ (equation in general form)}$$

2. Find the coordinates of the center and the radius of a circle whose equation is $x^2 + y^2 + 10x - 4y - 7 = 0$.

   Solution:
   Express the given equation as $(x^2 + 10x + ?) + (y^2 - 4y + ?) = 7$
   Complete the squares: $(x^2 + 10x + 25) + (y^2 - 4y + 4) = 7 + 25 + 4$
   Express in standard form: $(x + 5)^2 + (y - 2)^2 = 36$
   Hence, the center of the circle is $(-5, 2)$ and the radius is 6.

   Answer: center is $(-5, 2)$; radius is 6.

Exercises

In 1–5, state the center and the radius of the circle whose equation is given.

1. $x^2 + (y - 1)^2 = 25$ **2.** $(x - 3)^2 + y^2 = 9$

3. $(x - 4)^2 + (y - 5)^2 = 100$ **4.** $(x - 1)^2 + (y + 3)^2 = 16$

5. $(x + 3)^2 + (y + 1)^2 = 49$

In 6–9, transform the equation into the standard form and state the center and the radius of the circle.

6. $x^2 + 4x + 4 + y^2 = 25$ **7.** $x^2 + y^2 + 10y + 25 = 100$

8. $x^2 + 2x + y^2 + 6y = 46$ **9.** $4x^2 + 8x + 4y^2 + 16y = 44$

In 10–15, draw the graph of the equation.

10. $(x - 3)^2 + y^2 = 9$ **11.** $x^2 + (y + 4)^2 = 16$

12. $(x - 2)^2 + (y + 2)^2 = 25$ **13.** $x^2 + y^2 + 4y = 12$

14. $x^2 + y^2 - 6x = 16$ **15.** $3x^2 + 3y^2 + 6x = 24$

7. The Ellipse as a Graph

THE ELLIPSE WHOSE CENTER IS THE ORIGIN: INTERCEPT FORM OF ITS EQUATION

The equation of an ellipse (Fig. 1) whose center is the origin, whose x-intercepts are 5 and -5, and whose y-intercepts are 3 and -3, is

$$\frac{x^2}{25} + \frac{y^2}{9} = 1$$

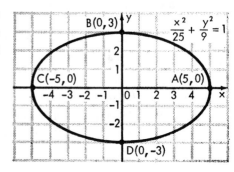

Fig. 1

If 0 is substituted for y, then $\frac{x^2}{25} = 1$. Hence, $x^2 = 25$ and $x = \pm 5$, verifying that the x-intercepts are $+5$ and -5. If 0 is substituted for x, then $\frac{y^2}{9} = 1$. Hence,

$y^2 = 9$ and $y = \pm 3$, verifying that the y-intercepts are 3 and -3. In Fig. 1, the length of the major axis, $CA = 10$. The length of the semimajor axis, $OA = 5$. The length of the minor axis, $BD = 6$. The length of the semiminor axis, $BO = 3$.

The equation $\frac{x^2}{25} + \frac{y^2}{9} = 1$ is the *intercept form* of the ellipse in Fig. 1.

In general, the intercept form of any ellipse whose center is the origin, whose x-intercepts are p and $-p$, and whose y-intercepts are q and $-q$, is

$$\frac{x^2}{p^2} + \frac{y^2}{q^2} = 1$$

Verify this by substituting 0 for y to obtain the x-intercepts, and by substituting 0 for x to obtain the y-intercepts.

THE ELLIPSE WHOSE CENTER IS THE ORIGIN: GENERAL FORM OF ITS EQUATION

The equation $\frac{x^2}{25} + \frac{y^2}{9} = 1$ is transformed into $9x^2 + 25y^2 = 225$ by multi-

plying each side by 225. The resulting equation, $9x^2 + 25y^2 = 225$, is an illustration of the *general form* of the equation of an ellipse whose center is the origin.

The general form of any ellipse whose center is the origin, and where a, b, and c are positive, is

$$ax^2 + by^2 = c$$

Thus, as shown in the following model problem, the graph of $4x^2 + 9y^2 = 36$ is an ellipse whose center is at the origin. Dividing each side by 36, we obtain $\frac{x^2}{9} + \frac{y^2}{4} = 1$, showing that the intercepts of the ellipse in the model problem are ± 3 for the x-intercepts and ± 2 for the y-intercepts.

The following model problem shows how the ellipse $4x^2 + 9y^2 = 36$ can be graphed by means of a table of values:

～～～～～～～ *MODEL PROBLEM* ～～～～～～～

Using a table of values, graph the equation $4x^2 + 9y^2 = 36$.

How To Proceed	*Solution*
1. Solve the equation for y in terms of x.	$9y^2 = 36 - 4x^2$
	$y^2 = \dfrac{36 - 4x^2}{9}$
	$y = \pm\tfrac{1}{3}\sqrt{36 - 4x^2}$

2. Prepare a table of ordered pairs by assigning consecutive integral values for x and finding the corresponding values of y. (Use a table of square roots to obtain or to check values correct to the nearest tenth.)

TABLE OF VALUES FOR
$4x^2 + 9y^2 = 36$

x	0	± 1	± 2	± 3
y	± 2	± 1.9	± 1.5	0

3. Plot the points representing the ordered pairs in the table and join them with a smooth curve.

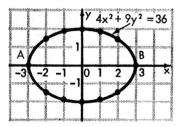

In the preceding figure, note in the graph of $4x^2 + 9y^2 = 36$ that the x-coordinates of points on the ellipse include the values between -3 and 3 inclusive; that is, $-3 \le x \le 3$. Similarly, the y-coordinates are restricted to the range, $-2 \le y \le 2$. If $|x| > 3$, then the corresponding values of y are imaginary; if $|y| > 2$, then the corresponding values of x are imaginary. For example, if $x = \pm 4$, then $y = \frac{2}{3}i\sqrt{7}$.

Note further that the graph of the ellipse does not pass the vertical line test of a function. Indeed, the equation $y = \pm\frac{2}{3}\sqrt{9 - x^2}$ shows that each real value of x for $-3 < x < 3$ is not associated with a unique value of y. Hence, the relation, $\{(x, y) \mid 4x^2 + 9y^2 = 36\}$ is not a function.

DRAWING AN ELLIPSE BY A MECHANICAL METHOD

An ellipse may be drawn by the following mechanical method:

Tack a piece of string at two points, F and F', on a sheet of paper. In Fig. 2, $FF' = 6$ units. Keeping FP and $F'P$ taut, pass a pencil along the string, drawing the ellipse in the process. P represents the changing point of contact as the pencil passes along the string. Be sure that the length of string is greater than the distance between F and F'.

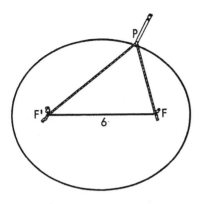

Fig. 2

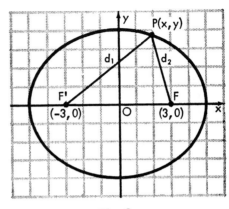

Fig. 3

The drawing of the ellipse is made in accordance with the following definition:

An *ellipse* is a set of points, the sum of whose distances from two fixed points is a constant distance greater than the distance between the fixed points.

Each of the points F and F' is a *focus* of the ellipse. Note in the resulting graph, Fig. 3, that the center of the ellipse is the midpoint of FF'. If the center is placed at the origin, then the coordinates of F and F' are $(3, 0)$ and $(-3, 0)$ respectively. The length of string, $d_1 + d_2$, is a constant, which may be shown to equal the length of the major axis.

THE CIRCLE IS A SPECIAL CASE OF AN ELLIPSE

Previously, we noted that the equation of a circle whose center is the origin may be written in the general form $ax^2 + by^2 = c$, where a, b, and c are all positive and $a = b$. The added condition, $a = b$, means that the circle is a special case of an ellipse. Note in the diagram of the circle and ellipse as conic sections, Fig. 4, that the ellipse is the section formed when the plane intersects all the elements of the cone. This is true for the circle with the added condition that the plane be parallel to the base of the cone.

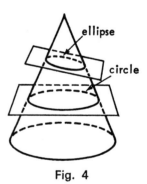

Fig. 4

Exercises

In 1–9, state the center, x-intercepts, and y-intercepts of the graph of the equation.

1. $x^2 + \dfrac{y^2}{4} = 1$ **2.** $\dfrac{x^2}{9} + y^2 = 1$ **3.** $\dfrac{x^2}{4} + \dfrac{y^2}{9} = 1$

4. $x^2 + 9y^2 = 36$ **5.** $25x^2 + y^2 = 25$ **6.** $25x^2 + 4y^2 = 100$

7. $x^2 = 1 - \dfrac{y^2}{49}$ **8.** $y^2 = 1 - \dfrac{x^2}{64}$ **9.** $y^2 = \dfrac{100 - x^2}{100}$

In 10–21, draw the graph of the equation.

10. $x^2 + 4y^2 = 36$ **11.** $4x^2 + y^2 = 36$ **12.** $x^2 + 9y^2 = 36$

13. $9x^2 + y^2 = 36$ **14.** $4x^2 + 25y^2 = 100$ **15.** $25x^2 + 4y^2 = 100$

16. $x^2 + 3y^2 = 12$ **17.** $3x^2 + y^2 = 12$ **18.** $4x^2 + 3y^2 = 48$

19. $x^2 + \dfrac{y^2}{4} = 1$ **20.** $\dfrac{x^2}{9} + y^2 = 1$ **21.** $\dfrac{x^2}{4} + \dfrac{y^2}{9} = 1$

In 22–27, by transforming the equation into the form $ax^2 + by^2 = c$, show that the graph of the equation is an ellipse whose center is the origin.

22. $2x^2 + 3y^2 - 10 = 0$ **23.** $2y^2 = 20 - 3x^2$

24. $10 - 4x^2 = 8y^2$ **25.** $\dfrac{x^2}{9} = 1 - \dfrac{y^2}{25}$

26. $\dfrac{x^2 - 16}{y} = -2y$ **27.** $2x(x + 5) = 10x + 8 - y^2$

8. The Hyperbola as a Graph

THE HYPERBOLA WHOSE CENTER IS THE ORIGIN: GENERAL FORM OF ITS EQUATION

In Fig. 1, the equation $x^2 - 4y^2 = 16$ is an example of the *general form* of the equation of a hyperbola whose center is the origin.

The general form of any hyperbola intersecting the x-axis whose center is the origin, and where a, b, and c are positive, is

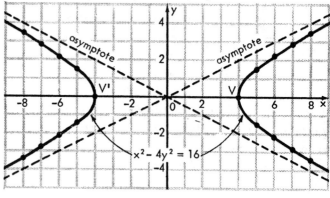

Fig. 1

$$ax^2 - by^2 = c$$

The graphing of the hyperbola is accomplished by following the same procedure as that used in the graphing of the ellipse. The first step is to solve the equation for y in terms of x. If $x^2 - 4y^2 = 16$, then $4y^2 = x^2 - 16$ and

$y = \pm\frac{1}{2}\sqrt{x^2 - 16}$. Note that when the absolute value of x is less than 4, the radicand, $x^2 - 16$, is negative and corresponding values of y are not real. Hence, in preparing a table of values for $x^2 - 4y^2 = 16$, avoid absolute values of x that are less than 4.

TABLE OF VALUES FOR $x^2 - 4y^2 = 16$

x	± 4	± 5	± 6	± 7	± 8
y	± 0	$\pm\frac{1}{2}\sqrt{9}$ or ± 1.5	$\pm\frac{1}{2}\sqrt{20} \approx \pm 2.2$	$\pm\frac{1}{2}\sqrt{33} \approx \pm 2.9$	$\pm\frac{1}{2}\sqrt{48} \approx \pm 3.5$

When the points representing the ordered pairs are plotted and joined with a smooth curve, the result is the hyperbola whose center is the origin, Fig. 1.

Note in Fig. 1 that the hyperbola is a discontinuous curve having two branches, unlike the other conic sections. The other conic sections—the circle, the ellipse, and the parabola—are continuous curves. Recall that the two branches of the hyperbola are obtained when a plane cuts a right circular cone and its extension.

TERMS USED WITH THE HYPERBOLA

The terms used with the hyperbola are illustrated in Fig. 1. The ***vertices*** of the hyperbola are $V(4, 0)$ and $V'(-4, 0)$. The ***transverse axis*** of the hyperbola is the line segment whose endpoints are the vertices of the hyperbola. The ***conjugate axis*** is a line segment that is the perpendicular bisector of the transverse axis. In Fig. 1, the transverse axis is a segment of the x-axis, whereas the conjugate axis is a segment of the y-axis. The ***center*** of the hyperbola, which is the origin in Fig. 1, is the point of intersection of the transverse and conjugate axes.

Note in Fig. 1 the two lines that are marked ***asymptote***. These are the lines $y = \frac{1}{2}x$ and $y = -\frac{1}{2}x$. The asymptotes are the lines that the branches of the hyperbola approach as they recede farther and farther from the origin, that is, as the absolute value of x becomes greater and greater. As the absolute value of x becomes very large, a value of y for the hyperbola, found by using $\pm\frac{1}{2}\sqrt{x^2 - 16}$, is very close to a value of y for the asymptotes, found by using $\pm\frac{1}{2}\sqrt{x^2}$, or $\pm\frac{1}{2}x$.

Another way of obtaining the equations of the asymptotes, $y = \frac{1}{2}x$ and $y = -\frac{1}{2}x$, is to replace 16 with 0 in the equation $x^2 - 4y^2 = 16$ to obtain $x^2 - 4y^2 = 0$. When the resulting equation is solved for y, we obtain $y = \frac{1}{2}x$ and $y = -\frac{1}{2}x$.

Figs. 2 and 3 show hyperbolas that either intersect the x-axis or intersect the y-axis. In Fig. 2, the hyperbola $4x^2 - y^2 = 16$ intersects the x-axis; in Fig. 3, the hyperbola $4y^2 - x^2 = 16$ intersects the y-axis.

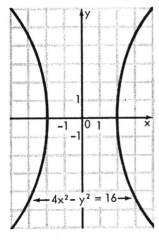

Fig. 2

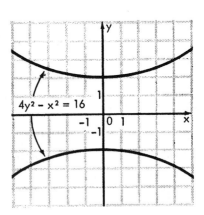

Fig. 3

TABLE OF ORDERED PAIRS FOR $xy = 4$

	Quadrant I					*Quadrant* III				
x	$\frac{1}{2}$	1	2	4	8	$-\frac{1}{2}$	-1	-2	-4	-8
y	8	4	2	1	$\frac{1}{2}$	-8	-4	-2	-1	$-\frac{1}{2}$

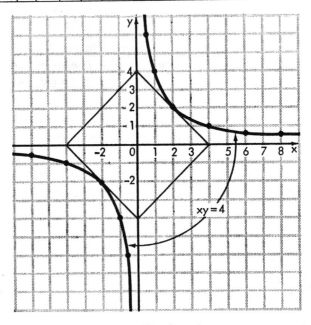

Fig. 4

Figs. 4 and 5 show hyperbolas that do not intersect either the x-axis or the y-axis. The x-axis and the y-axis are the asymptotes of each of the hyperbolas. The hyperbola $xy = 4$ (Fig. 4) has branches in quadrants I and III, while the hyperbola $xy = -4$ (Fig. 5) has branches in quadrants II and IV. It is left to the student to make a table of ordered pairs for $xy = -4$.

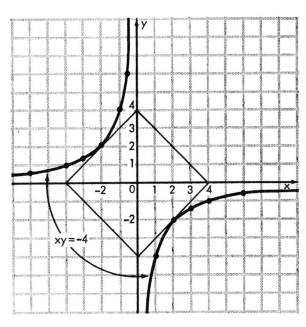

Fig. 5

Note in Figs. 4 and 5 that the x-axis and the y-axis are the extensions of the diagonals of a square or equilateral rectangle. Hence, the hyperbolas $xy = 4$ and $xy = -4$ are called **equilateral hyperbolas**. The general equations of equilateral hyperbolas whose asymptotes are the x-axis and the y-axis are $xy = k$ and $xy = -k$, where k is positive.

Exercises

In 1–9, state the coordinates of the points where the hyperbola crosses the x-axis or the y-axis.

1. $x^2 - y^2 = 9$ **2.** $y^2 - x^2 = 25$ **3.** $x^2 - 4y^2 = 36$

4. $4y^2 - x^2 = 36$ **5.** $4x^2 - 25y^2 = 100$ **6.** $25y^2 - x^2 = 25$

7. $4y^2 - 9x^2 = 36$ **8.** $\dfrac{x^2}{25} - \dfrac{y^2}{36} = 1$ **9.** $\dfrac{y^2}{25} - \dfrac{x^2}{64} = 1$

In 10–13, state the quadrants of the branches of the hyperbola.

10. $xy = 12$ **11.** $xy = -12$ **12.** $xy + 10 = 15$ **13.** $xy + 10 = 6$

In 14–17, by transforming the equation into the form $ax^2 - by^2 = c$ or $xy = k$, show that the graph of the equation is a hyperbola.

14. $x^2 = 25 + y^2$ **15.** $2y^2 = 20 + x^2$

16. $\dfrac{16}{x} = \dfrac{y}{2}$ **17.** $(x - y)(x + y) = 36$

In 18–23, draw the graph of the equation.

18. $xy = 6$ **19.** $xy = -6$ **20.** $x^2 - 4y^2 = 16$

21. $\dfrac{x^2}{25} - y^2 = 1$ **22.** $y^2 - \dfrac{x^2}{25} = 1$ **23.** $25x^2 - 4y^2 = 100$

SKETCHING AN ELLIPSE AND ITS RELATED HYPERBOLAS

Note in Fig. 6 that the sides of rectangle $ABCD$ are tangent to an inscribed ellipse. Also, sides $\overline{AD}$ and $\overline{BC}$ are tangent to the two branches of hyperbola 1, while sides $\overline{AB}$ and $\overline{CD}$ are tangent to the branches of hyperbola 2. The midpoints of the sides of the rectangle are the points of contact with the three curves.

In Fig. 6, the ellipse and its related hyperbolas are defined by equations that have many elements in common. Note these common elements and differences in the Table of Equations of an Ellipse and Its Related Hyperbolas (on the following page).

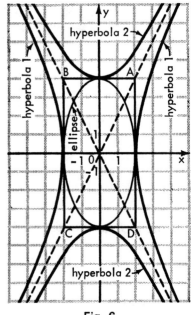

Fig. 6

Also, the same table lists the equations of the ellipse and its related hyperbolas, Fig. 6, in the second column. The last column sets forth the general equations for any set of an ellipse and its related hyperbolas that can be sketched or constructed in the same manner.

In the table, note the following differences in the equations:
1. The only difference between the general equations of the ellipse and hyperbola 1 is the change of sign between ax^2 and by^2, the terms in x and y.
2. The only difference between the general equations of the two hyperbolas is the interchanging of the terms containing x and y, the terms ax^2 and by^2.

TABLE OF EQUATIONS OF AN ELLIPSE
AND ITS RELATED HYPERBOLAS

Conic Section	*Equations of Curves in Fig. 6*	*General Equation Forms* $(a, b, c > 0)$
ellipse	$\dfrac{x^2}{4} + \dfrac{y^2}{16} = 1$ or $4x^2 + y^2 = 16$	$ax^2 + by^2 = c$
hyperbola 1 (crossing x-axis)	$\dfrac{x^2}{4} - \dfrac{y^2}{16} = 1$ or $4x^2 - y^2 = 16$	$ax^2 - by^2 = c$
hyperbola 2 (crossing y-axis)	$\dfrac{y^2}{16} - \dfrac{x^2}{4} = 1$ or $y^2 - 4x^2 = 16$	$by^2 - ax^2 = c$

GENERAL INTERCEPT FORMS OF AN ELLIPSE AND ITS RELATED HYPERBOLAS

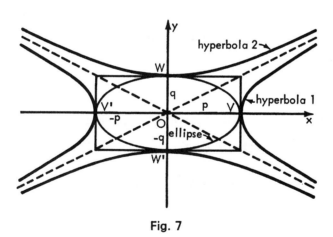

Fig. 7

If the x-intercepts are p and $-p$ and the y-intercepts are q and $-q$, as shown in Fig. 7, then the general equations of the ellipse and its related hyperbolas are:

ellipse: $\dfrac{x^2}{p^2} + \dfrac{y^2}{q^2} = 1$ hyperbola 1: $\dfrac{x^2}{p^2} - \dfrac{y^2}{q^2} = 1$ hyperbola 2: $\dfrac{y^2}{q^2} - \dfrac{x^2}{p^2} = 1$

Exercises

In 1–3, state the equations of the hyperbolas that are related to the ellipse whose equation is:

1. $x^2 + y^2 = 16$ **2.** $2x^2 + 5y^2 = 50$ **3.** $\dfrac{x^2}{9} + \dfrac{y^2}{49} = 1$

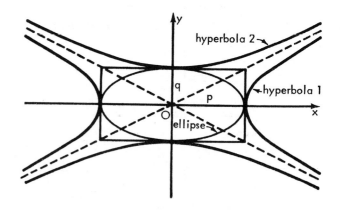

In the figure at the right, an ellipse and its related hyperbolas have been sketched. The lengths of the segments p and q are one-half the length of the axes of the ellipse. In 4–7, state the equations of the ellipse and its related hyperbolas if:

4. $p = 1$ and $q = 1$

5. $p = 6$ and $q = 5$

6. $p = 10$ and $q = 3$

7. $p = 1$ and $q = 9$

9. Identifying the Graphs of Second-Degree Equations

Following are the general second-degree equations of the parabola, circle, ellipse, and hyperbola:

Parabola

$y = ax^2 + bx + c$, $a \neq 0$ (axis of symmetry is parallel to the y-axis)

$x = ay^2 + by + c$, $a \neq 0$ (axis of symmetry is parallel to the x-axis)

Circle with center at the origin and radius r

$x^2 + y^2 = r^2$, $r > 0$

$ax^2 + by^2 = c$, where a, b, and c are all positive and $a = b$

Note. If $r = 0$ or $c = 0$, the graph is a point.

Ellipse with center at the origin

$ax^2 + by^2 = c$, where a, b, and c are all positive

Hyperbola with center at the origin

     1. Hyperbola whose transverse axis is on the x-axis:

         $ax^2 - by^2 = c$, where a, b, and c are all positive

     2. Hyperbola whose transverse axis is on the y-axis:

         $ay^2 - bx^2 = c$, where a, b, and c are all positive

     3. Equilateral hyperbola whose asymptotes are the x-axis and the y-axis:

         $xy = k$, $k \neq 0$

             If k is positive, the branches are in quadrants I and III.

             If k is negative, the branches are in quadrants II and IV.

Note. The general equation of hyperbolas 1 and 2 (see Fig. 7) may be written as $ax^2 - by^2 = c$, where a and b are positive and $c \neq 0$. If $c = 0$, then the graph consists of two lines intersecting at the origin. (The two lines, $ax^2 - by^2 = 0$, are the asymptotes of the hyperbola $ax^2 - by^2 = c$.)

In the model problems and in the exercises that follow, answer "circle" if the curve is an ellipse that is a circle.

〰〰〰〰〰〰〰〰 *MODEL PROBLEMS* 〰〰〰〰〰〰〰〰

In each of the following, name the curve that is the graph of the equation:

Equation	Answer	Equation	Answer
1. $x^2 + y^2 = 16$	circle	9. $(x - 3)^2 = 0$	line
2. $2x^2 + y^2 = 16$	ellipse	10. $4xy = 16$	hyperbola
3. $2x^2 - y^2 = 16$	hyperbola	11. $4xy = -16$	hyperbola
4. $2x^2 - y = 16$	parabola	12. $4xy = 0$	two lines
5. $2x - y^2 = 16$	parabola	13. $4x + y^2 = 25$	parabola
6. $2x - y^2 = 0$	parabola	14. $4y^2 + x^2 = 25$	ellipse
7. $x^2 - y^2 = 0$	two lines	15. $4y^2 - 9x^2 = 25$	hyperbola
8. $x^2 + y^2 = 0$	point		

〰〰〰〰〰〰〰〰〰〰〰〰〰〰〰〰〰〰〰〰〰〰〰

Exercises

In 1–15, name the curve that is the graph of the equation.

1. $y^2 = 5x$
2. $x^2 + y^2 = 7$
3. $x^2 - y^2 = 9$
4. $x^2 + y^2 - 17 = 0$
5. $x^2 - y^2 = 3$
6. $3x^2 + y^2 = 15$
7. $x^2 + 3x + 2 - y = 0$
8. $xy = 14$
9. $xy = -6$
10. $y^2 = 19 - x^2$
11. $2x + 3y - 6 = 0$
12. $y = x^2 - 5$
13. $y^2 = 20 - 2x^2$
14. $x = y^2 + 6y$
15. $y = 2x^2 + 6$

16. The graph of $2x^2 + 2y^2 = 50$ is (1) a parabola (2) a circle (3) a hyperbola
17. The graph of $x^2 + 4y^2 = 25$ is (1) an ellipse (2) a hyperbola (3) a parabola (4) a circle.
18. The graph of $8x^2 + 8y^2 = 32$ is (1) a parabola (2) a circle (3) a hyperbola
19. What is the name of the graph of $y = x^2$?
20. The graph of $3x^2 + 3y^2 = 10$ is (1) a circle (2) a parabola (3) a hyperbola
21. The graph of $ax^2 + ay^2 = c$ is always a circle. (Answer *true* or *false*.)
22. An equation of an ellipse may be (1) $3x^2 + y = 4$ (2) $3x^2 + y^2 = 4$ (3) $3x^2 - y^2 = 4$
23. The graph of the equation $y^2 = 6x$ is (1) a circle (2) an ellipse (3) a hyperbola (4) a parabola

24. Which equation has a circle as its graph? (1) $3x^2 = 5 + 3y^2$ (2) $3x^2 = 5 - 3y^2$ (3) $3x^2 = 5 - y^2$ (4) $3x^2 = 5 + y^2$

25. Which of the following is the equation of an ellipse? (1) $9x^2 = 4y^2 + 36$ (2) $xy = -8$ (3) $x^2 = 25 - 4y^2$ (4) $x^2 = y - 16x + 4$

26. Which of the following is the equation of a hyperbola? (1) $x^2 = 10 - y^2$ (2) $x = y^2 - 9$ (3) $xy = -6$ (4) $4x^2 + y^2 = 9$

27. The graph of the equation $x^2 - 1 = 2x + 2y$ is a (1) parabola (2) straight line (3) circle (4) hyperbola

In 28–36, state whether the graph of the equation is a point, a line, or a double line.

28. $x^2 + y^2 = 0$ **29.** $x^2 - y^2 = 0$ **30.** $xy = 0$

31. $x^2 = 4y^2$ **32.** $(x - 3)(x + 2) = 0$ **33.** $x(y - 5) = 0$

34. $(x + 2y)(x - 2y) = 0$ **35.** $(x - 5)^2 = 0$ **36.** $x^2 + 49 = 14x$

10. Solving a System of Two Equations Graphically

SOLVING A SYSTEM OF TWO EQUATIONS GRAPHICALLY: ONE A FIRST-DEGREE EQUATION AND THE OTHER A SECOND-DEGREE EQUATION

To solve a system of equations consisting of a first-degree equation and a second-degree equation graphically, both equations must be graphed using the same set of axes. The coordinates of the points of intersection of the two graphs, if any, are the ordered number pairs of real numbers of the solution set of the system.

Recall that the graph of a first-degree equation in two variables is a straight line. The graph of a second-degree equation in two variables is a conic section, a point, a line, or a double line. The number of ordered pairs of real numbers in the solution set of the system may be 0, 1, or 2 depending on the number of intersections of the graphs. If there are no intersections, then the solution set of the system has no ordered pairs of real numbers.

If the straight line is *tangent* to a conic section, there is one intersection. Hence, in the following cases, the solution set of the system has one ordered number pair:

 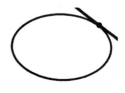 

    Parabola   Circle   Ellipse   Hyperbola

If the straight line intersects a conic section in two points, the solution set of the system has two ordered number pairs, as in the following cases:

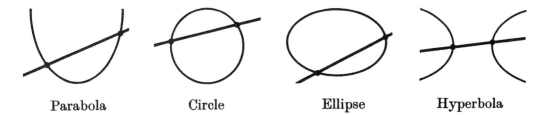

Parabola   Circle   Ellipse   Hyperbola

～～～～～～～ *MODEL PROBLEMS* ～～～～～～～

1. Solve graphically: $\begin{cases} y = 2x + 1 \\ y = x^2 - 4x + 9 \end{cases}$

How To Proceed	*Solution*
1. Graph the first equation.	The graph of $y = 2x + 1$ is a straight line.
2. Using the same set of axes, graph the second equation.	The graph of $y = x^2 - 4x + 9$ is a parabola.
3. Read the coordinates of the points of intersection of the two graphs.	The common solutions are: At A, $x = 2$, $y = 5$ At B, $x = 4$, $y = 9$

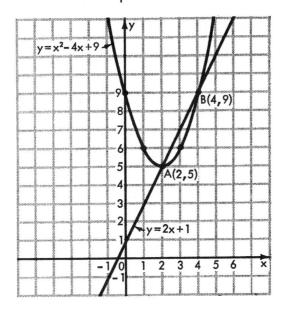

Answer: The solution set of the system is {(2, 5), (4, 9)}.

In 2–4, solve the system of equations algebraically; state whether the number of points of intersection will be 2, 1, or 0 if the graphs of the pair of equations in the system are plotted using the same set of axes.

2. $y = x^2 - 4x + 9$ **3.** $y = x^2 - 4x + 9$ **4.** $y = x^2 - 4x + 9$
    $y = 5$ $y = 2x$ $y = 4x - 7$

Solution:

2. If $y = 5$, then $5 = x^2 - 4x + 9$.
Hence, $(x - 2)^2 = 0$ and $x = 2$.
Ans. $\{(2, 5)\}$ 1 point

3. If $y = 2x$, then $2x = x^2 - 4x + 9$.
Hence, $(x - 3)^2 = 0$ and $x = 3$.
Ans. $\{(3, 6)\}$ 1 point

4. If $y = 4x - 7$, then $4x - 7 = x^2 - 4x + 9$.
Hence, $(x - 4)^2 = 0$ and $x = 4$.
Ans. $\{(4, 9)\}$ 1 point

It is left to the student to use the parabola in model problem 1 and show that each of the lines $y = 5$, $y = 2x$, and $y = 4x - 7$ is tangent to the parabola at a point whose coordinates are an ordered number pair of the solution set.

SOLVING A SYSTEM OF TWO EQUATIONS GRAPHICALLY: BOTH SECOND-DEGREE EQUATIONS

If the graph of each of the second-degree equations is a conic section, there may be as many as 4 ordered pairs of real numbers in the solution set of the system. The following figures, showing the relative positions of a circle and a parabola, indicate when the solution set of the system has either 4, 3, 2, or 1 ordered pairs of real numbers. (If the circle and parabola do not meet, there is no ordered pair of real numbers in the system.)

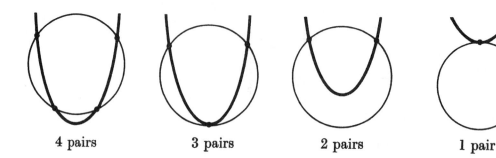

4 pairs   3 pairs   2 pairs   1 pair

~~~~~~~~~~~~~~~~~ *MODEL PROBLEM* ~~~~~~~~~~~~~~~~~

Find the solution set of the system and express irrational roots to the nearest tenth:

$$x^2 + y^2 = 25$$
$$xy = 10$$

How To Proceed	*Solution*
1. Graph the first equation.	The graph of $x^2 + y^2 = 25$ is a circle with its center at the origin and a radius of 5.
2. Using the same set of axes, graph the second equation.	The graph of $xy = 10$ is an equilateral hyperbola with branches in quadrants I and III.
3. Read the coordinates of the points of intersection of the two graphs.	The common solutions to the nearest tenth are: At A, $x = 4.5$, $y = 2.2$ At B, $x = 2.2$, $y = 4.5$ At C, $x = -4.5$, $y = -2.2$ At D, $x = -2.2$, $y = -4.5$

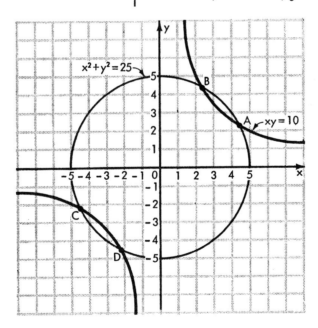

Answer: The solution set is $\{(4.5, 2.2), (2.2, 4.5), (-2.2, -4.5), (-4.5, -2.2)\}$.

~~~~~~

Note. In $\begin{cases} x^2 + y^2 = 1 \\ xy = 10 \end{cases}$ the graphs of the equations are a circle of radius 1 and the hyperbola of the previous model problem. These graphs do not intersect. When the graphs of the equations of a system do not intersect, the solutions of the system are not real numbers but complex numbers.

Exercises

In 1–9, solve the system of equations algebraically and state whether the number of points of intersection will be 2, 1, or 0 if the graphs of the pair of equations in the system are plotted using the same set of axes.

1. $x^2 + y^2 = 25$
$y = x - 1$

2. $x^2 + y^2 = 16$
$y = x$

3. $x^2 + y^2 = 9$
$y = 3$

4. $x^2 + y^2 = 9$
$x - 2y + 3 = 0$

5. $x^2 + y^2 = 36$
$x + y = 10$

6. $y = x^2 + 7x - 10$
$y = 4x$

7. $xy = 12$
$x - y = 5$

8. $xy = 16$
$x = y$

9. $x^2 - y^2 = 25$
$x = 3$

10. *a.* Draw the graph of the equation $y = x^2 - 2x - 5$ from $x = -2$ to $x = 4$ inclusive.
   b. On the same set of axes used in answer to *a*, draw the graph of the equation $x + 2y = 2$.
   c. From the graphs made in answer to *a* and *b*, estimate, correct to the nearest tenth, the values of x and y common to the two equations.

11. *a.* Using the same set of axes, draw the graphs of the equations $xy = 12$ and $x - y = 1$.
   b. From the graphs made in answer to *a*, determine the values of x and y common to the two equations.

In 12–20, solve the system of equations graphically.

12. $y = x^2 + 4$
$y = 4x + 1$

13. $y = x^2 + 3x$
$y - 3x = 4$

14. $x^2 - 2x - 4 = y$
$y + 4 = x$

15. $xy = 4$
$x = y + 3$

16. $xy = 24$
$x - 3y = 6$

17. $xy = 12$
$3x + 4y = 24$

18. $x^2 + y^2 = 16$
$x - y = 4$

19. $x^2 + y^2 = 25$
$y = 2x + 5$

20. $x^2 + y^2 = 50$
$2x = 15 - y$

21. The line $y = 2x - 4$ intersects a circle whose center is the origin at a point whose coordinates are represented by $(x, 8)$. Find the radius of the circle.

In 22–27, determine whether the number of points of intersection of the graphs of the pair of equations of the system is 4, 3, 2, 1, or 0.

22. $x^2 + y^2 = 25$
$xy = 10$

23. $x^2 + y^2 = 25$
$y^2 = x$

24. $y = x^2 - 3x$
$y = 2x^2$

25. $x^2 + y^2 = 9$
$x = y^2 + 5$

26. $x^2 - y^2 = 9$
$x^2 + y^2 = 9$

27. $x^2 + 4y^2 = 64$
$4y^2 - x^2 = 64$

28. *a.* Draw the graph of $y = x^2 - 4x$ from $x = -1$ to $x = 5$ inclusive.
   b. On the set of axes used in answer to *a*, draw the graph of $xy = 6$ from $x = 1$ to $x = 6$ inclusive.
   c. From the graphs, estimate, correct to the nearest tenth, a value of x and a value of y common to the two equations.

29. *a.* Draw the graph of the equation $x^2 + y^2 = 9$.
   b. On the axes used in answer to *a*, draw the graph of $x = y^2 - 5$.
   c. From the graphs, estimate the values of x and y common to the two equations.

11. Using the Parabola To Solve Problems Involving Maximum or Minimum

The graph of a parabola that opens downward can be used to solve a problem in which a maximum value is to be found at the turning point. Similarly, the graph of a parabola that opens upward can be used to solve a problem in which a minimum value is to be found.

~~~~~~~~~~~~~ *MODEL PROBLEM* ~~~~~~~~~~~~~

A man has 200 feet of wire fencing with which to enclose a vegetable garden. What are the dimensions and the area of the largest rectangular garden which he can fence off with this wire?

Solution:

Let $x = $ length of the garden in feet.
Then $100 - x = $ width of the garden.
Then $x(100 - x) = $ area of the garden.
$A = x(100 - x)$ [$A = $ area of garden]
$A = -x^2 + 100x$
Make a graph of $A = -x^2 + 100x$.

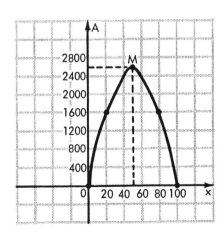

x	A
0	0
20	1600
40	2400
50	2500
60	2400
80	1600
100	0

From the graph, we observe that the garden will have a maximum area of 2500 square feet when the length of the rectangle is 50 feet. The width of the rectangle will be $100 - 50$, or 50 feet.

Answer: The length of the largest rectangle is 50 feet; the width is 50 feet; the area is 2500 square feet.

~~~~~~

Exercises

1. Separate 40 into two parts such that their product will be a maximum. Find the maximum product.
2. Find two numbers whose sum is 10 and whose product is as large as possible.
3. The perimeter of a rectangle is 24 feet.
   a. If x represents the length of the rectangle, express the width in terms of x.
   b. If A represents the area of the rectangle, express A in terms of x.
   c. Make a graph of the formula obtained in answer to *b*.
   d. From the graph, determine the maximum value of A.
4. Find the maximum area of a rectangle whose perimeter is 80 feet.
5. a. Show that the largest rectangle whose perimeter is represented by P is a square.
   b. Find the area of the square.
6. A rectangular flower garden is to be enclosed on three sides by wire fencing and on the fourth side by the side of a building; 12 yards of wire fencing are to be used for this purpose.
   a. Express the area, A, in terms of the width w.
   b. Draw the graph of the equation written in answer to *a*.
   c. From the graph, determine the value of w that will give the greatest area.
7. Find the number which when added to its square gives the smallest possible sum.
8. Find two numbers such that their sum is 10 and the sum of their squares is a minimum.

12. Solving Quadratic Inequalities in One Variable Graphically and Algebraically

Do you know what number has a square less than 4? You probably know many such numbers, such as 1, $\frac{1}{2}$, or $\frac{1}{4}$.

A more difficult question is: "Do you know the set of all the numbers whose square is less than 4?" Either of the following answers is correct:

1. The set of all numbers whose absolute value is less than 2.
2. The set of all numbers greater than -2 and less than 2.

Thus, 0, $\frac{1}{2}$, $-\frac{1}{2}$, $\frac{1}{4}$, $-\frac{1}{4}$, and -1 are numbers whose square is less than 4.

Let us consider the graphic and algebraic methods that can be used to solve such problems. If we begin by letting x represent any number whose square is less than 4, then $x^2 < 4$. When 4 is subtracted from each side, then we obtain the equivalent inequality $x^2 - 4 < 0$. The inequality $x^2 - 4 < 0$ is in the standard form of a *quadratic inequality,* which is $ax^2 + bx + c \neq 0$ where $a \neq 0$. This standard form of a quadratic inequality may be written either as $ax^2 + bx + c < 0$ or as $ax^2 + bx + c > 0$, where $a \neq 0$.

The model problems that follow show how quadratic inequalities in one variable can be solved graphically by means of a parabola, and how such inequalities can be solved algebraically. If a quadratic inequality is factorable, it can be solved by applying the following rules:

Rule 1. If the product of two factors is positive, then the factors denote numbers that have the same sign; that is, either (1) both numbers are positive or (2) both numbers are negative.

Rule 2. If the product of two factors is negative, then the factors denote numbers that differ in sign; that is, either (1) the first number is positive and the second number is negative or (2) the first number is negative and the second number is positive.

The rules may be better understood if the numbers denoted by the factors are represented by a and b, as follows:

Rule 1. If $ab > 0$, then either (1) $a > 0$ and $b > 0$ or (2) $a < 0$ and $b < 0$.

Rule 2. If $ab < 0$, then either (1) $a > 0$ and $b < 0$ or (2) $a < 0$ and $b > 0$.

~~~~~~~~~~~ *MODEL PROBLEMS* ~~~~~~~~~~~

Graphic Solution of the Inequality $x^2 < 4$

1. Find the solution set of $x^2 < 4$ graphically

How To Proceed	Solution
1. Transform the given inequality into the standard form $ax^2 + bx + c < 0$.	Transform $x^2 < 4$ into $x^2 - 4 < 0$.
2. Graph the parabola $y = ax^2 + bx + c$.	Graph the parabola $y = x^2 - 4$.
3. If $y < 0$, then $ax^2 + bx + c < 0$ and the required solution set is the set of the x-coordinates of those points whose y-coordinate is negative.	Points on the parabola whose y-coordinate is *negative* are shown in heavy black. Their x-coordinates agree with those points in

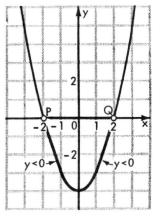

$\overline{PQ}$, not including points P and Q. Hence, $-2 < x < 2$.

The solution set is shown on the number line below.

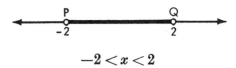

$$-2 < x < 2$$

Answer: The solution set is $\{x \mid -2 < x < 2\}$.

Note. If x has any real value between -2 and 2, then the absolute value of x is less than 2.

Using the Same Parabola To Solve the Inequality $x^2 > 4$

2. Find the solution set of $x^2 > 4$ graphically.

Solution: Suppose we wish to know which numbers have squares that are greater than 4. The same parabola, $y = x^2 - 4$, can be used to solve the inequality $x^2 > 4$ or $x^2 - 4 > 0$. However, we now look for the points on the parabola whose y-coordinate is *positive*, shown in heavy black. Their x-coordinates agree with the points on the x-axis to the left of P or to the right of Q, but not including P and Q. Hence, $x < -2$ or $x > 2$.

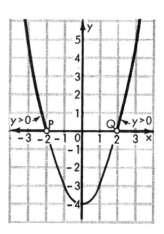

Answer: Numbers whose square exceeds 4 are those numbers that are less than -2 or greater than 2; or, numbers whose absolute value exceeds 2.

$$x < -2 \quad \text{or} \quad x > 2$$

Algebraic Solution of the Inequality $x^2 < 4$

3. Find the solution set of $x^2 < 4$ algebraically.

How To Proceed	Solution
1. Transform the given inequality into the form $ax^2 + bx + c < 0$.	Transform $x^2 < 4$ into $x^2 - 4 < 0$.
2. Factor $ax^2 + bx + c$.	$$x^2 - 4 < 0$$ $$(x + 2)(x - 2) < 0$$
3. Since the product of the two factors is negative, apply the following rule. *If the product of two factors is negative, then the factors denote numbers that differ in sign.*	Either (1) $x + 2 > 0$ and $x - 2 < 0$ or (2) $x + 2 < 0$ and $x - 2 > 0$.
4. Solve each of the first-degree inequalities.	**(1)** $\quad\quad$ **(2)** If $x + 2 > 0$, $\quad$ If $x + 2 < 0$, $\quad x > -2$. $\quad\quad$ $x < -2$. If $x - 2 < 0$, $\quad$ If $x - 2 > 0$, $\quad x < 2$. $\quad\quad\quad$ $x > 2$.
5. Obtain the required solution set by combining the results of step 4.	If $x > -2$ and $x < 2$, then $-2 < x < 2$. $\quad$ Reject, since x cannot be less than -2 and also greater than 2.

Answer: The solution set is $\{x \mid -2 < x < 2\}$.

Algebraic Solution of the Inequality $x^2 > 4$

4. Find the solution set of $x^2 > 4$ algebraically.

How To Proceed	Solution
1. Transform the given inequality into the standard form $ax^2 + bx + c > 0$.	Transform $x^2 > 4$ into $x^2 - 4 > 0$.
2. Factor $ax^2 + bx + c$.	$$x^2 - 4 > 0$$ $$(x + 2)(x - 2) > 0$$

3. Since the product of the two factors is positive, apply the following rule: *If the product of two factors is positive, then the factors denote numbers that have the same sign.*	Either (1) $x + 2 > 0$ and $x - 2 > 0$ or (2) $x + 2 < 0$ and $x - 2 < 0$.

(1)	(2)
4. Solve each of the resulting first-degree inequalities.	If $x + 2 > 0$, $x > -2$. If $x - 2 > 0$, $x > 2$.	If $x + 2 < 0$, $x < -2$. If $x - 2 < 0$, $x < 2$.
5. Obtain the solution set by combining the results of step 4.	If $x > -2$ and $x > 2$, then $x > 2$.	If $x < -2$ and $x < 2$, then $x < -2$.

Answer: The solution set is $\{x \mid x > 2 \quad \text{or} \quad x < -2\}$.

Solution of a Combination Equality and Inequality

5. The graph of the solution of the inequality $x^2 - 4x \geq 0$ is to be shown by a heavy line or lines on the real number axis. Which graph is correct?

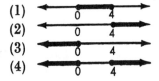

(1)
(2)
(3)
(4)

Solution: It is given that $x^2 - 4x \geq 0$.

Since $x^2 - 4x$ is the product of x and $(x - 4)$, then $x(x - 4) \geq 0$.

$x(x - 4) \geq 0$ is a combination of the equality $x(x - 4) = 0$ and the inequality $x(x - 4) > 0$.

If $x(x - 4) = 0$, then $x = 0$ or $x = 4$.

If $x(x - 4) > 0$, then either x and $(x - 4)$ are both positive or x and $(x - 4)$ are both negative.

1. If both x and $(x - 4)$ are positive, then x must be greater than 4.
2. If both x and $(x - 4)$ are negative, then x must be less than 0.

Hence, the solution has the form $x \leq 0$ or $x \geq 4$.

Answer: The correct choice is (4).

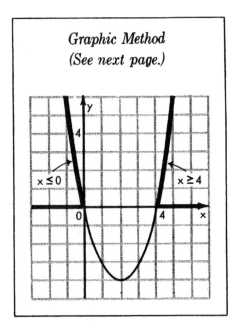

Graphic Method
(See next page.)

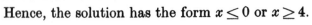

Graphic Method

By graphing $y = x^2 - 4x$, as shown, it can be seen that $y \geq 0$ for values of x such that $x \leq 0$ or $x \geq 4$.

Exercises

1. Graph the parabola $y = x^2 - 1$. Using the parabola, find the solution set of (*a*) $x^2 - 1 < 0$ and (*b*) $x^2 > 1$.
2. Graph the parabola $y = x^2 - 4x$. Using the parabola, find the solution set of (*a*) $x^2 - 4x > 0$ and (*b*) $x^2 < 4x$.
3. Graph the parabola $f(x) = x^2 - 3x - 4$. Using the parabola, find the solution set of (*a*) $x^2 - 3x \leq 4$ and (*b*) $x^2 \geq 3x + 4$.

In 4–12, find graphically the solution set and use a number line to represent the solution set.

4. $x^2 - 5x > 0$ 5. $x^2 < 5x$ 6. $x^2 - 5x + 6 < 0$
7. $x^2 > 5x - 6$ 8. $6 - 5x < x^2$ 9. $2x^2 - 18 < 0$
10. $3x^2 \geq 12$ 11. $3x^2 - 12x \geq 0$ 12. $3x^2 \leq 12x$

In 13–27, find algebraically the solution set and use a number line to represent the solution set.

13. $(x - 3)(x + 3) < 0$ 14. $(x - 3)(x + 3) > 0$ 15. $(x - 3)(x - 5) < 0$
16. $(x + 3)(x - 4) > 0$ 17. $x(x - 6) < 0$ 18. $y(y + 6) > 0$
19. $x^2 < 25$ 20. $y^2 - 49 > 0$ 21. $y(2y - 5) < 0$
22. $5n^2 \geq 45$ 23. $20 \geq 5n^2$ 24. $x^2 - 8x + 12 \leq 0$
25. $x^2 - 11x \geq 12$ 26. $2x^2 + 3x \leq 0$ 27. $2x^2 + 5x + 3 \leq 0$

In 28–31, use the same graph of a parabola to find the solution sets in parts *a*, *b*, and *c*. Then represent each solution set on a number line.

Part a	*Part b*	*Part c*
28. $x^2 - 4x - 5 = 0$	$x^2 - 4x - 5 > 0$	$x^2 - 4x - 5 < 0$
29. $x^2 - 25 = 0$	$x^2 - 25 > 0$	$x^2 - 25 < 0$
30. $-x^2 - 2x + 3 = 0$	$-x^2 - 2x + 3 > 0$	$-x^2 - 2x + 3 < 0$
31. $2x^2 - 5x + 2 = 0$	$2x^2 - 5x + 2 > 0$	$2x^2 - 5x + 2 < 0$

In 32–34, find the set of numbers such that in the case of any number in the set:

32. the square of the number is less than 64.
33. the square of the number exceeds 4 less than 5 times the number.
34. twice the square of the number exceeds 3 more than 5 times the number.

13. Systems of Inequalities

GRAPHS OF INEQUALITIES OF THE SECOND DEGREE

The graph of a second-degree equation in two variables divides the Cartesian plane into three sets of points. In Fig. 1, for example, the graph of $x^2 + y^2 = 16$ divides the plane into the following sets of points:

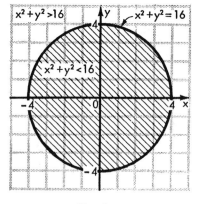

Fig. 1

1. The set of the points of the circle, which consists of the graphs of the ordered pairs that satisfy the equation $x^2 + y^2 = 16$. Examples of such points are $(0, 4)$ and $(-4, 0)$.

2. The set of points in the shaded interior region of the circle, which consists of the graphs of the ordered pairs that satisfy the inequality $x^2 + y^2 < 16$. Examples of such points are $(0, 2)$ and $(-3, 1)$.

3. The set of points in the nonshaded region outside the circle, which consists of the graphs of the ordered pairs that satisfy the inequality $x^2 + y^2 > 16$. Examples of such points are $(0, 6)$ and $(-5, 5)$.

In this case, the circle acts as a ***plane divider*** since it divides the Cartesian plane into two regions, the interior region within the circle and the region exterior to the circle.

In Fig. 2, the graph of $y = x^2$ divides the plane into three sets of points: (1) the set of points on the parabola $y = x^2$, (2) the set of points in the shaded interior region, which consists of the graphs of ordered pairs satisfying $y > x^2$, and (3) the set of points in the non-shaded exterior region, which consists of the graphs of ordered pairs satisfying $y < x^2$. In this case, the parabola is a plane divider.

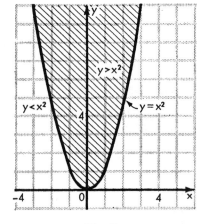

Fig. 2

GRAPHS OF SOLUTION SETS OF SYSTEMS OF INEQUALITIES OF THE SECOND DEGREE

Fig. 3 shows how the circle $x^2 + y^2 = 16$ and the parabola $y = x^2$ divide the Cartesian plane into four regions, A, B, C, and D. Since the points *on* the circle and *on* the parabola do not satisfy the inequalities, the curves are shown as dashed.

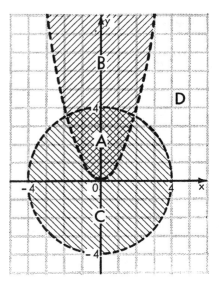

Fig. 3

$$\text{Region } A \quad \begin{cases} x^2 + y^2 < 16 \\ y > x^2 \end{cases}$$

The points in crosshatched region A, such as $(1, 2)$ and $(0, 3)$ are both in the interior region of the circle and in the interior region of the parabola.

$$\text{Region } B \quad \begin{cases} x^2 + y^2 > 16 \\ y > x^2 \end{cases}$$

The points in region B, such as $(1, 6)$ and $(-1, 7)$, are both in the exterior region of the circle and in the interior region of the parabola.

$$\text{Region } C \quad \begin{cases} x^2 + y^2 < 16 \\ y < x^2 \end{cases}$$

The points in region C, such as $(1, -2)$ and $(-2, -1)$, are both in the interior region of the circle and in the exterior region of the parabola.

$$\text{Region } D \quad \begin{cases} x^2 + y^2 > 16 \\ y < x^2 \end{cases}$$

The points in region D, such as $(5, 5)$ and $(5, -3)$, are both in the exterior region of the circle and in the exterior region of the parabola.

Procedure. To graph the solution set of a system of two inequalities:
1. **Graph each inequality, using the same set of coordinate axes.**
2. **The solution set of the system of inequalities is the region common to both graphs.**

〰〰〰〰〰〰〰〰〰 **MODEL PROBLEM** 〰〰〰〰〰〰〰〰〰

Find graphically the solution set of $\begin{cases} y > x + 2 \\ \dfrac{x^2}{16} + \dfrac{y^2}{4} \leq 1 \end{cases}$

Solution:

1. The graph of the inequality $y > x + 2$ is the shaded region above the line, $y = x + 2$. Since the points on the line do not satisfy the inequality, the line is shown as dashed. For example, the pair of coordinates of $A(-4, 3)$ satisfies $y > x + 2$.

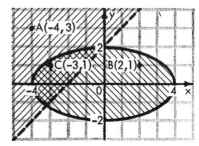

2. The graph of the combination inequality and equation $\dfrac{x^2}{16} + \dfrac{y^2}{4} \leq 1$ is the union

of the ellipse and its shaded interior region.
Since the points on the ellipse satisfy the equation, the ellipse is shown as a solid line. For example, the pair of coordinates of $B(2, 1)$ satisfies $\dfrac{x^2}{16} + \dfrac{y^2}{4} \leq 1$.

3. The crosshatched region that is the intersection of both graphs made in steps 1 and 2 is the graph of the solution set of the given system. For example, the pair of coordinates of $C(-3, 1)$ satisfies both sentences of the system.

Answer: The crosshatched region is the graph of the required solution set.

〰〰〰〰〰〰〰〰〰〰〰〰〰〰〰〰〰〰〰〰〰〰〰〰〰〰〰〰〰〰〰〰

Exercises

In 1–15, find graphically the solution set of the system.

1. $x^2 + y^2 < 25$
 $x + y < 5$

2. $x^2 + y^2 \leq 25$
 $x - y > 5$

3. $x^2 + y^2 \leq 25$
 $y \geq 2x$

4. $y < x^2 - 4$
 $y < 2$

5. $y \leq x^2 - 4$
 $x > -1$

6. $y > x^2 - 4$
 $y > 2x - 1$

7. $xy > 8$
 $y > x$

8. $xy \geq 12$
 $y < x - 3$

9. $xy < -6$
 $x + 2y \leq 6$

10. $x^2 + 4y^2 < 16$
 $y - x < 3$

11. $4x^2 + 9y^2 \leq 36$
 $2x + 3y \leq 6$

12. $x^2 + y^2 < 16$
 $xy > 8$

13. $x^2 + y^2 > 9$
 $xy \geq 4$

14. $y < x^2 + 1$
 $x^2 - y^2 < 1$

15. $y \geq x^2 + 2$
 $xy \geq -4$

━━━━━━ *CALCULATOR APPLICATIONS* ━━━━━━

On a hand-drawn graph, it may be very difficult to determine nonintegral values of the roots of a quadratic equation. The graphing calculator makes it possible to determine these roots exactly or find a close approximation.

━━━━━━━━ *MODEL PROBLEMS* ━━━━━━━

1. Solve $2x^2 - 4x - 1 = 0$ to the nearest thousandth.

   Solution: Begin by graphing the related function in the standard viewing window.

   Enter: $\boxed{\text{Y=}}$ 2 $\boxed{\text{X,T,}\theta,n}$ $\boxed{x^2}$ $\boxed{-}$ 4 $\boxed{\text{X,T,}\theta,n}$ $\boxed{-}$ 1 $\boxed{\text{GRAPH}}$

   Display:

   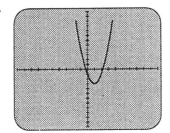

   The ZERO feature on the CALC menu can be used to find the roots. This requires providing a left bound and a right bound to define an interval that includes the root. *Enter* $\boxed{\text{2nd}}$ $\boxed{\text{CALC}}$ 2 and use $\boxed{\blacktriangle}$ or $\boxed{\blacktriangledown}$ to move the cursor onto the function for which the root is to be found. Use $\boxed{\blacktriangleleft}$ or $\boxed{\blacktriangleright}$ to select the bounds. For the left bound, locate a point a little to the left of the root and press $\boxed{\text{ENTER}}$. For the right bound, locate a point a little to the right and press $\boxed{\text{ENTER}}$. If these steps are done correctly, the y-values on the screen for the two bounds should have the opposite signs. Finally, after the Guess? prompt, press $\boxed{\text{ENTER}}$.

   Display:

   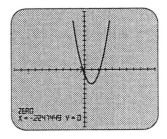

   Repeat this process to find the second root to be 2.2247449.

   Answer: To the nearest thousandth, $x = -.225$ and $x = 2.225$.

2. Graph the equation $4x^2 + 9y^2 = 36$ on a graphing calculator.

Solution: The equation is identifiable as that of an ellipse with center (0, 0). To graph this relation on a graphing calculator, we must first solve for y in terms of x. This was done on page 407.

$$y = \pm\tfrac{1}{3}\sqrt{36 - 4x^2}$$

Since for each y there are two values of x, one positive and one negative, we must enter the equation of the ellipse in two parts.

Enter: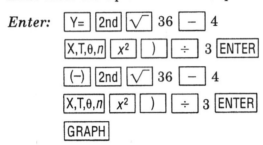

Display: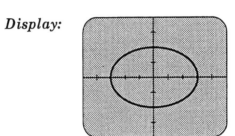

3. Graph the equation $x^2 + y^2 = 25$ on a graphing calculator.

Solution: The equation is identifiable as that of a circle with center (0, 0) and a radius of 5. The graph can be obtained using the same steps as for the ellipse above; that is, solving for y in terms of x and then entering the result in two parts. However, on the TI-83, there is another way to obtain the graph using the DRAW menu. The Circle(h, k, r) command draws a circle with center (h, k) and radius r.

With this command, use ZSquare *before* drawing the circle to adjust the window variables so the circle will not be distorted.

Enter: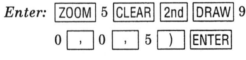

Display: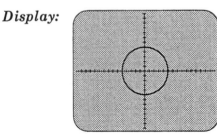

CHAPTER XII

Exponents and Logarithms

1. Exponents That Are Positive Integers

Recall (page 56) the following relationship:

$$\text{BASE}^{\text{EXPONENT}} = \text{POWER}$$

Thus, the expression x^m involves (1) the *base*, whice is x, (2) the *exponent*, which is m, and (3) the *power*, which is x^m.

EXPONENTS

By definition, $x^1 = x$; $x^2 = x \cdot x$; $x^3 = x \cdot x \cdot x$; $x^m = \underbrace{x \cdot x \cdot x \cdot \ldots \cdot x}_{m \text{ factors}}$, if m is a positive integer greater than 1.

Hence, if m is a positive integral exponent greater than 1, then m indicates the number of times the base is a factor. For example, $x^4 = x \cdot x \cdot x \cdot x$, $3^x = 3 \cdot 3 \cdot 3 \cdot \ldots \cdot 3$ (x factors), and $2^6 = 2 \cdot 2 \cdot 2 \cdot 2 \cdot 2 \cdot 2 = 64$.

Note, in 2^6, that 2 is the base, 6 is the exponent, and 2^6, or 64, is the power. We say that 2^6, or 64, is the 6th power of 2.

LAWS OF EXPONENTS FOR POSITIVE INTEGRAL EXPONENTS

The following six laws involving positive integral exponents can be derived by applying the definitions of exponent.

1. The Multiplication of Powers Law

$$x^m \cdot x^n = x^{m+n}$$

Rule. To multiply two powers having the same base, keep the base and *add* the exponents.

Thus, $x^3 \cdot x^2 = x^5$, $10^5 \cdot 10^4 = 10^9$, $(-3)^3(-3)(-3)^2 = (-3)^4(-3)^2 = (-3)^6$, and $2^x \cdot 2^y = 2^{x+y}$.

2. The Division of Powers Law

$$\text{If } m > n, \frac{x^m}{x^n} = x^{m-n}, x \neq 0 \qquad \text{If } m < n, \frac{x^m}{x^n} = \frac{1}{x^{n-m}}, x \neq 0$$

Rule. To divide two powers having the same base, keep the base and *subtract* the exponents.

$$\text{Thus, } \frac{x^5}{x^2} = x^3, \frac{x^2}{x^5} = \frac{1}{x^3}, \frac{3^9}{3^5} = 3^4, \text{ and } \frac{3^5}{3^9} = \frac{1}{3^4}.$$

3. The Power of a Power Law

$$(x^m)^n = x^{mn}$$

Rule. To find the power of a power, keep the base and *multiply* the exponents.
Thus, $(x^2)^3 = x^6$, $[(\pi^2)^3]^4 = (\pi^6)^4 = \pi^{24}$, and $(5^a)^b = 5^{ab}$.

4. The Root of a Power Law

$$\text{If } m \text{ is a multiple of } n, \sqrt[n]{x^m} = x^{\frac{m}{n}}, n \neq 0$$

Rule. To find the root of a power, keep the base and *divide* the exponent by the index.

$$\text{Thus, } \sqrt{x^6} = x^{\frac{6}{2}} = x^3, \sqrt[3]{2^{12}} = 2^{\frac{12}{3}} = 2^4, \text{ and } \sqrt[a]{5^{4a}} = 5^{\frac{4a}{a}} = 5^4.$$

5. The Power of a Product Law

$$(xy)^m = x^m y^m$$

Rule. To find the power of a product, make the exponent of the product the exponent of each of the separate factors and multiply the resulting powers. (This rule is part of the **extended law of distribution.** The power operation is distributive over multiplication.)

Thus, $(x^2y^3)^2 = (x^2)^2(y^3)^2 = x^4y^6$ and $(2x^2)^5 = (2)^5(x^2)^5 = 2^5x^{10} = 32x^{10}$.

6. The Power of a Quotient Law

$$\left(\frac{x}{y}\right)^m = \frac{x^m}{y^m}, y \neq 0$$

Rule. To find the power of a quotient, make the exponent of the quotient the exponent of both the dividend (numerator) and the divisor (denominator). Then divide the resulting powers. (This rule is part of the extended law of distribution. The power operation is distributive over division.)

Thus, $\left(\dfrac{a}{b}\right)^7 = \dfrac{a^7}{b^7}$, $\left(\dfrac{x^2}{y^3}\right)^2 = \dfrac{(x^2)^2}{(y^3)^2} = \dfrac{x^4}{y^6}$, and $\left(\dfrac{-3}{a}\right)^3 = \dfrac{(-3)^3}{a^3} = \dfrac{-27}{a^3}$.

Exercises

In 1–64, simplify. Use the set of positive integers as the domain of each literal exponent.

1. $x^5 \cdot x^7$
2. $y^3 \cdot y$
3. $r^4 \cdot r^6$
4. $t \cdot t^5$
5. $x^a \cdot x^b$
6. $a^{2b} \cdot a^{4b}$
7. $y^{2n} \cdot y^5$
8. $C^2 \cdot C^d$
9. $x^{a+2} \cdot x^{2a}$
10. $y^3 \cdot y^{2b-1}$
11. $y^{2c} \cdot y$
12. $z^{2x} \cdot z^{3y}$
13. $2^3 \cdot 2^2$
14. $5^4 \cdot 5^2$
15. $8^3 \cdot 8$
16. $10^4 \cdot 10^2$
17. $x^2 x^3 x^4$
18. $x^2 y^3 x^4 y$
19. $(\sin^2 x)(\sin^3 x)$
20. $(x+y)^3 (x+y)^4$
21. $y^8 \div y^3$
22. $y^2 \div y^8$
23. $A^5 \div A^2$
24. $A^2 \div A^5$
25. $a^{3n} \div a^n$
26. $c^{2n+2} \div c^2$
27. $x^{n+1} \div x^n$
28. $y^{3n+4} \div y^{n+1}$
29. $5^6 \div 5^2$
30. $2^5 \div 2$
31. $9^5 \div 9^7$
32. $10^4 \div 10^5$
33. $(y^3)^5$
34. $(a^2)^3$
35. $(b^4)^4$
36. $(C^2)^5$
37. $(-2)^3$
38. $(-4)^4$
39. $(-10)^2$
40. $(-3)^5$
41. $(ab)^4$
42. $(x^2 y)^7$
43. $(c^2 d^3)^5$
44. $(abc^2)^6$
45. $(-5a)^3$
46. $(-3y^3)^2$
47. $(2rs)^4$
48. $(-2x^2 y^3)^3$
49. $(\tfrac{1}{2}x)^4$
50. $(-\tfrac{2}{3}c^2)^3$
51. $(\tfrac{3}{4}c^2 d^3)^2$
52. $(-\tfrac{1}{2}xy^2)^4$
53. $(a^d)^2$
54. $(x^n y^r)^3$
55. $(r^2 s^3)^b$
56. $(c^r d^s)^t$
57. $\left(\dfrac{a}{b}\right)^c$
58. $\left(\dfrac{r^2 s^3}{t^3}\right)^a$
59. $\left(\dfrac{2x^2}{3y^3}\right)^3$
60. $\left(-\dfrac{3c}{d^3}\right)^4$
61. $\sqrt{x^8}$
62. $\sqrt[3]{y^9}$
63. $\sqrt[4]{r^4}$
64. $\sqrt[3]{a^3 b^6}$

65. Express 64 as a power of (a) 2, (b) 4, (c) 8.
66. Express x^{12} as a power of (a) x^2, (b) x^3, (c) x^4.
67. Express as a power of 10: (a) 100 (b) 1000 (c) 10,000.
68. Express as a power of 3: (a) 9 (b) 81 (c) 9^4 (d) $27^x \cdot 3^{2x}$.
69. Express as a power of 2: (a) 8 (b) 32 (c) 4^s (d) $4^x \cdot 8^x$.
70. Express as a power of 5: (a) 125 (b) 25^{4x} (c) $25^{4x} \div 5^x$ (d) $125^{2x} \cdot 5^x$.
71. The expression $(x^4)^2$ is equal to (1) x^{16} (2) x^8 (3) x^6

In 72–74, transform $a^2 + b^3$ into an equivalent expression in terms of x and y if:

72. $a = x^2, b = y^3$
73. $a = \dfrac{x}{3}, b = \dfrac{y}{2}$
74. $a = \sqrt{x}, b = \sqrt[3]{y}$

2. Exponents as Rational Numbers: Zero, Negative, and Fractional Exponents

Rational numbers, such as 0, $\frac{1}{2}$, $-\frac{1}{4}$, and 2.3456, may be used as exponents. To make this possible, new definitions are needed for powers having zero, negative, and fractional exponents. These definitions will not only provide meaning for powers having these new types of exponents, but will also preserve the laws, previously studied, that govern powers having the positive integral exponents.

DEFINITION OF A POWER HAVING A ZERO EXPONENT

By definition,

$$x^0 = 1, \qquad x \neq 0$$

We know that $\dfrac{x^5}{x^5} = 1$. By extending the division of powers law, $\dfrac{x^5}{x^5} = x^{5-5} = x^0$. Since $\dfrac{x^5}{x^5}$ equals both 1 and x^0, then x^0 must be defined as 1, provided $x \neq 0$.

Note that the new definition, $x^0 = 1$, $x \neq 0$, enables us to extend the division of powers law to cover the case involving division of powers that have equal exponents. That is,

$$\frac{x^m}{x^m} = x^{m-m} = x^0$$

This result agrees with the definition, $x^0 = 1$.

Rule. The zero power of any number, except 0, equals 1. Thus, $5^0 = 1$, $(-5)^0 = 1$, and $-5x^0 = -5$; $(5x)^0 = 1$ and $5x^0 = 5$; $(x+5)^0 = 1$ and $x + 5^0 = x + 1$.

DEFINITION OF A POWER HAVING A NEGATIVE EXPONENT

By definition,

$$x^{-n} = \frac{1}{x^n}, \qquad x \neq 0$$

We know that $\dfrac{x^5}{x^9} = \dfrac{1}{x^4}$. By extending the division of powers law, $\dfrac{x^5}{x^9} = x^{5-9} = x^{-4}$. Since $\dfrac{x^5}{x^9}$ equals both x^{-4} and $\dfrac{1}{x^4}$, then x^{-4} must be defined as

$\dfrac{1}{x^4}$, provided $x \neq 0$. This definition of x^{-4} conforms to the general definition of a negative exponent set forth above.

The definition of a power having a negative exponent may be stated in the following rule:

Rule. A power of any nonzero number having a negative exponent equals the reciprocal of the corresponding positive power of the same number.

Thus, $a^{-10} = \dfrac{1}{a^{10}}$, $2^{-5} = \dfrac{1}{2^5} = \dfrac{1}{32}$, $x^{-y} = \dfrac{1}{x^y}$, and $(x+2)^{-1} = \dfrac{1}{x+2}$.

Note in each of the illustrations that when the exponents of the same base are additive inverses of each other, the powers themselves are multiplicative inverses of each other.

DEFINITION OF A POWER HAVING A RATIONAL NUMBER AS AN EXPONENT

A power can have a rational number as an exponent. For $x^{\frac{1}{2}}$ to have a meaning consistent with the laws of exponents, it should be true that $(x^{\frac{1}{2}})^2 = x^{\frac{1}{2} \cdot 2} = x^1 = x$. Since $x^{\frac{1}{2}}$ denotes a number whose square is x, then $x^{\frac{1}{2}}$ must be defined as $\sqrt{x}$ rather than $-\sqrt{x}$ so that the inequality $x^0 < x^{\frac{1}{2}} < x^1$ is true for all values of x greater than 1. Similarly, $x^{\frac{1}{4}}$ must be defined as $\sqrt[4]{x}$.

By definition: $x^{\frac{1}{n}} = \sqrt[n]{x}$, $n \neq 0$.

Hence, $x^{\frac{1}{2}} = \sqrt{x}$, the principal square root of x; also, $x^{\frac{1}{3}} = \sqrt[3]{x}$, the principal cube root of x; and so on.

Applying the definition of $x^{\frac{1}{n}}$, then $\sqrt[n]{x^m} = (x^m)^{\frac{1}{n}} = x^{\frac{m}{n}}$; also, $(\sqrt[n]{x})^m = (x^{\frac{1}{n}})^m = x^{\frac{m}{n}}$.

By definition: $x^{\frac{m}{n}} = \sqrt[n]{x^m}$ or $x^{\frac{m}{n}} = (\sqrt[n]{x})^m$, $n \neq 0$.

Thus, $x^{\frac{3}{2}}$ is equivalent to either $\sqrt[3]{x^2}$ or $(\sqrt[3]{x})^2$. Hence, the value of $9^{\frac{3}{2}}$ may be shown to be 27 in the following ways:

(1) $(\sqrt{9})^3 = 3^3 = 27$,   (2) $\sqrt{9^3} = \sqrt{729} = 27$,   (3) $(3^2)^{\frac{3}{2}} = 3^3 = 27$.

Note. In extending powers to include all rational exponents, the root of a power law does not apply to cases *where n is even and x is negative.* The familiar laws of exponents do not hold in such cases.

Thus, if the root of a power law were to be applied to the evaluation of $(-8)^{\frac{2}{6}}$, we would obtain $(-8)^{\frac{2}{6}} = \sqrt[6]{(-8)^2} = \sqrt[6]{64} = 2$. This result would not be correct because $(-8)^{\frac{2}{6}} = (-8)^{\frac{1}{3}} = -2$ and not 2.

~~~~~~~~~~~~ **MODEL PROBLEMS** ~~~~~~~~~~~~

1. Find the value of
$$2(4^0) + 4^{\frac{1}{2}} - 4^{-\frac{1}{2}}.$$

Solution:

$$2(1) + \sqrt{4} - \frac{1}{\sqrt{4}}$$

$$2 + 2 - \tfrac{1}{2}$$

$3\tfrac{1}{2}$ *Ans.*

2. Find the value of
$$3x^0 + x^{\frac{2}{3}} \text{ when } x = 27.$$

Solution:

$$3(27^0) + 27^{\frac{2}{3}}$$

$$3(1) + (\sqrt[3]{27})^2$$

$$3 + 3^2$$

$$3 + 9 = 12 \quad Ans.$$

3. Find the value of
$$(x+2)^0 + (x+1)^{-\frac{2}{3}} \text{ if } x = 7.$$

Solution:

$$(7+2)^0 + (7+1)^{-\frac{2}{3}}$$

$$9^0 + 8^{-\frac{2}{3}}$$

$$1 + (2^3)^{-\frac{2}{3}}$$

$$[(2^3)^{-\frac{2}{3}} = 2^{3(-\frac{2}{3})} = 2^{-2}]$$

$$1 + 2^{-2}$$

$$1 + \frac{1}{2^2}$$

$$1 + \tfrac{1}{4} = 1\tfrac{1}{4} \quad Ans.$$

4. Find the value of
$$3x^0 - 54x^3 + 64^x \text{ if } x = -\tfrac{1}{3}.$$

Solution:

$$3\left(-\frac{1}{3}\right)^0 - 54\left(-\frac{1}{3}\right)^3 + 64^{-\frac{1}{3}}$$

$$3(1) - 54\left(-\frac{1}{27}\right) + \frac{1}{\sqrt[3]{64}}$$

$$3 + 2 + \tfrac{1}{4} = 5\tfrac{1}{4} \quad Ans.$$

5. Simplify the expression $\left(\dfrac{Z^6}{27}\right)^{\frac{1}{3}}$.

Solution:

$$\left(\frac{x}{y}\right)^m = \frac{x^m}{y^m} \quad \text{Apply the power of a quotient law.}$$

Hence, $\left(\dfrac{Z^6}{27}\right)^{\frac{1}{3}} = \dfrac{(Z^6)^{\frac{1}{3}}}{27^{\frac{1}{3}}}$ In the numerator, multiply exponents to obtain Z^2. In the denominator, the cube root of 27 is 3.

$$= \frac{Z^2}{3} \quad Ans.$$

Exercises

In 1–16, use a radical sign to indicate the expression.

1. $5^{\frac{1}{2}}$
2. $7^{\frac{1}{3}}$
3. $c^{\frac{1}{2}}$
4. $(2b)^{\frac{1}{2}}$

5. $(-8)^{\frac{1}{3}}$
6. $(-32)^{\frac{1}{5}}$
7. $(-5a)^{\frac{1}{2}}$
8. $(-3bc)^{\frac{1}{3}}$

9. $x^{\frac{1}{2}}$
10. $y^{\frac{1}{2}}$
11. $(3a)^{\frac{2}{3}}$
12. $(4b)^{\frac{2}{3}}$

13. $(-x)^{\frac{2}{3}}$
14. $(2y)^{\frac{3}{2}}$
15. $(ab)^{\frac{3}{4}}$
16. $(-8b)^{\frac{1}{4}}$

In 17–24, use a fractional exponent to express the radical.

17. $\sqrt{7}$
18. $\sqrt[3]{8}$
19. $\sqrt[3]{x}$
20. $\sqrt[4]{2y}$

21. $\sqrt{x^3}$
22. $\sqrt[3]{a^4}$
23. $\sqrt[4]{x^2y^3}$
24. $\sqrt[5]{r^2s^3}$

In 25–40, transform the expression into an equivalent expression involving only positive exponents.

25. y^{-5}
26. d^{-2}
27. $3a^{-4}$
28. $5b^{-3}$

29. $a^{-2}b^3$
30. $5c^{-1}d^2$
31. $(8r)^{-2}$
32. $(3s)^{-4}$

33. $\dfrac{1}{a^{-3}}$
34. $\dfrac{1}{c^{-2}}$
35. $\dfrac{5}{x^{-1}}$
36. $\dfrac{3}{y^{-5}}$

37. $\dfrac{r^{-2}}{t^{-3}}$
38. $\dfrac{m^2}{n^{-3}}$
39. $\dfrac{a^{-4}b}{cd^{-3}}$
40. $\dfrac{5x^{-4}}{7y^3}$

In 41–44, transform the fraction into an equivalent expression without a denominator.

41. $\dfrac{x^2}{y^3}$
42. $\dfrac{5ab^3}{c^4}$
43. $\dfrac{c^2}{d^{-1}}$
44. $\dfrac{8rs^2}{mn^3}$

In 45–113, simplify, using the set of rational numbers as the domain of each literal exponent.

45. 5^0
46. c^0
47. $(5x)^0$
48. $(a+b)^0$

49. $4x^0$
50. $8c^0$
51. 2^0x
52. 5^0y^a

53. $25^{\frac{1}{2}}$
54. $49^{\frac{1}{2}}$
55. $81^{\frac{1}{2}}$
56. $16^{\frac{1}{2}}$

57. $(8)^{\frac{1}{3}}$
58. $(-8)^{\frac{1}{3}}$
59. $(64)^{\frac{1}{3}}$
60. $(-27)^{\frac{1}{3}}$

61. $(16)^{\frac{1}{4}}$
62. $(a^8)^{\frac{1}{4}}$
63. $(32)^{\frac{1}{5}}$
64. $(y^{10})^{\frac{1}{5}}$

65. $\left(\dfrac{x^2}{25}\right)^{\frac{1}{2}}$
66. $\left(\dfrac{a^{3a}}{27}\right)^{\frac{1}{3}}$
67. $\left(\dfrac{x^{4c}}{y^8}\right)^{\frac{1}{2}}$
68. $\left(\dfrac{x^{3a}}{y^{6b}}\right)^{\frac{1}{3}}$

69. $(.64)^{\frac{1}{2}}$
70. $(.008)^{\frac{1}{3}}$
71. $(.04x^4)^{\frac{1}{2}}$
72. $(.125y^6)^{\frac{1}{3}}$

73. $(25)^{\frac{3}{2}}$
74. $(16)^{\frac{3}{2}}$
75. $(-8)^{\frac{2}{3}}$
76. $(-32)^{\frac{2}{5}}$

77. $(64)^{\frac{2}{3}}$
78. $(-27)^{\frac{2}{3}}$
79. $(625)^{\frac{3}{4}}$
80. $(-64)^{\frac{1}{3}}$

81. $(.16)^{\frac{3}{2}}$
82. $(.027)^{\frac{2}{3}}$
83. $(.0001)^{\frac{3}{4}}$
84. $(-.125)^{\frac{2}{3}}$

85. $4(16)^{\frac{1}{2}}$
86. $3(\tfrac{1}{9})^{\frac{3}{2}}$
87. $\tfrac{1}{2}(16)^{\frac{1}{2}}$
88. $\tfrac{1}{3}(81)^{\frac{1}{2}}$

89. $(5)^{-2}$
90. 10^{-5}
91. $(-2)^{-3}$
92. $(-4)^{-1}$

93. $(100)^{-\frac{1}{2}}$
94. $(25)^{-\frac{1}{2}}$
95. $(27)^{-\frac{1}{3}}$
96. $(-8)^{-\frac{1}{3}}$

97. $(.04)^{-\frac{1}{2}}$ **98.** $(.027)^{-\frac{1}{3}}$ **99.** $(.81)^{-\frac{1}{2}}$ **100.** $(.0001)^{-\frac{1}{4}}$

101. $(49)^{-\frac{3}{2}}$ **102.** $(8)^{-\frac{4}{3}}$ **103.** $(-32)^{-\frac{1}{5}}$ **104.** $(-27)^{-\frac{2}{3}}$

105. $4^0 - 4^{\frac{1}{2}}$ **106.** $8^{\frac{2}{3}} + 8x^0$ **107.** $(64)^{\frac{1}{3}} \cdot 2^{-2}$

108. $16 \cdot 2^{-3}$ **109.** $64^{\frac{1}{2}} \cdot 4^{-\frac{3}{2}}$ **110.** $4^{-\frac{1}{2}} - 8^{-\frac{1}{3}}$

111. $4 \times (8)^{\frac{2}{3}}$ **112.** $(81)^{-\frac{1}{2}} \times 3$ **113.** $9^0 \times 64^{-1}$

114. Find the value of $2a^0 + 3a^{-\frac{1}{2}}$ when $a = 9$.

115. Find the value of $2x^{\frac{1}{2}} + x^{-1}$ when $x = 4$.

116. Find the value of $x^{\frac{1}{2}} - x^0$ when $x = 4$.

117. When $x = 4$, the value of $x^{\frac{1}{2}} + 2x^0$ is   (1) 8   (2) 9   (3) 10

118. Find the value of $(a + 1)^0 + (4a)^{-\frac{1}{2}}$ when $a = 1$.

119. The expression $x^{-1} + y^{-1}$ is equal to   (1) $(x + y)^{-2}$ (2) $\dfrac{1}{x} + \dfrac{1}{y}$ (3) $\dfrac{1}{x + y}$

3. Applying the Laws of Exponents to All Rational Exponents Including Zero, Negative, and Fractional Exponents

It can be shown that the six laws of exponents, which were used previously for only positive integral exponents, may be extended to apply to any exponent which is a rational number, including exponents that are zero, negative, or fractional. The proof of such extensions is beyond the scope of this book.

The division of powers law may now be stated as follows: For all rational values of m and n,

$$\frac{x^m}{x^n} = x^{m-n}, \; x \neq 0$$

The root of a power law becomes: For all rational values of m and n,

$$\sqrt[n]{x^m} = x^{\frac{m}{n}}, \qquad n \neq 0$$

Thus, $x^{-2} \cdot x^3 \cdot x^{\frac{1}{2}} = x^{-2+3+\frac{1}{2}} = x^{1\frac{1}{2}}$, $a^5 \div a^{-3} = a^{5-(-3)} = a^8$,   and $(b^{\frac{2}{3}})^6 = b^{(\frac{2}{3} \cdot 6)} = b^4$.

Exercises

In 1–72, simplify. Use the set of rational numbers as the domain of each literal exponent.

1. $x^5 \cdot x^{-3}$ **2.** $a^{-1} \cdot a^4$ **3.** $b \cdot b^{-5}$ **4.** $s^{-4} \cdot s^{-2}$

5. $y^4 \cdot y^0$ **6.** $c^4 \cdot c^{-3}$ **7.** $r^3 s^2 \cdot r^{-2}$ **8.** $a^{-3} b^2 \cdot a^2 b^{-4}$

9. $a^2 \cdot a^{\frac{2}{3}}$ **10.** $b \cdot b^{\frac{1}{2}}$ **11.** $d^2 \cdot d^{2.5}$ **12.** $r^{-1} \cdot r^{.5}$

13. $x^{\frac{1}{2}} \cdot x^{\frac{1}{2}}$ **14.** $y^{\frac{1}{2}} \cdot y^{-\frac{1}{2}}$ **15.** $m^{\frac{3}{4}} \cdot m^{-\frac{1}{4}}$ **16.** $n^{-\frac{1}{4}} \cdot n^{-\frac{1}{4}}$

17. $3^{-2} \cdot 9$ **18.** $4^{\frac{1}{2}} \cdot 2^3$ **19.** $25 \cdot 5^{\frac{3}{2}}$ **20.** $16^{-1} \cdot 4^{-\frac{1}{2}}$

21. $x^3 \div x^5$ **22.** $y^2 \div y^{-3}$ **23.** $a^{-5} \div a$ **24.** $a^2 \div a^{-6}$

25. $x^{\frac{1}{2}} \div x^{\frac{1}{2}}$ **26.** $a^{-1} \div a^{\frac{1}{2}}$ **27.** $b^{-5} \div b^0$ **28.** $m^{\frac{1}{2}} \div m^{\frac{1}{4}}$

29. $r^{\frac{1}{2}} \div r^{-\frac{1}{2}}$ **30.** $s^{-\frac{1}{2}} \div s^{\frac{1}{2}}$ **31.** $c^{-\frac{1}{2}} \div c^{-\frac{1}{4}}$ **32.** $d^{-1} \div d$

33. $2^5 \div 2^7$ **34.** $8^{-\frac{2}{3}} \div 8^{-\frac{1}{3}}$ **35.** $9^{2\frac{1}{2}} \div 9^{-\frac{1}{2}}$ **36.** $5^{-\frac{3}{2}} \div 5^{\frac{1}{2}}$

37. $10^{-3} \div 10^2$ **38.** $10^4 \div 10^{-1}$ **39.** $10^{-2} \div 10^{-3}$ **40.** $100 \div 10^{-4}$

41. $(a^{\frac{1}{2}})^2$ **42.** $(b^{-3})^3$ **43.** $(c^{\frac{2}{3}})^{-2}$ **44.** $(d^{-4})^{-\frac{1}{2}}$

45. $(m^4)^{\frac{3}{2}}$ **46.** $(c^{-6})^{-\frac{1}{3}}$ **47.** $(c^6)^{-\frac{1}{2}}$ **48.** $(y^5)^{-\frac{2}{5}}$

49. $(x^2 y^{-3})^3$ **50.** $(x^{\frac{1}{2}} y)^{-4}$ **51.** $(a^{\frac{1}{2}} b^{-\frac{1}{3}})^6$ **52.** $(2c^{\frac{1}{2}} d^{\frac{1}{2}})^2$

53. $(10^2)^3$ **54.** $(10^{-1})^2$ **55.** $(10^{3 \cdot 6})^{\frac{1}{2}}$ **56.** $(10^4)^{-\frac{1}{2}}$

57. $3^x \cdot 3^y$ **58.** $2^{2a} \cdot 2^{-a}$ **59.** $5^{2x} \cdot 25$ **60.** $4^x \cdot 8^{-x}$

61. $7^{2x} \div 7^{3x}$ **62.** $4^2 \div 4^x$ **63.** $3^{2a} \div 3^{-a}$ **64.** $9^{-3x} \div 9^{-2x}$

65. $(2^{2a})^{\frac{1}{2}}$ **66.** $(9^{3x})^{\frac{1}{2}}$ **67.** $(3^c)^{-2}$ **68.** $(6^{4d})^{-\frac{1}{2}}$

69. $\sqrt{10^{-4}}$ **70.** $\sqrt[3]{x^3}$ **71.** $\sqrt{a^{\frac{1}{2}}}$ **72.** $\sqrt[4]{a^{\frac{1}{2}}}$

73. The value of $5^{\frac{1}{2}} \times 5^{\frac{1}{2}}$ is   (1) $5^{\frac{1}{4}}$ (2) 25   (3) 5

74. The expression $(a^2)^{-3}$ equals (1) a^{-6} (2) a^{-5} (3) a^{-1} (4) a^{-8}

75. Answer *true* or *false*: $(x^2)^{-\frac{2}{3}} = \sqrt[3]{x^{-4}}$.

76. If $3^y = x$, then 3^{y+1} equals (1) $x + 3$ (2) $3x$ (3) $x + 1$

4. Scientific Notation

In **scientific notation,** a number is expressed as the product of two factors: One factor is a number between 1 and 10, the other factor is an integral power of 10.

Thus, 630,000,000, the approximate number of seconds in 20 years, may be expressed in scientific notation as 6.3×10^8. Also, .00000000053, the approximate number of millimeters in the diameter of the orbit of an electron of a certain atom, may be expressed in scientific notation as 5.3×10^{-10}.

Scientific notation is used:

1. To provide a concise notation for very large and very small numbers, such as those shown above.
2. To simplify computations by applying the laws of exponents to numbers expressed in scientific notation.
3. To estimate or approximate the results of computation. Thus,

$$\frac{255,000}{.00125} = \frac{2.55 \times 10^5}{1.25 \times 10^{-3}} = \frac{2.55}{1.25} \times 10^8 \approx 2 \times 10^8 \approx 200,000,000$$

Procedure. **To write numbers in scientific notation:**

1. Place an apostrophe (') immediately after the first nonzero digit of the given number.
2. Starting from the apostrophe, count the number of places to the decimal point in the given number.
3. Write the first factor of the number in scientific notation by placing a decimal point at the position of the apostrophe, the result being a number between 1 and 10.
4. Write the second factor of the number in scientific notation by using the number of places counted in step 2 as the exponent of the power of 10. If the count is to the right, the exponent is positive; if the count is to the left, the exponent is negative.

In the following table, the arrow indicates the direction of counting.

$Number$ =	$\begin{pmatrix} Number\ between \\ 1\ and\ 10 \end{pmatrix}$	$\times (Power\ of\ 10)$
$\overrightarrow{7'340.}$ =	7.34	$\times$ 10^3
$\overleftarrow{.0007'34}$ =	7.34	$\times$ 10^{-4}
$5'.7$ =	5.7	$\times$ 10^0
$\overrightarrow{5'7000000.}$ =	5.7	$\times$ 10^7
$\overleftarrow{.5'7}$ =	5.7	$\times$ 10^{-1}

~~~~~~~~~~ *MODEL PROBLEMS* ~~~~~~~~~~

1. Write the number 4.86×10^{-3} in ordinary decimal notation.

   Solution:

   $$10^{-3} = \frac{1}{10^3} = \frac{1}{1000} \qquad \text{Hence, } 4.86 \times 10^{-3} = 4.86 \times \frac{1}{1000} = .00486. \quad Ans.$$

   Note. To multiply a decimal by 10^{-3}, simply move the decimal point 3 places to the left. Think of -3 as "3 to the left."

2. If each of the following numbers is expressed in scientific notation as 3.28×10^n, what is the value of n in each case?   *a.* 328,000   *b.* 0.000328

Solution: In each case, to find n, place an apostrophe immediately after the first nonzero digit of the given number; then count from the apostrophe to the decimal.

a. In the case of 3'28,000, the count from the apostrophe to the decimal is five places to the right. Hence, $n = 5$.   *Ans.*

b. In the case of 0.0003'28, the count from the apostrophe to the decimal is four places to the left. Hence, $n = -4$.   *Ans.*

Exercises

In 1–10, complete the table.

	Number	=	$\begin{pmatrix} Number\ between \\ 1\ and\ 10 \end{pmatrix} \times (10^n)$	
1.	750	=	7.5	$\times 10^?$
2.	.075	=	7.5	$\times 10^?$
3.	?	=	7.5	$\times 10^6$
4.	?	=	7.5	$\times 10^{-6}$
5.	80000	=	?	$\times 10^4$
6.	.00008	=	?	$\times 10^{-5}$
7.	?	=	3.25	$\times 10^{10}$
8.	?	=	3.25	$\times 10^{-10}$
9.	32500000	=	?	$\times$?
10.	.000000325	=	?	$\times$?

In 11–16, express the number as a power of 10.

11. 1,000,000

12. .000001

13. $\dfrac{1}{1,000,000}$

14. 1,000,000,000

15. .000000001

16. $\dfrac{1}{1,000,000,000}$

In 17–25, write the number in scientific notation.

17. 83,000

18. 83,000,000

19. 83,000,000,000

20. .0075

21. .0000075

22. .00000000075

23. $\dfrac{9}{1000}$

24. $\dfrac{9}{1,000,000}$

25. $\dfrac{9}{1,000,000,000,000}$

In 26–31, write the number as an integer or in ordinary decimal notation.

26. 5.8×10^3 **27.** 5.8×10^{-3} **28.** 5.8×10^{-6}

29. 9.3×10^{12} **30.** 9.3×10^{-12} **31.** 9.356×10^{-12}

In 32–39, applying the laws of exponents, calculate the result and express the result in both scientific notation and as an integer or in ordinary decimal notation.

32. $(2.5 \times 10^2)(4 \times 10^3)$

33. $(1.5 \times 10^{-2})(1.5 \times 10^{-3})$

34. $\dfrac{5 \times 10^2}{2.5 \times 10^3}$

35. $\dfrac{7.5 \times 10^{-3}}{1.5 \times 10^3}$

36. $(2.5 \times 10^5)^2$

37. $\sqrt{1.44 \times 10^8}$

38. $\dfrac{(9.3 \times 10^6)(5 \times 10^{-4})}{1.25 \times 10^{-3}}$

39. $\dfrac{(4 \times 10^{-5})^3(0.8 \times 10^{-4})^2}{(6.25 \times 10^{-6})^{\frac{1}{2}}}$

In 40–45, find the value of n.

40. $4000 = 4 \times 10^n$

41. $4{,}000{,}000 = 4 \times 10^n$

42. $.000004 = 4 \times 10^n$

43. $54{,}000{,}000 = 5.4 \times 10^n$

44. $.00000054 = 5.4 \times 10^n$

45. $.00000000054 = 5.4 \times 10^n$

In 46–49, express the number in scientific notation.

46. The approximate distance between our solar system and its nearest known star, Alpha Centauri, is 25,000,000,000,000 miles.

47. The number of atoms in a gram of hydrogen is approximately 600,000,000,000,000,000,000,000.

48. A light year is the distance light travels in a year. A light year is approximately 5,900,000,000,000 miles.

49. The diameter of the smallest visible particle is .0002 inch.

In 50–54, express the number as an integer, or in ordinary decimal notation.

50. The earth's mass is 1.32×10^{25} lb.

51. The distance from the earth to the sun is 9.3×10^7 miles.

52. The age of the earth's crust is 5×10^9 years.

53. The velocity of light is 3×10^{10} cm. per sec.

54. The average density of matter in metagalactic space is 10^{-29} gm. per cu. cm.

5. The Exponential Function

POWERS HAVING IRRATIONAL NUMBERS AS EXPONENTS

Until now we have defined all numbers b^x where b is a positive real number and x is a rational number. For example, we have defined 2^2 as 4, 8^0 as 1,

10^{-1} as $\frac{1}{10}$, and $27^{\frac{2}{3}}$ as 9. Now we will consider the meaning of b^x when b is a positive real number and x is an *irrational number*. We begin by investigating the meaning of 2^x, the special case of b^x when $b = 2$.

The following table contains ordered pairs of numbers which satisfy the equation $y = 2^x$:

x	-3	-2	-1	0	1	2	3
$y = 2^x$	$\frac{1}{8}$	$\frac{1}{4}$	$\frac{1}{2}$	1	2	4	8

To draw the graph of $y = 2^x$, we plot the points whose ordered pairs are included in the above table and then join them with a smooth curve as shown in the figure. In order to draw the smooth curve, we assume that, for every real value of x, there corresponds a unique real value of y such that $y = 2^x$. Note that the curve intersects the y-axis at $P(0, 1)$, but does not intersect the x-axis. Since y is positive for any real value of x, the curve lies entirely above the x-axis. As we proceed from right to left, the values of x become smaller and smaller and the curve draws closer and closer to the x-axis but does not touch the x-axis. The x-axis is an asymptote of the graph of $y = 2^x$.

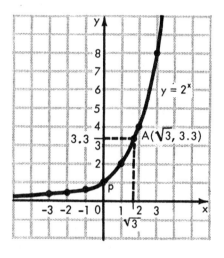

On the graph of $y = 2^x$, we can locate points whose x-coordinates are irrational numbers. For example, at point A, whose x-coordinate is $\sqrt{3}$, the y-coordinate is found to approximate 3.3. Hence, $2^{\sqrt{3}} \approx 3.3$.

Note that the curve meets the vertical line test, since no vertical line may intersect the curve in more than one point. Hence, the relation defined by $y = 2^x$ is a function. This function is called an *exponential function*.

Another method can be used to approximate the value of $2^{\sqrt{3}}$. First we approximate $\sqrt{3}$, the approximation being a rational number; then we use the result to approximate $2^{\sqrt{3}}$. For example, if we use 1.732 as an approximation of $\sqrt{3}$, then, using a method to be studied later in this chapter, we find that $2^{1.732} \approx 3.32$.

The definition of an exponent can be extended to include any irrational number as well as any rational number. Hence 2^x and, in general, b^x when $b > 0$ and $b \neq 1$, have meaning for all real numbers. With these definitions and mean·

ings, the laws of exponents that we have used previously can be extended to include exponents that may be any real number, irrational as well as rational. The proof of such extensions is beyond the scope of this book.

DEFINITION OF THE EXPONENTIAL FUNCTION

When $y = 2^x$, to each real value of x there corresponds a unique real value of y. Hence, the equation $y = 2^x$ defines a function. We call this function an *exponential function* in accordance with the following general definition:

The function $\{(x, y) \mid y = b^x, b > 0, b \neq 1\}$ is an exponential function whose base is b. The domain of the exponential function, the set of values of x, is the set of real numbers; the range of the exponential function, the set of the corresponding values of y, is the set of positive numbers.

~~~~~~~~~~ *MODEL PROBLEMS* ~~~~~~~~~~

1. The graph of $(\frac{1}{2})^x = y$ lies only in (1) quadrant I   (2) quadrant II   (3) quadrants I and II   (4) quadrants I and IV

    Solution: For any real values of x, positive, negative, or zero, $(\frac{1}{2})^x$ is always positive. Hence, y, which equals $(\frac{1}{2})^x$, is always positive for all real values of x. Since y is positive only in quadrants I and II, the graph of $y = (\frac{1}{2})^x$ lies only in these quadrants.

    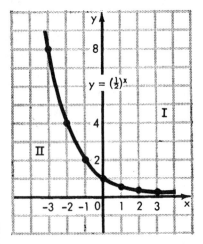

    Answer: The correct choice is (3).

    Note. The following table was used to sketch the graph of $y = (\frac{1}{2})^x$ for values of x from $x = -3$ to $x = 3$:

x	-3	-2	-1	0	1	2	3
y	8	4	2	1	$\frac{1}{2}$	$\frac{1}{4}$	$\frac{1}{8}$

    Since $y = (\frac{1}{2})^x = (2^{-1})^x = 2^{-x}$, then the graph of $y = (\frac{1}{2})^x$ is also the graph of $y = 2^{-x}$.

2. The graphs of the functions defined by $y = 2^x$ and $y = 2^{-x}$ are (1) symmetric to each other with respect to the x-axis   (2) symmetric to each other with respect to the y-axis   (3) symmetric to each other with respect to the origin   (4) not symmetric to each other

Solution: From the graphs of $y = 2^x$ and $y = 2^{-x}$, note that the curves are symmetric with respect to the y-axis; that is, the y-axis serves as the folding line or axis of symmetry with respect to the two curves. If the entire graph is folded along the y-axis, either curve can be made to coincide with the other.

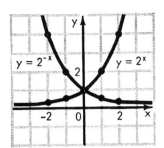

The following symmetry test can be applied to show that the curves are symmetric with respect to the y-axis:

In $y = 2^x$, if $-x$ replaces x, the other equation, $y = 2^{-x}$, is obtained. Similarly, in $y = 2^{-x}$, if x replaces $-x$, then the other equation, $y = 2^x$, is obtained.

Such interchangeability of x and $-x$ indicates that the graphs of equations in x and y will be symmetric with respect to the y-axis.

Answer: The correct choice is (2).

~~~~~~

Exercises

1. *a.* Plot the graph of $y = 3^x$, using integral values of x from -3 to 3, inclusive.
   b. Using the graph made in answer to part *a*, find the approximate value of:

      1. $3^{\frac{1}{2}}$ 2. $3^{1\frac{1}{2}}$ 3. $3^{\sqrt{2}}$ 4. $3^{\sqrt{3}}$

2. *a.* Plot the graph of $\{(x, y) \mid y = 4^x\}$, using integral values of x from -3 to 3.
   b. Using the graph made in answer to part *a*, find the approximate value of:

      1. $4^{\frac{1}{2}}$ 2. $4^{1\frac{1}{2}}$ 3. $4^{\sqrt{2}}$ 4. $4^{\sqrt{5}}$

In 3–6, sketch the graph of the function.

3. $y = 2^{-x}$ 4. $y = 3^{-x}$

5. $\{(x, y) \mid y = 5^x\}$ 6. $\{(x, y) \mid y = 10^x\}$

In 7–10, use an appropriate graph to find an approximation of the number.

7. $3^{2\frac{1}{2}}$ 8. $4^{1.25}$ 9. $5^{\sqrt{2}}$ 10. $\left(\frac{1}{2}\right)^{-\sqrt{2}}$

11. Name the point or points that the graphs of $y = 2^x$, $y = 3^x$, $y = \left(\frac{1}{2}\right)^x$, and $y = \left(\frac{1}{3}\right)^x$ have in common.

12. *a.* Sketch the graphs of $y = 3^x$ and $y = 3^{-x}$ and state how the graphs are related.
    b. Are the graphs of $y = 4^x$ and $y = \left(\frac{1}{4}\right)^x$ related in the same way? If so, explain.

13. *a.* Is $\{(x, y) \mid y = a^x, a > 0, a \neq 1\}$ an exponential function?
    b. State the reason for the answer given in part *a*.

6. Solving Equations Containing Fractional and Negative Exponents

The definitions and laws of exponents can be used to solve equations in which the variable base has a fractional or a negative exponent.

Thus, to solve $x^{\frac{1}{4}} = 10$, raise both sides to the fourth power: $(x^{\frac{1}{4}})^4 = 10^4$. Apply the power of a power law to multiply exponents. The product of the exponents is $\frac{1}{4}(4)$, or 1: $x^1 = 10,000$, or $x = 10,000$.

Observe that we raised both sides of the equation to a power whose exponent is the reciprocal of the exponent of the variable so that the exponent of the variable in the resulting equation became 1.

~~~~~~~~~~ *MODEL PROBLEMS* ~~~~~~~~~~

In 1 and 2, solve for the positive value of x.

     1. $x^{\frac{3}{2}} - 125 = 0$ **2.** $x^{-2} - 9 = 0$

How To Proceed	*Solution*	*Solution*
1. Transform the equation into the form $x^{\frac{m}{n}} = k$, where k is a constant.	$x^{\frac{3}{2}} = 125$	$x^{-2} = 9$
2. Raise both sides to the same power, using as an exponent $\dfrac{n}{m}$, the reciprocal of the exponent of the variable base x.	$(x^{\frac{3}{2}})^{\frac{2}{3}} = 125^{\frac{2}{3}}$ $x^1 = (5^3)^{\frac{2}{3}}$ $x = 5^2$ $= 25$ *Ans.*	$(x^{-2})^{-\frac{1}{2}} = 9^{-\frac{1}{2}}$ $x^1 = (3^2)^{-\frac{1}{2}}$ $x = 3^{-1}$ $= \frac{1}{3}$ *Ans.*
3. Check in the original equation.	$x^{\frac{3}{2}} - 125 = 0$ $25^{\frac{3}{2}} - 125 \overset{?}{=} 0$ $125 - 125 \overset{?}{=} 0$ $0 = 0$	$x^{-2} - 9 = 0$ $(\frac{1}{3})^{-2} - 9 \overset{?}{=} 0$ $\dfrac{1}{(\frac{1}{3})^2} - 9 \overset{?}{=} 0$ $9 - 9 \overset{?}{=} 0$ $0 = 0$

~~~~~~

Exercises

In 1–18, solve the equation for the positive value of the variable and check.

1. $x^{\frac{1}{2}} = 10$
2. $x^{\frac{1}{4}} = 2$
3. $x^{\frac{1}{3}} = 3$
4. $x^{\frac{1}{2}} - 4 = 2$
5. $4x^{\frac{1}{2}} = 20$
6. $2x^{\frac{1}{2}} + 7 = 11$
7. $x^{-\frac{1}{2}} = 6$
8. $x^{-\frac{1}{4}} = 2$
9. $x^{-\frac{1}{2}} + 1 = 2$
10. $x^{\frac{2}{3}} = 25$
11. $x^{-\frac{3}{4}} = 16$
12. $x^{\frac{2}{3}} - 27 = 0$
13. $x^{\frac{2}{3}} = 27$
14. $y^{-2} = 49$
15. $y^{-3} - 8 = 0$
16. $y^{\frac{2}{3}} = 125$
17. $y^{\frac{1}{2}} = \frac{4}{9}$
18. $z^{\frac{2}{3}} - 12 = 20$

7. Solving Problems Involving Exponential Equations

An *exponential equation* is an equation in which the variable appears as an exponent or a part of an exponential expression. Examples of exponential equations are $2^x = 8$, $2^{x+1} = 8^{2x}$, and $5^x = 25^{x-2}$.

Exponential equations, such as the illustrations given above, are readily solvable when both sides of the equation are expressed as powers of the same base. Exponential equations in which both sides cannot be expressed as powers of the same base will be considered later in the chapter.

In an exponential equation, if both sides are expressed as powers of the same base, then the exponents of the powers are equal.

Thus, if $3^{2x} = 81$, then $3^{2x} = 3^4$. Hence, $2x = 4$ and $x = 2$.

Also, if $2^{4+y} = 4^y$, then $2^{4+y} = (2^2)^y$. Hence, $4 + y = 2y$ and $y = 4$.

~~~~~~~~~~~~~ *MODEL PROBLEMS* ~~~~~~~~~~~~~

1. Solve for t: $27^{6-t} = 9^{t-1}$

Solution:

To solve the given equation for t, express both sides as powers of the same base. Do this by substituting 3^3 for 27 and 3^2 for 9.

$$27^{6-t} = 9^{t-1}$$
$$(3^3)^{6-t} = (3^2)^{t-1}$$
$$3^{18-3t} = 3^{2t-2}$$

Since both sides now have the same base, equate the exponents.

$$18 - 3t = 2t - 2$$
$$20 = 5t$$
$$t = 4 \quad Ans.$$

2. In the equation $y = 3^x$, if the variable x is decreased by 3, then y is
(1) decreased by 3   (2) divided by 3   (3) decreased by 27   (4) divided by 27

Solution:

If x is decreased by 3, the expression 3^x becomes 3^{x-3}.

Since $3^{x-3} = 3^x \cdot 3^{-3} = \dfrac{3^x}{3^3} = \dfrac{3^x}{27}$ and $y = 3^x$, then $3^{x-3} = \dfrac{y}{27}$.

Hence, y is divided by 27.

Answer: The correct choice is (4).

Exercises

In 1–18, solve the equation and check.

1. $2^x = 64$ **2.** $2^{-x} = 32$ **3.** $2^{-x} = \frac{1}{16}$

4. $3^x = \frac{1}{27}$ **5.** $3^{x-1} = 81$ **6.** $3^x + 3 = 30$

7. $9 \cdot 3^x = \frac{1}{27}$ **8.** $4^x = 8^{x-1}$ **9.** $8^x = 4^{2x-2}$

10. $10^{3y} = 100^{y+2}$ **11.** $100^{x+1} = 1000^{x-1}$ **12.** $100^x = .0001$

13. $8^{2x-3} = 16^{x+2}$ **14.** $4^{2x} = (\frac{1}{16})^{x-1}$ **15.** $16^{3x} = 8^{x+4}$

16. $8^{x-3} = 16^{\frac{x}{2}}$ **17.** $25^{6-x} = 125^x$ **18.** $27^{t+3} = (\frac{1}{3})^{2-t}$

19. If, in the equation $y = 2^x$, the variable x is increased by 2, then the value of y is (1) divided by 2   (2) multiplied by 2   (3) squared   (4) multiplied by 4

20. If $7^{-x} = 10$, then 7^{2x} is equal to (1) $\dfrac{1}{100}$ (2) $\dfrac{1}{20}$ (3) 20   (4) 100

21. If $a^x = b$, then a^{x+3} is equal to (1) $3 + b$ (2) $a^3 b$ (3) b^3 (4) $3b$

In 22–25, if $a^x = b$, transform the expression into a power of a.

22. $a^2 b^2$ **23.** $\dfrac{a}{b}$ **24.** $\dfrac{b^2}{a^3}$ **25.** $(ab)^3$

In 26–28, find the solution set of the system of equations.

26. $2^x \cdot 2^y = 8$ **27.** $3^x \cdot 3^y = 27$ **28.** $100^x \cdot 10^y = \frac{1}{10}$

$\qquad \dfrac{2^x}{2^y} = 32$ $\dfrac{3^x}{27^y} = \dfrac{1}{3}$ $\dfrac{10^x}{10^{2y}} = 100$

8. Using Exponents To Simplify Calculations: Powers of 2

The numbers in the following table are powers of the same base, 2. Operations upon numbers that can be expressed as powers of the same base are simplified through the application of the laws of exponents. Laborious calculations involving the numbers are replaced by mental operations involving exponents.

TABLE OF POWERS OF 2

$2^1 = 2$	$2^{11} = 2048$	$2^{21} = 2{,}097{,}152$
$2^2 = 4$	$2^{12} = 4096$	$2^{22} = 4{,}194{,}304$
$2^3 = 8$	$2^{13} = 8192$	$2^{23} = 8{,}388{,}608$
$2^4 = 16$	$2^{14} = 16{,}384$	$2^{24} = 16{,}777{,}216$
$2^5 = 32$	$2^{15} = 32{,}768$	$2^{25} = 33{,}554{,}432$
$2^6 = 64$	$2^{16} = 65{,}536$	$2^{26} = 67{,}108{,}864$
$2^7 = 128$	$2^{17} = 131{,}072$	$2^{27} = 134{,}217{,}728$
$2^8 = 256$	$2^{18} = 262{,}144$	$2^{28} = 268{,}435{,}456$
$2^9 = 512$	$2^{19} = 524{,}288$	$2^{29} = 536{,}870{,}912$
$2^{10} = 1024$	$2^{20} = 1{,}048{,}576$	$2^{30} = 1{,}073{,}741{,}824$

~~~~~~~~~~ *MODEL PROBLEM* ~~~~~~~~~~

Using the table of powers of 2, evaluate $\dfrac{(131{,}072)^2 \sqrt[3]{262{,}144}}{\sqrt{268{,}435{,}456}}$.

How To Proceed	*Solution*
1. Express each number as a power of 2.	1. $\dfrac{(2^{17})^2 (2^{18})^{\frac{1}{3}}}{(2^{28})^{\frac{1}{2}}}$
2. Apply the laws of exponents.	2. $\dfrac{2^{34}(2^6)}{2^{14}}$
	$2^{34+6-14} = 2^{26}$
3. Write the number which equals the resulting power of 2.	3. $\qquad = 67{,}108{,}864 \quad Ans.$

~~~~~~

Exercises

In 1–21, evaluate, using the table of powers of 2.

1. 32×16 **2.** $4 \times 64 \times 1024$ **3.** $2 \times 8 \times 128 \times 256$

4. $4096 \div 32$ **5.** $524{,}288 \div 65{,}536$ **6.** $1{,}073{,}741{,}824 \div 64$

7. 512^2 **8.** 1024^3 **9.** 128^4

10. $\sqrt{1{,}048{,}576}$ **11.** $\sqrt[3]{2{,}097{,}152}$ **12.** $\sqrt[5]{33{,}554{,}432}$

13. $67{,}108{,}864^{\frac{1}{3}}$ **14.** $16{,}777{,}216^{\frac{1}{4}}$ **15.** $134{,}217{,}728^{\frac{1}{3}}$

16. $4096^{\frac{3}{2}}$ **17.** $4096^{\frac{1}{3}}$ **18.** $4096^{\frac{1}{2}}(4096^{\frac{1}{4}})$

19. $\dfrac{524{,}288}{16{,}384^{\frac{1}{2}}}$ **20.** $\dfrac{1024^2(131{,}072)}{134{,}217{,}728}$ **21.** $\dfrac{67{,}108{,}864^{\frac{1}{3}}}{16^3(256^{\frac{1}{4}})}$

In 22–27, solve for x, using the table of powers of 2.

22. $2^{x+1} = 512$ **23.** $2^{3x} = 262{,}144$ **24.** $2^{4x-3} = 2{,}097{,}152$

25. $4^x = 1{,}048{,}576$ **26.** $2^x(4^x)(8) = 512$ **27.** $16^{x+2} = 32{,}768$

9. Understanding Logarithms: Writing Equations in Exponential and Logarithmic Form

The previous unit showed how laborious calculations involving numbers can be replaced by mental operations involving exponents—if the numbers can be expressed as powers of the same base. You can understand that the table of powers of 2 can be used to simplify calculations only if the numbers involved are the few numbers listed in the table.

In the beginning of the 17th century, a Scotsman named John Napier designed a table of logarithms by means of which *all numbers* could be expressed as powers of the same base. This new tool not only enabled mathematicians to simplify calculations of an extensive and laborious kind, but also made it possible to solve problems that could not have been done before the invention of the table. An understanding of logarithms is indispensable in advanced mathematics, science, and technology.

In $3^4 = 81$, the exponent 4 is called the *logarithm* of 81 to the base 3. The statement "4 is the logarithm of 81 to the base 3" is written as "$4 = \log_3 81$." Read "$4 = \log_3 81$" as "4 is the log of 81 to the base 3."

In the equation $3^4 = 81$, the number 4 plays a dual role, because 4 is the exponent of the base 3; and, at the same time, 4 is the logarithm of the power 81. Keep in mind that the same number is both an exponent and a logarithm.

By definition, the *logarithm* of a number, to a given base, is the exponent that is used with the base to obtain the number.

$$\textbf{If } \boldsymbol{b^e = n,} \textbf{ then } \boldsymbol{e = \log_b n.}$$

The equation $3^4 = 81$ is in **exponential form**. The equivalent equation, $\log_3 81 = 4$, is in **logarithmic form**. Note these two forms of equivalent equations in the following table:

Exponential Form	Logarithmic Form
$10^3 = 1000$	$\log_{10} 1000 = 3$
$16^{\frac{1}{2}} = 4$	$\log_{16} 4 = \frac{1}{2}$
$5^{-2} = \frac{1}{25}$	$\log_5 \frac{1}{25} = -2$
$3^4 = x$	$\log_3 x = 4$
$x^2 = 49$	$\log_x 49 = 2$

If no base is indicated when we find the logarithm of a number, the base is understood to be 10. Thus, $\log 1000 = 3$ means $\log_{10} 1000 = 3$. Logarithms which have 10 as their base are called **common logarithms**.

MODEL PROBLEMS

1. Express in logarithmic form $5^3 = 125$.   *Answer:* $\log_5 125 = 3$
2. Express in exponential form $\log_4 64 = 3$.   *Answer:* $4^3 = 64$
3. Express in exponential form $\log 100 = 2$.   *Answer:* $10^2 = 100$

4. Find the positive value of x: $\log_x 16 = 4$.
Solution:
$$\log_x 16 = 4$$
Express in exponential form: $x^4 = 16$
Find the fourth root of 16: $x = 2$ *Ans.*

Exercises

In 1–10, express in logarithmic form.
1. $2^5 = 32$ 2. $10^0 = 1$ 3. $3^5 = 243$ 4. $4^3 = 64$ 5. $36^{\frac{1}{2}} = 6$
6. $8^{\frac{1}{3}} = 2$ 7. $8^{\frac{2}{3}} = 4$ 8. $10^{-1} = \frac{1}{10}$ 9. $5^{-2} = \frac{1}{25}$ 10. $4^{-\frac{3}{2}} = \frac{1}{8}$

In 11–18, express in exponential form.

11. $\log_2 16 = 4$ **12.** $\log_9 81 = 2$ **13.** $\log 10 = 1$ **14.** $\log_3 1 = 0$

15. $\log_8 \frac{1}{8} = -1$ **16.** $\log .01 = -2$ **17.** $\log_9 3 = \frac{1}{2}$ **18.** $\log_8 2 = \frac{1}{3}$

In 19–27, solve for x.

19. $\log_x 81 = 2$ **20.** $\log_x 64 = 3$ **21.** $\log_2 x = 5$

22. $\log_x .1 = -1$ **23.** $\log_x .01 = -2$ **24.** $\log_7 49 = x$

25. $\log 100 = x$ **26.** $\log .001 = x$ **27.** $\log .1 = x$

In 28–30, evaluate.

28. $\log_3 27 + \log_2 16$ **29.** $5 \log 1 - 2 \log 10$ **30.** $\log_4 2 + \log_8 64$

In 31–33, express in exponential form.

31. $\log_a b = c$ **32.** $\log_c a = b$ **33.** $\log_b c = a$

In 34–36, express in logarithmic form.

34. $p^q = r$ **35.** $r^p = q$ **36.** $q^r = p$

10. The Logarithmic Function

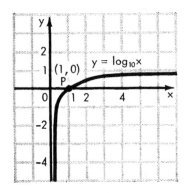

The inverse of the exponential function $y = 10^x$, obtained by interchanging x and y, is $x = 10^y$. When $x = 10^y$ is expressed in logarithmic form, we obtain the equation $y = \log_{10} x$.

The graph of $y = \log_{10} x$ is shown at the right. Note the following on the graph of $y = \log_{10} x$:

1. Since $x = 10^y$, if $y = 0$ then $x = 1$. Hence, the curve intersects the x-axis at $P(1, 0)$.
2. For any real values of y, positive, negative, or zero, 10^y must be positive; that is, the graph of $y = \log_{10} x$ is always to the right of the y-axis.
3. As we proceed from top to bottom, the values of y become smaller and smaller and the curve draws closer and closer to the y-axis but does not touch the y-axis. The y-axis is an asymptote to the graph of $y = \log_{10} x$.
4. Note that the curve meets the vertical line test since no vertical line may intersect the curve in more than one point. Hence, the relation defined by $y = \log_{10} x$ is a function. This function is called the *logarithmic function*.

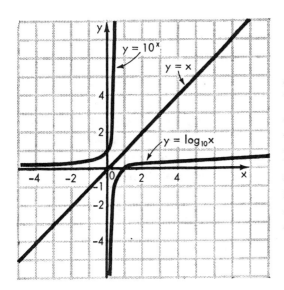

In the figure at the left, the graph of the logarithmic function defined by $y = \log_{10} x$ and also the graph of its inverse function, the exponential function defined by $y = 10^x$, have both been drawn. Note that the graph of the logarithmic function is symmetric to the graph of the exponential function with respect to the line $y = x$.

DEFINITION OF THE LOGARITHMIC FUNCTION

When $y = \log_{10} x$, to each real value of x there corresponds a unique real value of y. Hence, the equation $y = \log_{10} x$ defines a function. We call this function a **logarithmic function** in accordance with the following general definition:

The function $\{(x, y) \mid y = \log_b x, b > 0, b \neq 1\}$ is a logarithmic function whose base is b. The domain of the logarithmic function, the set of values of x, is the set of positive numbers. The range of the logarithmic function, the set of the corresponding values of y, is the set of real numbers.

Exercises

1. Sketch the graph of $y = \log_3 x$, using the following values of x: $\frac{1}{9}$, $\frac{1}{3}$, 1, 3, and 9.
2. Name the point or points that the graphs of $y = \log_2 x$, $y = \log_3 x$, and $y = \log_{10} x$ have in common.
3. To answer the following questions, refer to the preceding graphs of $y = \log_{10} x$.
   a. If $x > 1$, is the logarithm of x a positive or a negative number?
   b. If $x = 1$, what is the logarithm of x to any base?
   c. If $0 < x < 1$, is the logarithm of x a positive or a negative number?
   d. If $x < 0$, is the logarithm of x a real number?

11. Finding Common Logarithms

Since 10 is the basis of the decimal system of numbers, computations with logarithms will be performed with *common logarithms*, logarithms whose base is 10. To evaluate a numerical expression using logarithmic rules, we obtain the common logarithm of numbers in the expression. Let us begin with the integral powers of 10, the ones contained in the following table, since these are the numbers whose common logarithms are most easily determined.

TABLE OF POWERS OF 10

$(10^e = N)$	$(\text{Log } N = e)$
Exponential Form	*Logarithmic Form*
$10^4 = 10,000$ $10^3 = 1,000$ $10^2 = 100$ $10^1 = 10$	$\log 10,000 = 4$ $\log 1000 = 3$ $\log 100 = 2$ $\log 10 = 1$
$10^0 = 1$	$\log 1 = 0$
$10^{-1} = .1$ $10^{-2} = .01$ $10^{-3} = .001$ $10^{-4} = .0001$	$\log .1 = -1$ $\log .01 = -2$ $\log .001 = -3$ $\log .0001 = -4$

Note in the table, illustrations of the following principles:

Principle 1. The common logarithm of a number is *positive* if the number is greater than 1; the common logarithm of a number is *negative* if the number is less than 1 and greater than 0. Also, the common logarithm of 1 is 0.

Thus, $\log 100 = +2$ and $\log .01$ is -2.

Principle 2. Numbers and logarithms vary in the same sense; that is, the greater a number, the greater the logarithm, whereas the smaller a number, the smaller the logarithm.

Thus, since $100 > 10$, then $\log 100 > \log 10$. Also, since $.01 < .1$, then $\log .01 < \log .1$.

Principle 3. If a number is multiplied by 10, its common logarithm is increased by 1, whereas if a number is divided by 10, its common logarithm is decreased by 1.

For example, $\log 100 = 2$, $\log (100 \cdot 10) = 2 + 1$, and $\log 1000 = 3$. Also, $\log 100 = 2$, $\log (100 \div 10) = 2 - 1$, and $\log 10 = 1$.

Note. The common logarithm of a negative number is not a real number. It can be shown that the common logarithm of a negative number is a complex number.

Whether a number is an integral power of 10, such as 100 or 1000, or is not an integral power of 10, such as 3.25 or 31.7, *the common logarithm of any number is expressible as a combination of two parts: an integral part called the* **character-istic** *and a decimal part called the* **mantissa**.

Thus, log 1000 = 3.0000. Here, 3 is the characteristic and .0000 is the mantissa. The characteristic indicates the number's magnitude.

However, most numbers are not exact integral powers of 10. For example, $3.25 = 10^{0.5119}$. Hence, log 3.25 = 0.5119, approximately. Here, 0 is the characteristic and .5119 is the mantissa. Let us now see what happens when 3.25 is multiplied or divided by an integral power of 10.

If 3.25 is multiplied by 100, then $3.25 \times 100 = 10^{0.5119} \times 10^2 = 10^{2.5119}$. Since $325 = 3.25 \times 100$, then $325 = 10^{2.5119}$ and log 325 = 2.5119. Observe that both log 325 and log 3.25 have the same mantissa, which is .5119. However, the characteristic of log 325 is 2 more than the characteristic of log 3.25 due to the multiplication of 3.25 by 100.

If 3.25 is divided by 100, then $3.25 \div 100 = 10^{0.5119} \div 10^2 = 10^{.5119-2}$. Since $3.25 \div 100 = .0325$, then log .0325 = .5119 − 2. Observe that both log 3.25 and log .0325 have the same mantissa, which is .5119. However, the characteristic of log .0325 is 2 less than the characteristic of log 3.25 due to the division of 3.25 by 100.

The following table shows the results obtained when 3.25 is multiplied or divided by 100, as above, and also when 3.25 is multiplied or divided by 10.

Logarithm	Characteristic	Mantissa
log (3.25 × 100) = log 325 = 2.5119	2	.5119
log (3.25 × 10) = log 32.5 = 1.5119	1·	.5119
log 3.25 = 0.5119	0	.5119
log(3.25 ÷ 10) = log .325 = .5119 − 1	−1	.5119
log(3.25 ÷ 100) = log .0325 = .5119 − 2	−2	.5119

Observe in the table that the effect of multiplying a number by an integral power of 10 is to change the characteristic of its logarithm, but that the mantissa remains unchanged. Note also that multiplying a number by an integral power of 10 changes the position of the decimal point in the number, but the significant digits of the number remain unchanged. For example, 3, 2, and 5, the significant digits of 3.25, are unchanged.

On a graphing or scientific calculator, the LOG key is used to display the common logarithm of a number. Depending on the calculator, the LOG key may need to be pressed either before or after the number is entered. On a calculator, the logarithm of a number less than 1 is displayed with a negative mantissa. For example,

Method 1:

    Enter: [LOG] .0325 [ )] [ENTER]

Method 2:

    Enter: .0325 [LOG]

    Display: [- 1.488116639]

The value of log .0325 is approximately −1.4881. To write the logarithm with a positive mantissa, add and subtract 10.

    (−1.4881 + 10) − 10 = 8.5119 − 10.

The characteristic of the logarithm is 8 − 10 or −2 and the mantissa is .5119. This result agrees with what was found above. In the Model Problems, calculator *Method* 1 will be used.

MODEL PROBLEMS

1. *a.* Use a calculator to find the logarithm of 37.8 rounded to four decimal places.

    b. Use the result from part *a* to find log 37,800.

Solution:

    a. *Enter:* [LOG] 37.8 [ )] [ENTER]

        Display: [1.5774918]

Answer: log 37.8 is approximately 1.5775.

b. $37,800 = 37.8 \times 10^3$

Increase the logarithm from part *a* by 3:

$1.5775 + 3 = 4.5775$

Answer: log 37,800 = 4.5775

2. Find log .9. Write the result with a positive mantissa and identify the characteristic and mantissa.

Solution:

Enter: $\boxed{\text{LOG}}$ .9 $\boxed{\;)\;}$ $\boxed{\text{ENTER}}$

Display: $\boxed{\text{-.0457574906}}$

The value of log .9 is approximately –.0458.

Add and subtract 10.

$(-.0458 + 10) - 10 = 9.9542 - 10$

Answer: log .9 = 9.9542 – 10. The characteristic is –1 and the mantissa is .9542.

3. If $10^x = 5.41$, find x rounded to four decimal places.

Solution: Transform $10^x = 5.41$ into the logarithmic form log 5.41 = x.

log 5.41 = .7332 (rounded)

Answer: x = .7332

Exercises

In 1–15, use a calculator to find the logarithm of the given number.

1. 732	**2.** 546	**3.** 6.38	**4.** 400	**5.** 94
6. 8.7	**7.** 5	**8.** 4950	**9.** 72,600	**10.** .137
11. .00643	**12.** .0654	**13.** .2	**14.** .05	**15.** .0038

In 16–19, if the log 37.8 = 1.5775, state the log of the given number without using a calculator.

16. 3780	**17.** .378	**18.** 3.78	**19.** .000378

In 20–23, if the log $900 = 2.9542$, state the log of the given number without using a calculator.

20. 90   **21.** 9   **22.** .009   **23.** .09

In 24–26, if log $N = 2.3456$, find the indicated logarithm.

24. $\log 100\, N$ **25.** $\log 10{,}000\, N$ **26.** $\log \dfrac{N}{100}$

In 27–30, if $10^{0.4771} = 3$, find the value of the given powers of 10.

27. $10^{2.4771}$ **28.** $10^{4.4771}$ **29.** $10^{.4771-2}$ **30.** $10^{7.4771-10}$

In 31–33, solve the given equation for x, correct to four decimal places.

31. $10^x = 75.24$ **32.** $10^x = 180.6$ **33.** $10^x = 5.632$

12. Finding a Number Whose Logarithm Is Given: Finding the Antilogarithm

In general, if $a = \log N$, then N is the **antilogarithm** of a. That is, N is the number whose logarithm is a. For example, if $2.5119 = \log 325$, then 325 is the antilogarithm of 2.5119.

The process of finding a number whose logarithm is given is called *finding the antilogarithm*. To find the antilogarithm with a calculator, use the $\boxed{10^x}$ key and one of the following methods.

Method 1:

   Enter: $\boxed{\text{2nd}}$ $\boxed{10^x}$ 2.5119 $\boxed{\ )\ }$ $\boxed{\text{ENTER}}$

Method 2:

   Enter: 2.5119 $\boxed{\text{2nd}}$ $\boxed{10^x}$

   Display: $\boxed{325.0124519}$

To the *nearest integer*, $N = 325$.

Determine the method that is right for your calculator. In this book, *Method* 1 will be used.

———————————— *MODEL PROBLEMS* ————————————

1. Find the antilogarithm of 4.6759 to the nearest integer.

Solution: Use the calculator.

   Enter: $\boxed{\text{2nd}}$ $\boxed{10^x}$ 4.6759 $\boxed{\ )\ }$ $\boxed{\text{ENTER}}$

Display: $\boxed{47413.27996}$

Answer: 47,413

Note. Since log N = 4.6759 becomes log 47,413 = 4.6759, it follows that $10^{4.6759}$ = 47,413. A sense of estimation verifies this answer because $10^4 < 10^{4.6759} < 10^5$ or 10,000 $< 10^{4.6759} <$ 100,000.

2. If log n = 8.9367 − 10, find n.

Solution:

Enter: $\boxed{\text{2nd}}$ $\boxed{10^x}$ 8.9367 $\boxed{-}$ 10 $\boxed{)}$ $\boxed{\text{ENTER}}$

Display: $\boxed{.0864370626}$

Answer: n is approximately .08644, to four significant digits.

Note that 8.9367 − 10 = −1.0633. To check:

Enter: $\boxed{\text{LOG}}$ .08644 $\boxed{)}$ $\boxed{\text{ENTER}}$

Display: $\boxed{.-1.063285242}$

So, log .08644 = −1.0633.

~~~~~~

Exercises

In 1−6, if the log 5.9 = 0.7709, state the number whose log is:

1. 2.7709   **2.** 5.7709   **3.** 8.7709−10
4. 7.7709−10   **5.** .7709−1   **6.** 5.7709−10

In 7−21, use a calculator to find the number whose logarithm is given, correct to four significant digits.

7. 2.8162   **8.** 1.7767   **9.** 0.5391
10. 3.2900   **11.** 0.6532   **12.** 9.3962−10
13. −1.4724   **14.** 1.8612   **15.** 8.9579−10
16. 0.7303   **17.** −.2180   **18.** 3.8558
19. 1.8043   **20.** 0.0248   **21.** −1.9235

In 22−24, find the antilogarithm of:

22. 2.4752   **23.** 3.4060   **24.** 9.7353−10

In 25–27, find n if log n equals:

25. 1.7718 **26.** 0.6732 **27.** −.3209

28. Find, to the nearest tenth, the number whose logarithm is 2.6687.
29. Find, to the nearest hundredth, the antilogarithm of 1.7060.
30. Find, to the nearest thousandth, the antilogarithm of 0.7111.
31. Find the four-place decimal whose logarithm is 9.6060 − 10.
32. If $x = 10^{2.6609}$, find x. **33.** If $n = 10^{1.9540}$, find n.

In 34–37, if the antilogarithm of 2.5977 is 396, state the antilogarithm of:

34. 0.5977 **35.** 4.5977 **36.** .5977 − 2 **37.** −2.4023

In 38–40, if $10^{0.6487} = 4.453$, find:

38. $10^{3.6487}$ **39.** $10^{.6487-3}$ **40.** $10^{-1.3513}$

13. Laws of Logarithms: Using Logarithms To Find Products

Since logarithms are exponents, each of the following laws of logarithms follows from a corresponding law of exponents. In these laws, A and B are positive real numbers.

Name of Logarithmic Law	Statement of Law
logarithm of a product law	$\log (A \times B) = \log A + \log B$
logarithm of a quotient law	$\log \dfrac{A}{B} = \log A - \log B$
logarithm of a power law	$\log A^b = b \log A$
logarithm of a root law	$\log \sqrt[b]{A} = \log A^{\frac{1}{b}} = \dfrac{\log A}{b}$ or $\dfrac{1}{b} \log A$

Beginning with this unit, a separate unit will be devoted to each of the four logarithmic laws. In this unit, we consider the logarithm of a product law.

LOGARITHM OF A PRODUCT LAW

$$\log (A \times B) = \log A + \log B$$

EXTENSION OF THE LOGARITHM OF A PRODUCT LAW

$$\log (A \times B \times C \times D \ldots) = \log A + \log B + \log C + \log D + \ldots$$

Rule. The logarithm of a product is equal to the sum of the logarithms of its factors.

Thus, $\log \frac{1}{2}bh = \log .5 + \log b + \log h$.

Proving the Logarithm of a Product Law

Prove: $\log(A \times B) = \log A + \log B$

1. Let $\quad \log A = x$ and $\log B = y$.
2. Then $\quad A = 10^x \qquad$ Changing $\log A = x$ to exponential form.
3. And $\quad B = 10^y \qquad$ Changing $\log B = y$ to exponential form.
4. $\quad A \times B = 10^{x+y} \qquad$ Multiplying powers with the same base.
5. $\log (A \times B) = x + y \qquad$ Changing $A \times B = 10^{x+y}$ to logarithmic form.
6. $\log (A \times B) = \log A + \log B \quad$ Substituting.

〰〰〰〰〰〰 *MODEL PROBLEM* 〰〰〰〰〰〰

Using logarithms, find the value of 32.5×14.

How To Proceed	*Solution*
1. Let the product equal N.	$N = 32.5 \times 14$
2. Apply the logarithm of a product law.	$\log N = \log 32.5 + \log 14$
3. Find the needed logarithms and do the operations indicated in step 2.	$\log 32.5 = 1.5119$ $+ \log 14 = 1.1461$ $\overline{\log N = 2.6580}$
4. Find the antilogarithm of N.	$N = 455 \quad$ *Ans.*

〰〰〰〰〰〰〰〰〰〰〰〰〰〰〰〰〰〰〰〰〰〰〰〰〰〰〰〰〰

Using the product law, we can discover the important relationship that exists between the exponent of 10 when a number is written in scientific notation and the characteristic of the logarithm of a number.

If we know that, to the nearest ten-thousandth, $\log 2.53 = 0.4031$, then we can write the following:

$$\log 25.3 = \log (2.53 \times 10^1) = \log 2.53 + \log 10^1 = 0.4031 + 1 = 1.4031$$
$$\log 253 = \log (2.53 \times 10^2) = \log 2.53 + \log 10^2 = 0.4031 + 2 = 2.4031$$
$$\log 2{,}530 = \log (2.53 \times 10^3) = \log 2.53 + \log 10^3 = 0.4031 + 3 = 3.4031$$

In general:

If $1 < a < 10$, then $0 < \log a < 1$ and $\log (a \times 10^n) = n + \log a$.

Since $1 < a < 10$, $a \times 10^n$ represents a number written in scientific notation. In each case given, the mantissa of $\log (a \times 10^n)$ is $\log a$ and the characteristic is n, the exponent of 10 when the number is written in scientific notation.

Exercises

In 1–9, find the value of the product by using logarithms.

1. 15.6×5.84 2. 59×3.14 3. 550×1.41
4. 75.25×3.65 5. $65 \times .4245$ 6. $.875 \times .9063$
7. $8.16 \times 1.72 \times 3.14$ 8. $2.7 \times 5.6 \times .4384$ 9. $162 \times 53.6 \times .2391$

10. $C = 2\pi r$. Find C when $\pi = 3.14$, $r = 46.8$.
11. If $\log x = 1.4814$, find $\log 10x$.
12. If $\log y = 8.3010 - 10$, find $\log 1000y$.
13. Find the value of $\log 100y - \log y$.
14. If $\log b = x$, then $\log 100b$ equals (1) $100x$ (2) $2x$ (3) $x + 2$
15. If $\log 4.72$ equals m, then $\log 472$ equals (1) $100m$ (2) $2m$ (3) $m + 2$
16. If $\log x = \log a + \log b$, express x in terms of a and b.
17. If $\log y = \log r + \log s + \log t$, express y in terms of r, s, and t.
18. If $\log n = a$, then $\log np$ equals (1) ap (2) $a + p$ (3) $a + \log p$

In 19–22, if $\log 2 = 0.3010$, $\log 3 = 0.4771$, and $\log 6 = 0.7782$, find the indicated logarithm using the product law.

19. $\log 12$ 20. $\log 18$ 21. $\log 36$ 22. $\log 3600$

14. Using Logarithms To Find Quotients

LOGARITHM OF A QUOTIENT LAW

$$\log \frac{A}{B} = \log A - \log B$$

Rule. The logarithm of the quotient of two numbers is equal to the logarithm of the dividend minus the logarithm of the divisor. If the quotient is in fraction form, the logarithm of a fraction is equal to the logarithm of the numerator minus the logarithm of the denominator.

Thus, if $H = \dfrac{V}{B}$, then $\log H = \log V - \log B$.

Proving the Logarithm of a Quotient Law

Prove: $\log \dfrac{A}{B} = \log A - \log B$

1. Let $\log A = x$ and $\log B = y$.
2. Then $A = 10^x$ Changing $\log A = x$ to exponential form.
3. And $B = 10^y$ Changing $\log B = y$ to exponential form.

4. $\dfrac{A}{B} = 10^{x-y}$ Dividing powers with the same base.

5. $\log \dfrac{A}{B} = x - y$ Changing $\dfrac{A}{B} = 10^{x-y}$ to logarithmic form.

6. $\log \dfrac{A}{B} = \log A - \log B$ Substituting.

～～～～～ *MODEL PROBLEMS* ～～～～～

1. Using logarithms, find the value of $690 \div 3.75$.

How To Proceed	*Solution*
1. Let the quotient equal N.	$N = 690 \div 3.75$
2. Apply the logarithm of a quotient law.	$\log N = \log 690 - \log 3.75$
3. Find the needed logarithms and do the operations indicated in step 2.	$\log 690 = 2.8388$ $-\log 3.75 = 0.5740$ $\overline{\log N = 2.2648}$
4. Find the antilogarithm of N.	$N = 184$ *Ans.*

2. Find the value of $\dfrac{38.4 \times 6.78}{536 \times .471}$, correct to the nearest hundredth.

Solution: $N = \dfrac{38.4 \times 6.78}{536 \times .471}$

$\log N = \log \text{numerator} - \log \text{denominator}$

$\log N = (\log 38.4 + \log 6.78) - (\log 536 + \log .471)$

$\begin{array}{ll} \log 38.4 = 1.5843 & \log 536 = 2.7292 \\ +\log 6.78 = 0.8312 & +\log .471 = -.3270 \\ \hline \log \text{numerator} = 2.4155 & \log \text{denominator} = 2.4022 \\ -\log \text{denominator} = 2.4022 & \end{array}$

$\log N = 0.0133$

$N = 1.03$ *Ans.*

3. The expression $2 - \log a$ is equivalent to (1) $\log \dfrac{100}{a}$ (2) $\dfrac{2}{\log a}$ (3) $\log \dfrac{2}{a}$ (4) $\sqrt{\log a}$

Solution:

Since $2 = \log 100$, then the expression $2 - \log a$ is equivalent to the expression $\log 100 - \log a$.

If the logarithm of a quotient law is applied, then $\log 100 - \log a = \log \dfrac{100}{a}$.

Answer: The correct choice is (1).

Exercises

In 1–12, find the value, using logarithms.

1. $594 \div 237$ **2.** $9.63 \div 4.56$ **3.** $8.29 \div .974$

4. $.459 \div .84$ **5.** $756.8 \div 49.7$ **6.** $9.608 \div .746$

7. $\dfrac{.537}{.0469}$ **8.** $\dfrac{3.05}{.00008}$ **9.** $\dfrac{.06814}{.000053}$

10. $\dfrac{8.34 \times 74.8}{53.9}$ **11.** $\dfrac{46.35}{.82 \times 980}$ **12.** $\dfrac{.85 \times 1300}{160 \times .056}$

13. If $\log n = \log a - \log b$, express n in terms of a and b.

14. If $\log n = \log r + \log s - \log t$, express n in terms of r, s, and t.

In 15–18, if $\log 2 = 0.3010$, $\log 5 = 0.6990$, and $\log 11 = 1.0414$, find the indicated logarithm using the quotient law.

15. $\log 2.5$ **16.** $\log 5.5$ **17.** $\log 2.2$ **18.** $\log \dfrac{50}{11}$

15. Using Logarithms To Find Powers

LOGARITHM OF A POWER LAW

$$\log A^b = b \log A$$

Rule. The logarithm of a power of any number is equal to the logarithm of the number multiplied by the exponent of the power.

Thus, if $A = 1.05^{20}$, then $\log A = 20 \log 1.05$.

Proving the Logarithm of a Power Law

Prove: $\log A^b = b \log A$

1. Let $\log A = x$.
2. Then $A = 10^x$ Changing $\log A = x$ to exponential form.
3. $A^b = (10^x)^b$ Raising both sides of (2) to the bth power.
4. $A^b = 10^{bx}$ Power of a power rule.
5. $\log A^b = bx$ Changing $A^b = 10^{bx}$ to logarithmic form.
6. $\log A^b = b \log A$ Substituting.

~~~~~~~~~~ *MODEL PROBLEMS* ~~~~~~~~~~

1. Using logarithms, find 1.02^{20}, correct to the nearest hundredth.

How To Proceed	*Solution*
1. Let the power equal N.	$N = 1.02^{20}$
2. Apply the logarithm of a power law.	$\log N = 20 \log 1.02$
3. Find the needed logarithm and do the operation indicated in step 2.	$\log 1.02 = 0.0086$
	$\times 20$
	$\log N = \overline{0.1720}$
4. Find the antilogarithm of N.	$N = 1.49$ *Ans.*

2. If $\log x = n$, then $\log \dfrac{x^3}{10}$ is (1) $3n - 1$ (2) $\dfrac{3n}{10}$ (3) $\dfrac{n^3}{10}$ (4) $3 \log n - 1$

Solution:

First express $\log \dfrac{x^3}{10}$ in terms of $\log x$.

$$\log \frac{x^3}{10} = \log x^3 - \log 10 \qquad \text{(logarithm of a quotient law)}$$

$$= 3 \log x - \log 10 \qquad \text{(logarithm of a power law)}$$

Substitute n for log x and 1 for log 10.

$$\log \frac{x^3}{10} = 3n - 1$$

Answer: The correct choice is (1).

~~~~~~

Exercises

In 1–11, find the value using logarithms.
1. $(21)^2$ 2. $(1.04)^{20}$ 3. $(1.025)^{10}$ 4. $(.896)^2$ 5. $(.075)^3$
6. $3.14 \times (9.25)^2$ 7. $16 \times (2.5)^2$ 8. $4 \times 3.14 \times (6.5)^2$
9. $5000(1.02)^{10}$ 10. $4000(1.06)^{15}$ 11. $3500(1.015)^{20}$

12. If log $x = 0.5692$, find log x^2 and log x^3.
13. If log $x^3 = 3.3624$, find log x, log x^2, and log x^4.
14. If log $x = -.6990$, find log x^2 and log $100x^3$.
15. If log $x^2 = 0.8762$, then log $10x$ is (1) 4.3810   (2) 1.4381   (3) 1.7524

16. Express $\log \frac{u^-}{b}$ in terms of log a and log b.

17. If $50 = 10^{1.6990}$, find log $(50)^2$.
18. If $x = 10^{1.6990}$, find log x^2.

In 19–21, if log $a = 2.3000$ and log $b = 1.7000$, find the value of the indicated logarithm using the power law.

19. $\log \frac{a}{b}$ 20. $\log \frac{10a}{b^2}$ 21. $\log \frac{a^3}{1000b}$

In 22–24, if log $a = r$ and log $b = s$, express the indicated logarithm in terms of r and s.

22. $\log \frac{a^3}{b^2}$ 23. $\log \frac{a^2}{b^3}$ 24. $\log \frac{10a}{b^2}$

25. If log $a = p$, express log $10a^2$ in terms of p.
26. If log $n = 4$ log $r -$ log s, express n in terms of r and s.
27. If log $n = $ log $4 + $ log $\pi + 2$ log r, express n in terms of π and r.
28. If log $s = $ log $g + 2$ log $t - $ log 2, express s in terms of g and t.
29. If log $n - 3$ log $x = $ log y, express n in terms of x and y.
30. The expression log $r + $ log r^2 is equal to (1) $\log(r + r^2)$ (2) 3 log r (3) r^3

In 31–33, if log $y = a$, express the indicated logarithm in terms of a.

31. log $\dfrac{y^2}{10}$ **32.** log $\dfrac{100}{y^2}$ **33.** log $1000y^3$

34. If $T = 10x^2$, then log T equals (1) $1 + 2$ log x (2) $1 + 2x$ (3) $10 + 2$ log x
(4) 20 log x

16. Using Logarithms To Find Roots

LOGARITHM OF A ROOT LAW

$$\log \sqrt[b]{A} = \log A^{\frac{1}{b}} = \frac{1}{b} \log A = \frac{\log A}{b}$$

Rule. The logarithm of the root of a number is equal to the logarithm of the number divided by the index of the root.

Thus, if $A = \sqrt[3]{100}$, then log $A = \dfrac{\log 100}{3} = \dfrac{2}{3}$.

Proving the Logarithm of a Root Law

Prove: $\log \sqrt[b]{A} = \dfrac{\log A}{b}$

1. Let log $A = x$.
2. Then $A = 10^x$ Changing log $A = x$ to exponential form.

3. $A^{\frac{1}{b}} = (10^x)^{\frac{1}{b}}$ Finding the bth root of both sides of (2).

4. $A^{\frac{1}{b}} = 10^{\frac{x}{b}}$ Power of a power rule.

5. $\log A^{\frac{1}{b}} = \dfrac{x}{b}$ Changing $A^{\frac{1}{b}} = 10^{\frac{x}{b}}$ to logarithmic form.

6. $\log \sqrt[b]{A} = \dfrac{\log A}{b}$ Substituting.

MODEL PROBLEMS

1. Using logarithms, find $\sqrt[3]{.358}$, correct to the nearest hundredth.

How To Proceed	*Solution*
1. Let the root equal N.	$N = \sqrt[3]{.358}$
2. Apply the logarithm of a root law.	$\log N = \frac{1}{3} \log .358$
3. Find the needed logarithm and do the operation indicated in step 2.	$\log N = \frac{1}{3}(-.4461)$
	$\log N = -.1487$
4. Find the antilogarithm of N.	$N = .71$ *Ans.*

2. Express $\log \dfrac{a^2}{\sqrt[3]{b}}$ in terms of $\log a$ and $\log b$.

Solution:

$$\log \frac{a^2}{\sqrt[3]{b}} = \log \frac{a^2}{b^{\frac{1}{3}}} = \log a^2 - \log b^{\frac{1}{3}} \qquad \text{(logarithm of a quotient law)}$$

$$= 2 \log a - \frac{1}{3} \log b \quad \textit{Ans.} \qquad \begin{array}{l}\text{(logarithm of a power law; logarithm}\\ \text{of a root law)}\end{array}$$

Exercises

In 1–8, find the value, using logarithms.

1. $\sqrt{625}$ **2.** $\sqrt[3]{59.3}$ **3.** $\sqrt{.144}$ **4.** $\sqrt[3]{.476}$

5. $\dfrac{\sqrt[3]{536}}{3.2}$ **6.** $\dfrac{\sqrt{1764}}{14.8}$ **7.** $\dfrac{137}{\sqrt[3]{.8964}}$ **8.** $\sqrt{\dfrac{9800}{76.3}}$

9. If $\log x = 2.7186$, find (a) $\log \sqrt{x}$,   (b) $\log \sqrt[3]{x}$.
10. If $\log x = -.6988$, find (a) $\log \sqrt{x}$,   (b) $\log \sqrt[3]{x}$.
11. If $\log \sqrt{x} = 0.1526$, find (a) $\log x$,   (b) $\log x^2$.
12. If $50 = 10^{1.6990}$, find $\log \sqrt{50}$.   **13.** If $x = 10^{1.6990}$, find $\log \sqrt{x}$.
14. Find $\log \sqrt{10^{1.6990}}$.

15. If $\log x = a$ and $\log y = b$, express $\log \sqrt{xy}$ in terms of a and b.

16. Express the logarithm of $\dfrac{\sqrt[3]{a}}{b}$ in terms of $\log a$ and $\log b$.

17. Express $\log \dfrac{a}{\sqrt{b}}$ in terms of $\log a$ and $\log b$.

18. If $\log n = \dfrac{\log a + \log b - \log c}{3}$, express n in terms of a, b, and c.

In 19–21, if $\log n = a$, express the indicated logarithm in terms of a.

19. $\log \sqrt[3]{\dfrac{1000}{n}}$ **20.** $\log \sqrt[3]{1000n}$ **21.** $\log \sqrt[3]{100n^2}$

17. Applying Logarithmic Laws to More Difficult Problems

~~~~~~~~~~ *MODEL PROBLEM* ~~~~~~~~~~

The velocity u of a bullet in flight is given by the formula $u = kd \sqrt{\dfrac{g}{R}}$.

Using logarithms, find the value of u, correct to three significant digits, if $k = 834$, $d = 19.8$, $g = 980$, and $R = 295$.

Solution:

$$u = kd \sqrt{\dfrac{g}{R}}$$

Substitute the given values.

$$u = 834 \times 19.8 \sqrt{\dfrac{980}{295}}$$

$$\log u = \log 834 + \log 19.8 + \tfrac{1}{2}(\log 980 - \log 295)$$

$\log 834 = 2.9212$	$\log 980 = 2.9912$
$+\log 19.8 = 1.2967$	$-\log 295 = 2.4698$
$\overline{4.2179}$	$2\overline{\smash{)}0.5214}$
$+ 0.2607 \longleftarrow$	0.2607
$\log u = \overline{4.4786}$	
$u = 30{,}100$ *Ans.*	

~~~~~~

Exercises

In 1–8, write the logarithmic equation for the given equation.

1. $C = \pi D$ **2.** $S = 2\pi r h$ **3.** $S = 4\pi r^2$ **4.** $V = \tfrac{1}{3}BH$

5. $V = \tfrac{4}{3}\pi R^3$ **6.** $A = \dfrac{S^2}{4}\sqrt{3}$ **7.** $t = 2\pi\sqrt{\dfrac{l}{g}}$ **8.** $R = \sqrt{\dfrac{3V}{\pi H}}$

9. Match the expressions in Column I with those in Column II.

Column I	*Column II*
1. $\log xy$	$a.\ \log x + 2\log y$
2. $\log \dfrac{x}{y}$	$b.\ \dfrac{\log x + \log y}{2}$
3. $\log xy^2$	$c.\ \tfrac{1}{2}\log x - \log y$
4. $\log x^2 y$	$d.\ \log x + \log y$
5. $\log (xy)^2$	$e.\ \log y + \tfrac{1}{2}\log x$
6. $\log \sqrt{xy}$	$f.\ 2(\log x + \log y)$
7. $\log\sqrt{\dfrac{x}{y}}$	$g.\ \dfrac{\log x + 2\log y}{2}$
8. $\log\sqrt{\dfrac{y}{x}}$	$h.\ \log x - \log y$
9. $\log \dfrac{x}{\sqrt{y}}$	$i.\ \dfrac{\log y - \log x}{2}$
10. $\log \dfrac{\sqrt{x}}{y}$	$j.\ \dfrac{\log x - \log y}{2}$
11. $\log x\sqrt{y}$	$k.\ \dfrac{\log x - 2\log y}{2}$
12. $\log y\sqrt{x}$	$l.\ 2\log x + \log y$
13. $\log \sqrt{xy^2}$	$m.\ \log x + \tfrac{1}{2}\log y$
14. $\log \sqrt{\dfrac{x}{y^2}}$	$n.\ \log x - \tfrac{1}{2}\log y$

In 10–15, find the value, using logarithms.

10. $\dfrac{7.34 \times (87)^2}{155}$ **11.** $\dfrac{83\sqrt{521}}{437}$ **12.** $\dfrac{5.94 \times 86.68}{\sqrt[3]{824}}$

13. $\sqrt[4]{\dfrac{82 \times 61.7}{.016}}$

14. $\sqrt[3]{\dfrac{7.98}{0.586 \times 84}}$

15. $\sqrt{\dfrac{17.4 \times 96.3}{27.8 \times 156}}$

16. Using logarithms, find, to the nearest tenth, the value of L from the formula $L = \dfrac{t^2 g}{4\pi^2}$ when $t = 3.50$, $g = 32.2$, and $\pi = 3.14$.

17. If $V = \frac{1}{3}\pi r^2 h$, find V when $\pi = 3.14$, $r = 11.5$, and $h = 9.3$.

18. The volume, V, of a circular cylinder whose altitude is h and whose radius is r is given by the formula $V = \pi r^2 h$. Using logarithms, find, to the nearest tenth, the value of r if $V = 906$ and $h = 14.6$. (Use $\pi = 3.14$.)

19. Using logarithms, find, to the nearest integer, the value of $\dfrac{3.84^3 \times (1.82)^2}{\sqrt[3]{0.0870}}$.

20. Using logarithms, find, to the nearest hundredth, the value of $\dfrac{762 \times \sqrt[3]{0.364}}{94.4}$.

21. Given the formula $t = \pi \sqrt{\dfrac{L}{g}}$. Using logarithms, find, to the nearest hundredth, the value of t when $L = 1.38$, $g = 32.2$, and $\pi = 3.14$.

In 22–24, find c, using logarithms, if $c = \sqrt{a^2 + b^2}$ and:

22. $a = 23.8$ and $b = 16.2$
23. $a = 350$ and $b = 485$
24. $a = 2340$ and $b = 5764$

18. Using Logarithms To Solve Exponential Equations

Earlier in this chapter, exponential equations were solved by expressing both sides of the equation as powers of the same base. Exponential equations which cannot be solved by this method may be solved using logarithms. The following model problems show how the exponential equations are transformed into logarithmic equations by equating the logarithms of both sides of the given equation. The resulting equation is then solved for the unknown.

~~~~~~~ *MODEL PROBLEMS* ~~~~~~~

1. Solve for x, correct to the nearest tenth.
$$6^x = 19$$

2. Solve for x, correct to the nearest integer.
$$1.01^{2x} = 2$$

How To Proceed	*Solution*	*Solution*
1. Equate the logarithms of both sides of the given equation.	$6^x = 19$ $\log(6^x) = \log 19$	$1.01^{2x} = 2$ $\log(1.01^{2x}) = \log 2$
2. Solve the resulting equation for the unknown.	$x \log 6 = \log 19$ $0.7782x = 1.2788$ $x = \dfrac{1.2788}{.7782}$ $x = 1.6 \quad Ans.$	$2x \log 1.01 = \log 2$ $2x(0.0043) = 0.3010$ $.0086x = .3010$ $x = \dfrac{.3010}{.0086}$ $x = 35 \quad Ans.$

Exercises

In 1–16, solve the equation, giving the answer correct to the nearest tenth.

1. $2^x = 42$
2. $3^y = 15$
3. $5^y = 29$
4. $12^x - 87 = 500$
5. $22^m = 629$
6. $5^y = 37.6$
7. $15^x = 96.3$
8. $1.5^x + .4 = 4$
9. $1.04^x = 2$
10. $1.02^x = 3$
11. $5^{2x} = 72.5$
12. $1.2^{3x} - 25 = 27.25$
13. $1.01^{4x} = 2$
14. $3^{x+1} = 85.6$
15. $5^{3x-1} = 57.9$
16. $4^{3x} = 5^{x+1}$

In 17–19, the formula $A = e^{rt}$ has values of $A = 1.5$ and $e = 2.718$. Find t, to the nearest hundredth, when r has the given value.

17. $r = .01$
18. $r = .025$
19. $r = .0225$

19. Using Logarithms To Solve Problems Involving Compound Interest

If a sum of P dollars is placed at compound interest for a period of n years, interest being compounded annually at the rate of $r\%$, the sum of money to which it will amount at the end of this period of time, the *compound amount*, or *amount*, is given by the formula: $A = P(1 + r)^n$.

If interest is compounded semiannually (twice a year), $A = P\left(1 + \dfrac{r}{2}\right)^{2n}$; if quarterly (four times a year), $A = P\left(1 + \dfrac{r}{4}\right)^{4n}$; if t times a year, $A = P\left(1 + \dfrac{r}{t}\right)^{nt}$.

The interest which the principal has accumulated during the given period of years can be found by finding the difference between the compound amount and the original principal.

In problems, the expression "invested at $r\%$" means "invested at an *annual rate* of $r\%$."

〜〜〜〜〜〜〜 **MODEL PROBLEMS** 〜〜〜〜〜〜〜

1. How much money, to the nearest ten dollars, must be invested at 6% interest, compounded semiannually, to yield $1250 at the end of 10 years?

   Solution:

   $$A = P\left(1 + \frac{r}{2}\right)^{2n}$$

   $A = 1250$, $r = .06$, $n = 10$

   $$1250 = P\left(1 + \frac{.06}{2}\right)^{2 \times 10}$$

   $$1250 = P(1.03)^{20}$$
   $$\log 1250 = \log P + 20 \log 1.03$$
   $$\log P = \log 1250 - 20 \log 1.03$$
   $$\log 1250 = 3.0969$$
   $$-20 \log 1.03 - 0.2560$$
   $$\overline{\log P = 2.8409}$$
   $$P = \$690, \text{ rounded to tens} \quad Ans.$$

2. Find the number of years, correct to the nearest year, it will take $450 to amount to $730 if the principal of $450 is invested at $5\frac{1}{2}\%$, compounded annually.

   Solution:

   $$A = P(1 + r)^n$$

   $A = 730$, $P = 450$, $r = .055$

   $$730 = 450(1.055)^n$$
   $$\log 730 = \log 450 + n \log 1.055$$
   $$n \log(1.055) = \log 730 - \log 450$$
   $$.0233n = 2.8633 - 2.6532$$
   $$.0233n = .2101$$

   $$n = \frac{.2101}{.0233} = 9, \text{ to the nearest integer} \quad Ans.$$

Exercises

In 1–6, find to the nearest dollar the amount and the interest, if interest is compounded annually on a principal of:
1. $250 for 5 years at 7%   2. $375 for 10 years at 6%
3. $600 for 20 years at 4%   4. $1500 for 15 years at 4%
5. $400 for 5 years at $5\frac{1}{2}$%   6. $750 for 10 years at $4\frac{1}{2}$%

In 7–12, find the amount and the interest, if interest is compounded semi-annually on a principal of:
7. $300 for 5 years at 4%   8. $550 for 10 years at 8%
9. $2500 for 6 years at 4%   10. $3000 for 4 years at 6%
11. $900 for 5 years at 6%   12. $1200 for 3 years at 5%

13. Find the principal that must be invested at 3%, interest compounded annually, in order that it amount to $800 in 5 years.
14. Find the principal that must be invested at 7%, interest compounded semi-annually, to yield $10,000 in 10 years.

In 15–19, find, to the nearest year, the time in which:
15. $750 will amount to $900 at 6% compounded annually.
16. $580 will amount to $760 at $5\frac{1}{2}$% compounded annually.
17. $300 will amount to $600 at 4% compounded annually.
18. $400 will double itself at 4% compounded annually.
19. $600 will amount to $910 at 7% compounded semiannually.

In 20–22, use $A = P(1 + r)^n$ and give your answer to the nearest year.
20. In how many years, n, will $570 amount to $965 if interest is compounded annually at 3%?
21. In how many years, n, will $350 amount to $1030 if the money earns interest at the rate of 8% compounded annually?
22. Find n, to the nearest integer, when $r = .04$, $P = 1$, and $A = 3$.

23. For $500 it is possible to buy a bond that will be worth $1000 in 10 years. Using the formula $A = P(1 + r)^n$, find, to the nearest tenth of a per cent, the rate of interest on this investment if interest is compounded annually.

~~~~~~~~ *CALCULATOR APPLICATIONS* ~~~~~~~~

Most graphing calculators will display the result of a computation in scientific notation when the number of digits needed to display the number in ordinary (normal mode) notation exceeds the display capability of the calculator or if the number is very small. In scientific mode, a number entered in ordinary notation is displayed in scientific notation.

~~~~~~~~~~ *MODEL PROBLEMS* ~~~~~~~~~~

1. Write each number in scientific notation:

    362,000,000   .0000159

    Solution: Set the calculator to Sci mode.

    Enter: | MODE | | ▶ | | ENTER | | CLEAR | 362000000 | ENTER | .0000159 | ENTER |

    Display:
3.62 E 8
1.59 E -5

    Answer: 3.62×10^8 and 1.59×10^{-5}.

2. Express $\dfrac{(7.5 \times 10^5) \times (2.8 \times 10^{-8})}{1.5 \times 10^{-5}}$ as a single number in scientific notation.

    Solution: Use Sci mode to be sure the answer will be displayed in scientific notation. Otherwise, the answer may appear in ordinary notation and need to be converted. Numbers can be entered in scientific notation using the | EE | key, the | ∧ | key, or the | 10ˣ | key. Two methods are shown below.

    Method 1:

    Enter: | MODE | | ▶ | 7.5 | 2nd | | EE | 5 | × | 2.8 | 2nd | | EE | | (−) | 8
    | ENTER | | ÷ | 1.5 | 2nd | | EE | | (−) | 5 | ENTER |

    Method 2:

    Enter: | MODE | | ▶ | 7.5 | × | 10 | ∧ | 5 | × | 2.8 | × | 10 | ∧ | | (−) | 8
    | ÷ | | (| 1.5 | × | 10 | ∧ | | (−) | 5 |) | | ENTER |

    Display: | 1.4 E 3 |

    Answer: 1.4×10^3

On a graphing calculator we can draw the graphs of several exponential functions and compare them.

3. Graph 8^x, 3^x, 2^x, and 1.5^x.

    Solution: Set the ⌊WINDOW⌋ for Xmin = −3, Xmax = 3, Xscl = 1, Ymin = −1, Ymax = 9, and Yscl = 1.

    Enter: ⌊Y=⌋ 8 ⌊∧⌋ ⌊X,T,θ,n⌋ ⌊ENTER⌋

               3 ⌊∧⌋ ⌊X,T,θ,n⌋ ⌊ENTER⌋

               2 ⌊∧⌋ ⌊X,T,θ,n⌋ ⌊ENTER⌋

            1.5 ⌊∧⌋ ⌊X,T,θ,n⌋ ⌊ENTER⌋ ⌊GRAPH⌋

    Display:

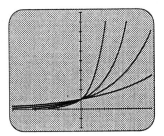

The calculator display shows us that each graph intersects the *y*-axis at (0, 1) and that the larger the value of *b*, the more rapidly the graph rises as the positive values of *x* increase. For negative values of *x*, as the value of *x* decreases, each graph approaches but does not intersect the *x*-axis.

If you use the ⌊LOG⌋ key on the calculator and graph $y = \log x$, then the graph displayed is that of $y = \log_{10} x$. To graph a logarithmic function for a base other than 10, a special formula that is not proved in this course must be used:

$$\log_a x = \frac{\log_{10} x}{\log_{10} a}$$

4. Graph $y = \log_2 x$.

    Solution: In the above formula, substitute 2 for *a* to get

$$\log_2 x = \frac{\log_{10} x}{\log_{10} 2}$$

Use the standard ⌊WINDOW⌋.

    Enter: ⌊Y=⌋ ⌊LOG⌋ ⌊X,T,θ,n⌋ ⌊)⌋ ⌊÷⌋ ⌊LOG⌋ 2 ⌊)⌋ ⌊GRAPH⌋

    Display:

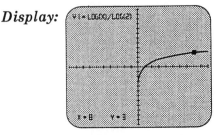

To verify that $\log_2 8 = 3$, enter ⌊2nd⌋ ⌊CALC⌋ 1 8 ⌊ENTER⌋.

CHAPTER XIII

Types of Variation

In solving problems in mathematics and science, we very often apply formulas that have the same general structure, $z = xy$, a relationship among three variables in which one variable is the product of the other two.

Mathematical Formula	Scientific Formula	
Motion formula: $D = RT$	Density $\times$ Volume $=$ Mass	$DV = M$
Interest formula: $I = PR$	Height $\times$ Density $=$ Pressure	$HD = P$
Cost formula: $C = NP$	Force $\times$ Distance $=$ Work	$FD = W$
Area formula: $A = LW$	Weight $\times$ Distance $=$ Moment	$WD = M$
Percentage formula: $P = BR$	Current $\times$ Resistance $=$ Voltage	$IR = V$

In this chapter, we shall study the manner in which the three basic types of variation are associated with formulas having the structure of $z = xy$. The basic types are *direct variation*, *inverse variation*, and *joint variation*.

1. Understanding Direct Variation: $y = kx$ or $\frac{y}{x} = k$

The values in the table at the right are those of the perimeters (p) and sides (s) of four squares, I, II, III, and IV. From the four pairs of corresponding values of p and s, we note that $\frac{p}{s} = \frac{4}{1} = \frac{8}{2} = \frac{24}{6} = \frac{240}{60}$. For the four squares, the formula $p = 4s$ states the relationships between the perimeter and a side of any one of the squares. By a transformation, $p = 4s$ becomes

VALUES OF SIDES
AND PERIMETERS
OF SQUARES

	I	II	III	IV
p	4	8	24	240
s	1	2	6	60

$\frac{p}{s} = 4$. The equations $p = 4s$ and $\frac{p}{s} = 4$ illustrate **direct variation**. If, instead of squares, we were to use equilateral triangles, our equations would be $p = 3s$ and $\frac{p}{s} = 3$; for regular hexagons, our equations would be $p = 6s$ and $\frac{p}{s} = 6$.

EXPRESSING DIRECT VARIATION IN EQUATION FORM BY USING A CONSTANT OF VARIATION

If $y = kx$ where x and y are variables and k is a constant, then we say that x and y vary *directly* as each other; that is, x varies directly as y, or y varies directly as x. The constant, k, $k \neq 0$, is called the **constant of variation.**

Note that k cannot have a value of 0. In this chapter, in all cases, assume that the constant of variation does not equal 0.

Thus, if $p = 4s$, then p and s vary directly as each other and the constant of variation is 4. In a motion problem, when there is a uniform rate of 30 miles per hour, then $D = 30T$. If $D = 30T$, then D and T vary directly as each other and 30 is the constant of variation.

If $y = kx$, then $\dfrac{y}{x} = k$. The equation $\dfrac{y}{x} = k$ also expresses direct variation.

The equation $\dfrac{y}{x} = k$ is exemplified in $\dfrac{p}{s} = 6$, $\dfrac{D}{T} = 30$, and $\dfrac{A}{L} = 35$. In this form of the equation, the constant of variation stands alone on one side and the two variables are on the other.

KEEP IN MIND

If $y = kx$ or $\dfrac{y}{x} = k$, where x and y are variables and k

is the constant, then x and y vary directly as each other and k is the constant of variation.

EXPRESSING DIRECT VARIATION IN PROPORTION FORM

Direct variation is expressible in proportion form. In the figure, Square I and Square II are any two squares whose perimeters are p_1 and p_2 and whose corresponding sides are s_1 and s_2. Since the ratio of the perimeter of a square to its side is 4, then $\dfrac{p_2}{s_2} = \dfrac{p_1}{s_1}$. This proportion may be transformed

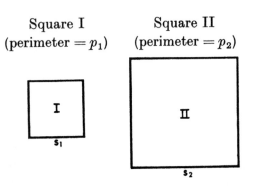

Square I
(perimeter $= p_1$)

Square II
(perimeter $= p_2$)

into $\dfrac{p_2}{p_1} = \dfrac{s_2}{s_1}$. In this form, we note that *the ratio of the perimeters directly equals the ratio of the corresponding sides.*

In general, if x and y vary directly as each other, then when x varies from x_1 to x_2 and y varies from y_1 to y_2, $\dfrac{x_2}{x_1} = \dfrac{y_2}{y_1}$.

Thus, in the table at the right, if two travelers A and B are traveling at 20 mph, A for 6 hours and B for 4 hours, the ratio of their times of travel, $\frac{6}{4}$, equals the ratio of their distances traveled, $\frac{120}{80}$. Note in the table that the time-ratio and the distance-ratio, each equal to $\frac{3}{2}$, have been placed alongside the values of time and distance.

TRAVELERS TRAVELING AT A CONSTANT RATE OF SPEED

| | (mph) | (hr.) | (mi.) |
	Rate $\times$	Time $=$	Distance
A	20	$6\big)_{\frac{3}{2}}$	$120\big)_{\frac{3}{2}}$
B	20	$4\big)$	$80\big)$

If two travelers are traveling at the same rate of 20 miles per hour, then $D = 20T$ and D and T vary directly as each other. The constant of variation is 20.

MULTIPLICATION AND DIVISION OPERATIONS IN DIRECT VARIATION

If a man travels at a constant rate of speed, then when he travels *twice* as long, he will cover a distance *twice* as far. If the length of a rectangle is constant, dividing the width by 3, or taking a width that is one-third as long, will result in a rectangle whose area is one-third as great.

In general, if y varies directly as x:

(1) When x is *multiplied* by a number, y will be *multiplied* by the same number.

(2) When x is *divided* by a nonzero number, y will be *divided* by the same number.

THE GRAPH OF DIRECT VARIATION

If x and y vary directly as each other and k is the constant of variation, then $y = kx$. Recall that the graph of $y = kx$ is a straight line passing through the origin. In the figure are shown the graphs of $y = 4x$ and $y = -4x$, using 4 and -4 as constants of variation. Note in the case of $y = -4x$ that y and x vary directly; yet, when x increases, y decreases.

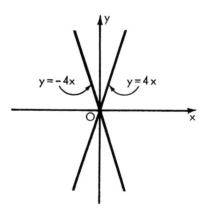

─── KEEP IN MIND ───

If y varies directly as x, then:

1. In equation form, $y = kx$ or $\dfrac{y}{x} = k$, where k is the constant of variation.

2. In proportion form, $\dfrac{y_2}{y_1} = \dfrac{x_2}{x_1}$ when x varies from x_1 to x_2 and y varies from y_1 to y_2.

3. Whenever x is multiplied by a number, y is multiplied by the same number.

4. Whenever x is divided by a nonzero number, y is divided by the same number.

5. The graph of the equation $y = kx$ is a straight line passing through the origin.

﹏﹏﹏ MODEL PROBLEMS ﹏﹏﹏

1. *a.* If $D = RT$, when will D vary directly as R?
   b. If $V = IR$, when will V vary directly as R?

Solution:
   a. D will vary directly as R when T is constant.
   b. V will vary directly as R when I is constant.

2. If y varies directly as x and if $y = 15$ when $x = 3$, find y when $x = 9$.

Solution:

Method 1	Method 2
($y = kx$ method)	(direct proportion method)
Since y varies directly as x,	$\dfrac{y_2}{y_1} = \dfrac{x_2}{x_1}$
$y = kx$	
If $y = 15$ when $x = 3$, $15 = 3k$.	$\dfrac{y}{15} = \dfrac{9}{3}$
Hence, k, the constant of variation, equals 5, and $y = 5x$. Therefore, when $x = 9$, $y = 5(9) = 45$. *Ans.*	$3y = 135$ $y = 45$ *Ans.*

	y	x
2nd values	y	9
1st values	15	3

3. *a.* If a man triples his rate of speed, how many times as far will he go if his time of travel is the same?
   b. If the length of a rectangle of constant width is divided by 5, what happens to the area of the rectangle?

Solution:

a. If $D = RT$ and T is constant, then D varies directly as R. Hence, if the rate is multiplied by 3, the distance is also multiplied by 3. *Answer:* 3 times as far

b. If $A = LW$ and W is constant, then A varies directly as L. Hence, if the length is divided by 5, the area is divided by 5. *Answer:* The area is divided by 5.

Exercises

In 1–3, show that one variable in the table varies directly as the other and write an equation or formula that expresses the relationship between the variables.

1.
y	2	4	6
x	1	2	3

2.
C	5	10	15
n	1	2	3

3.
A	5	10	15
h	2	4	6

In 4–7, in the table, the ratio $\dfrac{y}{x}$ is constant. Find the missing numbers.

4.
y	3	12	?	27
x	1	4	5	?

5.
y	7	14	?	56
x	2	4	8	?

6.
y	-5	-10	?	-25
x	-1	-2	-3	?

7.
y	-8	-12	?	20
x	2	3	4	?

In 8–12, write an equation or formula that expresses the relationship between the variables, using k as the constant of variation.

8. The circumference of a circle, C, varies directly as the diameter, D.
9. The perimeter of an equilateral triangle, P, varies directly as a side, s.
10. The resistance, R, of a copper wire varies directly as its length, l.
11. The weight, W, of a circular pipe is directly proportional to its length, l.
12. Under certain conditions, the volume, V, of a gas varies directly as the absolute temperature, T.

13. If y varies directly as x and $y = 50$ when $x = 5$, find y when $x = 10$.
14. If C varies directly as n and $C = 6$ when $n = 4$, find C when $n = 12$.
15. If C varies directly as r and $C = 6.28$ when $r = 1$, find r when $C = 25.12$.
16. If A varies directly as h and $A = 72$ when $h = 8$, find A when $h = 5$.
17. If D varies directly as t and $D = 30$ when $t = \frac{1}{2}$, find t when $D = 20$.
18. If x varies directly as y and if $y = 27$ when $x = 6$, find y when $x = 8$.
19. If y varies directly as x and if $y = 8$ when $x = 4$, find the value of y when $x = -50$.

20. If r varies directly as s and if $r = 13$ when $s = 52$, find s when $r = 100$.

21. If 5 hats cost $60, how much will 11 hats of the same kind cost?

22. If 3 men earn $180 in a week, what will 15 men working at the same rate of pay earn?

23. If a boat travels 132 miles in 11 hours, how far can it travel in $38\frac{1}{2}$ hours, traveling at the same rate of speed?

24. A 20-acre field yields 300 bushels of wheat. At the same rate, how many bushels will a 50-acre field yield?

25. If one quantity varies directly as another, (1) their product is constant (2) their sum is constant (3) their ratio is constant

26. Tell what happens to the area of a rectangle if its width is fixed and its length is (*a*) doubled, (*b*) divided by 3, and (*c*) halved.

27. How many times as far will a man travel if his time of travel is fixed and his rate of speed is (*a*) tripled, (*b*) divided by 5, and (*c*) halved?

28. How many times as far will a man travel if his rate of speed is fixed and his time of travel is (*a*) quadrupled, (*b*) divided by 10, and (*c*) halved?

29. D varies directly as R. What change takes place in D if (*a*) R is tripled (*b*) R is halved?

30. $A = LW$. (*a*) When will A vary directly as L? (*b*) When will A vary directly as W?

In 31–34, select two variables that may vary directly and state when this will occur.

31. $C = NP$ **32.** $I = PR$ **33.** $A = \frac{1}{2}bh$ **34.** $V = \frac{1}{3}Bh$

35. $V = LWH$. (*a*) When will V and L vary directly? (*b*) When will V and W vary directly? (*c*) When will V and H vary directly?

2. Understanding Inverse Variation: $xy = k$ or $y = \dfrac{k}{x}$

The values in the table at the right are those of the lengths (l) and widths (w) of four rectangles, I, II, III, and IV, having the same area. From the table, we see that for rectangle III, when the length is 8, the width is 3; for rectangle IV, when the length is 6, the width is 4. For the four rectangles, the formula $lw = 24$ states the relationships between the length, l, and the width, w, of any of the rectangles. The equation $lw = 24$ illustrates *inverse variation.*

VALUES OF LENGTHS AND WIDTHS OF REC-TANGLES HAVING A CONSTANT AREA

	I	II	III	IV
l	24	12	8	6
w	1	2	3	4

EXPRESSING INVERSE VARIATION IN EQUATION FORM BY USING A CONSTANT OF VARIATION

If $xy = k$ where x and y are variables and k is a constant, then we say that x and y vary *inversely* as each other; that is, x varies inversely as y, and y varies inversely as x. The constant, k, is called the *constant of variation*.

Thus, if $lw = 24$, then l and w vary inversely as each other and the constant of variation is 24. In a motion problem where the distance traveled is 120 miles and this distance remains the same in several situations, then $RT = 120$. If $RT = 120$, then R and T vary inversely as each other and 120 is the constant of variation. In an interest problem, where $PR = I$ and I is fixed at 200 in several situations, then $PR = 200$. If $PR = 200$, then P and R vary inversely as each other and the constant of variation is 200.

KEEP IN MIND

If $xy = k$ or $y = \dfrac{k}{x}$, where x and y are variables and k is a constant, then x and y vary inversely as each other and k is the constant of variation.

EXPRESSING INVERSE VARIATION IN PROPORTION FORM

Inverse variation is expressible in proportion form. In the figure, Rectangle I and Rectangle II are any two rectangles whose lengths are l_1 and l_2 and whose corresponding widths are w_1 and w_2. When the areas of the rectangle are the same, $l_1 w_1 = l_2 w_2$, which, when transformed, becomes the proportion $\dfrac{l_2}{l_1} = \dfrac{w_1}{w_2}$.

In this form, note that *the ratio of the lengths is "inversely" equal to the ratio of the widths*; that is, the ratio of the lengths is the inverse or reciprocal of the corresponding ratio of the widths.

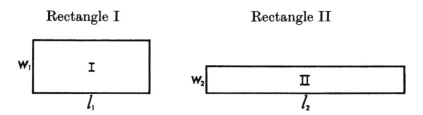

In general, if x and y vary inversely as each other, then $\dfrac{y_2}{y_1} = \dfrac{x_1}{x_2}$; that is, the ratio of two values of one variable *inversely* equals the ratio of the corresponding values of the other variable.

Thus, in the table at the right, if two travelers A and B are traveling the same distance, 120 miles, A at 20 miles per hour and B at 15 miles per hour, the ratio of their rates is $\frac{20}{15}$, and the ratio of their times of travel is $\frac{6}{8}$. Note in the table that the rate-ratio, $\frac{4}{3}$, and the time-ratio, $\frac{3}{4}$, have been placed alongside the values of rate and time.

TRAVELERS TRAVELING A CONSTANT DISTANCE			
	(mph) Rate	(hr.) × Time	(mi.) = Distance
A	20 $\Big)_{\frac{4}{3}}$	6 $\Big)_{\frac{3}{4}}$	120
B	15	8	120

If two travelers are traveling the same distance of 120 miles, then $RT = 120$ and R and T vary inversely as each other. The constant of variation is 120.

MULTIPLICATION AND DIVISION OPERATIONS IN INVERSE VARIATION

A man travels over a fixed distance in two trips. If he goes four times as fast, his time will be divided by 4; that is, he will travel only $\frac{1}{4}$ as long. If the price of an article is doubled, then if the cost of an entire purchase remains fixed, a purchaser can buy only one-half the original number of such articles.

In general, if x varies inversely as y:
(1) When x is *multiplied* by a nonzero number, y will be *divided* by the same number.
(2) When x is *divided* by a nonzero number, y will be *multiplied* by the same number.

THE GRAPH OF INVERSE VARIATION

If x and y vary inversely as each other and k is the constant of variation, then $xy = k$. Recall that the graph of $xy = k$ is an equilateral hyperbola. If k is positive, the branches of the hyperbola are in quadrants I and III, and if k is negative, the branches of the hyperbola are in quadrants II and IV, as shown in the figure. Note in the case of $xy = -4$ that x and y vary inversely; yet, when x increases, y increases also.

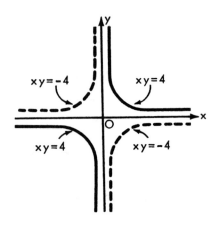

```
┌─────────────────── KEEP IN MIND ───────────────────┐
│                                                     │
│   If y varies inversely' as x, then:                │
│                                    k                │
│   1. In equation form, xy = k or y = -, where k is  │
│                                    x                │
│      the constant of variation.                     │
│                           y₂   x₁                   │
│   2. In proportion form,  -- = --, when x varies    │
│                           y₁   x₂                   │
│      from x₁ to x₂ and y varies from y₁ to y₂.      │
│   3. Whenever x is multiplied by a nonzero number,  │
│      y is divided by the same number.               │
│   4. Whenever x is divided by a nonzero number, y   │
│      is multiplied by the same number.              │
│                                                     │
└─────────────────────────────────────────────────────┘
```

~~~~~ MODEL PROBLEMS ~~~~~

1. Express as an equation: The measure of the central angle, C, of a regular polygon varies inversely as the number of sides, n. *Answer: $Cn = k$*

2. If $D = RT$, when will R and T vary inversely? *Answer:* when D is constant

3. If y varies inversely as x and if $y = 3$ when $x = 12$, find y when $x = 4$.

Solution:

Method 1
($xy = k$ method)

Since y varies inversely as x,

$$xy = k$$

If $y = 3$ when $x = 12$, $(12)(3) = k$

$$36 = k$$

Hence, $xy = 36$

When $x = 4$,

$$4y = 36 \text{ or } y = 9 \quad Ans.$$

Method 2
(inverse proportion method)

$$\frac{y_2}{y_1} = \frac{x_1}{x_2}$$

$$\frac{y}{3} = \frac{12}{4}$$

$$4y = 36$$

$$y = 9 \quad Ans.$$

	y	x
2nd values	y	4
1st values	3	12

4. The expression $xy = z$ represents (1) direct variation of x with respect to y when z is constant (2) inverse variation of x with respect to y when z is constant (3) inverse variation of x with respect to z when y is constant

Solution:

Make the correct choice by applying the following principle:

If the product of two variables is constant, then either of the variables varies inversely with respect to the other.

In choice (2), the product xy equals z, which is given as constant. Hence, x varies inversely with respect to y.

Answer: The correct choice is (2).

Exercises

In 1–3, show that one variable in the table varies inversely as the other and write an equation or formula that expresses the relationship between the variables.

1.

x	1	2	4
y	40	20	10

2.

R	2	4	10
T	50	25	10

3.

C	12	24	36
D	12	6	4

In 4–9, in the table, the product xy is constant. Find the missing numbers.

4.

y	36	18	8	9
x	2	4	?	?

5.

y	10	6	4	?
x	3	5	?	20

6.

y	2	$2\frac{1}{2}$	3	?
x	5	4	?	$2\frac{1}{2}$

7.

y	6	4	?	1
x	−2	−3	−6	?

8.

y	−4	−3	−2	?
x	−6	−8	?	4

9.

y	4	−2	−$\frac{1}{2}$	?
x	−1	2	?	−8

In 10–12, write an equation or formula that expresses the relationship between the variables, using k as the constant of variation where k is not given.

10. If the distance remains at 100, the rate, R, varies inversely as the time, T.

11. For a fixed sum of money, the number of articles, N, that can be bought varies inversely as the cost, C, of an article.

12. Under fixed conditions, the volume, V, of a gas varies inversely as the pressure, p.

13. If x varies inversely as y and $x = 4$ when $y = 5$, find x when $y = 10$.

14. If n varies inversely as c and $n = 20$ when $c = 10$, find c when $n = 50$.

15. If R varies inversely as T and $R = 40$ when $T = \frac{1}{2}$, find R when $T = 4$.

16. If y is inversely proportional to z and $y = 9$ when $z = 8$, find y when $z = 2$.

17. If x varies inversely as y and $x = 8$ when $y = 9$, find y when $x = 24$.

18. If s varies inversely as t, and if $s = 6$ when $t = 2$, find s when $t = 3$.

19. The speed of a gear varies inversely as the number of teeth. If a gear which has 36 teeth makes 30 revolutions per minute, how many revolutions per minute will a gear which has 24 teeth make?

20. If 8 printing presses can do a job in 6 hours, how many hours would it take 3 printing presses to do the same job, assuming the rate of work of each press is the same?

In 21–32, state whether the relationship is one of direct variation, inverse variation, or neither.

21. $bh = 40$ **22.** $RT = 80$ **23.** $P = \dfrac{60}{V}$ **24.** $b = 40h$

25. $R + T = 80$ **26.** $\dfrac{e}{i} = 20$ **27.** $h = 40b$ **28.** $R = 80T$

29. $t = \dfrac{100}{r}$ **30.** $b + h = 40$ **31.** $T = 80R$ **32.** $\dfrac{N}{C} = 4$

33. V varies inversely as P. (*a*) If P is tripled, what change takes place in V? (*b*) If P is divided in half, what change takes place in V?
34. $A = LW$. When will L and W vary inversely?

In 35–38, select two variables that may vary inversely and state when this will occur:
35. $C = NP$ **36.** $I = RP$ **37.** $A = \frac{1}{2}bh$ **38.** $V = \frac{1}{3}Bh$
39. $V = LWH$. (*a*) When will L and W vary inversely? (*b*) When will L and H vary inversely? (*c*) When will W and H vary inversely?

3. Understanding Joint Variation : $z = kxy$ or $\dfrac{z}{xy} = k$

The values in the table at the right are those of the areas (A), bases (b), and altitudes (h) of four triangles, I, II, III, and IV. From the table, we see that when the length of a base is 10 and the altitude is 4, then the area of triangle II is 20. For the four triangles, the formula $A = \frac{1}{2}bh$ states the relationships among the base, altitude, and area of any one of the triangles. By a transformation, $A = \frac{1}{2}bh$ becomes $\dfrac{A}{bh} = \frac{1}{2}$. Note in the table that the ratio of A to bh equals $\frac{1}{2}$ in each instance. The equations $A = \frac{1}{2}bh$ and $\dfrac{A}{bh} = \frac{1}{2}$ illustrate *joint variation*.

VALUES OF BASES, ALTITUDES, AND AREAS OF TRIANGLES

	I	II	III	IV
A	12	20	24	40
bh	24	40	48	80
h	3	4	4	5
b	8	10	12	16

EXPRESSING JOINT VARIATION IN EQUATION FORM BY USING A CONSTANT OF VARIATION

If $z = kxy$ where x, y, and z are variables and k is a constant, then we say that z varies jointly as the product of x and y, or, in simpler language, "z varies jointly as x and y." The constant, k, is called the *constant of variation*.

Thus, if $A = \frac{1}{2}bh$, we say that A varies jointly as b and h and the constant of variation is $\frac{1}{2}$. In the motion formula $D = RT$, D varies jointly as R and T and the constant of variation is 1.

The equation $z = kxy$, which expresses joint variation, can be transformed into $\dfrac{z}{xy} = k$. The latter equation is exemplified in $\dfrac{A}{bh} = \frac{1}{2}$, $\dfrac{C}{NP} = 1$, and $\dfrac{V}{BH} = \dfrac{1}{3}$. In this form, the constant of variation stands alone on one side of the equation and the three variables are on the other side.

KEEP IN MIND

If $z = kxy$ or $\dfrac{z}{xy} = k$, where z, x, and y are variables and k is a constant, then z varies jointly as x and y and k is the constant of variation.

Note. The equation $z = kxy$ may be transformed into an equivalent equation $y = \dfrac{z}{kx}$. If $\dfrac{1}{k}$ is replaced by k', we obtain $y = k'\dfrac{z}{x}$. If $y = k'\dfrac{z}{x}$ where x, y, and z are variables, the expression is used, "y varies directly as z and inversely as x." It should be kept in mind that in such cases, the type of variation that is involved is that of joint variation.

EXPRESSING JOINT VARIATION IN PROPORTION FORM

Joint variation is expressible in proportion form. In the figure Triangle I and Triangle II are any two triangles whose bases are b_1 and b_2, whose altitudes are h_1 and h_2, and whose areas are A_1 and A_2. Since $A_2 = \frac{1}{2}b_2h_2$ and $A_1 = \frac{1}{2}b_1h_1$, then by division, $\dfrac{A_2}{A_1} = \dfrac{b_2h_2}{b_1h_1}$, eliminating the constant, $\frac{1}{2}$. This proportion may be transformed into $\dfrac{A_2}{A_1} = \left(\dfrac{b_2}{b_1}\right)\left(\dfrac{h_2}{h_1}\right)$. In this form, we note that *the ratio of the areas equals the product of the ratio of the bases and the ratio of the corresponding altitudes drawn to these bases.*

Triangle I   Triangle II

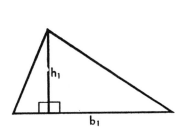 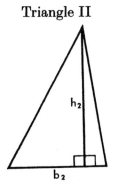

In general, if z varies jointly as x and y, then when z varies from z_1 to z_2 and x from x_1 to x_2 and y from y_1 to y_2, $\dfrac{z_2}{z_1} = \dfrac{x_2 y_2}{x_1 y_1}$ or $\dfrac{z_2}{z_1} = \left(\dfrac{x_2}{x_1}\right)\left(\dfrac{y_2}{y_1}\right)$; that is, the ratio of two values of z equals the product of the ratios of the corresponding values of x and y. In this case, (x_1, y_1, z_1) and (x_2, y_2, z_2) are ordered triplets of numbers that satisfy $z = kxy$.

Thus, in the table at the right, if traveler B goes twice as fast and three times as long as A, then B will go six times as far as A. The ratio of the distances, 6, equals the product of the ratio of the rates, 2, and the ratio of the times, 3. Note in the table that the three ratios, $\frac{2}{1}$, $\frac{3}{1}$, and $\frac{6}{1}$, are next to the values.

In the above example of two travelers, the motion formula, $D = RT$ applies. Hence, D varies jointly as R and T.

TRAVELERS TRAVELING AT VARYING RATES, TIMES, AND DISTANCES

	(mph)		(hr.)		(mi.)	
	Rate	$\times$	Time	$=$	Distance	
B	30	$\tfrac{2}{1}$	6	$\tfrac{3}{1}$	180	$\tfrac{6}{1}$
A	15		2		30	

MULTIPLICATION AND DIVISION OPERATIONS IN JOINT VARIATION

In a problem involving the area of a triangle, if the base is multiplied by 4 and the altitude is multiplied by 3, then the area of the new triangle is 12 times the area of the old triangle. In the motion problem shown in the preceding table, if a man travels twice as fast and three times as long, then he will go six times as far; that is, if the ratio of the speeds is 2 and the ratio of the times of travel is 3, then the ratio of distances traveled is 6.

In general, if z varies jointly as x and y:

When x is multiplied by a number a and y is multiplied by a number b, then z is multiplied by the product of these numbers ab. This is another way of saying that when the ratio of the x-values is a and the ratio of the y-values is b, then the ratio of the z-values is ab.

─── KEEP IN MIND ───

If z varies jointly as x and y, then:

1. In equation form, $z = kxy$ or $\dfrac{z}{xy} = k$, where k is the constant of variation.

2. In proportion form, $\dfrac{z_2}{z_1} = \dfrac{x_2 y_2}{x_1 y_1}$ or $\dfrac{z_2}{z_1} = \left(\dfrac{x_2}{x_1}\right)\left(\dfrac{y_2}{y_1}\right)$, where x varies from x_1 to x_2, y varies from y_1 to y_2, and z varies from z_1 to z_2.

3. When x is multiplied by a number a and y is multiplied by a number b, then z is multiplied by ab.

Note. If a number is divided by a nonzero number a, the number may be considered to be multiplied by $\dfrac{1}{a}$. Hence, the division operation may be replaced by a multiplication operation in order to apply this principle. For example, if $z = kxy$, x is divided by 4, and y is multiplied by 3, then z is multiplied by $\frac{1}{4}(3)$ or $\frac{3}{4}$.

∿∿∿∿∿∿∿∿ MODEL PROBLEMS ∿∿∿∿∿∿∿∿

1. If $V = LWH$, when will V vary jointly as (a) W and H (b) L and H (c) L and W?

Solution:

(a) when L is constant   (b) when W is constant   (c) when H is constant

2. If z varies jointly as x and y, and $z = 120$ when $x = 8$ and $y = 5$, find z when $x = 15$ and $y = 4$.

Solution:

Since z varies jointly as x and y, $z = kxy$

If $z = 120$ when $x = 8$ and $y = 5$, $120 = k(8)(5)$

$$120 = 40k$$
$$3 = k$$

Substitute $x = 15$ and $y = 4$. $z = 3xy = 3(15)(4) = 180$ *Ans.*

Exercises

In 1–3, write an equation or formula that expresses the relationship between the variables, using k as the constant of variation.

1. The area of a parallelogram, A, varies jointly as its base, b, and altitude, h.
2. The volume of a pyramid, V, varies jointly as its base, B, and altitude, h.
3. The area of a rhombus, A, varies jointly as its diagonals, D and d.

4. If V varies jointly as B and h, and $V = 100$ when $B = 30$ and $h = 10$, find V when $B = 60$ and $h = 30$.
5. If I varies jointly as p and r, and $I = 8$ when $p = 100$ and $r = .04$, find I when $p = 400$ and $r = .06$.
6. D varies jointly as R and T. If R is doubled and T is trebled, what change takes place in D?
7. V varies jointly as l, w, and h. If l is multiplied by 4, w is multiplied by 5 and h is halved, what change takes place in V?
8. $I = PRT$. (a) When will I vary jointly as P and R? (b) When will I vary jointly as P and T?
9. If the base of a triangle is quadrupled and the height is tripled, then its area is multiplied by _____.

In 10–12, complete the statement, using $A = \frac{1}{2}bh$.
10. If the base and height are each doubled, the area of a triangle will be _____.
11. If the base and height are each halved, the area of a triangle will be _____.
12. If the area of a triangle is to be multiplied by 6 and the base is doubled, then its height must be _____.

13. $A = \frac{1}{2}nsr$. (a) When will A vary jointly as s and r? (b) When will A vary jointly as n and r? (c) If n and s are constant, how will A and r vary? (d) If A and n are constant, how will s and r vary?

4. Understanding Direct and Inverse Square and Square Root Variation

DIRECT SQUARE VARIATION: $y = kx^2$ or $\frac{y}{x^2} = k$

The values in the table at the right are those of the edges (e) and surfaces (S) of four cubes, I, II, III, and IV. From the four pairs of corresponding values of S and e^2, we note that $\frac{S}{e^2} = \frac{6}{1} = \frac{24}{4} = \frac{54}{9} = \frac{150}{25}$. For each of the four cubes, the formula, $S = 6e^2$ states the relationships between its surface, S, and its edge, e.

VALUES OF EDGES AND SURFACES OF CUBES

	I	II	III	IV
S	6	24	54	150
e^2	1	4	9	25
e	1	2	3	5

EXPRESSING DIRECT SQUARE VARIATION IN EQUATION FORM BY USING A CONSTANT OF VARIATION

--- KEEP IN MIND ---

If $y = kx^2$ or $y = \dfrac{k}{x^2}$, where x and y are variables and k is a constant, then y varies directly as the square of x, and k is the constant of variation.

In general, if y varies directly as the square of x, then when x varies from x_1 to x_2 and y varies from y_1 to y_2,

$$\frac{y_2}{y_1} = \left(\frac{x_2}{x_1}\right)^2 \text{ or } \frac{y_2}{y_1} = \frac{x_2{}^2}{x_1{}^2}$$

Thus, if $S = 6e^2$, then S varies directly as the square of e, and the constant of variation is 6. Also, if $A = \pi r^2$, then A varies directly as the square of r, and the constant of variation is π.

If $y = kx^2$, then $\dfrac{y}{x^2} = k$. The equation, $\dfrac{y}{x^2} = k$ also expresses direct square variation. The equation $\dfrac{y}{x^2} = k$ is illustrated by $\dfrac{S}{e^2} = 6$ and $\dfrac{A}{r^2} = \pi$. In this form, the constant of variation stands alone on one side of the equation and the two variables are on the other.

The graph that is associated with direct square variation is the parabola, $y = kx^2$.

DIRECT SQUARE ROOT VARIATION: $y = k\sqrt{x}$ or $\dfrac{y}{\sqrt{x}} = k$

If $y = k\sqrt{x}$ or $\dfrac{y}{\sqrt{x}} = k$ where x and y are variables and k is a constant, then y varies directly as the square root of x, and k is the constant of variation.

If y varies directly as the square root of x, then when x varies from x_1 to x_2 and y from y_1 to y_2, $\dfrac{y_2}{y_1} = \sqrt{\dfrac{x_2}{x_1}}$ or $\dfrac{y_2}{y_1} = \dfrac{\sqrt{x_2}}{\sqrt{x_1}}$. To solve problems involving direct square root variation, procedures are used that are similar to those used in direct square variation.

INVERSE SQUARE VARIATION: $yx^2 = k$ or $y = \dfrac{k}{x^2}$

In the table of values at the right, h represents the altitude and s represents the side of the base of four prisms having a square base and a constant volume. From the four pairs of corresponding values of h and s^2, we note that $hs^2 = 18 \times 4 = 8 \times 9 = 2 \times 36 = \frac{1}{2} \times 144$. In each case, $hs^2 = 72$. For prisms having a square base and a constant volume, $hs^2 = k$ states the relationship between the altitude, h, and the side of the base, s.

VALUES OF BASES AND ALTITUDES OF PRISMS

	I	II	III	IV
h	18	8	2	$\frac{1}{2}$
s^2	4	9	36	144
s	2	3	6	12

EXPRESSING INVERSE SQUARE VARIATION IN EQUATION FORM, USING A CONSTANT OF VARIATION

KEEP IN MIND

In $yx^2 = k$ or $y = \dfrac{k}{x^2}$ where x and y are variables and k is a constant, then y varies inversely as the square of x and k is the constant of variation.

In general, if y varies inversely as the square of x, then when x varies from x_1 to x_2 and y varies from y_1 to y_2, $\dfrac{y_2}{y_1} = \left(\dfrac{x_1}{x_2}\right)^2$ or $\dfrac{y_2}{y_1} = \dfrac{x_1^2}{x_2^2}$. Thus, if $hs^2 = 72$, then h varies inversely as the square of s, and 72 is the constant of variation. Also, if $r^2h = \dfrac{100}{\pi}$, then h varies inversely as the square of r and the constant of variation is $\dfrac{100}{\pi}$.

INVERSE SQUARE ROOT VARIATION: $y\sqrt{x} = k$ or $y = \dfrac{k}{\sqrt{x}}$

If $y\sqrt{x} = k$ or $y = \dfrac{k}{\sqrt{x}}$ where x and y are variables and k is a constant, then y varies inversely as the square root of x, and k is the constant of variation.

If y varies inversely as the square root of x, then when x varies from x_1 to x_2 and y varies from y_1 to y_2, $\dfrac{y_2}{y_1} = \sqrt{\dfrac{x_1}{x_2}}$ or $\dfrac{y_2}{y_1} = \dfrac{\sqrt{x_1}}{\sqrt{x_2}}$. To solve problems involving inverse square root variation, procedures are used that are similar to those used in inverse square variation.

MULTIPLICATION AND DIVISION OPERATIONS IN DIRECT AND INVERSE SQUARE AND SQUARE ROOT VARIATION

If y varies directly as the square of x, then when x is multiplied by a number a, y is multiplied by a^2.

Thus, if S varies directly as the square of e, then when e is multiplied by 3, S is multiplied by 9.

If y varies inversely as the square of x, then when x is multiplied by a number a, y is divided by a^2.

Thus, if h varies inversely as the square of s, then when s is multiplied by 3, h is divided by 9.

If y varies directly as the square root of x, then when x is multiplied by a number a, y is multiplied by $\sqrt{a}$.

Thus, if the radius of a circle, r, varies directly as the square root of the area, A, then if A is multiplied by 9, r is multiplied by 3.

If y varies inversely as the square root of x, then when x is multiplied by a number a, y is divided by $\sqrt{a}$.

Thus, if r varies inversely as the square root of h, then when h is multiplied by 9, r is divided by 3.

~~~~~~~~~~ *MODEL PROBLEMS* ~~~~~~~~~~

1. Express each sentence as an equation.
   a. The distance that a freely falling body falls, D, varies directly as the square of the time it falls, T.
   b. The force of attraction, F, of two bodies having constant masses varies inversely as the square of the distance between them, d.

Solution:

$$a. \ D = kT^2 \text{ or } \frac{D}{T^2} = k$$

$$b. \ F = \frac{k}{d^2} \text{ or } Fd^2 = k$$

2. Express each sentence as an equation.

    a. The radius of a circle, r, varies directly as the square root of the area, A.

    b. The radius, r, of a cylinder of constant volume varies inversely as the square root of the altitude, h.

Solution:

$$a.\ r = k\sqrt{A} \text{ or } \frac{r}{\sqrt{A}} = k$$

$$b.\ r\sqrt{h} = k$$

3. If x varies inversely as the square of y and if $x = 2$ when $y = 6$, find x when $y = 3$.

Solution:

Since x varies inversely as the square of y, then $xy^2 = k$.

Since $x = 2$ when $y = 6$, substitute 2 for x and 6 for y: $\quad 2(6^2) = k$

$$2(36) = k$$
$$72 = k$$

Hence, $xy^2 = 72$

To find x when $y = 3$, substitute 3 for y: $\quad x(3^2) = 72$

$$9x = 72$$
$$x = 8 \quad Ans.$$

4. Indicate the number by which T is to be multiplied if L is multiplied by 4 and:

    a. T varies directly as the square of L

    b. T varies inversely as the square of L

    c. T varies directly as the square root of L

    d. T varies inversely as the square root of L

Solution:

    a. If T varies directly as the square of L, then if L is multiplied by a, T is multiplied by a^2. Hence, if L is multiplied by 4, T is multiplied by 4^2 or 16. *Ans.*

    b. If T varies inversely as the square of L, then if L is multiplied by a ($a \neq 0$), T is divided by a^2. Hence, if L is multiplied by 4, T is divided by 4^2 or 16. Since multiplying by $\frac{1}{16}$ is equivalent to dividing by 16, T is multiplied by $\frac{1}{16}$. *Ans.*

    c. If T varies directly as the square root of L, then if L is multiplied by a, T is multiplied by $\sqrt{a}$. Hence, if L is multiplied by 4, T is multiplied by $\sqrt{4}$ or 2. *Ans.*

d. If T varies inversely as the square root of L, then if L is multiplied by a ($a \neq 0$), T is divided by $\sqrt{a}$. Hence, if L is multiplied by 4, T is divided by $\sqrt{4}$ or 2. Since multiplying by $\frac{1}{2}$ is equivalent to dividing by 2, T is multiplied by $\frac{1}{2}$. *Ans.*

Exercises

In 1–4, express the statement as an equation, using k as the constant of variation.

1. The area of a circle, A, varies directly as the square of its radius, r.

2. The area of a square, A, varies directly as the square of its edge, e.

3. The surface area of a sphere, S, varies directly as the square of its radius, r.

4. The intensity of illumination, I, upon a surface varies inversely as the square of the distance, d, between the surface and the source of light.

5. If d varies directly as the square of t and if $d = 24$ when $t = 2$, find the value of d when $t = 5$.

6. If x varies directly as y^2 and if $x = 2$ when $y = 1$, find the value of x when $y = 2$.

7. If v varies directly as b^2 and $v = 45$ when $b = 3$, find the value of v when $b = 4$.

8. If x varies directly as the square of y and if $x = 200$ when $y = 5$, find the value of x when $y = 4$.

9. If x varies directly as the square root of y and if $x = 10$ when $y = 100$, find x when $y = 900$.

10. If S varies directly as the square of e and $S = 150$ when $e = 5$, find S when $e = 10$.

11. If I varies inversely as the square of d and $I = 400$ when $d = 5$, find I when $d = 20$.

12. If x varies inversely as the square root of y and $x = 4$ when $y = 25$, find x when $y = 4$.

13. If t varies directly as $\sqrt{x}$ and if $t = 2$ when $x = 25$, what is the value of t when $x = 100$?

14. If $y = 10$ when $x = 5$, find the value of y when $x = 20$ and:
a. y varies directly as the square of x.
b. y varies inversely as the square of x.
c. y varies directly as the square root of x.
d. y varies inversely as the square root of x.

15. If F varies inversely as the square of d, what change takes place in F when d is (a) doubled (b) trebled?

In 16 and 17, express the statement as an equation, using k as the constant of variation.

16. The volume of a sphere, V, varies directly as the cube of its radius, r.
17. The velocity, v, of sound in air varies directly as the square root of the absolute temperature, t, of the air.
18. $K = \frac{1}{2}mv^2$. (a) How do K and v vary when m is constant? (b) How do m and v vary when K is constant? (c) How do K and m vary when v is constant?
19. The surface area of a sphere varies directly as the square of its radius. If the surface area is 64π square inches when the radius is 4 inches, what is the surface area, in square inches, when the radius is 8 inches? (1) 128π (2) 32π (3) 64π (4) 256π
20. Indicate the number by which A is to be multiplied if B is multiplied by 9 and:
    a. A varies directly as the square of B.
    b. A varies inversely as the square of B.
    c. A varies directly as the square root of B.
    d. A varies inversely as the square root of B.
21. What is the graph of the relation between y and x when (a) y varies directly as x (b) y varies inversely as x (c) y varies directly as the square of x?

SUMMARY OF THE VARIATION FORMULAS

Type of Variation	Equation	Equation with Subscripts
direct	$y = kx$ or $$\frac{y}{x} = k$$	$$\frac{y_2}{y_1} = \frac{x_2}{x_1}$$
inverse	$y = \frac{k}{x}$ or $$xy = k$$	$\frac{y_2}{y_1} = \frac{x_1}{x_2}$ or $$\frac{y_2}{y_1} = \frac{1}{\frac{x_2}{x_1}}$$
joint	$z = kxy$ or $$\frac{z}{xy} = k$$	$$\frac{z_2}{z_1} = \left(\frac{x_2}{x_1}\right)\left(\frac{y_2}{y_1}\right)$$
direct square	$y = kx^2$ or $$\frac{y}{x^2} = k$$	$$\frac{y_2}{y_1} = \left(\frac{x_2}{x_1}\right)^2$$
inverse square	$y = \frac{k}{x^2}$ or $$yx^2 = k$$	$\frac{y_2}{y_1} = \left(\frac{x_1}{x_2}\right)^2$ or $$\frac{y_2}{y_1} = \frac{1}{\left(\frac{x_2}{x_1}\right)^2}$$
direct square root	$y = k\sqrt{x}$ or $$\frac{y}{\sqrt{x}} = k$$	$$\frac{y_2}{y_1} = \sqrt{\frac{x_2}{x_1}}$$
inverse square root	$y = \frac{k}{\sqrt{x}}$ or $$y\sqrt{x} = k$$	$\frac{y_2}{y_1} = \sqrt{\frac{x_1}{x_2}}$ or $$\frac{y_2}{y_1} = \frac{1}{\sqrt{\frac{x_2}{x_1}}}$$

CALCULATOR APPLICATIONS

Some of the important features of the graph of an inverse variation can be explored with a graphing calculator.

MODEL PROBLEMS

1. Graph $xy = 6$ in the standard viewing window.

   Solution: First, solve the equation of variation for y.

   $$xy = 6$$
   $$y = \frac{6}{x}$$

   Enter:

   Display: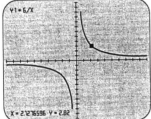

   The graph shows the two branches of the equilateral hyperbola. Since $k = 6$ is positive, the branches are in quadrants I and III. Press the $\boxed{\text{TRACE}}$ key and use the $\boxed{\blacktriangleleft}$ and $\boxed{\blacktriangleright}$ arrow keys to move the cursor along one branch. Note that as the x-values increase, the y-values decrease. If you trace along the graph to $x = 0$, you will see there is no corresponding y-value shown on the screen. This is because when $x = 0$, the value of y is undefined. This can also be observed by displaying a table of values.

   Enter: $\boxed{\text{2nd}}$ $\boxed{\text{TABLE}}$

   Display: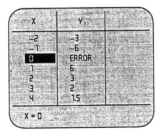

2. Suppose you needed to find the values of y when $x = 2.3$, 2.5, and 2.7. Use the $\boxed{\text{TBLSET}}$ key and enter a starting value of 2.2. Since x is in tenths, set ΔTbl for .1. Use the $\boxed{\text{TABLE}}$ key to display the table:

All three required values are easily read off the table.

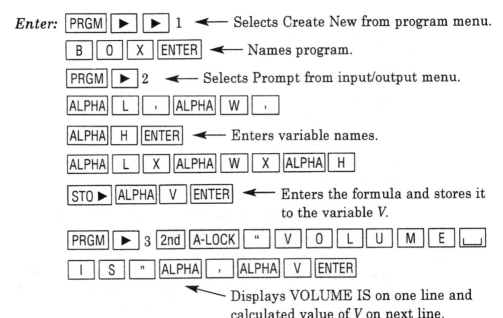

3. Formulas involving joint variation can be programmed into a graphing calculator for repeated execution. To create a program that prompts for the length *L*, width *W*, and height *H* of a box and then computes its volume, follow these steps.

Enter: [PRGM] [▶] [▶] 1 ◀—— Selects Create New from program menu.

[B] [O] [X] [ENTER] ◀—— Names program.

[PRGM] [▶] 2 ◀—— Selects Prompt from input/output menu.

[ALPHA] [L] [,] [ALPHA] [W] [,]

[ALPHA] [H] [ENTER] ◀—— Enters variable names.

[ALPHA] [L] [X] [ALPHA] [W] [X] [ALPHA] [H]

[STO ▶] [ALPHA] [V] [ENTER] ◀—— Enters the formula and stores it to the variable *V*.

[PRGM] [▶] 3 [2nd] [A-LOCK] ["] [V] [O] [L] [U] [M] [E] [␣]

[I] [S] ["] [ALPHA] [,] [ALPHA] [V] [ENTER]

◀—— Displays VOLUME IS on one line and calculated value of *V* on next line.

Display:

```
PROGRAM:BOX
:PROMPT L,W,H
:L*W*H→V
:DISP "VOLUME IS
",V
:▮
```

Now you are ready to execute the program for specific values of the variables. Let the length of the box be 8 inches, the width be 4.5 inches, and the height be 12 inches. If necessary, press [2nd] [QUIT] to display the home screen.

Enter: [PRGM] 1 [ENTER]

8 [ENTER] 4.5 [ENTER] 12 [ENTER]

Display:

```
PRGMBOX
L=?8
W=?4.5
H=? 12
VOLUME IS
            432
          DONE
▮
```

CHAPTER XIV

Statistics and Probability

1. The Summation Symbol

In statistics, we often work with sums. The **summation symbol** is the Greek capital letter sigma, Σ, used to indicate the sum of related terms. For example, $\sum_{i=1}^{6} i$ means the sum of the integers, symbolized by i, from $i = 1$ to $i = 6$. Therefore:

$$\sum_{i=1}^{6} i = 1 + 2 + 3 + 4 + 5 + 6 = 21$$

In the preceding example, the letter i is called the **index**; and when the indicated sum is evaluated, i is replaced by a series of consecutive integers, starting with the lower limit of summation and ending with the upper limit of summation.

The **lower limit of summation** is the value of the index placed below the summation symbol, and the **upper limit of summation** is the value of the index placed above the summation symbol. In $\sum_{i=1}^{6} i$, the lower limit is 1, and the upper limit is 6.

Any letter can serve as an index; but i, j, k, and n are the most frequently used letters. Since an index acts as a placeholder, an index of i is *not* equal to the imaginary unit, $\sqrt{-1}$.

The meaning of the summation symbol is shown in the following equations:

$$\sum_{k=0}^{5} 3^k = 3^0 + 3^1 + 3^2 + 3^3 + 3^4 + 3^5 = 1 + 3 + 9 + 27 + 81 + 243 = 364$$

$$\sum_{j=3}^{6} j^2 = 3^2 + 4^2 + 5^2 + 6^2 = 9 + 16 + 25 + 36 = 86$$

USE OF SUBSCRIPTS

If a variable quantity is denoted by x, then successive values of that variable can be indicated by using a **subscript,** as in x_1, x_2, x_3 (read as x sub-1, x sub-2, x sub-3). The use of sigma notation and subscripted variables is shown in the following examples.

❑ Mr. Cook teaches five classes. The number of students absent from his classes today are 3, 1, 2, 0, and 1. Mr. Cook records the number of student absences using subscripted variables:

$$x_1 = 3, \qquad x_2 = 1, \qquad x_3 = 2, \qquad x_4 = 0 \qquad x_5 = 1$$

The total number of absentees from Mr. Cook's classes for the day can be written as follows in sigma notation:

$$\sum_{i=1}^{5} x_i = x_1 + x_2 + x_3 + x_4 + x_5 = 3 + 1 + 2 + 0 + 1 = 7$$

When the summation symbol is used without an index and without upper and lower limits of summation, Σ designates the sum of *all* values of the given variable under consideration. Summation symbols without an index are used in the following example.

Reports x	Frequency f	xf
5	1	5
4	5	20
3	8	24
2	7	14
1	4	4
0	2	0

❑ Mrs. Gallagher, a science teacher, has completed five laboratory sessions with her class. In the frequency distribution shown here, she has recorded the number of students who have completed 0, 1, 2, 3, 4, or 5 lab reports. The sum of the frequencies equals the number of students in the class. Therefore:

Number of students in class = $\Sigma f = 1 + 5 + 8 + 7 + 4 + 2 = 27$

To find the total number of reports completed, we first, for each row, multiply the number of completed reports, x, by the number of students who have completed that many reports, f. The sum of these products, Σxf, is the total number of reports completed.

$$\Sigma xf = 5(1) + 4(5) + 3(8) + 2(7) + 1(4) + 0(2)$$
$$= 5 \quad + \quad 20 + \quad 24 + \quad 14 + \quad 4 \quad + \quad 0 \quad = \quad 67$$

Therefore, 67 lab reports have been completed.

～～～～～ MODEL PROBLEMS ～～～～～

1. In parts *a* and *b*, in each case, evaluate the sum.

a. $\displaystyle\sum_{k=1}^{5} 3k$ *b.* $\displaystyle\sum_{k=1}^{5} k$

Solution:

a. $\displaystyle\sum_{k=1}^{5} 3k = 3(1) + 3(2) + 3(3) + 3(4) + 3(5)$

$\qquad = 3 + 6 + 9 + 12 + 15 = 45$ *Answer*

b. $3 \displaystyle\sum_{k=1}^{5} k = 3(1 + 2 + 3 + 4 + 5) = 3(15) = 45$ *Answer*

Note. $\displaystyle\sum_{k=1}^{5} 3k = 3 \sum_{k=1}^{5} k.$

Model Problem 1 illustrates a general statement that is proved in Model Problem 2.

2. If c is a constant, show that $\displaystyle\sum_{j=1}^{n} cj = c \sum_{j=1}^{n} j.$

Solution: $\displaystyle\sum_{j=1}^{n} cj = c(1) + c(2) + c(3) + \cdots + c(n)$

$\qquad = c(1 + 2 + 3 + \cdots + n)$ by the distributive property

$\qquad = c \displaystyle\sum_{j=1}^{n} j$

3. Using the summation symbol, write and expression that equals $1 + \dfrac{1}{4} + \dfrac{1}{9} + \dfrac{1}{16}.$

Solution: $1 + \dfrac{1}{4} + \dfrac{1}{9} + \dfrac{1}{16} \;=\; \dfrac{1}{1^2} + \dfrac{1}{2^2} + \dfrac{1}{3^2} + \dfrac{1}{4^2} \;=\; \cdot\displaystyle\sum_{i=1}^{4} \dfrac{1}{i^2}$ *Answer*

Note. Other answers also exist. For example:

$$1 + \frac{1}{4} + \frac{1}{9} + \frac{1}{16} = \sum_{n=0}^{3} \frac{1}{(n+1)^2}$$

Exercises

In 1–12, find the value indicated by each summation symbol.

1. $\displaystyle\sum_{k=0}^{6} 2k$

2. $\displaystyle\sum_{i=1}^{5} (i + 1)$

3. $\displaystyle\sum_{j=0}^{4} j^2$

4. $\displaystyle\sum_{n=2}^{5} (n - 2)^2$

5. $\displaystyle\sum_{j=1}^{3} j^3$

6. $\displaystyle\sum_{n=0}^{3} 2^n$

7. $\displaystyle\sum_{n=1}^{4} \frac{1}{n}$

8. $\displaystyle\sum_{k=0}^{5} (10 - k)$

9. $\displaystyle\sum_{i=3}^{7} (2 - i)^2$

10. $5 \displaystyle\sum_{n=1}^{4} (n - 1)$

11. $\displaystyle\sum_{k=1}^{3} 3k^2$

12. $\displaystyle\sum_{k=1}^{3} k^{k-1}$

In 13–18, in each case, use the summation symbol to write an expression to indicate the sum.

13. $2(1) + 2(2) + 2(3) + 2(4) + 2(5)$ **14.** $0 + 1 + 4 + 9 + 16$

15. $\dfrac{1}{3} + \dfrac{1}{6} + \dfrac{1}{9} + \dfrac{1}{12} + \dfrac{1}{15}$ **16.** $\dfrac{1}{2} + \dfrac{2}{3} + \dfrac{3}{4} + \dfrac{4}{5}$

17. $1^1 + 2^2 + 3^3 + 4^4 + 5^5 + 6^6$ **18.** $2(1) + 3(2) + 4(3)$

In 19–22, find the value of each indicated sum when $x_1 = 12$, $x_2 = 5$, $x_3 = 4$, $x_4 = 8$, and $x_5 = 7$.

19. $\displaystyle\sum_{i=1}^{5} x_i$ **20.** $\displaystyle\sum_{k=3}^{5} x_k$ **21.** $3\displaystyle\sum_{j=1}^{3} x_j$ **22.** $\displaystyle\sum_{n=2}^{4} 5x_n$

In 23 and 24, select the *numeral* preceding the choice that best completes each sentence.

23. The value of $3\displaystyle\sum_{k=1}^{5} (k - 1)$ is

     (1) 10   (2) 15   (3) 30   (4) 45

24. The sum $1 + 8 + 27 + 64$ is *not* equal to

     (1) $\displaystyle\sum_{i=1}^{4} i^3$ (2) $\displaystyle\sum_{k=0}^{4} (k + 1)^3$ (3) $\displaystyle\sum_{n=0}^{3} (n + 1)^3$ (4) $\displaystyle\sum_{j=2}^{5} (j - 1)^3$

25. Represent the sum $5 + 10 + 15 + 20 + 25 + 30$ by *three* different expressions, each involving the summation symbol.

26. In *a*–*d*, evaluate each expression.

     a. $\displaystyle\sum_{n=4}^{7} (3n + 2)$ *b.* $3\displaystyle\sum_{n=4}^{7} (n + 2)$ *c.* $\displaystyle\sum_{n=4}^{7} (3n + 6)$ *d.* $3\displaystyle\sum_{n=4}^{7} (n + 6)$

     e. Which expression, if any, in parts *a*–*d*, represent the same sum?

27. Show that $\displaystyle\sum_{k=1}^{n} 7k = 7\displaystyle\sum_{k=1}^{n} k$.

28. If b is a constant, show that $\displaystyle\sum_{n=1}^{6} bn = b\displaystyle\sum_{n=1}^{6} n$.

29. Show that $\displaystyle\sum_{k=1}^{n} (k - 1) = \displaystyle\sum_{k=0}^{n-1} k$.

30. Evaluate $\displaystyle\sum_{n=1}^{100} n$. (*Hint:* Add $1 + 100$, add $2 + 99$, etc.)

2. Measures of Central Tendency

In analyzing data, we often find it useful to represent all collected values by a single number called a *measure of central tendency*. Each measure of central tendency is a number that in some way can be used to designate all of the numbers in a particular set of data. Three important measures of central tendency are the *mean*, the *median*, and the *mode*.

THE MEAN

The *mean,* or *arithmetic mean,* of a set of n numbers is the sum of the numbers divided by n. The symbol for mean is $\bar{x}$, read as x-bar. Therefore, for a set of data, $x_1, x_2, x_3, \ldots, x_n$:

$$\text{Mean} = \bar{x} = \frac{\sum_{i=1}^{n} x_i}{n} = \frac{x_1 + x_2 + x_3 + \cdots + x_n}{n}$$

For example, a cabdriver collected the following fares one afternoon: $7.50, $6.00, $9.50, $8.75, $14.00, $10.50, $6.25, $8.75, $9.25, and $11.00. To find the mean of these 10 fares, we first find their sum and then divide by 10.

$$\bar{x} = \frac{\sum_{i=1}^{10} x_i}{10} = \frac{\$91.50}{10} = \$9.15$$

The mean is often called the *average.* If the cabdriver had collected $9.15 from each of the 10 fares, the total amount collected would have been the same, that is, $91.50.

THE MEDIAN

When a set of data values is arranged in numerical order, the middle value is called the *median.* Therefore, in an ordered set, the number of values that precede the median is equal to the number of values that follow it. We will consider two cases: one in which there is an odd number of values in the set of data, and the other in which there is an even number of scores.

CASE 1: Kurt's grades on his report card are 88, 81, 91, 83, and 86. To find Kurt's median grade, we follow these two steps:

1. Arrange the grades in numerical order.   81, 83, 86, 88, 91

2. Mark off equal numbers of grades from
   the top and from the bottom.

81, 83, 86, 88, 91

The middle number is the median.   Median = 86

Alternative Method: If the number of data values, n, is an odd number, the median is the value that is $\frac{n+1}{2}$ from the bottom or from the top when the values are arranged in order. Since there are five grades on Kurt's report card, find $\frac{n+1}{2} = \frac{5+1}{2} = 3$. The median grade is therefore third from the top and third from the bottom. This grade is 86.

CASE 2: Sarah's grades on her report card are 88, 90, 85, 84, 83, and 88. To find Sarah's median grade, we follow these three steps:

1. Arrange the grades in numerical order.   90, 88, 88, 85, 84, 83

2. Mark off equal numbers of grades   90, 88, 88, 85, 84, 83
   from the top and from the bottom,
   leaving two in the middle.

3. Find the mean of the two middle grades. $\frac{88 + 85}{2} = \frac{173}{2} = 86.5$

   The mean of the middle numbers is the   Median = 86.5
   median.

Alternative Method: If the number of data values, n, is an even number, the median is the mean (average) of the values that are $\frac{n}{2}$ and $\frac{n+2}{2}$ from the bottom or from the top when the values are arranged in order. Since there are six grades on Sarah's report card, find $\frac{n}{2} = \frac{6}{2} = 3$ and $\frac{n+2}{2} = \frac{6+2}{2} = 4$. The median grade is therefore the mean of the third and fourth grades from the top or from the bottom. This grade is 86.5.

THE MODE

The *mode* is the value that appears most often in a set of data. For example, in Case 2 above, Sarah's grades were 90, 88, 88, 85, 84, and 83. Since the grade of 88 occurs twice and every other grade only once, the mode for Sarah's grades is 88.

When a set of data is arranged in a table as a frequency distribution, the mode is the entry with the highest frequency. The table at the right shows the number of children in each of 20 families that answered a survey. The mode

Number of Children	Frequency
6	1
5	0
4	2
3	4
2	8
1	5

for this set of data is 2, the entry with the highest frequency.

For some sets of data, there may be more than one mode, and, for some other set, no mode whatsoever.

1. The ages of employees at a fast-food restaurant are 17, 17, 17, 18, 18, 19, 19, 19, 21, 23, and 37. This set of data contains *two* modes: 17 and 19. When two modes appear, the set of data is said to be *bimodal*, and both modes are reported. We do *not* take an average of these modes, since a mode tells us where most values occur.

2. Mrs. Mangold found the following numbers of spelling errors on compositions that she graded: 0, 0, 0, 1, 2, 2, 3, 4, 4, 4, 5, 5, 5, 6, 7, 7, 8, 9, 9, and 12. This set of data has *three* modes: 0, 4, and 5.

3. On his last five trips to Sound Beach, Mr. Fernandes caught the following numbers of fish: 3, 1, 5, 0, 2. Since no number appears more often than others, this set of data has *no* mode.

QUARTILES

When a set of data is arranged in numerical order, the median (also called the *second quartile*) separates the data into two equal parts. The data values below the mean are separated into two equal parts by a value called the *lower* or *first quartile*. Similarly, the data values above the mean are separated into two equal parts by a value called the *upper* or *third quartile*. In this way, quartiles separate a set of data into four equal parts.

To find the quartiles, we begin by finding the median. For example, the heights, in inches, of 17 children are given below. The median is the ninth height in the ordered list.

$$45, 47, 47, 48, 49, 50, 50, 50, 51, 51, 52, 53, 53, 55, 55, 56, 58$$

median

The median separates the data into two equal groups with eight heights in each group. The lower quartile is the middle value of the heights that are below the median, and the upper quartile is the middle value of the heights that are above the median. Since there are eight heights in each of these sets of data, the middle values are the averages of the fourth and fifth heights in each group.

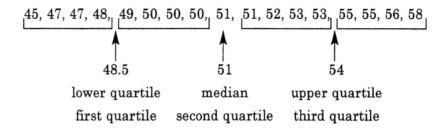

lower quartile   median   upper quartile

first quartile   second quartile   third quartile

A *box-and-whisker plot* is a graph that displays the quartile values as well as the smallest and largest numbers in a set of data. To draw a box-and-whisker plot for the set of heights given above, we choose a scale that includes all of the heights, 45 to 58, in the set of data. Then we place dots above the numbers that are the lowest value, 45; the first quartile, 48.5; the median, 51; the third quartile, 54; and the largest value, 58.

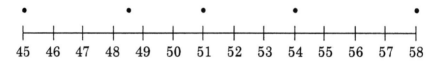

Next, we draw a box between the dots that represent the lower and the upper quartiles, and a vertical line in the box at the median.

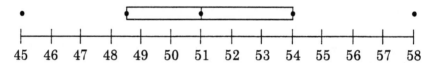

Finally, we draw the whiskers, a line segment connecting the lowest value and the first quartile, and a line segment connecting the upper quartile and the largest value.

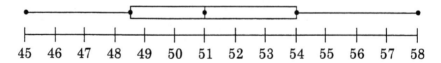

MODEL PROBLEMS

1. Mrs. Taggart bought a new car. She kept a record of the number of miles that she drove per gallon of gas for each of the first six times she filled the tank. Her record showed the following numbers of miles per gallon: 29, 32, 32, 33, 35, 37. Find: *a.* the mean   *b.* the median   *c.* the mode   *d.* the first and third quartiles

Solution:

a. $\bar{x} = \dfrac{\sum\limits_{i=1}^{n} x_i}{n} = \dfrac{29 + 32 + 32 + 33 + 35 + 37}{6} = \dfrac{198}{6} = 33$

b. There are six entries in this set of data. The median is the mean (average) of the entries that fall into positions 3 and 4 from the top or the bottom.

29, 32, 32, 33, 35, 37

$\text{median} = \dfrac{32 + 33}{2}$

$= 32.5$

c. The mode is the entry that occurs most often: 32.

median

$\boxed{32.5}$

d. The median separates the data into two sets, each of which contains three numbers:

{29, 32, 32} and {33, 35, 37}

The first quartile is the middle number in the lower half, 32, and the third quartile is the middle number in the upper half, 35.

29, $\boxed{32}$, 32, 33, $\boxed{35}$, 37

first   third
quartile   quartile

Answer: a. Mean = 33   *b.* Median = 32.5   *c.* Mode = 32   *d.* First quartile = 32, Third quartile = 35

2. The ages of 25 students in a senior high school mathematics class are recorded in the frequency distribution table at the right. For these ages, find: *a.* the mean *b.* the median *c.* the mode

Age in years x	Frequency f
18	2
17	11
16	12

Solution:

a. 1. The total number of students in the class is $\Sigma f = n = 25$.

2. To find the sum of the ages of the 25 students, add two 18's, eleven 17's, and twelve 16's. Each product of a frequency, f, and an age, x, is fx, the sum of the ages of all students of the same age. The sum of these products, Σfx, equals the sum of the ages of all 25 students. Here $\Sigma fx = 415$.

x	f	fx
18	2	36
17	11	187
16	12	192
—	$\Sigma f = 25$	$\Sigma fx = 415$

3. The mean $= \bar{x} = \dfrac{\Sigma fx}{\Sigma f} = \dfrac{415}{25} = 16.6$

b. The median is the middle value.

For $n = 25$, the median is the value that is $\frac{n + 1}{2} = \frac{25 + 1}{2} = 13$, that is, 13th from the top or 13th from the bottom. Therefore, the median age lies in the interval 17.

c. The age that appears most frequently is 16 because this interval has the highest frequency.

Answer: a. Mean = 16.6   b. Median = 17   c. Mode = 16

Exercises

1. Find the mean grade for each of the following students. If necessary, round the mean to the *nearest tenth.*
   a. Peter: 90, 70, 88, 82, 70   b. Maria: 80, 82, 93, 91, 94
   c. Elizabeth: 82, 75, 100, 83   d. Thomas: 92, 91, 75, 93, 98
   e. Al: 80, 70, 92, 78, 78, 98   f. Joanna: 90, 90, 61, 90

2. a. Mr. Katzel will give a grade of A on the report card to any student whose mean average is 90.0 or higher. Which student(s) listed in Exercise 1, if any, will receive a grade of A?
   b. If Mr. Katzel omits the lowest test grade for each student listed in Exercise 1, which student(s) will then receive a grade of A?

3. For each set of student grades in Exercise 1, find the median grade.

In 4–11, find the mode for each distribution. If no mode exists, write "None."

4. 3, 3, 4, 5, 9   5. 4, 4, 5, 9, 9   6. 4, 4, 6, 6, 6
7. 3, 4, 7, 8, 9   8. 3, 8, 3, 8, 3   9. 5, 2, 2, 5, 2, 5
10. 1, 7, 4, 3, 2, 4, 3, 1, 7, 1   11. 5, 2, 7, 2, 8, 5, 9, 3

12. What is the median for the digits 0, 1, 2, . . . , 9?

13. The median age of four children is 9.5 years. If Jeanne is 11, Debbie is 8, and Jimmy is 5 years old, then Kathy's age *cannot* be
    (1) 10   (2) 11   (3) 13   (4) 16

14. The set of data 6, 8, 9, x, 9, 8 is given. Find all possible values of x such that:
    a. There is no mode because all scores appear an equal number of times.
    b. There is only one mode.
    c. There are two modes.

In 15–18, in each case, the scores of a student on four tests are recorded. Find the score needed by each student on a fifth test so that the mean average of all five tests is exactly 80, or explain why such an average is not possible.

15. Edna: 80, 75, 92, 85   **16.** Rosemary: 77, 81, 76, 83
17. Joe: 78, 72, 70, 75   **18.** Jerry: 68, 82, 79, 71

19. Alice Garr typed a seven-page report. She made the following numbers of typing errors, reported for successive pages of the report: 2, 0, 2, 1, 4, 5, 7.
   a. For the number of errors per page, find: (1) the mean   (2) the median   (3) the mode   (4) the first quartile   (5) the third quartile
   b. Draw a box-and-whisker plot for this set of data.

20. David enters bicycle races. His times to complete the last 10 races, each covering a distance of 20 miles, were recorded to the nearest minute as follows: 55, 58, 53, 50, 52, 50, 54, 55, 59, 55. For these recorded times, find: *a.* the mean   *b.* the median   *c.* the mode   *d.* the first quartile   *e.* the third quartile

21. In the last 12 times that Mary ran the 100-yard dash, her times to the nearest tenth of a second were 13.5, 13.2, 13.1, 13.3, 13.2, 13.0, 12.8, 13.1, 13.0, 13.1, 13.4, 13.1.   *a.* For these 12 times, find: (1) the mean   (2) the median   (3) the mode   (4) the first quartile   (5) the third quartile
   b. Draw a box-and-whisker plot for this set of data.

22. The ages of 30 students enrolled in a health class are shown in the table.
   a. Find the mean age to the *nearest tenth* of a year.
   b. Find the median age.
   c. For this set of data, which statement is true?
     (1) median > mode
     (2) median = mode
     (3) median < mode
     (4) median > mean

Ages x_i	Frequency f_i
18	2
17	12
16	14
15	2

In 23–27, for each given frequency distribution, find: *a.* the mean   *b.* the median   *c.* the mode.

23.

Measure x_i	Frequency f_i
10	3
11	5
12	2

24.

Index i	Measure x_i	Frequency f_i
1	100	3
2	90	3
3	80	4

25.

x_i	f_i
10	1
20	5
30	2
40	4

26.

x_i	f_i
5	5
4	5
3	2
2	3

27.

x_i	f_i
15	12
10	4
5	4
0	4

28. Eddie Dunn scored the following numbers of goals in his last seven hockey games: 3, 4, 0, 2, 3, 1, and 2. What is the least number of goals that Eddie must score in his next game to claim that his mean average number of goals per game is greater than 2?

3. Measures of Dispersion: The Range and The Mean Absolute Deviation

The mean is the measure of central tendency most frequently used to describe statistical data. The mean, however, does not give us sufficient information about the data to draw conclusions.

Frequency diagrams for four sets of data are given in Figures 1–4. The pictures clearly show us that these sets of data are very different.

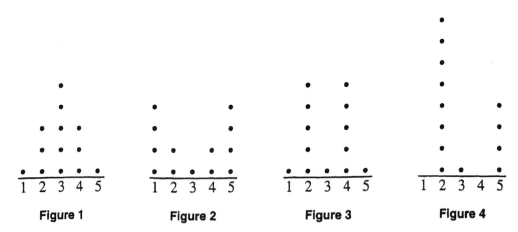

Figure 1   Figure 2   Figure 3   Figure 4

Nevertheless, for each set of data displayed here, *the mean is 3*. For example, using the data in Figure 4, we can complete the frequency table at the right and find sums. It follows that:

x_i	f_i	$f_i x_i$
1	0	0
2	8	16
3	1	3
4	0	0
5	4	20
—	$n = 13$	$\Sigma f_i x_i = 39$

$$\text{Mean} = \bar{x} = \frac{\Sigma f_i x_i}{n} = \frac{39}{13} = 3$$

Since each of these sets of data has a mean of 3, we need another measure to show how the sets are different. The new measure should indicate how individual values are scattered, or distributed, about the mean. A number that indicates the spread, or variation, of data values about the mean is called a ***measure of dispersion.***

In this section, we will study two measures of dispersion: the *range* and the *mean absolute deviation.* In the next section, we will study two other measures of dispersion, *variance* and *standard deviation.*

THE RANGE

The simplest of the measures of dispersion is the range. The *range* is the difference between the highest value and the lowest value in a set of data.

For example, if Eve Lucano's grades for this marking period are 97, 94, 92, 89, and 87, the range of Eve's grades is 97 − 87 or 10.

Since the range is dependent on only the highest and lowest values in a distribution, the range is *often unreliable* as a measure of dispersion. Let us consider the following cases in which we compare data.

Comparison 1. For each set of data in Figures 5 and 6, the mean is 6 and the range = 10 − 2 = 8.

Although these frequency diagrams are very different, the range does not help us in comparing the distribution of data in these sets.

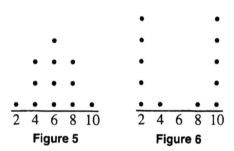

Figure 5   Figure 6

Comparison 2. The test scores for two students are shown below.

Heidi: 75, 88, 91, 92, 95, 99   Range = 99 − 75 = 24
Eric: 87, 88, 91, 92, 95, 99   Range = 99 − 87 = 12

While the range of Heidi's scores is twice the range of Eric's scores, the individual scores of these two students are exactly the same, except for one extreme score, 75, in Heidi's set, In fact, both students have averages in the low nineties. Here, the sets of data are very much alike, but we have been led to believe that they are different because of the differences in the ranges.

THE MEAN ABSOLUTE DEVIATION

Let us study a more useful measure of dispersion, based on the information in the accompanying table.

x_i	$\bar{x}$	$x_i - \bar{x}$	$\lvert x_i - \bar{x} \rvert$
93	86	7	7
90	86	4	4
89	86	3	3
87	86	1	1
85	86	−1	1
72	86	−14	14
Σx_i 516	—	$\Sigma x_i - \bar{x}$ 0	$\Sigma \lvert x_i - \bar{x} \rvert$ 30

1. The data values in column 1 are used to find the mean:

$$\bar{x} = \frac{\Sigma x_i}{n} = \frac{516}{6} = 86$$

2. The mean of 86 is entered in every row of column 2.

3. The difference between each entry, x_i, in the sample and the mean, $\bar{x}$, is recorded in column 3. We note that the sum of these differences is 0; that is, $\Sigma(x_i - \bar{x}) = 0$. If the differences had been reversed, it would still be true that $\Sigma(\bar{x} - x_i) = 0$. This example illustrates the fact that *the sum of the differences between each entry in a sample and the mean of that sample is always equal to* 0.

4. In column 4, however, by taking the absolute value of the deviation of each entry from the mean, we see that the sum of these absolute values is usually a number other than 0. Here, $\Sigma \lvert x_i - \bar{x} \rvert = 30$.

5. By definition, the *mean absolute deviation* $= \dfrac{\displaystyle\sum_{i=1}^{n} \lvert x_i - \bar{x} \rvert}{n} = \dfrac{30}{6} = 5.$

If $\bar{x}$ is the mean of a set of numbers denoted by x_i, then the **mean absolute deviation,** or simply the **mean deviation,** is

$$\frac{\displaystyle\sum_{i=1}^{n} \lvert x_i - \bar{x} \rvert}{n} \quad \text{or} \quad \frac{1}{n}\sum_{i=1}^{n} \lvert x_i - \bar{x} \rvert$$

—————————— *MODEL PROBLEMS* ——————————

1. Two students in a computer course have the same average, based on the grades received by each student for six computer programs.

       Thomas: 90, 70, 85, 100, 80, 85   (mean = 85)
       Robert: 100, 90, 65, 90, 65, 100   (mean = 85)

a. For each set of grades, find the mean absolute deviation to the *nearest tenth.*

b. Which student has more widely dispersed grades? Explain why.

Solution:

a. For each student, organize the data in a table, find $\Sigma|x_i - \bar{x}|$, and find the mean absolute deviation.

Thomas

x_i	$\bar{x}$	$\|x_i - \bar{x}\|$
100	85	15
90	85	5
85	85	0
85	85	0
80	85	5
70	85	15

$\Sigma|x_i - \bar{x}| = 40$

Robert

x_i	$\bar{x}$	$\|x_i - \bar{x}\|$
100	85	15
100	85	15
90	85	5
90	85	5
65	85	20
65	85	20

$\Sigma|x_i - \bar{x}| = 80$

$$\text{mean deviation} = \frac{\Sigma|x_i - \bar{x}|}{n}$$

$$= \frac{40}{6} = 6.66$$

$$= 6.7 \quad Answer$$

$$\text{mean deviation} = \frac{\Sigma|x_i - \bar{x}|}{n}$$

$$= \frac{80}{6} = 13.33$$

$$= 13.3 \quad Answer$$

b. Robert has more widely dispersed grades because the mean deviation of his grades, 13.3, is about twice as great as the mean deviation, 6.7, for Thomas. *Answer*

2. George Goldstein is a business student. His last six scores on tests in accounting were 87, 76, 93, 83, 84, and 81. For these scores, find:
a. the range   b. the mean deviation

Solution:

a. The range is the difference between the highest and the lowest scores. Here, the range = 93 – 76 = 17.

b. 1. Organize the data in column 1 of a table.

2. Find the mean:

$$\bar{x} = \frac{\Sigma x_i}{n} = \frac{504}{6} = 84$$

3. After the mean is entered in every row of column 2, find the values of $x_i - \bar{x}$ in column 3 and $|x_i - \bar{x}|$ in column 4.

x_i	$\bar{x}$	$x_i - \bar{x}$	$|x_i - \bar{x}|$
93	84	9	9
87	84	3	3
84	84	0	0
83	84	−1	1
81	84	−3	3
76	84	−8	18
Σx_i 504	—	—	$\Sigma|x_i - \bar{x}|$ 24

4. Find the mean deviation:

$$\frac{\Sigma|x_i - \bar{x}|}{n} = \frac{24}{6} = 4$$

Answer: a. Range = 17
          b. Mean deviation = 4

Exercises

In 1–6, for each set of student grades, find the range.

1. Ann: 83, 87, 92, 92, 95

2. Barbara: 94, 90, 86, 86, 85

3. Bill: 78, 97, 82, 86, 90

4. Cathy: 88, 81, 90, 74, 72

5. Stephen: 91, 65, 92, 94, 98

6. Tom: 90, 90, 90, 90, 90

7. If each student in Mr. Pedersen's class is either 16 or 17 years old, what is the range of student ages in the class?

8. The most expensive item in Dale Singer's shopping basket is meat at $5.60, and the least expensive item is fruit at $0.39. What is the range of the prices of items in the shopping basket?

9. *a.* Anna has 14 grandchildren. If the oldest is 18 and the youngest is 3, what is the range of ages of the grandchildren?
   b. If Andrea, Anna's fifteenth grandchild, is born today, what is now the range of ages of Anna's grandchildren?

10. For the data given in the accompanying table:

a. Find the mean.

b. Copy and complete the table.

c. Find the mean absolute deviation.

x_i	$\bar{x}$	$x_i - \bar{x}$	$|x_i - \bar{x}|$
27			
26			
25			
22			
20			

11. "Curveball" Klopfer is a baseball pitcher. In his last six games, he struck out the following numbers of batters: 16, 20, 14, 13, 21, 12. For this set of data, find:

a. the range b. the mean c. the mean deviation

12. *True* or *False:* For any set of data, $\frac{1}{n}\Sigma|x_i - \bar{x}| = \frac{1}{n}\Sigma|\bar{x} - x_i|$. Explain your answer.

13. Frances owns and manages a printing business. Over the last 6 days, she has processed the following numbers of jobs: 5, 8, 12, 7, 3, 4. For the number of jobs processed, find: a. the mean (*Hint:* The mean is not an integer.)

b. the mean absolute deviation

14. Last week, Florence kept a log of the number of hours that she watched television each day. The times, recorded to the nearest half hour, are 5, 4, 3, 5, 2, 1.5, and 0. For these times: a. Find the mean to the *nearest tenth.*

b. Using the mean from part a, find the mean deviation to the *nearest hundredth.*

15. Three sets of data (Set I, Set II, and Set III) are displayed in the following frequency diagrams.

a. For each set of data, find: (1) the range (2) the mean (3) the mean deviation

b. By using the mean deviations found, tell: (1) which set of data is most closely grouped about the mean (2) which set of data is most widely dispersed

16. Two students in a mathematics class are comparing their grades.

Mary Murray: 87, 98, 82, 96, 99, 84
Thea Olmstead: 95, 92, 79, 94, 90, 96

    a. For Mary's grades, find: (1) the range   (2) the mean   (3) the mean deviation to the *nearest tenth*

    b. For Thea's grades, find: (1) the range   (2) the mean   (3) the mean deviation to the *nearest tenth*

    c. Which student has more widely dispersed grades? Explain why.

17. The data sets below show the highest daily temperatures, recorded in degrees Celsius, for two weeks in the summer. Each week had the same mean daily reading.

Week 1:   37, 35, 34, 30, 32, 36, 34   (mean = 34)
Week 2:   37, 36, 40, 33, 31, 30, 31   (mean = 34)

    a. For each week's data, find the mean deviation to the *nearest tenth*.

    b. Using part *a*, tell which week had the more consistent readings.

18. The set of data 3, 4, 4, 5, 5, 5, 6, 6, 7 contains nine numbers and has a range of 4 and a mean of 5. Write three more sets of data for which $n = 9$, the range $= 4$, and the mean $= 5$.

4. Measures of Dispersion: The Variance and the Standard Deviation

To find the mean absolute deviation, we used the absolute value of the deviation of each element of the data set from the mean to change negative differences to positive numbers. Negative differences can also be changed to positive values by squaring the differences. For example:

$$(5 - 7)^2 = (-2)^2 = 4 \qquad \text{and} \qquad (1 - 7)^2 = (-6)^2 = 36$$

THE VARIANCE

A measure of dispersion that uses the squares of the deviations from the mean gives greatest weight to the scores that are farthest from the mean.

The *variance, v,* of a set of data is the average of the squares of the deviations from the mean.

$$v = \frac{\sum_{i=1}^{n}(x_i - \bar{x})^2}{n} \qquad \text{or} \qquad v = \frac{1}{n}\sum_{i=1}^{n}(x_i - \bar{x})^2$$

A method for finding variance can be demonstrated by the following example:

☐ On five mathematics tests taken this month, Fred earned grades of 92, 86, 95, 84, and 78. Find the variance of this set of grades.

To solve this problem, follow steps 1–4 to organize the data in a table, and then step 5 to use the entries in the table to find the variance.

1. In column 1, write the data in numerical order.

2. Find the mean of the data:

$$\bar{x} = \frac{\Sigma x_i}{n} = \frac{435}{5} = 87$$

Write the mean in each row of column 2.

x_i	$\bar{x}$	$x_i - \bar{x}$	$(x_i - \bar{x})^2$
95	87	8	64
92	87	5	25
86	87	−1	1
84	87	−3	9
78	87	−9	81
435 Σx_i	—	—	180 $\Sigma(x_i - \bar{x})^2$

3. In column 3, write the difference of each entry, x_i, minus the mean, $\bar{x}$.

4. In column 4, write the square of each difference shown in column 3.

5. Add the entries in column 4, and divide this sum, $\Sigma(x_i - \bar{x})^2$, by 5, the number of grades.

$$v = \frac{\sum_{i=1}^{5}(x_i - \bar{x})^2}{5} = \frac{180}{5} = 36$$

Answer: The variance of Fred's grades is 36.

THE STANDARD DEVIATION

Some people object to variance as a measure of dispersion because it contains a distortion that results from squaring the differences. To overcome this objection, we can find the square root of the variance and thus obtain a measure of dispersion that has the same units as the given data. This measure, called the *standard deviation*, is the most important and widely used measure of dispersion in the world today.

The *standard deviation*, σ, of a set of data is equal to the square root of the variance.

The symbol for standard deviation is σ, the lowercase Greek letter sigma.

$$\sigma = \sqrt{v} = \sqrt{\dfrac{\displaystyle\sum_{i=1}^{n}(x_i - \bar{x})^2}{n}} \quad \text{or} \quad \sigma = \sqrt{\dfrac{1}{n}\sum_{i=1}^{n}(x_i - \bar{x})^2}$$

❑ Find the standard deviation of Fred's five mathematics grades: 92, 86, 95, 84, and 78.

1–5. Find the variance, using steps 1–5 of the preceding example.

6. Find the square root of the variance.

$$v = \dfrac{\displaystyle\sum_{i=1}^{5}(x_i - \bar{x})^2}{5} = 36$$

$$\sigma = \sqrt{v} = \sqrt{36} = 6$$

Answer: The standard deviation of Fred's grades is 6.

──── MODEL PROBLEM ────

The times, in minutes, required by five students to complete a test were as follows: 35, 27, 30, 25, and 38. For this set of data, find:

a. the mean   *b.* the standard deviation to the *nearest tenth*

Solution:

a. $\bar{x} = \dfrac{x_i}{n} = \dfrac{38 + 35 + 30 + 27 + 25}{5}$

$= \dfrac{155}{5} = 31$

b. To find the standard deviation, construct a table.
1. In column 1, write the data, x_i, in numerical order.
2. In each row of column 2, write the mean, $\bar{x}$.
3. In column 3, write the deviations from the mean, $x_i - \bar{x}$.
4. In column 4, write the squares of the deviations and find the sum, $\Sigma(x_i - \bar{x})^2$.
5. Use the rule for standard deviation and round to the *nearest tenth*, 4.9.

x_i	$\bar{x}$	$x_i - \bar{x}$	$(x_i - \bar{x})^2$
38	31	7	49
35	31	4	16
30	31	−1	1
27	31	−4	16
25	31	−6	36
155 Σx_i	—	—	118 $\Sigma(x_i - \bar{x})^2$

$$\sigma = \sqrt{\dfrac{1}{n}\sum_{i=1}^{n}(x_i - \bar{x})^2}$$

$$= \sqrt{\dfrac{1}{5}(118)}$$

$$= \sqrt{23.6} = 4.85$$

Exercises

1. For the data given in the accompanying table:
   a. Find the mean.
   b. Copy and complete the table.
   c. Find the standard deviation to the *nearest tenth*.

x_i	$\overline{x}$	$x_i - \overline{x}$	$(x_i - \overline{x})^2$
30			
29			
26			
23			
12			

2. The scores of six students on an IQ test are listed below.

Fred: 130	Toni: 127	Lee: 125
Paul: 122	Lynn: 128	John: 118

   For these scores, find: *a.* the mean   *b.* the standard deviation

3. The highest number of points that a student can score in a mathematics competition is 5. In the last six competitions, Jennifer's scores were 2, 1, 4, 2, 4, 5. For these scores, calculate:
   a. the mean   *b.* the standard deviation to the *nearest tenth*

4. On his last five fishing trips, Jim caught the following numbers of fish: 6, 5, 12, 3, 9. For this set of data, find: *a.* the mean   *b.* the standard deviation to the *nearest tenth*

5. On a test, five students received scores of 63, 60, 59, 57, and 56. For these scores, find: *a.* the mean   *b.* the standard deviation to the *nearest tenth*

6. The heights, in centimeters, of five players on the basketball team are listed at the right. For these heights, find: *a.* the mean   *b.* the standard deviation to the *nearest tenth*

Player	Height (cm)
R. Melendy	192
C. Cronin	189
D. Schmeling	187
M. Natale	184
R. Weinrich	183

7. Two students are comparing their grades on their report cards.

Sean O'Brien:	83, 92, 79, 65, 82, 85
Eddy Capobianco:	83, 75, 78, 86, 77, 87

   a. Which student, if either, has the higher mean average?
   b. For Sean's grades, calculate the standard deviation to the *nearest tenth*.
   c. For Eddy's grades, calculate the standard deviation to the *nearest tenth*.
   d. Which student has the greater dispersion in grades?

8. During a 5-day work week, Helen worked the following numbers of hours per day: 8, 10, 9, 11, 10. For this set of data, calculate: *a.* the mean   *b.* the standard deviation to the *nearest tenth*

9. Dave worked the following numbers of hours at a fast-food restaurant over a period of 7 days: 5, 4, 0, 9, 7, 0, 3. Using $n = 7$, find: *a.* the mean number of hours worked per day   *b.* the standard deviation to the *nearest tenth*

10. Over the last 7 days, Mr. Kavanagh spent the following number of hours reading a novel: 3, 1.5, 2, 1.5, 4, 2.5, 3. For these times, find: *a.* the mean   *b.* the standard deviation to the *nearest tenth*

11. In a statistical study, if the variance is 64, the standard deviation is
    (1) 6.4   (2) 16   (3) 8   (4) 64

12. If the standard deviation for a set of data is 2.5, find the value of the variance.

13. In a statistical study, variance and standard deviation are usually not equal. For what two numerical values would these measures be equal?

5. Permutations and Combinations

PERMUTATIONS OF THE FORM $_nP_n$; FACTORIALS

A bank contains 4 coins: a penny, a nickel, a dime, and a quarter, represented by P, N, D, and Q, respectively. If 1 coin is taken out of the bank at random, there are 4 possible outcomes. The **sample space, s,** or the set of all possible outcomes, is $\{P, N, D, Q\}$.

If a coin is tossed, there are 2 possible outcomes, heads and tails, represented by H and T, respectively. This sample space is $\{H, T\}$.

Therefore, if 1 of the 4 coins is taken out of the bank and tossed, the sample space becomes

$$\{(P, H), (P, T), (N, H), (N, T), (D, H), (D, T), (Q, H), (Q, T)\}$$

This sample space is represented on the graph at the right. We see that there are 4 · 2 or 8 possible outcomes. The result illustrates the **counting principle.**

- **The Counting Principle.** If one activity can occur in any of m ways and, following this, a second activity can occur in any of n ways, then both activities can occur in the order given in $m \cdot n$ ways.

The 4 coins in the bank are to be drawn out 1 at a time without replacement. There are 4 possible outcomes for the first draw. Then, since 3 coins remain in the bank, there are 3 possible coins to be drawn and $4 \cdot 3$ ways in which the first 2 coins could be removed. On the third draw there are 2 coins left to be selected and $4 \cdot 3 \cdot 2$ ways in which the first 3 coins could be removed. On the last draw 1 coin is left to be selected, and there are $4 \cdot 3 \cdot 2 \cdot 1$ or 24 possible orders in which the coins could have been drawn.

A *permutation* is an arrangement of objects in some specific order. The symbol $_4P_4$ is read as "the number of permutations of 4 things taken 4 at a time."

$$_4P_4 = 4 \cdot 3 \cdot 2 \cdot 1 = 24$$

There are 24 permutations of the 4 coins.

We have learned that, for any natural number n, we define *n factorial,* or *factorial n,* as

$$n! = n(n-1)(n-2) \cdots 3 \cdot 2 \cdot 1$$

Therefore, $4! = 4 \cdot 3 \cdot 2 \cdot 1$ and $_4P_4 = 4!$

- **In general, the number of permutations of n things taken n at a time is**

$$_nP_n = n! = (n-1)(n-2) \cdots 2 \cdot 1$$

PERMUTATIONS OF THE FORM $_nP_r$ $(r \le n)$

Let us imagine that a bank contains 5 coins (a penny, a nickel, a dime, a quarter, and a half-dollar), and 2 of these coins are to be drawn. There are 5 ways in which the first coin can be selected and 4 ways in which the second coin can be selected. Therefore, there are $5 \cdot 4$ or 20 orders in which these coins can be selected. The symbol $_5P_2$ is read as "the number of permutations of 5 things taken 2 at a time."

$$_5P_2 = 5 \cdot 4 = 20$$

An alternative approach to evaluating $_5P_2$ makes use of factorials:

$$_5P_2 = \frac{5!}{(5-2)!} = \frac{5!}{3!} = \frac{5 \cdot 4 \cdot 3 \cdot 2 \cdot 1}{3 \cdot 2 \cdot 1} = 20$$

- **In general, the number of permutations of n things taken r at a time is**

$$_nP_r = \underbrace{n(n-1)(n-2) \cdots}_{r \text{ factors}} \qquad \text{or} \qquad _nP_r = \frac{n!}{(n-r)!}$$

Algebra II

PERMUTATIONS WITH REPETITION

A bank contain 6 coins: a penny, a nickel, a dime, and 3 quarters. In how many different orders can the 6 coins be removed from the bank?

If the coins are considered to be all different, then the number of orders is

$$_6P_6 = 6! = 6 \cdot 5 \cdot 4 \cdot 3 \cdot 2 \cdot 1 = 720$$

But within any given order, such as $P N Q D Q Q$, there are 3! or 6 permutations of the 3 quarters that produce this arrangement of coins. These permutations are indicated by using subscripts:

$$P N Q_1 D Q_2 Q_3 \qquad P N Q_2 D Q_1 Q_3 \qquad P N Q_3 D Q_1 Q_2$$
$$P N Q_1 D Q_3 Q_2 \qquad P N Q_2 D Q_3 Q_1 \qquad P N Q_3 D Q_2 Q_1$$

We can divide the 720 arrangements of coins into groups of 6 that are the same. Therefore, the number of different orders of the 6 coins, 3 of which are quarters, is

$$\frac{6!}{3!} = \frac{6 \cdot 5 \cdot 4 \cdot 3 \cdot 2 \cdot 1}{3 \cdot 2 \cdot 1} = \frac{720}{6} = 120$$

- **In general, the number of permutations of n things taken n at a time when r are identical is $\frac{n!}{r!}$.**

COMBINATIONS, $_nC_r$ $(r \le n)$

A bank contains 4 coins: a penny, a nickel, a dime, and a quarter. Two coins are to be drawn from the bank, and the sum of the values noted. The number of ways in which 2 coins can be drawn from the bank, one after another, is a permutation: $_4P_2 = 4 \cdot 3 = 12$. Here, order is important, and the 12 permutations can be written as 12 ordered pairs.

Let us now find the number of possible *sums of the values* of the 2 coins selected. If the penny is drawn first and then the dime, the sum is 11 cents. If the dime is drawn first and then the penny, the sum is again 11 cents. Here, the order of the coins drawn is *not* important in finding the sum. A selection in which order is not important is called a **combination.**

For any 2 coins selected, such as the penny and the dime, there are $2! = 2 \cdot 1 = 2$ orders. Therefore, to find the number of combinations of 4 things taken 2 at a time, we divide the number of permutations of 4 things taken 2 at a time by the 2! orders:

Permutations		Combinations
$(P, D)(D, P)$	$\longrightarrow$	$\{P, D\}$
$(P, N)(N, P)$	$\longrightarrow$	$\{P, N\}$
$(P, Q)(Q, P)$	$\longrightarrow$	$\{P, Q\}$
$(N, D)(D, N)$	$\longrightarrow$	$\{N, D\}$
$(N, Q)(Q, N)$	$\longrightarrow$	$\{N, Q\}$
$(D, Q)(Q, D)$	$\longrightarrow$	$\{D, Q\}$

$$_4C_2 = \frac{_4P_2}{2!} = \frac{4 \cdot 3}{2 \cdot 1} = 6$$

In the box shown above, each of the six combinations listed has a unique sum; these sums are 11¢, 6¢, 26¢, 15¢, 30¢, and 35¢, respectively. By comparing the listed permutations and combinations, we see that:

1. *Permutations* are regarded as *ordered* elements, such as the ordered pairs (P, D) and (D, P).

2. *Combinations* are regarded as *sets*, such as $\{P, D\}$, in which order is not important.

If, from a class of 23 students, 4 are to represent the class in a science contest, the order in which the students are chosen is not important. This is a combination, $_{23}C_4$, which may also be written as $\binom{23}{4}$.

$$_{23}C_4 = \frac{_{23}P_4}{4!} = \frac{23 \cdot 22 \cdot 21 \cdot 20}{4 \cdot 3 \cdot 2 \cdot 1} = 8,855$$

- In general, the number of combinations of n things taken r at a time is

$$_nC_r = \frac{_nP_r}{r!} \quad \text{or} \quad \binom{n}{r} = \frac{_nP_r}{r!}$$

Note. The alternative symbol for the combination, $\binom{n}{r}$, is *not* the fraction $\left(\frac{n}{r}\right)$.

RELATIONSHIPS INVOLVING COMBINATIONS

Certain relationships with combinations can be shown to be true:

1. There is only 1 way to select a 3-person committee from a group of 3 people, and a 4-person committee from a group of 4 people.

$$_3C_3 = \frac{_3P_3}{3!} = \frac{3 \cdot 2 \cdot 1}{3 \cdot 2 \cdot 1} = 1 \quad \text{and} \quad \binom{4}{4} = \frac{_4P_4}{4!} = \frac{4 \cdot 3 \cdot 2 \cdot 1}{4 \cdot 3 \cdot 2 \cdot 1} = 1$$

- **In general, for any counting number *n*:**

$$_nC_n = 1 \quad \text{or} \quad \binom{n}{n} = 1$$

2. There is only 1 way to take 0 object from a set of *n* objects.

- **In general, for any counting number *n*:**

$$_nC_0 = 1 \quad \text{or} \quad \binom{n}{0} = 1$$

3. Since $_7C_2 = \dfrac{7 \cdot 6}{2 \cdot 1}$ and $_7C_5 = \dfrac{7 \cdot 6 \cdot \cancel{5} \cdot \cancel{4} \cdot \cancel{3}}{\cancel{5} \cdot \cancel{4} \cdot \cancel{3} \cdot 2 \cdot 1} = \dfrac{7 \cdot 6}{2 \cdot 1}$, then $_7C_2 = {_7C_5}$.

- **In general, for any whole numbers *n* and *r*, when $r \le n$:**

$$_nC_r = {_nC_{n-r}} \quad \text{or} \quad \binom{n}{r} = \binom{n}{n-r}$$

MODEL PROBLEMS

1. In how many different orders can the program for a music recital be arranged if 7 students are to perform?

   Solution: This is a permutation of 7 things taken 7 at a time.

   $$_7P_7 = 7 \cdot 6 \cdot 5 \cdot 4 \cdot 3 \cdot 2 \cdot 1 = 5,040$$

   Answer: 5,040 orders

2. In how many ways can 1 junior and 1 senior be selected from a group of 8 juniors and 6 seniors?

   Solution: Use the counting principle. The junior can be selected in 8 ways, and the senior can be selected in 6 ways. Therefore, there are $8 \cdot 6 = 48$ possible selections.

   Answer: 48 ways

3. How many 5-letter arrangements can be made from the letters in the word BOOKS?

Solution: This is a permutation of 5 things taken 5 at a time when 2 are identical.

$$\frac{{}_5P_5}{2!} = \frac{5!}{2!} = \frac{5 \cdot 4 \cdot 3 \cdot \cancel{2} \cdot \cancel{1}}{\cancel{2} \cdot \cancel{1}} = 60$$

Answer: 60 arrangements

4. A reading list gives the titles of 20 novels and 12 biographies from which each student is to choose 3 novels and 2 biographies to read. How many different combinations of titles can be chosen?

How To Proceed	*Solution*
1. Find the number of ways in which 3 novels can be chosen.	$\binom{20}{3} = \dfrac{20 \cdot 19 \cdot 18}{3 \cdot 2 \cdot 1}$ $= 1{,}140$
2. Find the number of ways in which 2 biographies can be chosen.	$\binom{12}{2} = \dfrac{12 \cdot 11}{2 \cdot 1}$ $= 66$
3. Use the counting principle to find the number of possible choices of novels and biographies.	$1{,}140 \cdot 66 = 75{,}240$

Answer: 75,240 combinations

Exercises

In 1–18, evaluate each expression.

1. ${}_6P_6$
2. ${}_8P_4$
3. ${}_{12}P_2$
4. ${}_8C_5$
5. ${}_8C_3$
6. ${}_{12}C_{12}$

7. ${}_{12}C_0$
8. $4!$
9. $\binom{13}{4}$
10. $\binom{17}{15}$
11. $\dfrac{{}_5P_3}{3!}$
12. $\dfrac{{}_6P_2}{2!}$

13. $\dfrac{30!}{26!4!}$
14. ${}_{18}C_5$
15. $\dfrac{{}_{18}P_5}{5!}$
16. ${}_{14}P_{14}$
17. ${}_{91}C_{89}$
18. ${}_{50}C_{48}$

19. In how many different ways can 6 runners be assigned to 6 lanes at the start of a race?

20. In how many different ways can a hand of 5 cards be dealt from a deck of 52 cards?

21. There are 7 students in a history club. *a.* In how many ways can a president, a vice-president, and a treasurer be elected from the members of the club? *b.* In how many ways can a committee of 3 club members be selected to plan a visit to a museum?

22. There are 5 red and 4 white marbles in a jar. A marble is drawn from the jar and is not replaced. Then, a second marble is drawn. *a.* In how many ways can a red marble and a white marble be drawn in that order? *b.* In how many ways can a red marble and a white marble be drawn in either order?

In 23–28, in each case, find the number of "words" (arrangements of letters) that can be formed from the letters of the given word, using all letters in each arrangement.

23. axis   24. circle   25. identity

26. parabola   27. abscissa   28. minimum

29. From a standard deck of 52 cards, 2 cards are drawn without replacement. *a.* How many combinations of 2 hearts are possible? *b.* How many combinations of 2 kings are possible?

30. Each week, Mark does the dishes on 4 days and Lisa does them on the remaining 3 days. In how many different orders can they choose to do the dishes? (*Hint:* This is an arrangement of 7 things with repetition.)

31. Each week, Albert does the dishes on 3 days, Rita does them on 2 days, and Marie does them on the remaining 2 days. In how many different orders can they choose to do the dishes?

32. There are 4 boys and 5 girls who are members of a chess club.
    a. How many games must be played if each member is to play every other member once?
    b. In how many ways can 1 boy and 1 girl be selected to play a demonstration game?
    c. In how many ways can a group of 3 members be selected to represent the club at a regional meet?
    d. In how many ways can 2 boys and 2 girls be selected to attend the state tournament?

6. Probability

A sample space is the set of all possible outcomes or results of an activity. An *event, E,* is a subset of a sample space. For example, if a die is rolled, the sample space is {1, 2, 3, 4, 5, 6}. The event of rolling a number less than 3 is {1, 2}.

If the die is fair, or unbiased, each outcome is equally likely to occur. The probability of rolling a number less than 3 is the ratio of the number of elements in the set, {1, 2}, to the number of elements in the sample space, {1, 2, 3, 4, 5, 6}:

$$P(\text{rolling a number less than 3}) = \frac{2}{6} = \frac{1}{3}$$

- The *theoretical probability* of an event is the number of ways that an event can occur divided by the total number of possible outcomes when each outcome is equally likely to occur.

If $P(E)$ represents the probability of an event E,
  $n(E)$ represents the number of ways that E can occur, and
  $n(S)$ represents the number of possible outcomes in the same space S, then:

$$P(E) = \frac{n(E)}{n(S)}$$

Consider the following examples:

❑ What is the probability of drawing a red queen from a standard deck of 52 cards?

Use counting.

1. The sample space S = {cards in the deck}. → $n(S) = 52$

2. The event E = {queen of hearts, queen of diamonds}. → $n(E) = 2$

3. Substitute these values in the formula: $P(\text{red queen}) = \dfrac{n(E)}{n(S)} = \dfrac{2}{52} = \dfrac{1}{26}$

Answer: $P(\text{red queen}) = \dfrac{1}{26}$

The number of outcomes in the sample space and in the event can often be determined by using permutations or combinations, as seen in the next example.

❑ Two cards are to be drawn from a standard deck of 52 cards without replacement. What is the probability that both cards will be red?

Since order is not required, use combinations.

1. The total number of outcomes is $_{52}C_2 = \dfrac{52 \cdot 51}{2 \cdot 1} = 26 \cdot 51$.

2. The number of favorable outcomes is $_{26}C_2 = \dfrac{26 \cdot 25}{2 \cdot 1} = 13 \cdot 25$.

3. Substitute these values in the formula: $P(E) = \dfrac{n(E)}{n(S)}$

$$P(2 \text{ red cards}) = \frac{n(E)}{n(S)} = \frac{_{26}C_2}{_{52}C_2} = \frac{\overset{1}{\cancel{13}} \cdot 25}{\underset{2}{\cancel{26}} \cdot 51} = \frac{25}{102}$$

Answer: $P(2 \text{ red cards}) = \dfrac{25}{102}$

Numerical values are assigned to probabilities such that:

1. The probability of an event that is ***certain*** is 1. For example, the probability of rolling a number less than 7 on a single roll of a die is $\dfrac{6}{6}$ or 1.

2. The probability of an ***impossible*** event is 0. For example, the probability of rolling a number greater than 7 on a single roll of a die is $\dfrac{0}{6} = 0$.

3. The probability of any event is greater than or equal to 0 and less than or equal to 1.

$$0 \le P(E) \le 1$$

4. Since $P(E) + P(\text{not } E) = 1$, it follow that:

$$\text{If } P(E) = p, \text{ then } P(\text{not } E) = 1 - p$$

For example, if $P(\text{rain}) = .40$, then $P(\text{not rain}) = 1 - .40 = .60$.

────────────── **MODEL PROBLEMS** ──────────────

1. A choral group is composed of 6 juniors and 8 seniors. If a junior and a senior are chosen at random to sing a duet at the spring concert, what is the probability that the choices are Emira, who is a junior, and Jean, who is a senior?

Solution:

1. Since there are 6 ways of choosing the junior and 8 ways of choosing the senior, there are $6 \cdot 8 = 48$ possible choices.

2. There is 1 choice that includes Emira and Jean.

3. $P(\text{Emira and Jean}) = \dfrac{n(E)}{n(S)} = \dfrac{1}{48}$.

Alternative Solution: Since the choices are independent events, use the counting principle for probabilities.

$$P(\text{choosing Emira}) = \frac{1}{6}, \text{ and } P(\text{choosing Jean}) = \frac{1}{8}$$

Therefore, $P(\text{choosing Emira and Jean}) = \frac{1}{6} \cdot \frac{1}{8} = \frac{1}{48}$.

Answer: $P(\text{choosing Emira and Jean}) = \frac{1}{48}$

2. From the set of two-digit numbers, $\{00, 01, 02, \ldots, 99\}$, a number is selected at random. What is the probability that both digits in the number are even?

Solution: Even digits are 0, 2, 4, 6, 8; odd digits are 1, 3, 5, 7, 9.

$$P(\text{first digit even}) = \frac{5}{10}, \text{ and } P(\text{second digit even}) = \frac{5}{10}$$

By the counting principle for probabilities:

$$P(\text{both even}) = P(\text{first even}) \cdot P(\text{second even}) = \frac{5}{10} \cdot \frac{5}{10} = \frac{1}{2} \cdot \frac{1}{2} = \frac{1}{4}$$

Answer: $P(\text{both digits even}) = \frac{1}{4}$ or .25

3. What is the probability that a 3-letter word formed from the letters of the word COINAGE consists of all vowels?

How To Proceed	*Solution*
1. Find $n(S)$, the number of 3-letter permutations of the 7 letters in the word COINAGE	$n(S) = {}_7P_3 = 7 \cdot 6 \cdot 5$ $= 210$
2. Find $n(E)$, the number of 3-letter permutations of the 4 vowels (O, I, A, E) in the word.	$n(E) = {}_4P_3 = 4 \cdot 3 \cdot 2$ $= 24$
3. Find the probability of event E.	$P(E) = \dfrac{n(E)}{n(S)} = \dfrac{24}{210} = \dfrac{4}{35}$

Answer: $P(\text{all vowels}) = \dfrac{4}{35}$

Exercises

1. What is the probability of getting a number less than 5 on a single throw of a fair die?

2. If 1 letter of the word ELEMENT is chosen at random, what is the probability that the vowel *e* is chosen?

3. A bag contains only 5 red marbles and 3 blue marbles. If 1 marble is drawn at random from the bag, what is the probability that it is blue?

4. If the letters of the word *equal* are rearranged at random, what is the probability that the first letter of the new arrangement is a vowel?

5. If 2 coins are tossed, what is the probability that both show heads?

6. If 2 coins are tossed, what is the probability the neither shows heads?

7. If a card is drawn from a standard deck of 52 cards, what is the probability that the card is a queen?

8. If 2 cards are drawn from a standard deck of 52 cards without replacement, what is the probability that both cards are queens?

9. A bank contains 4 coins: a penny, a nickel, a dime, and a quarter. One coin is removed at random and tossed. *a.* What is the probability that the dime is removed? *b.* What is the probability that the coin shows heads? *c.* What is the probability that the coin is a quarter that shows heads? *d.* What is the probability that the coin has a value less than 20 cents and shows heads?

10. The weather report gives the probability of rain on Saturday as 20% and the probability of rain on Sunday as 10%. *a.* What is the probability that it will *not* rain on Saturday? *b.* What is the probability that it will *not* rain on Sunday? *c.* What is the probability that it will *not* rain either day?

11. A seed company advertises that, if its geranium seed is properly planted, the probability that the seed will grow is 90%. *a.* What is the probability that a geranium seed that has been properly planted will fail to grow? *b.* If 5 geranium seeds are properly planted, what is the probability that all will fail to grow?

12. Of the 5 sandwiches that Mrs. Muth made for her children's lunches, 2 contain tuna fish and 3 contain peanut butter and jelly. Her son, Tim, took 2 sandwiches at random. What is the probability that these 2 sandwiches contain peanut butter and jelly?

13. Mrs. Gillis's small son, Brian, tore all the labels off the soup cans on the kitchen shelf. If Mrs. Gillis knows that she bought 4 cans of tomato soup and 2 cans of vegetable soup, what is the probability that the first 2 cans of soup she opens are both tomato?

14. Of the 15 students in Mrs. Barney's mathematics class, 10 take Spanish. If 2 students are absent from Mrs. Barney's class on Monday, what is the probability that both of these students take Spanish?

15. At a card party, 2 door prizes are to be awarded by drawing 2 names at random from a box. The box contains the names of 40 persons, including Patricia Sullivan and Joe Ramirez. *a.* What is the probability that Joe's name is *not* drawn for either prize? *b.* What is the probability that Patricia wins 1 of the prizes?

7. Probability with Two Outcomes; Binomial Probability

In Section 6, some exercises were solved by using the counting principle with probabilities. In this section, we will study applications of this principle in greater detail.

In the spinner at the right, the arrow can land on 1 of 3 equally likely regions, numbered 1, 2, and 3. If the arrow lands on a line, the spin is not counted, and the arrow is spun again.

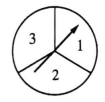

Let us define an experiment with 2 outcomes for this spinner as either obtaining an odd number or obtaining an even number. Therefore:

$$P(\text{odd}) = P(O) = \frac{2}{3} \quad \text{and} \quad P(\text{even}) = P(E) = \frac{1}{3}$$

When the arrow is spun several times, the result of each spin is independent of the results of the other spins. To find the probability of getting an odd number each time in several spins, we can use the *counting principle with probabilities.*

• **If E and F are independent events, and if the probability of E is $m(0 \leq m \leq 1)$ and the probability of F is $n(0 \leq n \leq 1)$, then the probability of E and F occurring jointly is $m \cdot n(0 \leq m \cdot n \leq 1)$.**

If the arrow is spun twice:

$$P(\text{2 odd numbers in 2 spins}) = \frac{2}{3} \cdot \frac{2}{3} = \frac{4}{9}$$

If the arrow is spun 3 times:

$$P(\text{3 odd numbers in 3 spins}) = \frac{2}{3} \cdot \frac{2}{3} \cdot \frac{2}{3} = \frac{8}{27}$$

If the arrow is spun 4 times:

$$P(4 \text{ odd numbers in 4 spins}) = \frac{2}{3} \cdot \frac{2}{3} \cdot \frac{2}{3} \cdot \frac{2}{3} = \frac{16}{81}$$

We can also apply the counting principle with probabilities to situations where the arrow is spun n times and an odd number is obtained *less than* n times, as shown in the following examples.

❑ Find the probability of obtaining *exactly* 1 *odd* number on 4 spins of the arrow.

Consider getting an odd number on the first spin and an even number on each of the other 3 spins.

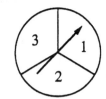

$$P(\text{odd on first spin only}) = \frac{2}{3} \cdot \frac{1}{3} \cdot \frac{1}{3} \cdot \frac{1}{3} = \frac{2}{81}$$

As seen in each row of the chart at the right, however, the odd number could appear on the second spin, or on the third spin, or on the fourth spin. Then:

O	E	E	E
E	O	E	E
E	E	O	E
E	E	E	O

$$P(\text{odd on second spin only}) = \frac{1}{3} \cdot \frac{2}{3} \cdot \frac{1}{3} \cdot \frac{1}{3} = \frac{2}{81}$$

$$P(\text{odd on third spin only}) = \frac{1}{3} \cdot \frac{1}{3} \cdot \frac{2}{3} \cdot \frac{1}{3} = \frac{2}{81}$$

$$P(\text{odd on fourth spin only}) = \frac{1}{3} \cdot \frac{1}{3} \cdot \frac{1}{3} \cdot \frac{2}{3} = \frac{2}{81}$$

Since there are 4 possible ways to spin exactly 1 odd number, each with a probability of $\left(\frac{2}{3}\right)^1 \cdot \left(\frac{1}{3}\right)^3$ or $\frac{2}{81}$, it follows that:

$$P(\text{exactly 1 odd on 4 spins}) = 4 \cdot \left(\frac{2}{3}\right)^1 \cdot \left(\frac{1}{3}\right)^3 = \frac{8}{81}$$

Answer: $P(\text{exactly 1 odd on 4 spins}) = \frac{8}{81}$

❑ Find the probability of obtaining *exactly* 2 *odd* numbers on 4 spins of the arrow.

Consider the case in which the first 2 spins are odd and the last 2 spins are even.

$$P(\text{odd on first 2 spins only}) = \frac{2}{3} \cdot \frac{2}{3} \cdot \frac{1}{3} \cdot \frac{1}{3} = \left(\frac{2}{3}\right)^2 \cdot \left(\frac{1}{3}\right)^2 = \frac{4}{81}$$

The chart at the right, however, shows that there are 6 possible ways to spin exactly 2 odd numbers. This is a *combination* of 2 odd numbers out of 4 spins, obtained by the formula $_4C_2 = \dfrac{4 \cdot 3}{2 \cdot 1} = 6$. Thus:

O	O	E	E
O	E	O	E
O	E	E	O
E	O	O	E
E	O	E	O
E	E	O	O

$$P(\text{exactly 2 odds on 4 spins}) = {}_4C_2 \cdot \left(\frac{2}{3}\right)^2 \cdot \left(\frac{1}{3}\right)^2$$

$$= 6 \cdot \frac{4}{9} \cdot \frac{1}{9} = \frac{24}{81} \text{ or } \frac{8}{27}$$

Answer: $P(\text{exactly 2 odds on 4 spins}) = \dfrac{24}{81} \text{ or } \dfrac{8}{27}$

❑ Find the probability of obtaining *exactly* 3 *odd* numbers on 4 spins of the arrow.

Consider the case in which the first 3 spins are odd and the last spin is even.

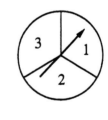

$$P(\text{odd on only the first 3 spins}) = \frac{2}{3} \cdot \frac{2}{3} \cdot \frac{2}{3} \cdot \frac{1}{3}$$

$$= \left(\frac{2}{3}\right)^3 \cdot \left(\frac{1}{3}\right)^1 = \frac{8}{81}$$

Since the number of possible ways to obtain exactly 3 odd numbers on 4 spins of the arrow is a combination, $_4C_3$, it follows that:

$$P(\text{exactly 3 odds on 4 spins}) = {}_4C_3 \cdot \left(\frac{2}{3}\right)^3 \cdot \left(\frac{1}{3}\right)^1$$

$$= \frac{4 \cdot \cancel{3} \cdot \cancel{2}}{\cancel{3} \cdot \cancel{2} \cdot 1} \cdot \frac{8}{27} \cdot \frac{1}{3} = \frac{32}{81}$$

Answer: $P(\text{exactly 3 odds on 4 spins}) = \dfrac{32}{81}$

The patterns seen in the previous solutions enable us to determine the probability of exactly r successes in n independent trials of an experiment with *exactly two outcomes*, called a ***Bernoulli experiment*** or ***binomial-probability.***

A Bernoulli experiment, such as tossing a coin, has two outcomes: heads and tails. Other experiments, such as tossing a die, can also be thought of as having two outcomes, for example, rolling a one and not rolling a one. If an event E, such as rolling a one on a die, is to occur exactly r times in n trials, then the event *not E*, that is, rolling a number that is *not* one, must occur $n - r$ times in n trials.

- **In general, for a given experiment, if the probability of success is p and the probability of failure is $1 - p = q$, then the probability of exactly r successes in n independent trials is**

$$_nC_r p^r q^{n-r}$$

This formula will be examined again in Chapter XV, Section 14, when we study the binomial theorem.

~~~~~~~~~~ *MODEL PROBLEMS* ~~~~~~~~~~

1. If a fair coin is tossed 10 times, what is the probability that it falls tails exactly 6 times?

Solution:

In 1 toss of a fair coin: probability of success $p = P(\text{tails}) = \frac{1}{2}$

and probability of failure $q = P(\text{heads}) = \frac{1}{2}$

Here, the number of trials, n, is 10, and the number of successes, r, is 6. Use the formula for a Bernoulli experiment:

$$_nC_r \quad p^r \quad q^{n-r}$$

$$_{10}C_6 \left(\frac{1}{2}\right)^6 \left(\frac{1}{2}\right)^{10-6} = \frac{10 \cdot 9 \cdot 8 \cdot 7 \cdot 6 \cdot 5}{6 \cdot 5 \cdot 4 \cdot 3 \cdot 2 \cdot 1} \left(\frac{1}{2}\right)^6 \left(\frac{1}{2}\right)^4$$

$$= 210 \cdot \frac{1}{64} \cdot \frac{1}{16} = \frac{210}{1{,}024} = \frac{105}{512}$$

Answer: P(exactly 6 tails in 10 tosses) $= \dfrac{105}{512}$

2. If 5 fair dice are tossed, what is the probability that they show exactly 3 fours?

Solution:

For this set of data:

$$p = P(4) = \frac{1}{6}$$

$$q = P(\text{not } 4) = \frac{5}{6}$$

$$n = 5 \text{ trials}$$

$$r = 3 \text{ successes}$$

$$n - r = 2 \text{ failures}$$

Use the formula: $_nC_r p^r q^{n-r}$

$$P(3 \text{ fours in 5 trials}) = {}_5C_3\left(\frac{1}{6}\right)^3\left(\frac{5}{6}\right)^2$$

$$= \frac{5 \cdot 4 \cdot 3}{3 \cdot 2 \cdot 1} \cdot \frac{1}{216} \cdot \frac{25}{36}$$

$$= 10 \cdot \frac{1}{216} \cdot \frac{25}{36}$$

$$= \frac{250}{7,776} = \frac{125}{3,888}$$

Answer: $P(3 \text{ fours in 5 trials}) = \dfrac{125}{3,888}$

Exercises

1. A fair coin is tossed 4 times.   *a.* Find the probability of tossing:
(1) exactly 4 tails   (2) exactly 3 tails   (3) exactly 2 tails
(4) exactly 1 tail   (5) exactly 0 tails
b. What is the sum of the probabilities found in part *a?*

2. A fair coin is tossed 5 times. Find the probability of tossing:
a. exactly 2 heads   *b.* exactly 3 heads   *c.* exactly 4 heads

3. If 4 fair dice are tossed, find the probability of getting:
a. exactly 3 fives   *b.* exactly 4 fives   *c.* exactly 2 fives

4. If 4 fair dice are tossed, find the probability of getting: *a.* exactly 3 even numbers   *b.* exactly 2 odd numbers   *c.* no odd numbers

5. Jan's record shows that her probability of success on a basketball free throw is $\frac{3}{5}$. Find the probability that Jan will be successful on 2 out of 3 shots.

6. The probability that the Wings will win a baseball game is $\frac{2}{3}$. State the probability that the Wings will win: *a.* exactly 2 out of 4 games   *b.* exactly 3 out of 4 games   *c.* their next 4 games

In 7 and 8, select the *numeral* preceding the choice that best answers each question.

7. What is the probability of getting a number less than 3 on 6 out of 10 tosses of a fair die?

   (1) $6\left(\frac{1}{3}\right)^6$ (2) $210\left(\frac{1}{3}\right)^6$ (3) $6\left(\frac{1}{3}\right)^6\left(\frac{2}{3}\right)^4$ (4) $210\left(\frac{1}{3}\right)^6\left(\frac{2}{3}\right)^4$

8. A coin is loaded so that the probability of heads is $\frac{1}{4}$. What is the probability of getting exactly 3 heads on 8 tosses of the coin?

   (1) $_8C_3\left(\frac{1}{4}\right)^3$ (2) $_8C_3\left(\frac{1}{4}\right)^3\left(\frac{3}{4}\right)^5$ (3) $_8C_3\left(\frac{1}{4}\right)^3\left(\frac{1}{4}\right)^5$ (4) $3\left(\frac{1}{4}\right)^3\left(\frac{3}{4}\right)^5$

In 9–13, answers may be expressed in exponential form, as in the choices for Exercises 7 and 8.

9. The probability that a flashbulb is defective is found to be $\frac{1}{20}$. What is the probability that a package of 6 flashbulbs has only 1 defective bulb?

10. A multiple-choice test gives 5 possible choices for each answer, of which 1 is correct. The probability of selecting the correct answer by guessing is $\frac{1}{5}$.

    a. What is the probability of getting 5 out of 10 questions correct by guessing?
    b. What is the probability of getting only 1 out of 10 questions correct by guessing?
    c. What is the probability of getting 9 correct answers out of 10 by guessing?

11. Mrs. Shusda gave a true-false test of 10 questions. The probability of selecting the correct answer by guessing is $\frac{1}{2}$.

    a. What is the probability that Fred, who guessed at every answer, will get 9 out of 10 correct?
    b. What is the probability that Fred will get 10 out of 10 correct?
    c. What is the probability that Fred will get either 9 or 10 out of 10 correct?

12. In a group of 100 persons who were born in June, what is the probability that exactly 2 were born on June 1? (Assume that a person born in June is equally likely to have been born on any 1 of the 30 days.)

13. In a box there are 4 red marbles and 5 white marbles. Marbles are drawn 1 at a time and replaced after each drawing. What is the probability of drawing:

    a. exactly 2 red marbles when 3 marbles are drawn?
    b. exactly 3 white marbles when 5 marbles are drawn?
    c. exactly 7 red marbles when 12 marbles are drawn?

8. *At Least* and *At Most*

When we are anticipating success on repeated trials of an experiment, we often require *at least* a given number. For example, in a game, David rolls 5 dice. To win, *at least* 3 of the 5 dice that David rolls must be "ones." Therefore, David will win if he rolls 3, 4, or 5 "ones." In general:

- At least r successes in n trials means $r, r + 1, r + 2, \ldots, n$ successes.

If a manufacturer considers *at most* 2 defective parts in a lot of 100 parts to be an acceptable standard, then there can be 2, 1, or 0 defective parts in every 100 parts.

- At most r successes in n trials means $r, r - 1, r - 2, \ldots, 0$ successes.

The model problems that follow show how these concepts are applied to probability.

─────────── *MODEL PROBLEMS* ───────────

1. Rose is the last person to compete in a basketball free-throw contest. To win, Rose must be successful in at least 4 out of 5 throws. If the probability that Rose will be successful on any single throw is $\frac{3}{4}$, what is the probability that Rose will win the contest?

   Solution: To be successful in at least 4 out of 5 throws means to be successful in 4 or in 5 throws.

   On 1 throw, $P(\text{success}) = \frac{3}{4}$ and $P(\text{failure}) = 1 - \frac{3}{4} = \frac{1}{4}$. Then:

   1. $P(4 \text{ out of } 5 \text{ successes}) = {}_5C_4\left(\frac{3}{4}\right)^4\left(\frac{1}{4}\right)^1$

   $$= \frac{5 \cdot 4 \cdot 3 \cdot 2}{4 \cdot 3 \cdot 2 \cdot 1} \cdot \frac{81}{4^4} \cdot \frac{1}{4^1} = \frac{405}{4^5} = \frac{405}{1,024}$$

   2. $P(5 \text{ out of } 5 \text{ successes}) = {}_5C_5\left(\frac{3}{4}\right)^5\left(\frac{1}{4}\right)^0 = 1 \cdot \frac{243}{4^5} \cdot 1 = \frac{243}{1,024}$

   3. $P(\text{at least 4 out of 5 successes})$:

   $$= P(4 \text{ out of } 5 \text{ successes}) + P(5 \text{ out of } 5 \text{ successes})$$

   $$= \frac{405}{1,024} + \frac{243}{1,024} = \frac{648}{1,024} = \frac{81}{128}$$

   Answer: $P(\text{at least 4 out of 5 successes}) = \frac{81}{128}$

2. A family of 5 children is chosen at random. What is the probability that there are at most 2 boys in this family of 5 children?

Solution: To have at most 2 boys means to have 0, 1, or 2 boys. Let us assume that $P(\text{boy}) = \frac{1}{2}$ and $P(\text{girl}) = \frac{1}{2}$. Then:

1. $P(0 \text{ boys in 5 children}) = {}_5C_0\left(\frac{1}{2}\right)^5\left(\frac{1}{2}\right)^0 = 1 \cdot \left(\frac{1}{2}\right)^5 \cdot 1 = \frac{1}{32}$

2. $P(1 \text{ boy in 5 children}) = {}_5C_1\left(\frac{1}{2}\right)^1\left(\frac{1}{2}\right)^4 = 5 \cdot \frac{1}{2} \cdot \frac{1}{16} = \frac{5}{32}$

3. $P(2 \text{ boys in 5 children}) = {}_5C_2\left(\frac{1}{2}\right)^2\left(\frac{1}{2}\right)^3 = \frac{5 \cdot 4}{2 \cdot 1} \cdot \frac{1}{4} \cdot \frac{1}{8} = \frac{10}{32}$

4. $P\left(\begin{array}{c}\text{at most 2 boys} \\ \text{in 5 children}\end{array}\right) = P(0 \text{ boys}) + P(1 \text{ boy}) + P(2 \text{ boys})$

$$= \frac{1}{32} \quad + \frac{5}{32} \quad + \frac{10}{32} \quad = \frac{16}{32} = \frac{1}{2}$$

Answer: $P(\text{at most 2 boys in family of 5 children}) = \frac{1}{2}$

3. A coin is loaded so that the probability of heads is 4 times the probability of tails.
a. What is the probability of heads on a single throw?
b. What is the probability of at least 1 tail in 5 throws?

Solution:

a. Let $P(\text{tails}) = q$ and $P(\text{heads}) = 4q$.

$$q + 4q = 1$$
$$5q = 1$$
$$q = \frac{1}{5}, \text{ or } P(\text{tails}) = \frac{1}{5}$$
$$\text{Then } 4q = 4 \cdot \frac{1}{5} = \frac{4}{5}, \text{ and } P(\text{heads}) = \frac{4}{5}$$

b. $P(\text{at least 1 tail in 5 throws})$

$= P(1 \text{ tail}) + P(2 \text{ tails}) + P(3 \text{ tails}) + P(4 \text{ tails}) + P(5 \text{ tails})$

$= {}_5C_1\left(\frac{1}{5}\right)^1\left(\frac{4}{5}\right)^4 + {}_5C_2\left(\frac{1}{5}\right)^2\left(\frac{4}{5}\right)^3 + {}_5C_3\left(\frac{1}{5}\right)^3\left(\frac{4}{5}\right)^2 + {}_5C_4\left(\frac{1}{5}\right)^4\left(\frac{4}{5}\right)^1 + {}_5C_5\left(\frac{1}{5}\right)^5$

$$= 5 \cdot \frac{256}{3,125} + 10 \cdot \frac{64}{3,125} + 10 \cdot \frac{16}{3,125} + 5 \cdot \frac{4}{3,125} + 1 \cdot \frac{1}{3,125}$$

$$= \frac{1,280}{3,125} + \frac{640}{3,125} + \frac{160}{3,125} + \frac{20}{3,125} + \frac{1}{3,125}$$

$$= \frac{2,101}{3,125}$$

Alternative Solution: There is one way to fail to get at least 1 tail in 5 throws, that is, to get 5 heads in 5 throws.

$P(\text{failure}) = P(5 \text{ heads})$

$$= {}_5C_0 \left(\frac{4}{5}\right)^5$$

$$= 1 \cdot \frac{1,024}{3,125}$$

$$= \frac{1,024}{3,125}$$

Then, $P(\text{success}) = 1 - P(\text{failure})$

$$= 1 - \frac{1,024}{3,125}$$

$$= \frac{3,125}{3,125} - \frac{1,024}{3,125}$$

$$= \frac{2,101}{3,125}$$

Answer: a. $P(\text{heads}) = \frac{4}{5}$ b. $P(\text{at least 1 tail in 5 throws}) = \frac{2,101}{3,125}$

Exercises

1. A fair coin is tossed 4 times. Find the probability of tossing:
   a. exactly 3 heads   b. exactly 4 heads   c. at least 3 heads
   d. exactly 0 heads   e. exactly 1 head   f. at most 1 head

2. If a fair coin is tossed 6 times, find the probability of obtaining at most 2 heads.

3. If 5 fair coins are tossed, what is the probability that at least 2 tails are obtained?

4. A fair die is rolled 3 times. Find the probability of rolling:
   a. exactly 1 five   b. no fives   c. at most 1 five
   d. at least 2 sixes   e. at most 2 fours   f. at most 1 even
   g. exactly 1 even   h. at most 1 number less than three
   i. at least 1 even   j. at least 1 number greater than four

5. If a fair die is tossed 5 times, find the probability of obtaining at most 2 ones.

6. If 4 fair dice are rolled, what is the probability that at least 2 are fives?

In 7–18, an arrow can land on one of 5 equally likely regions on a spinner, numbered 1, 2, 3, 4, and 5. If the arrow lands on a line, the spin is not counted and the arrow is spun again.

In 7–16, for the spinner described, find each probability.

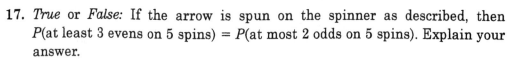

   7. P(even number)

   8. P(odd number)

   9. P(both even on 2 spins)

10. P(both odd on 2 spins)

11. P(at least 1 even on 2 spins)

12. P(at least 1 odd on 2 spins)

13. P(exactly 2 evens on 3 spins)   **14.** P(at least 2 evens on 3 spins)

15. P(exactly 2 odds on 4 spins)   **16.** P(at most 2 odds on 4 spins)

17. *True* or *False:* If the arrow is spun on the spinner as described, then P(at least 3 evens on 5 spins) = P(at most 2 odds on 5 spins). Explain your answer.

18. For the spinner described: *a.* Find $P(2)$.   *b.* If the arrow is spun 5 times, will a "2" appear exactly once? Support your answer by finding P(exactly one "2" on 5 spins).

19. A coin is weighted so that the probability of heads is $\frac{4}{5}$.

   a. What is the probability of getting at least 3 heads when the weighted coin is tossed 4 times?

   b. What is the probability of getting at most 2 tails when the weighted coin is tossed 5 times?

20. Each evening the members of the Sanchez family are equally likely to watch the news on any one of 3 possible TV channels: 5, 8, and 13. *a.* What is the probability that they watch the news on channel 8 on at least 3 out of 5 evenings? *b.* What is the probability that they watch the news on channel 13 on at most 2 out of 5 evenings?

21. In a game, the probability of winning is $\frac{1}{5}$ and the probability of losing is $\frac{4}{5}$. If 3 games are played, what is the probability of winning at least 2 games?

22. In each game that the school basketball team plays, the probability that the team will win is $\frac{2}{3}$. What is the probability that the team will win at least 3 of the next 4 games?

23. A die is loaded so that the probability of rolling a one is $\frac{3}{4}$. What is the probability of rolling at least 2 ones when the die is tossed 3 times?

24. If a family of 4 children is selected at random, what is the probability that at most 3 of the children are boys?

In 25 and 26, an electronic game contains 9 keys. As shown at the right, 5 keys have numbers, and 4 keys have colors. Each key is equally likely to be pressed on each move.

Red	1	Blue
2	3	4
Green	5	Yellow

25. If one key is pressed at random in the electronic game, find:
   a. *P*(number key)   b. *P*(color key)

26. If 3 keys are pressed at random in the electronic game, find the probability of selecting: a. exactly 1 color key   b. at least 2 color keys   c. at most 1 color key   d. all color keys   e. exactly 1 number key   f. at least 1 number key

In 27–31, answers may be expressed in exponential form.

27. Of last year's graduates, 3 out of 5 are enrolled in college. If the names of 10 of last year's graduates are chosen are random, what is the probability that at least 8 out of 10 are in college?

28. A coin is loaded so that the probability of heads on a single throw is 3 times the probability of tails.
   a. What is the probability of heads and the probability of tails on a single throw?
   b. What is the probability of at most 3 heads when the coin is tossed 6 times?

29. A license number consists of 5 letters of the alphabet selected at random. Each letter can be selected any number of times.
   a. What is the probability that the license number has at most 2 Q's?
   b. What is the probability that the license number has at least 4 X's?

30. A manufacturer tests her product and finds that the probability of a defective part is .02. What is the probability that out of 5 parts selected at random at most 1 will be defective?

31. A seed company advertises that, if its seeds are properly planted, 95% of them will germinate. What is the probability that, when 20 seeds are properly planted, at least 15 will germinate?

⎯⎯⎯⎯⎯ CALCULATOR APPLICATIONS ⎯⎯⎯⎯⎯

A graphing calculator can be used to sum frequencies and calculate the mean and standard deviation of a set of data.

⎯⎯⎯⎯⎯⎯ MODEL PROBLEMS ⎯⎯⎯⎯⎯⎯

1. *a.* Find the number of students, total number of reports, mean number of reports per student, and standard deviation for Mrs. Gallagher's class (see page 506).

   b. Draw a box plot of the data.

   Solution: On the TI-83, data sets are stored in lists. Before you enter any new data clear all previous list entries by pressing [2nd] [MEM] 4 [ENTER]. The number of reports will be listed in L_1 and the frequency, f, in L_2.

   Enter: [STAT] [ENTER] 5 [ENTER] 4 [ENTER] 3 [ENTER]
   2 [ENTER] 1 [ENTER] 0 [ENTER] [▶] 1 [ENTER]
   5 [ENTER] 8 [ENTER] 7 [ENTER] 4 [ENTER] 2 [ENTER]

   After the data entry is complete, press [STAT] and use the [▶] key to choose [CALC]; then press [ENTER]. The calculator will copy 1-Var Stats to the home screen. Then enter the list names L_1 and L_2; for the 1-Var Stats, the calculator interprets this instruction to mean that (first entry in L_1) occurs (first entry in L_2) times, and so on. (L_2 is the frequency list for L_1.)

   Enter: [STAT] [▶] [ENTER] [2nd] [L_1] [,] [2nd] [L_2] [ENTER]

   Display:

   ```
   1-VAR STATS
   X̄=2.481481481
   ΣX=67
   ΣX²=209
   SX=1.282136749
   σX=1.258169463
   ↓n=27
   ```

   ```
   1-VAR STATS
   ↑n=27
   minX=0
   Q₁=2
   MED=3
   Q₃=3
   maxX=5
   ■
   ```

   The first screen shows the mean, $\bar{x}$, on the first line: $\bar{x} = 2.48$ (nearest hundredth).
   The total number of reports, Σx, is shown on the second line to be 67.
   The number of students, n, is shown at the bottom to be 27.
   The standard deviation, σx, is shown on the fifth line; $\sigma x = 1.26$ (nearest hundredth).

The number of students, n, is shown at the bottom to be 27.

The standard deviation, σx, is shown on the fifth line; $\sigma x = 1.26$ (nearest hundredth).

By scrolling down, additional information can be obtained. The second screen shows that the first quartile, Q_1, is 2, the median, Med, is 3, and the third quartile, Q_3, is also 3.

b. Use $\boxed{\text{STAT PLOT}}$, the second function of the $\boxed{\text{Y=}}$ key. Specify L_1 for Xlist and L_2 for Freq.

Enter:

Display: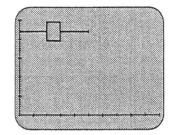

Note. There is no center box line because Med = Q_3 = 3.

The graph is shown in a $\boxed{\text{WINDOW}}$ with Xmin = 0, Xmax = 10, Ymin = 0, and Ymax = 5.

Many probability calculations can be performed easily with a calculator. Permutation and combination functions are on the $\boxed{\text{MATH}}$ PRB menu.

2. How many different ways can a committee of 4 students be selected from a group of 15 students?

   Solution: Order does not matter. Find the number of combinations of 15 students taken 4 at a time, $_{15}C_4$.

   Enter: 15 $\boxed{\text{MATH}}$ $\boxed{\blacktriangleright}$ $\boxed{\blacktriangleright}$ $\boxed{\blacktriangleright}$ 3 4 $\boxed{\text{ENTER}}$

   Display: $\boxed{1365}$

3. Find the number of permutations of 9 items taken 4 at a time.

   Solution: Find $_9P_4$.

   Enter: 9 $\boxed{\text{MATH}}$ $\boxed{\blacktriangleright}$ $\boxed{\blacktriangleright}$ $\boxed{\blacktriangleright}$ 2 4 $\boxed{\text{ENTER}}$

   Display: $\boxed{3024}$

CHAPTER XV

Sequences and Series

1. Understanding the Meaning of a Sequence

Observe that there is a simple pattern in the number arrangement 4, 8, 12, 16, ... If this pattern is extended, the next two terms are 20 and 24. A set of ordered numbers such as 4, 8, 12, 16, 20, 24 is called a *sequence*.

The sequence 4, 8, 12, 16 may be extended by setting up a one-to-one correspondence between the four terms of the given sequence and the first four members of the set of positive integers as follows:

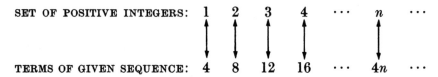

SET OF POSITIVE INTEGERS: 1   2   3   4   ··· n ···

TERMS OF GIVEN SEQUENCE: 4   8   12   16   ··· $4n$ ···

Observe that each term of the given sequence is 4 times the positive integer with which it is paired. If this pattern is applied to find new terms, the fifth term is 4(5) or 20; the sixth term is 4(6) or 24; and, in general, the nth term is $4n$.

A sequence such as 4, 8, 12, 16, which has a first term and a last term, is called a *finite sequence*. A sequence such as 4, 8, 12, 16, . . . in which, after each term of the sequence there is another term, is called an *infinite sequence*.

In the sequence whose terms are a_1, a_2, a_3, ..., a_n, ..., a_1 represents the first term, a_3 represents the third term, and a_n represents the nth term. The subscript in a term represents the place of the term in the sequence.

An infinite sequence may also be represented by $\{a_n\}$, the domain of n being the set of positive integers.

The expression that represents the general term of a sequence, a_n, is the rule that may be used to find any term of the sequence. For example, in the sequence 4, 8, 12, 16, . . . , $4n$, . . . , the nth term is represented by $4n$. Therefore, $a_n = 4n$ is the rule for forming the sequence. Each term of the sequence may be found by substituting a positive integer for n. For example, to find the 8th term, substitute 8 for n. Hence, the 8th term, $a_8 = 4(8)$ or 32. Since in the sequence 4, 8, 12, 16, . . . , the nth term $a_n = 4n$, the sequence may also be represented by $\{4n\}$.

550

~~~~~ *MODEL PROBLEMS* ~~~~~

1. *a.* Write a rule that can be used in forming a sequence 1, 4, 9, 16, ...
   b. Use the rule found in part *a* to write the next three terms of the sequence.

Solution:

a. Set up a one-to-one correspondence between the members of the set of positive integers and the terms of the given sequence.

SET OF POSITIVE INTEGERS:	1	2	3	4	. . .	n	. . .
TERMS OF GIVEN SEQUENCE:	1	4	9	16	. . .	n^2	. . .

Since each term of the sequence is the square of the positive integer with which it is paired, the rule $a_n = n^2$ is a rule that can be used to form the sequence.

Answer: $a_n = n^2$

b. Use the rule which forms the sequence $a_n = n^2$ to find the next three term of the sequence as follows:
The fifth term, $a_5 = 5^2$ or 25.
The sixth term, $a_6 = 6^2$ or 36.
The seventh term, $a_7 = 7^2$ or 49.

Answer: 25, 36, 49

2. Write the first five terms of the sequence represented by $\{n + 2\}$.

Solution:

To find the first five terms of the sequence represented by $\{n + 2\}$: In $n + 2$, replace n by the positive integers 1, 2, 3, 4, and 5 respectively.

$$a_1 = 1 + 2 \text{ or } 3 \quad a_2 = 2 + 2 \text{ or } 4 \quad a_3 = 3 + 2 \text{ or } 5$$
$$a_4 = 4 + 2 \text{ or } 6 \quad a_5 = 5 + 2 \text{ or } 7$$

Answer: 3, 4, 5, 6, 7

Exercises

In 1–16: *a.* Write a rule (that is, write a formula for the general term a_n) that can be used in forming a sequence. *b.* Use the rule found in part *a* to write the next three terms of the sequence.

1. 4, 5, 6, 7, . . .   **2.** 2, 4, 6, 8, . . .

3. $1, \frac{1}{2}, \frac{1}{3}, \frac{1}{4}, \ldots$

4. $-1, 0, 1, 2, \ldots$

5. $3, 6, 9, 12, \ldots$

6. $-5, -10, -15, -20, \ldots$

7. $6, 7, 8, 9, \ldots$

8. $-2, -1, 0, 1, \ldots$

9. $1, \frac{1}{4}, \frac{1}{9}, \frac{1}{16}, \ldots$

10. $\frac{1}{2}, \frac{2}{3}, \frac{3}{4}, \frac{4}{5}, \ldots$

11. $3, 5, 7, 9, \ldots$

12. $1, 4, 7, 10, \ldots$

13. $2 \cdot 3, 2 \cdot 3^2, 2 \cdot 3^3, 2 \cdot 3^4, \ldots$

14. $5, 5 \cdot 2, 5 \cdot 2^2, 5 \cdot 2^3, \ldots$

15. $4, 12, 36, 108, \ldots$

16. $-2, 0, 2, 4, \ldots$

In 17–25, write the first four terms of the sequence defined by the given rule.

17. $a_n = 5n$

18. $a_n = n + 5$

19. $a_n = 3n - 4$

20. $a_n = n^2$

21. $a_n = 3^n$

22. $a_n = 2^{n-1}$

23. $a_n = n(n + 1)$

24. $a_n = \dfrac{2n}{2n + 1}$

25. $a_n = (-1)^n \cdot 2n$

In 26–33, write the first four terms of the sequence that is defined.

26. $\{n + 6\}$

27. $\{4n - 3\}$

28. $\{n^2 + n\}$

29. $\dfrac{1}{\{n + 2\}}$

30. $\{n^3\}$

31. $\{4^n\}$

32. $\{3^{n-1}\}$

33. $\{2 \cdot 5^{n-1}\}$

2. Finding the General Term of an Arithmetic Sequence

In the sequence 2, 5, 8, 11, each term after the first term, 2, is the sum of the preceding term and a constant, 3. Such a sequence is an **arithmetic sequence**, or an **arithmetic progression (A.P.)**.

An arithmetic sequence is a sequence in which each term after the first is the sum of the preceding term and a constant.

Thus, in the sequence 2, 5, 8, 11, the term $5 = 2 + 3$ and the term $11 = 8 + 3$. Note that the constant 3 may be found by subtracting any term from its successor; for example, $5 - 2 = 3$, and $8 - 5 = 3$, also $11 - 8 = 3$. Hence, the constant 3 is the **common difference** in the arithmetic sequence 2, 5, 8, 11.

Procedure: To find the common difference in an arithmetic sequence, subtract any term from its successor.

Thus, in the decreasing arithmetic sequence 15, 12, 9, 6, the common difference is -3 since $12 - 15 = -3$.

In general, in an arithmetic sequence, if d represents the common difference, and a_{n-1} and a_n represent two consecutive terms, then $d = a_n - a_{n-1}$. Keep in mind that the common difference is positive in an increasing sequence such as 2, 5, 8, 11; it is negative in a decreasing sequence such as 15, 12, 9, 6; it is zero in the sequence 5, 5, 5, 5.

If a_1 represents the first term of an arithmetic sequence and d represents the common difference, then the sequence may be written as follows:

NUMBER OF TERM: 1   2   3   4   5 ..., n

TERMS OF SEQUENCE: $a_1, a_1 + d, a_1 + 2d, a_1 + 3d, a_1 + 4d, \ldots, a_1 + (n-1)d$.

Note that the fourth term of the sequence is $a_1 + 3d$ and the fifth term is $a_1 + 4d$. In general, the nth term $= a_1 + (n-1)d$; that is, the nth term is the sum of the first term and $(n-1)$ common differences.

In general, if the nth term of an arithmetic sequence is represented by a_n, then

$$a_n = a_1 + (n-1)d$$

For example, in the sequence $2, 5, 8, 11, \ldots$, the 51st term is $a_{51} = 2 + 50(3) = 2 + 150 = 152$. Also, in the sequence $15, 10, 5, \ldots$, the 25th term is $a_{25} = 15 + 24(-5) = 15 - 120 = -105$.

～～～～～ *MODEL PROBLEMS* ～～～～～

1. Find the 12th term of the arithmetic sequence $3, 8, 13, 18, \ldots$.

Solution:
First find the common difference by subtracting the first term from the second. Then find the 12th term.

$$d = 8 - 3 = 5$$
$$a_n = a_1 + (n-1)d \qquad a_1 = 3, \ n = 12, \ d = 5$$
$$a_n = 3 + (12-1)5$$
$$= 3 + (11)5$$
$$= 3 + 55 = 58$$

Answer: The 12th term is 58.

2. Write the first seven terms of an arithmetic sequence in which the third term is 7 and the seventh term is 15.

Solution:

$$a_n = a_1 + (n-1)d$$

When $n = 7$, $a_n = a_7 = 15$. Hence, $15 = a_1 + 6d$
When $n = 3$, $a_n = a_3 = 7$. Hence, $7 = a_1 + 2d$

Subtract: $8 = 4d$

$$2 = d$$

Since the third term is 7:	$7 = a_1 + 2d$
Substitute $d = 2$:	$7 = a_1 + 2(2)$
	$7 = a_1 + 4$
	$3 = a_1$

Answer: The required terms are 3, 5, 7, 9, 11, 13, 15.

Exercises

In 1–6, write the first four terms in an arithmetic sequence that has the given values for a_1 and d.

1. $a_1 = 2, d = 4$ **2.** $a_1 = -3, d = 2$ **3.** $a_1 = -5, d = -3$
4. $a_1 = 9, d = \frac{1}{3}$ **5.** $a_1 = 6, d = .5$ **6.** $a_1 = -2, d = -.7$

In 7–15, find the common difference of the arithmetic sequence. Then find the next two terms.

7. 5, 7, 9, ... **8.** $x + 1, x + 4, x + 7, ...$ **9.** 7, 2, −3, ...
10. $2\frac{1}{2}, 3, 3\frac{1}{2}, ...$ **11.** $2x, 3x, 4x, ...$ **12.** $5\frac{2}{3}, 5\frac{1}{3}, 5, ...$
13. .3, .5, .7, ... **14.** 1.02, 1.04, 1.06, ... **15.** $b - c, b, b + c, ...$

In 16–23, find the indicated term of the arithmetic sequence.

16. 7, 11, 15, ... (15th term) **17.** 10, 7, 4, ... (13th term)
18. −5, −7, −9, ... (18th term) **19.** $5\frac{1}{2}, 6, 6\frac{1}{2}, ...$ (40th term)
20. $9, 8\frac{3}{4}, 8\frac{1}{2}, ...$ (20th term) **21.** 1.05, 1.10, 1.15, ... (12th term)
22. 7, 6.9, 6.8, ... (35th term) **23.** $7x, 9x, 11x, ...$ (10th term)

24. Find the 17th term in the arithmetic sequence 3, 7, 11, ...
25. Find the 40th term of the arithmetic progression −26, −24, −22, ...
26. Write the 30th term of the arithmetic sequence $x - y, x, x + y, ...$.
27. Find a_7 in the arithmetic sequence 8, 5, 2, ...

In 28–31, find a_n in an arithmetic sequence which has the given values for a_1, d, and n.

28. $a_1 = 6, d = 4, n = 8$ **29.** $a_1 = 9, d = -2, n = 10$
30. $a_1 = \frac{5}{2}, d = -\frac{1}{2}, n = 7$ **31.** $a_1 = -\frac{4}{5}, d = -\frac{2}{5}, n = 28$

32. Find the first term of an arithmetic sequence in which $a_4 = 9$ and $d = 2$.
33. Find a_1 in an arithmetic progression in which $a_6 = 8$ and $d = -\frac{2}{3}$.
34. Find the first term in an arithmetic progression in which the sixth term is −2 and the seventh term is −7.
35. Which term of the arithmetic sequence 5, 8, 11, ... is 35?
36. Which term of the arithmetic sequence 15, 13, 11, ... is −13?

37. Which term of the arithmetic sequence 7, 13, 19, ... is 133?

38. If the first term of an arithmetic sequence is 7 and the 25th term is 79, what is the common difference?

39. The first term of an arithmetic progression is 10 and the thirtieth term is -77. Find the common difference.

40. In an arithmetic progression, $a_1 = \frac{2}{3}$ and $a_{12} = 19$. Find the common difference.

41. Find the number of terms in the arithmetic sequence 4, 7, 10, ... , 40.

42. Find the number of terms in the arithmetic sequence 84, 82, 80, ..., 20.

43. The second and fourth terms of an arithmetic sequence are 15 and 21. Find the third term.

44. Mr. Ward accepted a job which pays $38,000 the first year with yearly increases of $1600. Find his salary during his tenth year on the job.

45. In a theatre, the front row of the orchestra has 30 seats. Each row after the first row has 6 seats more than the previous row. Find the number of seats in the 15th row.

46. Miss Taylor saved $2400 during 1995. In each of the following years, she saved $350 more than she had saved the previous year. During which year did her savings for the year amount to $5200?

47. *a.* Find x so that the numbers $4x - 1$, $2x + 2$, and $2x - 3$ are in arithmetic progression.

  b. Find the fifth term of the progression.

In 48–50, solve $a_n = a_1 + (n - 1)d$ for the indicated variable.

48. a_1 **49.** d **50.** n

In 51–53, if $a_1, a_2, a_3, a_4, \ldots, a_n$ are in arithmetic progression, prove the stated relationship.

51. $a_1 + a_3 = 2a_2$ **52.** $a_1 + a_4 = a_2 + a_3$ **53.** $a_1 + a_n = a_2 + a_{n-1}$

3. Inserting Arithmetic Means

The terms between any two given terms of an arithmetic sequence are called the **arithmetic means** between those terms.

Thus, in the sequence 5, 7, 9, 11, 13, the terms 7, 9, 11 are the three arithmetic means between 5 and 13.

A single arithmetic mean between two given terms of an arithmetic sequence is called the **arithmetic mean** between the two terms.

Thus, in the sequence 5, 7, 9, the number 7 is the arithmetic mean between 5 and 9. Observe that the arithmetic mean, 7, is the average of the other terms 5 and 9; that is, $7 = \frac{1}{2}(5 + 9)$.

In general, if M represents the arithmetic mean between two real numbers a and b, then $M - a = b - M$

$$2M = a + b$$

$$M = \frac{a + b}{2}$$

Rule. The arithmetic mean between two real numbers is the average of the numbers.

~~~~~~~~~~ **MODEL PROBLEMS** ~~~~~~~~~~

1. Insert three arithmetic means between 19 and 11.

How To Proceed	*Solution*
1. Use spaces to represent the arithmetic means to be inserted between the given numbers.	1. 19, __, __, __, 11
2. Determine the values of a_1, a_n, and n.	2. $a_1 = 19$, $a_n = 11$ Since there are five terms, $n = 5$.
3. Find d by substituting the values of a_1, a_n, and n in the following formula: $a_n = a_1 + (n-1)d$.	3. $a_n = a_1 + (n-1)d$ $11 = 19 + (5-1)d$ $11 = 19 + 4d$ $-8 = 4d$ $-2 = d$
4. Find the required means by successively adding the value of d to the value of a_1.	4. $19 + (-2) = 17$ $17 + (-2) = 15$, etc. 19, 17, 15, 13, 11 *Ans.*

2. Find the arithmetic mean between 8 and 14.

Solution: If M represents the arithmetic mean between 8 and 14, then 8, M, 14 is an arithmetic sequence.

Method 1	*Method 2*
Using a common difference:	Using an average:
$M - 8 = 14 - M$	$M = \dfrac{a+b}{2}$ $a = 8, b = 14$
$2M = 22$	$M = \dfrac{8+14}{2} = \dfrac{22}{2} = 11$ *Ans.*
$M = 11$ *Ans.*	

Exercises

In 1–3, write the arithmetic means between the given terms of the sequence −12, −9, −6, −3, 0, 3, 6, 9.
1. −12 and 9   2. −9 and 0   3. −6 and 6

In 4–9, insert the indicated number of arithmetic means between the given numbers.
4. 4, ..., 16 (3 means)   5. −5, ..., 10 (4 means)
6. 25, ..., 13 (5 means)   7. 6, ..., −6 (5 means)
8. 3, ..., 5 (7 means)   9. 1, ..., −2 (8 means)

In 10–13, find the arithmetic mean between the two given numbers.
10. 7 and 15   11. −9 and −5   12. $5\frac{3}{4}$ and $7\frac{1}{4}$ 13. −6.4 and 8.4

14. Insert two arithmetic means between 8 and 23.
15. Find the arithmetic mean between −24 and 6.
16. The arithmetic mean between two numbers is $4\frac{1}{2}$. If one of the numbers is 3, find the other number.
17. Three numbers inserted between 4 and 6 form with these numbers an arithmetic progression. Find the common difference of this progression.
18. What four numbers inserted between 2 and 37 will form with these numbers an arithmetic progression of 6 terms?
19. The first term of an arithmetic progression is −4 and the fifth term is 20. Find the second term.
20. In a certain school system, a teacher's annual salary increases in arithmetic progression for six years. If a teacher's first-year salary is $36,400 and his sixth-year salary is $42,400, what is his salary during each of the other years?
21. Write an equation which shows that x is the arithmetic mean between a and b.

In 22–24, represent, in terms of x, the arithmetic mean between the given expressions.
22. $2x$ and $8x$ 23. $3x + 4$ and $5x - 2$ 24. $2x - 4$ and $7x + 4$

25. Prove that if M is the arithmetic mean between the real numbers p and q, then $M = \dfrac{p+q}{2}$.
26. Show that if a and b are both even integers, then their arithmetic mean, M, is an integer.
27. Show that if a and b are both odd integers, then their arithmetic mean, M, is an integer.
28. Show that if a is an odd integer and b is an even integer, then their arithmetic mean, M, cannot be an integer.

4. Finding the Sum of an Arithmetic Series

A series is the indicated sum of the terms of a sequence.

Thus, $2 + 5 + 8 + 11$ is an example of a *finite series*.

Also, $2 + 5 + 8 + 11 + \cdots$ is an example of an **infinite series**.

In general, $a_1 + a_2 + a_3 + \cdots + a_n$ represents a finite series.

Also, $a_1 + a_2 + a_3 + \cdots + a_n + \cdots$ represents an infinite series.

Be careful to note that a series is an *indicated sum* of the terms of a sequence, not the sum itself. Do not confuse the series $2 + 5 + 8 + 11$ with its sum, 26. The series $2 + 5 + 8 + 11$ cannot be represented by the numeral 26.

An **arithmetic series** is the indicated sum of the terms of an arithmetic sequence.

Hence, since 5, 8, 11, 14, 17 is an arithmetic sequence, then $5 + 8 + 11 + 14 + 17$ is an arithmetic series. Since $5 + 8 + 11 + 14 + 17 = 55$, we say that the *sum* of this series is 55.

In an arithmetic series, if a_1 is the first term, n is the number of terms, a_n is the nth term, and d is the common difference, then S_n, the sum of the arithmetic series, is given by the following formulas:

$$S_n = \frac{n}{2}(a_1 + a_n)$$

$$S_n = \frac{n}{2}[2a_1 + (n-1)d]$$

PROOF OF THE FORMULA FOR THE SUM, S_n, OF AN ARITHMETIC SERIES IN TERMS OF THE NUMBER OF TERMS, n, THE FIRST TERM, a_1, AND THE nTH TERM, a_n

1. $S_n = a_1 + (a_1 + d) + (a_1 + 2d) + \cdots + (a_n - d) + a_n$
2. $S_n = a_n + (a_n - d) + (a_n - 2d) + \cdots + (a_1 + d) + a_1$
3. $2S_n = (a_1 + a_n) + (a_1 + a_n) + (a_1 + a_n) + \cdots + (a_1 + a_n) + (a_1 + a_n)$

Since there are n terms in the series, there will be n addends in (3), each of which is $(a_1 + a_n)$.

4. $2S_n = n(a_1 + a_n)$

5. $S_n = \frac{n}{2}(a_1 + a_n)$

PROOF OF THE FORMULA FOR THE SUM, S_n, OF AN ARITHMETIC SERIES IN TERMS OF THE NUMBER OF TERMS, n, THE FIRST TERM, a_1, AND THE COMMON DIFFERENCE, d

1. $S_n = \dfrac{n}{2}(a_1 + a_n)$ (Formula for the sum of an arithmetic series.)

2. $a_n = a_1 + (n-1)d$ (Formula for the nth term of an arithmetic sequence.)

In (1), substitute $a_1 + (n-1)d$ for a_n.

3. $S_n = \dfrac{n}{2}[a_1 + a_1 + (n-1)d]$

4. $S_n = \dfrac{n}{2}[2a_1 + (n-1)d]$

~~~~~~~~~~ *MODEL PROBLEMS* ~~~~~~~~~~

1. Find the sum of ten terms of an arithmetic sequence whose first term is 5 and whose tenth term is -13.

Solution:

$S_n = \dfrac{n}{2}(a_1 + a_n)$ $a_1 = 5,\ a_{10} = -13,\ n = 10$

$S_n = \tfrac{10}{2}[5 + (-13)]$
$\quad = 5(-8)$
$\quad = -40$

Answer: The sum is -40.

2. In an arithmetic sequence, $S_n = 155$, $a_n = 28$, and $n = 10$. Find a_1.

Solution:

Using $S_n = \dfrac{n}{2}(a_1 + a_n)$:

First, substitute the given values. Then solve for a_1.

$$155 = \tfrac{10}{2}(a_1 + 28)$$
$$155 = 5(a_1 + 28)$$
$$31 = a_1 + 28$$
$$3 = a_1$$

Answer: $a_1 = 3$

3. Find the sum of the first 50 terms of the arithmetic series $3 + 5 + 7 + 9 + \cdots$

Solution:

Method 1	*Method 2*

Method 1

Using $S_n = \dfrac{n}{2}(a_1 + a_n)$:

First, find the 50th term.

$a_1 = 3,\ d = 2,\ n = 50$

$a_n = a_1 + (n-1)d$
$a_{50} = 3 + (50-1)2$
$\quad = 3 + (49)2$
$\quad = 101$

Then find the sum of 50 terms.

$S_n = \dfrac{n}{2}(a_1 + a_n)$

$S_{50} = \dfrac{50}{2}(3 + 101)$

$\quad = 25(104) = 2600 \quad Ans.$

Method 2

Using $S_n = \dfrac{n}{2}[2a_1 + (n-1)d]$:

$a_1 = 3,\ d = 2,\ n = 50$

$S_n = \dfrac{n}{2}[2a_1 + (n-1)d]$

$S_{50} = \frac{50}{2}[2(3) + (50-1)2]$
$\quad = 25[6 + (49)2]$
$\quad = 25(6 + 98)$
$\quad = 25(104)$
$\quad = 2600 \quad Ans.$

Exercises

In 1–6, find the sum of the terms of the arithmetic sequence which has the given data.

1. $a_1 = 7,\ a_{12} = 29,\ n = 12$

2. $a_1 = -14,\ a_5 = 30,\ n = 5$

3. $a_1 = 2,\ a_{14} = 54,\ n = 14$

4. $a_1 = 20,\ a_9 = -6,\ n = 9$

5. $a_1 = 5,\ d = 3,\ n = 15$

6. $a_1 = -12,\ d = -2,\ n = 10$

In 7–12, find the sum of the indicated number of terms of the arithmetic sequence.

7. 1, 8, 15, … (first ten terms)

8. 20, 17, 14, … (first 16 terms)

9. 8, $7\frac{1}{2}$, 7, … (first 20 terms)

10. 5, 1, −3, … (first 13 terms)

11. −3, −5, −7, … (first 18 terms)

12. −12, −10, −8, … (first 12 terms)

13. Find the sum of all the positive odd integers less than 100.

14. Find the sum of all the even integers from 2 to 100 inclusive.

15. Find the sum of all the positive numbers less than 200 which are exactly divisible by 3.

16. Find the sum of all integers between 1 and 100 that are exactly divisible by 7.

17. An object falls 16.08 feet, 48.24 feet, 80.40 feet in successive seconds.
  a. How far will it fall in the tenth second?
  b. How far will it fall in 15 seconds?
18. In an arithmetic sequence, $S_n = 210$, $a_n = 24$, and $n = 14$. Find a_1.
19. In an arithmetic progression, $d = 5$, $n = 14$, $S_n = 651$. Find a_1 and a_n.
20. Prove that the sum of the first n positive integers $1 + 2 + 3 + \cdots$ is $\dfrac{n(n+1)}{2}$.
21. Prove that the sum of the first n positive odd integers $1 + 3 + 5 + \cdots$ is n^2.
22. Prove that the sum of the first n positive even integers $2 + 4 + 6 + \cdots$ is $n(n+1)$.

5. Solving Verbal Problems Involving Arithmetic Sequences

~~~~~~~~~~~ *MODEL PROBLEMS* ~~~~~~~~~~~

1. Three numbers are in the ratio of $3 : 4 : 7$. If 4 is added to the middle number, that resulting number will be the second term of an arithmetic sequence of which the other two numbers are the first and third terms. Find the three numbers.

Solution: Let $3x$, $4x$, and $7x =$ the original numbers.
  Then $3x$, $4x + 4$, and $7x =$ the numbers which form an arithmetic sequence. In an arithmetic sequence of three terms, the middle term is the average of the first and the last terms. Hence,

$$4x + 4 = \frac{3x + 7x}{2}$$
$$4x + 4 = 5x$$
$$4 = x$$

If $x = 4$, then $3x = 12$, $4x = 16$, and $7x = 28$. *Answer:* 12, 16, 28

Note. The numbers in arithmetic sequence are 12, 20, 28.

2. A man wishes to pay off a debt of $1160 by making monthly payments in which each payment after the first is $4 more than that of the preceding month. According to this plan, how long will it take him to pay the debt if the first payment is $20 and no interest charge is made?

Solution: Let $n =$ the number of payments.
The payments form an arithmetic progression 20, 24, 28, ...

$$S_n = \frac{n}{2}[2a_1 + (n-1)d] \qquad S_n = 1160,\ a_1 = 20,\ d = 4$$

$$1160 = \frac{n}{2}[40 + (n-1)4]$$

$$1160 = \frac{n}{2}(40 + 4n - 4)$$

$$1160 = \frac{n}{2}(4n + 36)$$

$$1160 = 2n^2 + 18n$$
$$0 = 2n^2 + 18n - 1160$$
$$0 = n^2 + 9n - 580$$
$$0 = (n + 29)(n - 20)$$

$n + 29 = 0 \qquad\qquad n - 20 = 0$

$n = -29 \qquad\qquad n = 20$

Reject

Answer: 20 payments

Exercises

1. Three numbers are in the ratio of 4:5:7. If 3 is subtracted from the last number, the resulting number will be the third term in an arithmetic sequence of which the other two numbers are the first and second terms. Find the three numbers.
2. In an arithmetic sequence, the common difference is 2. If the sum of the first four terms of the sequence is divided by the third term, the quotient is 3 and the remainder is 5. Find the first term.
3. The sum of three numbers in arithmetic progression is 12. The sum of their squares is 56. Find the numbers.
4. Three numbers whose sum is 27 are in arithmetic progression. The square of the smallest number exceeds the largest by 12. Find the numbers.
5. In a certain school system, the salary scale for teachers starts at $36,800 and provides for a yearly increase of $1500 for the next 5 years. Andrea, starting at the minimum salary, plans to make $28,000 cover her entire expenses for each year. How much will she be able to save if she stays five years? (Use a formula in the solution.)
6. The sum of a certain number of consecutive odd numbers is 180. The largest is three times the smallest.
   a. Find the smallest number.   *b.* Find the number of numbers.

7. The balcony in a theatre has 780 seats. If the first row has 36 seats and each succeeding row has four seats more than the previous row, find the number of rows in the balcony.

8. A man desiring to pay a debt of $340 finds that he can pay $25 the first month and thereafter increase each monthly payment $2 over that of the previous month. How long will it take him to pay the debt if the interest is not included?

9. Mr. Jones brought a machine costing $240. He arranged to pay $60 down and the rest in monthly installments. If he paid $33 the first month and each payment thereafter was $3 less than the preceding one, in how many months did Mr. Jones complete payment for the machine?

10. A sum of money is to be distributed among 10 prize winning contestants in such a way that each one after the first receives $10 more than the preceding person. The largest prize awarded is three times the smallest. Let a represent the smallest prize.

  a. Express the largest prize in terms of a.

  b. Using the formula for the last term of an arithmetic progression, find a.

  c. Find the amount distributed in prizes.

6. Writing Series Using Summation Notation

The use of the summation symbol, Σ, was introduced in Chapter XIV, Section 1. In this chapter, summation notation will be used to generate both arithmetic and geometric series.

See how the summation notation can be used to generate the series that indicates the sum of the first 500 positive consecutive even integers.

$$\sum_{i=1}^{500} 2i = 2(1) + 2(2) + 2(3) + \cdots + 2(498) + 2(499) + 2(500)$$
$$= 2 + 4 + 6 + \cdots + 996 + 998 + 1000$$

MODEL PROBLEMS

1. *a.* Write the series generated by $\sum_{i=1}^{9} (2i - 1)$.

  b. Find the sum of the series found in answer to part *a.*

Solution:

a. Beginning with 1, the value of i below sigma, and ending with 9, the value above sigma, substitute consecutive integers for the variable i in the expression $(2i - 1)$, to generate the required series:

$$1 + 3 + 5 + 7 + 9 + 11 + 13 + 15 + 17$$

For example, 17, the last term of the generated series, is found by substituting 9 for i in $(2i-1)$: $2(9)-1=18-1=17$

Answer: $1+3+5+7+9+11+13+15+17$

b. Since the series generated in part *a* is an arithmetic series, apply the formula $S_n=\dfrac{n}{2}(a_1+a_n)$. Substitute 9 for n, 1 for a_1, and 17 for a_n:

$$S_9=\tfrac{9}{2}(1+17)=\tfrac{9}{2}(18)=81 \quad Ans.$$

2. Write the series $5+9+13+17+21$ using the summation notation.

Solution:
The sequence 5, 9, 13, 17, 21 is an arithmetic sequence in which the first term is 5 and the common difference is 4. Therefore, the representation of the general term, the nth term, found by using the formula $a_n=a_1+(n-1)d$, is $5+(n-1)4$ or $4n+1$. The sequence is obtained by varying n from a lower limit of 1 to an upper limit of 5. The given series may be represented in summation notation as $\displaystyle\sum_{n=1}^{5}(4n+1)$. *Ans.*

Exercises

In 1–8, write the series generated by the summation symbol.

1. $\displaystyle\sum_{i=1}^{7} i$ **2.** $\displaystyle\sum_{z=4}^{9} 3(z+1)$ **3.** $\displaystyle\sum_{w=5}^{8} (3w-1)$ **4.** $\displaystyle\sum_{x=15}^{20} \frac{7}{x}$

5. $\displaystyle\sum_{i=1}^{5} (i^2-1)$ **6.** $\displaystyle\sum_{i=4}^{6} (i^2+2i-1)$ **7.** $\displaystyle\sum_{x=0}^{4} \frac{x-1}{x+1}$ **8.** $\displaystyle\sum_{i=2}^{5} 4\cdot 3^{i-1}$

In 9–22, represent the series using the sigma notation for summation
9. $10+20+30+40$ **10.** $5+8+11+14$
11. $1^2+2^2+3^2+4^2+5^2$ **12.** $5+25+125+625$
13. $1\cdot3+2\cdot4+3\cdot5+4\cdot6$ **14.** $\frac{1}{2}+\frac{2}{3}+\frac{3}{4}+\frac{4}{5}+\frac{5}{6}$
15. $10(12)+11(13)+12(14)$ **16.** $1+\frac{1}{3}+\frac{1}{9}+\frac{1}{27}$
17. $\frac{2}{7}+\frac{3}{7}+\frac{4}{7}+\frac{5}{7}+\frac{6}{7}$ **18.** $\sqrt{5}+\sqrt{6}+\sqrt{7}+\sqrt{8}$
19. $1+2+3+\cdots+49+50$ **20.** $6+9+14+21+30+41$
21. $-1+2-3+4-5+6$ **22.** $2-4+8-16+32-64+128-256$

In 23–25, use sigma notation to represent the series of:
23. the first 25 positive integers.
24. the first ten positive even integers.
25. the consecutive integers from -5 to 10.

In 26–29, (*a*) write the arithmetic series generated by the summation symbol and (*b*) find the sum of the series generated in part (*a*).

26. $\displaystyle\sum_{i=1}^{4} 4i$ **27.** $\displaystyle\sum_{y=2}^{6} 2(y-1)$ **28.** $\displaystyle\sum_{z=7}^{10} (3z+2)$ **29.** $\displaystyle\sum_{w=12}^{20} (1\tfrac{1}{2})(w+4)$

7. Finding the General Term of a Geometric Sequence

In the sequence 2, 6, 18, 54, each term after the first term, 2, is the product of the preceding term and a constant 3. Such a sequence is a *geometric sequence*, or a *geometric progression* (**G.P.**).

A *geometric sequence* is a sequence in which each term after the first is the product of the preceding term and a constant.

Thus, in the sequence 2, 6, 18, 54, the term $6 = 2(3)$ and the term $54 = 18(3)$. Note that the constant 3 may be found by dividing any term by its predecessor. Thus, $\frac{6}{2} = 3$ and $\frac{18}{6} = 3$. Also, $\frac{54}{18} = 3$. Hence, the constant 3 is the **common ratio** in the geometric sequence 2, 6, 18, 54.

Procedure: To find the common ratio in a geometric sequence, divide any term by its predecessor, the term that immediately precedes it.

Thus, in the geometric sequence 5, −25, 125, −625, the common ratio is −5 since $\dfrac{-25}{5} = -5$.

In general, in a geometric sequence, if r represents the common ratio and a_{n-1} and a_n represent two consecutive terms, then

$$r = \frac{a_n}{a_{n-1}}$$

Note that the common ratio may be positive or negative, but not 0.

If a_1 represents the first term of a geometric sequence and r represents the common ratio, then the sequence may be written as follows:

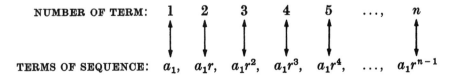

NUMBER OF TERM: 1   2   3   4   5   ..., n

TERMS OF SEQUENCE: a_1, $a_1 r$, $a_1 r^2$, $a_1 r^3$, $a_1 r^4$,   ..., $a_1 r^{n-1}$

Note that the fourth term of the sequence is $a_1 r^3$ and the fifth term is $a_1 r^4$. In general, the *n*th term $= a_1 r^{n-1}$; that is, the *n*th term is the product of the first term, and the common ratio raised to the $(n-1)$ power.

Thus, the 9th term of 2, 20, 200, 2000, ... is $2(10^8) = 200,000,000$. In general, if the nth term of a geometric sequence is represented by a_n, then

$$a_n = a_1 r^{n-1}$$

For example, in the sequence 8, 4, 2, 1, ..., $a_6 = 8(\frac{1}{2})^5 = 8(\frac{1}{32}) = \frac{1}{4}$. Also, in the sequence 12, -6, 3, $-1\frac{1}{2}$, ..., $a_6 = 12(-\frac{1}{2})^5 = 12(-\frac{1}{32}) = -\frac{3}{8}$.

~~~~~~~ MODEL PROBLEM ~~~~~~~

Find the seventh term of the geometric progression 32, -16, 8, ...

Solution: First find the common ratio by dividing the second term by the first. Then find the seventh term.

$$r = (-16) \div (32) = -\frac{1}{2}$$
$$a_n = a_1 r^{n-1}$$
$$= 32(-\frac{1}{2})^6$$
$$= 32(\frac{1}{64})$$
$$= \frac{1}{2} \quad Ans.$$

$a_1 = 32, n = 7, r = -\frac{1}{2}$

Exercises

In 1–9, find the common ratio of the geometric sequence and the next two terms.

1. 5, 10, 20, ...
2. -3, 6, -12, ...
3. -3, -9, -27, ...
4. 9, 3, 1, ...
5. $4x$, $16x$, $64x$, ...
6. $\frac{1}{2}$, $\frac{1}{2}y$, $\frac{1}{2}y^2$, ...
7. x^3, x^6, x^9, ...
8. 1, 1.05, $(1.05)^2$, ...
9. 2, 2, 2, ...

In 10–12, write the first four terms of a geometric progression in which a_1 and r have the given values.

10. $a_1 = 5, r = -2$
11. $a_1 = 6, r = \frac{1}{3}$
12. $a_1 = 4, r = x$

In 13–15, write the last four terms of a geometric progression in which a_n and r have the given values.

13. $a_n = 243, r = 3$
14. $a_n = 12, r = \frac{1}{2}$
15. $a_n = 368, r = -2$

In 16–21, find the indicated term of the geometric sequence.

16. 1, 2, 4, ... (7th term)
17. 12, 6, 3, ... (8th term)
18. -3, 6, -12, ... (6th term)
19. 4, 1, $\frac{1}{4}$, ... (5th term)
20. 3, $\frac{3}{2}$, $\frac{3}{4}$, ... (8th term)
21. $-\frac{3}{2}$, 3, -6, ... (9th term)

In 22–24, supply the missing term of the geometric sequence.

22. 4, 12, 36, ..., 324
23. 2, 3, $4\frac{1}{2}$, ..., $10\frac{1}{8}$
24. 6, -2, $\frac{2}{3}$, ..., $\frac{2}{27}$

25. What term of the sequence 4, 8, 16, ... is 128?

26. What term of the sequence 4, −12, 36, ... is 324?

27. What term of the sequence 36, 18, 9, ... is $\frac{9}{32}$?

28. Which of the following is *not* a geometric progression? (1) 16, 8, 4, ... (2) 3, −6, 12, ... (3) 2, 5, 8, ...

29. Find the missing term in the sequence 2, −6, 18, ..., 162.

30. Find the fifth term of the sequence 2, 3, $4\frac{1}{2}$, ...

31. What is the fifth term of the sequence 2, 6, 18, ...?

32. The first term of a geometric progression is 8 and the fourth term is 125. Find the common ratio.

33. The fourth term of a geometric progression is 125 and the sixth term is 3125. Find the common ratio and the first three terms.

34. In a culture, there are 200 bacteria. Every hour the number of bacteria in the culture doubles. Find the number of bacteria in the culture at the end of 8 hours.

35. In a lottery, the first ticket drawn paid a prize of $50,000. Every succeeding ticket paid half as much as the preceding one. Four tickets were drawn. How much did the fourth ticket pay?

36. A house which was built at a cost of $50,000 depreciates each year by an amount equal to 4% of its value at the beginning of the year. Find the value of this house at the end of 5 years.

8. Inserting Geometric Means

The terms between any two given terms of a geometric sequence are called the *geometric means* between those terms.

Thus, in the sequence 5, 10, 20, 40, 80, the terms 10, 20, 40 are the three geometric means between 5 and 80.

A single geometric mean between two given terms of a geometric sequence is called the *geometric mean* between the two terms.

Thus, in the sequence 5, 10, 20, the number 10 is the geometric mean between 5 and 20. Observe that the geometric mean 10 is the *mean proportional* between 5 and 20; that is, $\frac{20}{10} = \frac{10}{5}$.

In general, if G represents the geometric mean between two real numbers a and b, then $\frac{a}{G} = \frac{G}{b}$.

Rule. The geometric mean between two real numbers is their mean proportional.

Note. Since $\frac{a}{G} = \frac{G}{b}$, then $G^2 = ab$ and $G = \pm\sqrt{ab}$.

~~~~~~~~~~~~~ *MODEL PROBLEMS* ~~~~~~~~~~~~

1. Insert three geometric means between $\frac{1}{9}$ and 9.

How To Proceed	*Solution*
1. Use spaces to represent the geometric means to be inserted between the given numbers.	1. $\frac{1}{9}$, ___, ___, ___, 9
2. Determine the values of a_1, a_n, and n.	2. $a_1 = \frac{1}{9}$, $a_n = 9$. Since there are five terms, $n = 5$.
3. Substitute the values of a_1, a_n, and n in the formula $a_n = a_1 r^{n-1}$ and then find r.	3. $a_n = a_1 r^{n-1}$ $9 = \frac{1}{9} r^4$ $81 = r^4$ $\pm 3 = r$
4. Find the required means by successively multiplying the value of a_1 by the value of r.	4. If $r = 3$, the resulting series is $\frac{1}{9}, \frac{1}{3}, 1, 3, 9$. If $r = -3$, the resulting series is $\frac{1}{9}, -\frac{1}{3}, 1, -3, 9$.

Answer: $\frac{1}{9}, \frac{1}{3}, 1, 3, 9$ or $\frac{1}{9}, -\frac{1}{3}, 1, -3, 9$

2. Find the positive geometric mean between $\frac{1}{2}$ and $\frac{1}{32}$.

Solution: If G represents the geometric mean between $\frac{1}{2}$ and $\frac{1}{32}$, then $\frac{1}{2}, G, \frac{1}{32}$ is a geometric sequence.

Method 1	*Method 2*
Using $\dfrac{b}{G} = \dfrac{G}{a}$:	Using $G = \pm \sqrt{ab}$:
	Since G is positive, then
$\dfrac{\frac{1}{32}}{G} = \dfrac{G}{\frac{1}{2}}$	$G = \sqrt{\left(\dfrac{1}{2}\right)\left(\dfrac{1}{32}\right)}$
$G^2 = \left(\dfrac{1}{32}\right)\left(\dfrac{1}{2}\right)$	$= \sqrt{\dfrac{1}{64}}$
$= \dfrac{1}{64}$	$= \dfrac{1}{8}$ *Ans.*
Since G is positive, then	
$G = \sqrt{\dfrac{1}{64}} = \dfrac{1}{8}$ *Ans.*	

Exercises

In 1–3, find the common ratio of a geometric sequence which has the given terms.

1. $a_1 = 5, a_3 = 45$ **2.** $a_1 = 36, a_3 = 9$ **3.** $a_1 = 4, a_4 = 500$

In 4–7, insert the indicated number of geometric means between the given numbers. Give all possible real answers.

4. 2 means between 8 and 64   **5.** 4 means between 1 and 32

6. 3 means between 2 and 162   **7.** 3 means between -6 and $-\frac{3}{8}$

8. Find the positive geometric mean between 2 and 50.

9. Find the negative geometric mean between 16 and 25.

10. The geometric mean between two numbers is 10. If one of the numbers is 4, find the other number.

11. Insert two geometric means between 2 and 54.

12. Insert a positive geometric mean between 2 and 6.

13. Two geometric means between 2 and 128 are (1) 16 and 64   (2) 44 and 86   (3) -8 and 32   (4) 8 and 32

14. Find the fifth term of a geometric sequence whose first term is 4 and whose fourth term is 32.

9. Finding the Sum of a Geometric Series

A *geometric series* is the indicated sum of the terms of a geometric sequence.

Since 5, 10, 20, 40, 80 is a geometric sequence, $5 + 10 + 20 + 40 + 80$ is a geometric series. Since $5 + 10 + 20 + 40 + 80 = 155$, we say that the sum of this series is 155.

In general, a geometric series is represented by $a_1 + a_1r + a_1r^2 + \cdots + a_1r^{n-1}$, in which a_1 is the first term, n is the number of terms, a_n is the nth term, and r is the common ratio. The sum of the geometric series, S_n, is given by the formula:

$$S_n = \frac{a_1 - a_1 r^n}{1 - r}, r \neq 1$$

PROOF OF THE FORMULA FOR THE SUM, S_n, OF A GEOMETRIC SERIES IN TERMS OF THE FIRST TERM, a_1, THE COMMON RATIO, r, AND THE NUMBER OF TERMS, n

1. $\qquad\qquad S_n = a_1 + a_1r + a_1r^2 + a_1r^3 + \cdots + a_1r^{n-1}$

2. M_r: $\qquad\quad rS_n = \qquad\; a_1r + a_1r^2 + a_1r^3 + \cdots + a_1r^{n-1} + a_1r^n$

3. Subtract: $\overline{S_n - rS_n = a_1 \qquad\qquad\qquad\qquad\qquad\qquad\quad\; - a_1r^n}$

4. Factor: $(1 - r)S_n = a_1 - a_1r^n$

5. D_{1-r}: $\qquad\qquad S_n = \dfrac{a_1 - a_1 r^n}{1 - r}$

By replacing $a_1 r^{n-1}$ with a_n, we obtain another formula for the sum of a geometric series:

$$S_n = \frac{a_1 - a_n r}{1 - r}, \ r \neq 1$$

〰〰〰〰〰〰 *MODEL PROBLEM* 〰〰〰〰〰〰

Find the sum of the first five terms of the geometric sequence 5, 15, 45, ...

Solution:

1. Find the common ratio, r, by dividing the second term, 15, by the first term, 5. $r = 15 \div 5$, or $r = 3$.
2. Find the sum of the first five terms of the series $5 + 15 + 45 + \cdots$.

$$S_n = \frac{a_1 - a_1 r^n}{1 - r} \qquad a_1 = 5, \ r = 3, \ n = 5$$

$$S_5 = \frac{5 - 5(3)^5}{1 - 3} = \frac{5 - 1215}{-2} = \frac{-1210}{-2} = 605 \quad Ans.$$

Exercises

In 1–6, find the sum of the indicated number of terms of the geometric sequence. (In exercise 6, $y \neq 0$, $y \neq 1$.)

1. 1, 3, 9, ... (first 6 terms)
2. 8, 4, 2, ... (first 6 terms)
3. −3, −15, −75, ... (first 5 terms)
4. 3, −6, 12, ... (first 5 terms)
5. $\frac{1}{27}, \frac{1}{9}, \frac{1}{3}, \ldots$ (first 5 terms)
6. 1, y, y^2, ... (first 10 terms)

In 7–10, find the sum of the indicated number of terms of the geometric series.

7. $8 + 16 + 32 + \cdots$ (first 5 terms)
8. $81 + 27 + 9 + \cdots$ (first 6 terms)
9. $\frac{1}{8} + \frac{1}{4} + \frac{1}{2} + \cdots$ (first 7 terms)
10. $1 - 5 + 25 - 125 + \cdots$ (first 8 terms)

In 11–14, (a) write the geometric series and (b) find the sum of the series.

11. $\sum_{i=1}^{4} 2^i$
12. $\sum_{i=1}^{3} (\frac{1}{3})^i$
13. $\sum_{j=1}^{5} 2(3)^{j-1}$
14. $\sum_{k=-1}^{3} (-5)(4)^{k+1}$

In 15–18, the values of a_1, n, and r of a geometric sequence are given. Find the sum of the terms of the sequence.

15. $a_1 = 6, n = 4, r = 3$
16. $a_1 = 5, n = 6, r = 2$
17. $a_1 = -3, n = 5, r = -2$

18. $a_1 = 18, n = 5, r = \frac{2}{3}$

19. The first term of a geometric progression is 1 and the common ratio is 2. Find the sum of the first ten terms.

20. The sum of the first eight terms of the progression 3, 12, 48, ... is
(1) $4^8 + 1$ (2) 4^{10} (3) $4^8 - 1$

21. Find the first term of a geometric sequence of five terms in which the sum of the five terms is 363 and the common ratio is 3.

22. Find the number of terms of a geometric progression whose first term is 8, whose common ratio is 2, and the sum of whose terms is 120.

23. Find the common ratio of a geometric sequence in which $a_1 = 25$, $a_n = 400$, and $S_n = 775$.

24. One Sunday morning, Ted sent a letter to each of three different friends in which he asked each of them to send a similar letter to each of three of their friends on the following Sunday. If no one breaks the chain and there are no duplications among the friends, how many letters will have been sent in the first seven Sundays?

10. Solving Verbal Problems Involving Geometric Sequences

In the solution of verbal problems involving geometric sequences, the following may be helpful in writing the equations that can be used in solving the problems:

1. The ratio of any term of a geometric sequence to its preceding term is equal to the ratio of any other term to its preceding term.
2. Apply the formulas for a_n and S_n in a geometric sequence.

ᐡᐡᐡᐡᐡᐡ *MODEL PROBLEM* ᐡᐡᐡᐡᐡᐡ

There are three numbers such that the second is two more than the first and the third is nine times the first. The numbers form a geometric sequence. Find the numbers.

Solution:

Let x = the first number. Then $x + 2$ = the second number and $9x$ = the third number.

The numbers x, $x + 2$, $9x$ are numbers in a geometric sequence.

In a geometric sequence of three terms, the middle term is the mean proportional between the first and the last terms. Hence,

$$\frac{x}{x+2} = \frac{x+2}{9x}$$
$$9x^2 = (x+2)^2$$
$$9x^2 = x^2 + 4x + 4$$
$$8x^2 - 4x - 4 = 0$$
$$2x^2 - x - 1 = 0$$
$$(x-1)(2x+1) = 0$$

$$x - 1 = 0 \qquad\qquad 2x + 1 = 0$$
$$x = 1 \qquad\qquad x = -\tfrac{1}{2}$$

If $x = 1$, the geometric sequence is 1, 3, 9. If $x = -\tfrac{1}{2}$, the geometric sequence is $-\tfrac{1}{2}, 1\tfrac{1}{2}, -4\tfrac{1}{2}$.

Answer: 1, 3, 9   or $-\tfrac{1}{2}, 1\tfrac{1}{2}, -4\tfrac{1}{2}$

Exercises

1. Three numbers are in the ratio of $1 : 3 : 5$. If 12 is added to the third number, the resulting number will be the third term of a geometric sequence of which the other two numbers are the first and second terms. Find the three numbers.

2. Three integers are in the ratio of $2 : 5 : 8$. If 2 is subtracted from the middle integer, the resulting number will be the second term of a geometric sequence of which the other two integers are the first and third terms. Find the three integers.

3. How much must be added to each of the numbers 2, 9, 23 so that the resulting numbers should form a geometric progression?

4. How much must be subtracted from each of the numbers 6, 10, and 18 so that the resulting numbers will form a geometric sequence?

5. What number must be added to the first number of the sequence 15, 10, 5 so that the resulting numbers will form a geometric progression?

6. Three numbers are in arithmetic progression, the common difference being 3. If the first number is diminished by 2, the second is increased by 4, and the third is doubled and then increased by 2, the resulting numbers are in geometric progression. Find the original numbers.

7. The sum of three numbers in a geometric progression is 14. Their product is 64. Find the numbers.

8. The sum of three positive numbers in an arithmetic progression is 18. If the third number is increased by 8, the numbers then form a geometric progression. Find the numbers in the arithmetic progression.

11. Finding the Sum of an Infinite Geometric Series

We have learned that if the number of terms in a sequence increases without limit, the sequence is called an *infinite sequence*. Also, if the number of terms in a series increases without limit, we call the series an *infinite series*.

Thus, $1, \frac{1}{2}, \frac{1}{4}, \frac{1}{8}, \frac{1}{16}, \frac{1}{32}, \ldots$ is an infinite geometric sequence. Also, $1 + \frac{1}{2} + \frac{1}{4} + \frac{1}{8} + \frac{1}{16} + \frac{1}{32} + \cdots$ is an infinite geometric series.

Consider the infinite geometric series $1 + 2 + 4 + 8 + 16 + 32 + \cdots$
Let us find the sums of different numbers of terms of this series.

$$S_1 = 1 \text{ (We define } S_1 \text{ as the first term, 1.)}$$
$$S_2 = 1 + 2 = 3$$
$$S_3 = 1 + 2 + 4 = 7$$
$$S_4 = 1 + 2 + 4 + 8 = 15$$
$$S_5 = 1 + 2 + 4 + 8 + 16 = 31$$

In general, $S_n = 1 + 2 + 4 + 8 + 16 + \cdots$ to n terms.

We can see that as we increase the number of terms that are being added in the series, the individual terms become larger and larger. Furthermore, the sums $S_1, S_2, S_3, \ldots$ also become larger and larger. In fact, we can make S_n as large as we please by adding a sufficiently large number of terms. There is no limit to the value of S_n. We say that the sum of the terms of this infinite series is not finite. Such an infinite series is said to be a **divergent series.**

Other examples of infinite geometric series that are divergent are

$$-1 - 2 - 4 - 8 - 16 - 32 - \cdots \text{ and } 1 - 2 + 4 - 8 + 16 - 32 + \cdots$$

In general, an infinite geometric series is a divergent series if the absolute value of the common ratio, r, is greater than or equal to 1; that is, $|r| \geq 1$.

On the other hand, consider the infinite geometric series

$$1 + \frac{1}{2} + \frac{1}{4} + \frac{1}{8} + \frac{1}{16} + \frac{1}{32} + \cdots$$

Let us find the sums of different numbers of terms of this series.

$$S_1 = 1 \text{ (We define } S_1 \text{ as the first term, 1.)}$$
$$S_2 = 1 + \frac{1}{2} = 1\frac{1}{2}, \text{ or } 1.5$$
$$S_3 = 1 + \frac{1}{2} + \frac{1}{4} = 1\frac{3}{4}, \text{ or } 1.75$$
$$S_4 = 1 + \frac{1}{2} + \frac{1}{4} + \frac{1}{8} = 1\frac{7}{8}, \text{ or } 1.875$$
$$S_5 = 1 + \frac{1}{2} + \frac{1}{4} + \frac{1}{8} + \frac{1}{16} = 1\frac{15}{16}, \text{ or } 1.9375$$

In general, $S_n = 1 + \frac{1}{2} + \frac{1}{4} + \frac{1}{8} + \frac{1}{16} + \cdots$ to n terms.

We can see that as we increase the number of terms in the series, the individual terms are becoming smaller and smaller. Although the sums $S_1, S_2, S_3, \ldots$ are becoming larger, there is a limit to the value of S_n. The value of S_n gets closer and closer to 2, but is never equal to 2. We can obtain a value of S_n as close to 2 as we wish by adding a sufficiently large number of terms. For example, to obtain a value of S_n within $\frac{1}{1000}$ of 2, we would add 12 terms. However, there is no value of n for which the value of S_n is equal to 2. We say that the value of S_n "approaches" the number 2 as a limit, and for that reason we also say that the sum of an infinite number of terms of this series is 2. Such a series, in which the infinite sequence $S_1, S_2, S_3, \ldots$ approaches a limit, is called a **convergent series.**

Other examples of infinite geometric series that are convergent are

$$-1 - \tfrac{1}{2} - \tfrac{1}{4} - \tfrac{1}{8} - \tfrac{1}{16} - \tfrac{1}{32} - \cdots \text{ and } 1 - \tfrac{1}{2} + \tfrac{1}{4} - \tfrac{1}{8} + \tfrac{1}{16} - \tfrac{1}{32} + \cdots$$

In general, an infinite geometric series is a convergent series if the absolute value of the common ratio, r, is less than 1; that is, $|r| < 1$.

In order to find the limit of the sum of an infinite convergent geometric series, we can use the following formula:

$$S = \frac{a_1}{1 - r}, \; |r| < 1$$

For example, in the series $1 + \tfrac{1}{2} + \tfrac{1}{4} + \tfrac{1}{8} + \cdots$, $a_1 = 1$ and $r = \tfrac{1}{2}$. Hence,

$$S = \frac{a_1}{1 - r}$$

$$= \frac{1}{1 - \tfrac{1}{2}} = \frac{1}{\tfrac{1}{2}} = 2$$

PROOF OF THE FORMULA FOR THE SUM, S, OF AN INFINITE SERIES IN TERMS OF THE FIRST TERM, a_1, AND THE COMMON RATIO, r, WHOSE ABSOLUTE VALUE IS LESS THAN 1

1. The formula for the sum of a geometric series is $S_n = \dfrac{a_1 - a_1 r^n}{1 - r}$, or

$$S_n = \frac{a_1}{1 - r} - \frac{a_1 r^n}{1 - r}$$

2. When $|r| < 1$: As n becomes greater and greater, r^n becomes smaller and smaller and, in so doing, the value of r^n comes closer and closer to 0. A value of n can always be found that will make the difference between r^n and 0 smaller than any arbitrarily selected value, no matter how small this selected value may be. Hence, we say that r^n approaches 0 as a limit. If r^n approaches

0 as a limit, then $\dfrac{a_1 r^n}{1-r}$ must also approach 0 as a limit. Hence, we may replace

$\dfrac{a_1 r^n}{1-r}$ with 0 in step 1, if the geometric series is an infinite series.

3. Therefore, if $\dfrac{a_1 r^n}{1-r}$ is replaced by 0 in step 1, we see that

$$S = \frac{a_1}{1-r}$$

if $|r| < 1$ and the geometric series is an infinite series.

Note that we have replaced S_n with S for the reason that n can only be used to represent a finite number. The above formula applies only to an infinite series, the number of whose terms is not a finite number.

〰〰〰〰〰〰 MODEL PROBLEMS 〰〰〰〰〰〰

1. Find the sum of the infinite series $4 + \frac{4}{3} + \frac{4}{9} + \frac{4}{27} + \cdots$.

   Solution:
   First, find the ratio by dividing the second term by the first term. Then find the sum.

   $r = \left(\frac{4}{3}\right) \div (4) = \frac{4}{3} \times \frac{1}{4} = \frac{1}{3}$

   $S = \dfrac{a_1}{1-r}$ Substitute: $a_1 = 4$, $r = \dfrac{1}{3}$

   $= \dfrac{4}{1 - \frac{1}{3}} = \dfrac{4}{\frac{2}{3}} = 6$ *Ans.*

2. Find the sum of the terms of the infinite geometric sequence $2, -1, \frac{1}{2}, -\frac{1}{4}, \ldots$.

   Solution:

   $$r = (-1) \div (2) = -\tfrac{1}{2}$$

   $$S = \frac{a_1}{1-r} \qquad a_1 = 2, r = -\frac{1}{2}$$

   $$= \frac{2}{1 - (-\frac{1}{2})} = \frac{2}{1 + \frac{1}{2}} = \frac{2}{\frac{3}{2}} = \frac{4}{3} \quad Ans.$$

3. Find the sum of the infinite geometric progression in which a_1 and r have the given values.

   $a_1 = 15 \qquad r = -.2$

Solution:
$$S = \frac{a_1}{1-r} = \frac{15}{1-(-.2)}$$

$$= \frac{15}{1+.2} = \frac{15}{1.2} = \frac{150}{12} = 12.5$$

Answer: 12.5

4. Find the value of the repeating decimal .535353 . . . .

Solution:
$$.535353 \ldots = .53 + .0053 + .000053 + \cdots$$
In this infinite geometric series, the common ratio $r = .0053 \div .53 = .01$.

Since $a_1 = .53$ and $r = .01$, then $S = \dfrac{a_1}{1-r} = \dfrac{.53}{1-.01} = \dfrac{.53}{.99} = \dfrac{53}{99}$.

Answer: $\frac{53}{99}$

Exercises

In 1–4, find the first term and the common ratio of the infinite geometric series and state whether the series diverges or converges.

1. $20 + 10 + 5 + 2\frac{1}{2} + \cdots$ **2.** $3 + 12 + 48 + 192 + \cdots$

3. $8 - 2 + \frac{1}{2} - \frac{1}{8} + \cdots$ **4.** $2 - 6 + 18 - 54 + \cdots$

In 5–10, find the sum of the terms of the infinite geometric sequence.

5. $4, 2, 1, \ldots$ **6.** $5, \frac{5}{2}, \frac{5}{4}, \ldots$ **7.** $3, -1, \frac{1}{3}, \ldots$

8. $12, 8, 5\frac{1}{3}, \ldots$ **9.** $.4, .04, .004, \ldots$ **10.** $.23, .0023, .000023, \ldots$

In 11–19, find the sum of the infinite series.

11. $6 + 2 + \frac{2}{3} + \cdots$ **12.** $12 + 3 + \frac{3}{4} + \cdots$

13. $200 + 50 + 12\frac{1}{4} + \cdots$ **14.** $8 + 4 + 2 + \cdots$

15. $\frac{7}{10} + \frac{7}{100} + \frac{7}{1000} + \cdots$ **16.** $\frac{1}{5} + \frac{1}{10} + \frac{1}{20} + \cdots$

17. $-2 - \frac{1}{4} - \frac{1}{32} \cdots$ **18.** $9 - 6 + 4 - \cdots$

19. $\frac{1}{2} - \frac{1}{3} + \frac{2}{9} - \cdots$

In 20–22, find the sum of the infinite geometric progression in which a_1 and r have the given values.

20. $a_1 = 7, r = \frac{1}{2}$ **21.** $a_1 = 8, r = -\frac{1}{4}$ **22.** $a_1 = .7, r_1 = .1$

23. Find the common ratio in an infinite geometric progression whose first term is 4 and whose sum is 8.

24. Find the first term in an infinite geometric progression whose sum is 10 and whose common ratio is $\frac{1}{2}$.

25. A rubber ball is dropped from a height of 18 feet. It rebounds $\frac{2}{3}$ of the distance from which it fell previously. This process continues until the ball comes to rest. Find the total distance through which the ball moved.

26. The sum of the first two terms of an infinite geometric series is $\frac{8}{9}$ and the first term is $\frac{2}{3}$. Find the sum of the series.

In 27–38, represent the repeating decimal as an equivalent fraction.

27. .888888 . . .	**28.** .131313 . . .	**29.** .363636 . . .	**30.** .484848 . . .
31. .727272 . . .	**32.** .838383 . . .	**33.** .125125 . . .	**34.** .431431 . . .
35. .566666 . . .	**36.** .844444 . . .	**37.** .45151 . . .	**38.** .326363 . . .

12. Recursion

The arithmetic sequence 3, 5, 7, 9, . . . can be represented by using the general formula for its nth term:

$$a_n = a_1 + (n - 1)d$$
$$a_n = 3 + (n - 1)2 \qquad a_1 = 3, d = 2$$

This same arithmetic sequence can also be represented by using a *recursive formula* that describes how to find the nth term from the previous term:

$$a_n = a_{n-1} + d$$
$$a_n = a_{n-2} + 2, \qquad a_1 = 3$$

Note that a recursive formula has two parts: a rule or *recursion equation* that shows how to find each term from the term(s) before it and the value of the first term. The formula $a_n = a_1 + (n - 1)d$ is not recursive because each term is determined by the number of the term n rather than by the preceding term. With a recursive formula, you cannot calculate a term such as a_5 without first calculating a_2, a_3, and a_4.

Similarly, the geometric sequence 2, 6, 18, 54, . . . can be represented two ways:

nth term formula	recursive formula
$a_n = a_1 r^{n-1}$	$a_n = r a_{n-1}$
$a_n = 2 \cdot 3^{n-1}$	$a_n = 3 \cdot a_{n-1}, \quad a_1 = 2$

Depending on the information given in a problem, one type of formula may be more convenient to use than the other. Recursive formulas are often used to represent sequences that are neither arithmetic nor geometric.

~~~~~~~~~~~~~~~~~~ **MODEL PROBLEMS** ~~~~~~~~~~~~~~~~~

1. Find the first four terms of the sequence in which $a_1 = 5$ and $a_n = 2a_{n-1} + 1$.

Solution:

$$a_1 = 5 \text{ and } a_n = 2a_{n-1} + 1$$

$$a_2 = 2a_1 + 1 = 2(5) + 1 = 11$$
$$a_3 = 2a_2 + 1 = 2(11) + 1 = 23$$
$$a_4 = 2a_3 + 1 = 2(23) + 1 = 47$$

Answer: The first four terms of the sequence are 5, 11, 23, and 47. Note that the sequence is neither arithmetic nor geometric.

2. Find the first six terms of the sequence in which $a_1 = 5$, $a_2 = 2$, and $a_n = a_{n-1} + a_{n-2}$.

Solution:

Note that the recursion equation involves two previous values, a_1 and a_2.

$$a_n = a_{n-1} + a_{n-2}, a_1 = 5, a_2 = 2$$

$$a_3 = a_2 - a_1 = 2 - 5 = -3$$
$$a_4 = a_3 - a_2 = -3 - 2 = -5$$
$$a_5 = a_4 - a_3 = -5 - (-3) = -2$$
$$a_6 = a_5 - a_4 = -2 - (-5) = 3$$

Answer: The first six terms of the sequence are 5, 2, −3, −5, −2, and 3.

~~~~~~

ITERATION

To *iterate* a function means to use the output as a new input for the next step. The function rule or equation for $f(x)$ must be given, along with an initial value x_0 at which to begin the evaluation process. For example, we can iterate the function $f(x) = x^2$ for an initial value of $x_0 = 2$. The first four iterates are:

$$f(x_0) = f(2) = 2^2 = 4 = x_1$$
$$f(x_1) = f(4) = 4^2 = 16 = x_2$$
$$f(x_2) = f(16) = 16^2 = 256 = x_3$$
$$f(x_3) = f(256) = 256^2 = 65{,}536 = x_4$$

Iteration is a recursion that composes a function with itself repeatedly. In symbols, the iterative process can be represented as

$$x_n = f(x_{n-1}) = f(f(x_{n-2})) \ldots \text{ where } f(f(x)) = f \circ f(x)$$

Note that all iterations are recursions but not all recursions are iterations. Moreover, since iterating a function produces a sequence of numbers, some functions will generate arithmetic or geometric sequences.

~~~~~~~~~~~~ *MODEL PROBLEMS* ~~~~~~~~~~~~

1. Find the first four iterates (x_1, x_2, x_3, x_4) of the function $f(x) = 3x - 6$ for an initial value of $x_0 = 4$.

    Solution:

    $$f(x) = 3x - 6, \; x_0 = 4$$

    $$f(x_0) = 3(4) - 6 = 6 = x_1$$
    $$f(x_1) = 3(6) - 6 = 12 = x_2$$
    $$f(x_2) = 3(12) - 6 = 30 = x_3$$
    $$f(x_3) = 3(30) - 6 = 84 = x_4$$

    Answer: The first four iterates of the function $f(x) = 3x - 6$ for an initial value of $x_0 = 4$ are 6, 12, 30, and 84.

2. Find the first four iterates of $f(x) = x + 5$ for $x_0 = 2$.

    Solution:

    $$f(x) = x + 5, \; x_0 = 2$$

    $$f(x_0) = 2 + 5 = 7 = x_1$$
    $$f(x_1) = 7 + 5 = 12 = x_2$$
    $$f(x_2) = 12 + 5 = 17 = x_3$$
    $$f(x_3) = 17 + 5 = 22 = x_4$$

    Answer: The first four iterates of $f(x) = x + 5$ for $x_0 = 2$ are 7, 12, 17, and 22.

Note that the sequence of iterates in this case forms an arithmetic sequence.

3. Find the first four iterates of $f(x) = .3x$ for $x_0 = 10$.

    Solution:

    $$f(x) = .3x, \; x_0 = 10$$

    $$f(x_0) = .3(10) = 3 = x_1$$
    $$f(x_1) = .3(3) = .9 = x_2$$
    $$f(x_2) = .3(.9) = .27 = x_3$$
    $$f(x_3) = .3(.27) = .081 = x_4$$

Answer: The first four iterates of $f(x) = .3x$ for $x_0 = 10$ are 3, .9, .27, and .081.

Note that in this case the sequence of iterates is a geometric sequence.

~~~~~~

We can generalize the results of Model Problems 2 and 3 above.

• **When a linear function $y = mx + b$ is iterated, the sequence of iterates will be an arithmetic sequence if $m = 1$ ($y = x + b$); the sequence of iterates will be geometric if $b = 0$ ($y = mx$).**

Suppose that in Model Problem 1, the initial value was given as $x_0 = 3$. Look what happens.

$$f(x) = 3x - 6, x_0 = 3$$

$$f(x_0) = 3(3) - 6 = 3 = x_1$$
$$f(x_1) = 3(3) - 6 = 3 = x_2$$
$$f(x_3) = 3(3) - 6 = 3\ x_4$$

Each output is the same as the input. A value for which all iterates are the same is called a ***fixed point*** of the function. Some of the properties of fixed points will be explored in the exercises.

FRACTALS: GEOMETRIC RECURSION

A *fractal* is a geometric figure created by a recursive geometric process. When small parts of a fractal are examined, these parts look the same as the entire shape. This property is called *self-similarity*. Theoretically, the recursive process for the fractal continues without end.

To create a fractal, we start with a rule that describes a pattern and apply this rule over and over to each new part. As an example, we will construct a famous fractal called the Koch curve.

The initial figure is a line segment. The recursion rule is to construct an equilateral triangle on the middle third of each segment, deleting the base of the triangle.

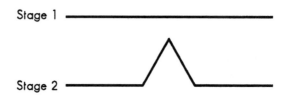

Stage 3

The next stage is left as an exercise. When the number and lengths of the segments in this curve are studied, many interesting patterns can be found.

Fractals are generated on a computer by iterating a function with an initial value that is a complex number and then plotting the points on the complex plane. The resulting figures are striking to look at and fascinating to examine. The name fractal was first used by the American mathematician Benoit Mandelbrot in the early 1970's.

Exercises

In 1–4, write a recursive formula to represent each sequence.

1. 4, 12, 20, 28, . . .   **2.** 37, 31, 25, 19, . . .

3. 1, 6, 36, 216, . . .   **4.** 5, −2, .8, −.32, . . .

In 5–8, find the first five terms of each sequence.

5. $a_1 = 7, a_n = 2a_{n-1} + 4$

6. $a_1 = 64, a_n = \frac{1}{2}a_{n-1}$

7. $a_1 = 1, a_2 = 2, a_n = 3a_{n-1} + 2a_{n-2}$

8. $a_1 = 100, a_n = .8a_{n-1}$

In 9–14, write the first four iterates of each function using the given initial values. Is the sequence of iterates arithmetic, geometric, or neither?

9. $f(x) = 4x - 1, x_0 = 2$

10. $f(x) = x - 6, x_0 = 5$

11. $f(x) = x^2 - 2x + 3, x_0 = 2$

12. $f(x) = \frac{3}{4}x, x_0 = 80$

13. $f(x) = 2x^2 + 1, x_0 = 0$

14. $f(x) = 9 - x, x_0 = 12$

In 15 and 16, find a fixed point for the given function.

15. $y = 7x - 12$ **16.** $y = 4x + 9$

17. Find a formula for the fixed point of the function $y = mx + b$.

18. The sequence of numbers below is called the Fibonacci sequence, named after the Italian mathematician Leonardo Fibonacci who devised it.

$$1, 1, 2, 3, 5, 8, 13, 21, 34, 55, \ldots$$

    a. Write the next three terms of the sequence.

    b. Write a recursive formula to represent the sequence.

19. Maria invested $1000 in a certificate of deposit that pays 5% interest compounded annually.
    a. Write a recursive formula that gives the balance b_n in the account after n years in terms of the previous year's balance.
    b. Find the end-of-year balance in the account for the first four years.
20. Draw Stage 4 of the Koch curve.
21. a. Complete the following table for the number of line segments in each stage of the Koch curve.

Stage	1	2	3	4
Number of Line segments	1	4	?	?

   b. Write a recursive formula for the number of line segments at stage n in terms of the number in the previous stage.
   c. Use the result from part b to predict the number of segments in Stage 6.
22. The fractal known as Sierpinski's triangle can be created as follows.
    · Start with an equilateral triangle.
    · Join the midpoints of the sides with line segments.
    · Remove (shade) the middle triangle.
    · Repeat the steps above for each new triangle.

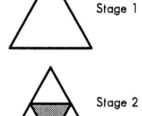

Stage 1

Stage 2

23. Draw the fifth figure in the sequence. Explain the rule in words.

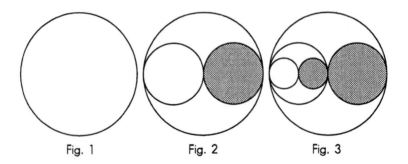

Fig. 1   Fig. 2   Fig. 3

13. Expanding a Binomial by the Binomial Theorem

The binomial theorem provides a method for *expanding* any power of a binomial, without repeated multiplication. We shall consider only those cases of binomials which have positive integral exponents.

If we begin with $(x + y)^1 = x + y$ and repeatedly multiply by $x + y$, we obtain the following:

$$(x + y)^1 = x + y$$
$$(x + y)^2 = x^2 + 2xy + y^2$$
$$(x + y)^3 = x^3 + 3x^2y + 3xy^2 + y^3$$
$$(x + y)^4 = x^4 + 4x^3y + 6x^2y^2 + 4xy^3 + y^4$$
$$(x + y)^5 = x^5 + 5x^4y + 10x^3y^2 + 10x^2y^3 + 5xy^4 + y^5$$

These results suggest the following rules for expanding $(x + y)^n$ when n is a positive integer:

1. The exponent of x in the first term is n. In each succeeding term, the exponent of x decreases by 1. The last term does not contain x.
2. The first term does not contain y. The exponent of y is 1 in the second term and increases by 1 in each succeeding term until it is n in the last term.
3. The coefficient of the first term is 1; of the second, n.
4. If the coefficient of any term is multiplied by the exponent of x in that term, and the product divided by the number of the term, the quotient is the coefficient of the next term.

In the expansion of $(x + y)^n$, when n is a positive integer, the following statements are always true:

1. The number of terms in the expansion is $n + 1$.
2. The sum of the exponents of x and y in each term in the expansion is n.
3. The coefficients of terms "equidistant" from the ends are the same.

The binomial theorem stated in general terms is :

$$(x + y)^n = x^n + \frac{n}{1}x^{n-1}y + \frac{n(n-1)}{1 \cdot 2} x^{n-2}y^2 + \frac{n(n-1)(n-2)}{1 \cdot 2 \cdot 3} x^{n-3}y^3 + \cdots + y^n$$

The symmetric coefficients form a pattern that can be displayed in a triangular formation. This display, called *Pascal's triangle,* is shown below.

$(x + y)^0$					1						
$(x + y)^1$				1		1					
$(x + y)^2$			1		2		1			**Pascal's triangle**	
$(x + y)^3$		1		3		3		1			
$(x + y)^4$		1		4		6		4		1	
$(x + y)^5$	1		5		10		10		5		1

In Pascal's triangle, elements inside the triangle can also be obtained by adding a pair of adjacent entries from the row above. Thus, the elements of the fifth row (1-4-6-4-1) are added to find the elements of the sixth row, which are 1-5-10-10-5-1.

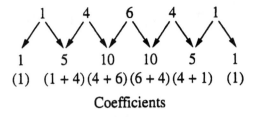

$$(1) \quad (1+4)\,(4+6)\,(6+4)\,(4+1) \quad (1)$$

Coefficients

~~~~~~~~ **MODEL PROBLEMS** ~~~~~~~~

1. Expand $(a+1)^4$ by using the binomial theorem.

Solution:
If we write the terms of the expansion of $(x+y)^4$ without their coefficients, we have

$$(x+y)^4 = (?)x^4 + (?)x^3y^1 + (?)x^2y^2 + (?)x^1y^3 + (?)y^4$$

In this example, $x = a$ and $y = 1$. Hence,

$$(a+1)^4 = (?)a^4 + (?)a^3(1)^1 + (?)a^2(1)^2 + (?)a^1(1)^3 + (?)(1)^4$$

To find the coefficients:

a. The coefficient of the first term is 1.
b. The coefficient of the second term is 4, the exponent of the binomial.
c. To find the coefficient of the third term, multiply 4, the coefficient of the second term, by 3, the exponent of a in the second term. Then divide the product, 12, by 2, the number of the second term. The result, 6, represents the coefficient of the third term.
d. To find the coefficient of the fourth term, multiply 6, the coefficient of the third term, by 2, the exponent of a in the third term. Then divide the product, 12, by 3, the number of the third term. The result, 4, represents the coefficient of the fourth term.
e. The coefficient of the last term, the fifth term, is 1.

Hence, $(a+1)^4 = (1)a^4 + 4a^3(1)^1 + 6(a^2)(1)^2 + 4(a)^1(1)^3 + (1)(1)^4$.

Answer: $(a+1)^4 = a^4 + 4a^3 + 6a^2 + 4a + 1$

2. Expand $(2c-d)^3$ using the binomial theorem.

Solution:
If we write the terms of the expansion of $(x+y)^3$ without their coefficients, we have

$$(x + y)^3 = (?)x^3 + (?)x^2y^1 + (?)x^1y^2 + (?)y^3$$

In this example, $x = 2c$ and $y = -d$. Hence,

$$(2c - d)^3 = (?)(2c)^3 + (?)(2c)^2(-d)^1 + (?)(2c)^1(-d)^2 + (?)(-d)^3$$

To find the coefficients:

a. The coefficient of the first term is 1.

b. The coefficient of the second term is 3, the exponent of the binomial.

c. To find the coefficient of the third term, multiply 3, the coefficient of the second term, by 2, the exponent of $(2c)$ in that term, and divide the product by 2, the number of the term. The result, 3, represents the coefficient of the third term.

d. The coefficient of the last term, the fourth term, is 1. Hence,

$$(2c - d)^3 = (1)(2c)^3 + (3)(2c)^2(-d)^1 + (3)(2c)^1(-d)^2 + (1)(-d)^3$$
$$= (1)(8c^3) + (3)(4c^2)(-d) + (3)(2c)(d^2) + (1)(-d^3)$$

Answer: $(2c - d)^3 = 8c^3 - 12c^2\,d + 6cd^2 - d^3$

3. Use the pattern in Pascal's triangle and the pattern of exponents to expand $(x + y)^7$.

Solution: Write the next two rows of Pascal's triangle.

$$1\ \ 6\ \ 15\ \ 20\ \ 15\ \ 6\ \ 1$$

$$1\ \ 7\ \ 21\ \ 35\ \ 35\ \ 21\ \ 7\ \ 1 \qquad (x + y)^7 \text{ has 8 terms}$$

$$(x + y)^7 = 1x^7 + 7x^6y + 21x^5y^2 + 35x^4y^3 + 35x^3y^4 + 21x^2y^5 + 7xy^6 + 1y^7$$

Answer: $(x + y)^7 = x^7 + 7x^6y + 21x^5y^2 + 35x^4y^3 + 35x^3y^4 + 21x^2y^5 + 7xy^6 + y^7$

Exercises

In 1–4, state the number of terms there are in the expansion of the expression.

1. $(a + b)^8$ **2.** $(3x - 5y)^5$ **3.** $\left(a + \dfrac{1}{2}\right)^7$ **4.** $\left(\dfrac{b}{4} + \dfrac{4}{b}\right)^{10}$

5. The number of terms in the expansion of $(a + b)^n$, when n is a positive integer, is (1) n (2) $n + 1$ (3) $2n$ (4) $n - 1$

In 6–21, expand the expression using the binomial theorem.

6. $(a + b)^5$ **7.** $(r - s)^4$ **8.** $(m + n)^6$ **9.** $(x + 2)^4$

10. $(d - 2)^5$ **11.** $(a^2 + b)^4$ **12.** $(c^2 - d^2)^5$ **13.** $(m^2 - 1)^6$

14. $(2a + b)^5$ **15.** $(r - 2s)^4$ **16.** $(2c - 3d)^3$ **17.** $(1 + x)^6$

18. $(2 - b)^5$ **19.** $(x + \frac{1}{2})^4$ **20.** $\left(x + \dfrac{1}{x}\right)^3$ **21.** $(\frac{1}{2} - x^2)^3$

In 22–25, write the first two terms in the expansion of the expression.

22. $(a + b)^8$ **23.** $(x + y^2)^4$ **24.** $(2a + b)^4$ **25.** $(x + \frac{1}{2})^6$

In 26–33, write the first three terms of the expansion of the expression.

26. $(a + b)^6$ **27.** $(a - b)^7$ **28.** $(1 + 2a)^4$ **29.** $(x^2 - y)^3$

30. $(c^2 + 1)^8$ **31.** $(a^2 - 2b)^{10}$ **32.** $\left(2 - \dfrac{a}{2}\right)^5$ **33.** $\left(x + \dfrac{y}{3}\right)^7$

In 34–37, write the required term in the expansion of the expression.

34. the second term of $(a + b)^{12}$ **35.** the second term of $(2x - y)^4$

36. the third term of $(2 - s)^5$ **37.** the fourth term of $\left(\dfrac{b}{3} + \dfrac{3}{b}\right)^6$

38. Find the middle term of $\left(\dfrac{x}{2} + \dfrac{2}{x}\right)^8$.

39. *a.* Write the coefficients in row 10 of Pascal's triangle.
   b. Expand $(p + q)^9$.

14. Alternate Form of the Binomial Theorem; Probability Formula

Since $(x + y)^5 = (x + y)(x + y)(x + y)(x + y)(x + y)$, we can think of the expansion of $(x + y)^5$ as the result of choosing all possible combinations of either x or y from each of 5 factors.

1. We can choose x from 5 factors and y from 0 factors: $_5C_0 x^5 y^0$.
   The number of ways of doing this is $_5C_0$.

2. We can choose x from 4 factors and y from 1 factor: $_5C_1 x^4 y^1$.
   The number of ways of doing this is $_5C_1$.

3. We can choose x from 3 factors and y from 2 factors: $_5C_2 x^3 y^2$.
   The number of ways of doing this is $_5C_2$.

4. We can choose x from 2 factors and y from 3 factors: $_5C_3 x^2 y^3$.
   The number of ways of doing this is $_5C_3$.

5. We can choose x from 1 factor and y from 4 factors: $_5C_4 x^1 y^4$.
   The number of ways of doing this is $_5C_4$.

6. We can choose x from 0 factors and y from 5 factors: $_5C_5 x^0 y^5$.
   The number of ways of doing this is $_5C_5$.

Therefore:

$$(x + y)^5 = {_5}C_0 x^5 y^0 + {_5}C_1 x^4 y^1 + {_5}C_2 x^3 y^2 + {_5}C_3 x^2 y^3 + {_5}C_4 x^1 y^4 + {_5}C_5 x^0 y^5$$

or

$$(x + y)^5 = \binom{5}{0}x^5 y^0 + \binom{5}{1}x^4 y^1 + \binom{5}{2}x^3 y^2 + \binom{5}{3}x^2 y^3 + \binom{5}{4}x^1 y^4 + \binom{5}{5}x^0 y^5$$

or

$$(x + y)^5 = 1x^5 + 5x^4 y + 10x^3 y^2 + 10x^2 y^3 + 5xy^4 + 1y^5$$

- **In general:**

$$(x + y)^n = {_n}C_0 x^n y^0 + {_n}C_1 x^{n-1} y^1 + {_n}C_2 x^{n-2} y^2 + \cdots + {_n}C_{n-1} x^1 y^{n-1} + {_n}C_n x^0 y^n$$

or

$$(x + y)^n = \binom{n}{0}x^n y^0 + \binom{n}{1}x^{n-1} y^1 + \binom{n}{2}x^{n-2} y^2 + \cdots + \binom{n}{n-1}x^1 y^{n-1} + \binom{n}{n}x^0 y^n$$

The model problems that follow show how this general form is used to expand any binomial. Other observations based on patterns tell us that:

1. For any binomial expansion $(x + y)^n$, there are $n + 1$ terms.

2. In general, the rth term of the expansion is

$$_nC_{r-1} x^{n-r+1} y^{r-1}$$

The formula for a Bernoulli experiment or binomial probability given in Chapter XIV is closely related to the powers of a binomial. For example, if a binomial experiment has 3 trials with a probability of success p and a probability of failure $1 - p = q$, then the probabilities of 3, 2, 1, or 0 successes are shown in the table below.

x	3	2	1	0
$Pr(x)$	${_3}C_3 p^3 q^0$	${_3}C_2 p^2 q$	${_3}C_1 pq^2$	${_3}C_0 p^0 q^3$
	$= p^3$	$= 3p^2 q$	$= 3pq^2$	$= q^3$

Now consider the expansion of $(p + q)^3$:

$$(p + q)^3 = p^3 + 3p^2 q + 3pq^2 + q^3$$

The formula given for r successes in n trials,

$$_nC_r p^r q^{n-r}$$

corresponds to the $(n - r + 1)$th term of the binomial expansion of $(p + q)^n$.

Algebra II

~~~~~~~~~~~~~~~~~~~ **MODEL PROBLEMS** ~~~~~~~~~~~~~~~~~~

1. Write the expansion of $(b^2 - 3)^3$.

Solution: Write the general expansion of $(x + y)^3$ and let $x = b^2$ and $y = -3$.

$$(x + y)^3 = {_3C_0}x^3y^0 + {_3C_1}x^2y^1 + {_3C_2}x^1y^2 + {_3C_3}x^0y^3$$

Then:

$$(b^2 - 3)^3 = 1(b^2)^3(-3)^0 + 3(b^2)^2(-3)^1 + 3(b^2)^1(-3)^2 + 1(b^2)^0(-3)^3$$

$$= 1 \cdot b^6 \cdot 1 + 3 \cdot b^4 \cdot (-3) + 3 \cdot b^2 \cdot 9 + 1 \cdot 1 \cdot (-27)$$

$$= \quad b^6 \qquad\quad - 9b^4 \qquad\quad + 27b^2 \qquad\quad -27$$

Answer: $(b^2 - 3)^3 = b^6 - 9b^4 + 27b^2 - 27$

2. Write the *eleventh* term of the expansion of $(2a - 1)^{12}$.

Solution:

1. Write the general formula for the rth term of a binomial expansion.

$${_nC_{r-1}}x^{n-r+1}y^{r-1}$$

2. To find the 11th term of $(2a - 1)^{12}$, let $n = 12$, $r = 11$, $x = 2a$, and $y = -1$. Substitute these values in the formula, and simplify.

$${_{12}C_{11-1}}(2a)^{12-11+1}(-1)^{11-1}$$
$$= {_{12}C_{10}} \cdot (2a)^2 \cdot (-1)^{10}$$
$$= {_{12}C_2} \cdot (2a)^2 \cdot (-1)^{10}$$
$$= 66 \cdot 4a^2 \cdot 1$$
$$= 264a^2$$

Answer: $264a^2$

3. A die is tossed 10 times. Which term of the expansion of $(p + q)^n$ represents the probability that exactly 7 faces with a 1 on them will result? Write the term.

Solution: In this case, $n = 10$, $r = 7$, and $n - r + 1 = 4$.

The fourth term of the expansion of $(p + q)^{10}$ is

$${_{10}C_3}p^7q^3$$

Since $p = \frac{1}{6}$ and $q = \frac{5}{6}$, this gives

$$120\left(\frac{1}{6}\right)^7\left(\frac{5}{6}\right)^3$$

Answer: The fourth term, $120\left(\frac{1}{6}\right)^7\left(\frac{5}{6}\right)^3$

Exercises

In 1–4, write the expansion of each binomial. Use combinations.

1. $(3a - 1)^3$ **2.** $(x - 2)^4$ **3.** $(1 - b^3)^5$ **4.** $(2a^2 - b^3)^3$

In 5–8, in each case, write in simplest form the *third term* of the expansion. Use combinations.

5. $(k - 5)^3$ **6.** $(2 + y)^7$ **7.** $(2x - 1)^6$ **8.** $(4x + 3)^7$

In 9–12, in each case, write in simplest form the *fourth term* of the expansion. Use combinations.

9. $(5 - b)^4$ **10.** $(2a - 3)^5$ **11.** $(k^2 + 1)^7$ **12.** $\left(2x - \frac{1}{2}\right)^6$

13. Write in expanded form the volume of a cube if the measure of each edge is represented by $(2x - 3)$.

In 14–17, evaluate each power by using the expansion of a binomial.

14. $(1.01)^5$ **15.** $(1.2)^3$ **16.** $(1.02)^4$ **17.** $(1.05)^3$

In 18–22, select the *numeral* preceding the choice that best completes each sentence.

18. The third term of the expansion of $(a + 2b)^4$ is
  (1) $2a^2b^2$ (2) $4a^2b^2$ (3) $12a^2b^2$ (4) $24a^2b^2$

19. The eighth term of the expansion of $(2r - 1)^8$ is
  (1) $16r$ (2) $-16r$ (3) 1 (4) -1

20. The middle term of the expansion of $(x - 2y)^4$ is
  (1) $12x^2y^2$ (2) $-12x^2y^2$ (3) $24x^2y^2$ (4) $-24x^2y^2$

21. The last term in the expansion of $(6 - y)^9$ is
  (1) $54y^9$ (2) $-54y^9$ (3) y^9 (4) $-y^9$

22. A die is tossed 20 times. Which term of the expansion of $(p + q)^{20}$ gives the probability of getting exactly 13 faces that show 5?
  (1) 7th (2) 8th (3) 15th (4) 18th

---------- *CALCULATOR APPLICATIONS* ----------

Graphing calculators have several features that can be used to generate sequences, find the sums of series, and iterate functions.

-------------------- *MODEL PROBLEMS* --------------------

1. Find the first seven terms of the arithmetic sequence represented by $a_n = 9 + (n - 1)3$.

   Solution: Use the seq(command that is found on the ⎹LIST⎸ OPS menu. Enter the expression for the *n*th term, the variable to be incremented, starting value, ending value, and increment. In this case, the expression is $9 + (n - 1)3$, the variable is *n*, the starting value is 1, the ending value is 7, and the increment is 1.

   Enter: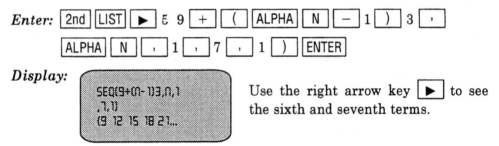

   Display:
   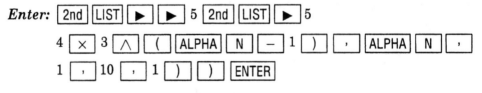

   Use the right arrow key ⎹▶⎸ to see the sixth and seventh terms.

   Answer: The first seven terms are 9, 12, 15, 18, 21, 24, 27.

2. Find the sum of the first ten terms of the geometric sequence 3, 12, 36, ...

   Solution: Since $a_1 = 4$ and $r = 3$, the general term is

   $$a_n = a_1 r^{n-1} = 4(3)^{n-1}$$

   Use both the sum(and seq(commands. The sum(command is on the ⎹LIST⎸ MATH menu.

   Enter:
   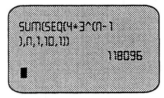

   Display:

   SUM(SEQ(4*3^(n-1
   ),n,1,10,1)
                    118096
   ■

   Answer: The sum of the first ten terms of the sequence 4, 12, 36, . . . is 118,096.

3. *a.* Find the first four iterates of the function $y = 3x + 5$ for an initial value of $x_0 = 1$.

  b. Determine the fixed point of $y = 3x + 5$ graphically.

Solution:

a. Start by storing the initial value as x.

Enter: 1 |STO ▶| |X, T, θ, n| |ENTER|

Then, to evaluate the function for the initial value, store $3x + 5$ as x.

Enter: 3 |X, T, θ, n| |+| 5 |STO ▶| |X, T, θ, n| |ENTER|

Continue to press |ENTER| to iterate the function. Each time |ENTER| is pressed, the calculator finds $3x + 5$ for the current value of x.

Display:

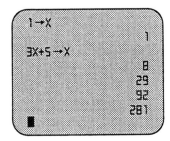

Answer: The first four iterates of $y = 3x + 5$ for $x_0 = 1$ are 8, 29, 92, and 281.

  b. For a fixed point, the output is the same as the input, so $y = x$. Therefore, the fixed point of $y = 3x + 5$ can be found by determining the point at which its graph intersects the graph of $x = y$. Use a standard window.

Enter: |Y=| 3 |X, T, θ, n| |+| 5 |ENTER| |X, T, θ, n| |GRAPH|
  |2nd| |CALC| 5

Use the arrow keys to move the cursor as close as possible to the intersection point. Then press |ENTER| |ENTER| |ENTER|.

Display:

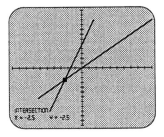

Answer: The x-coordinate, -2.5, is the fixed point of $y = 3x + 5$. To check, note that $3(-2.5) + 5 = -2.5$.

CHAPTER XVI

Trigonometry of the Right Triangle

1. Defining Three Trigonometric Functions

There are many situations where the measure of a distance or of an angle must be found even though the measure may prove difficult or impossible to obtain directly. By applying the principles of trigonometry, such measures may be obtained by indirect measurement.

In the figure, triangle ABC is a right triangle in which C is the right angle. $\overline{AB}$, the side opposite $\angle C$, is the **hypotenuse** of the right triangle; the length of $\overline{AB}$ is represented by c. The other two sides of $\triangle ABC$, $\overline{BC}$ and $\overline{AC}$, are called the **legs** of the triangle. We call $\overline{BC}$ the leg opposite $\angle A$; the length of $\overline{BC}$ is represented by a. We call $\overline{AC}$ the leg opposite $\angle B$; the length of $\overline{AC}$ is represented by b. We may also call $\overline{AC}$ the leg adjacent to (next to) $\angle A$. We may also call $\overline{BC}$ the leg adjacent to (next to) $\angle B$.

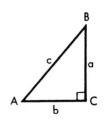

Now we will define three ratios, each of which involves two sides of the right triangle. These ratios are called **trigonometric ratios**.

For either acute angle in a right triangle:

$$\text{sine (sin) of the angle} = \frac{\text{length of leg opposite the angle}}{\text{length of hypotenuse}}$$

$$\text{cosine (cos) of the angle} = \frac{\text{length of leg adjacent to the angle}}{\text{length of hypotenuse}}$$

$$\text{tangent (tan) of the angle} = \frac{\text{length of leg opposite the angle}}{\text{length of leg adjacent to the angle}}$$

Note. Next to the name of each ratio, we find in parentheses the abbreviation for that name.

592

Using these definitions, we can represent the trigonometric ratios involving acute angles A and B in right triangle ABC as follows:

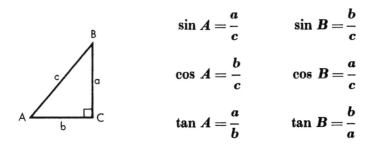

$$\sin A = \frac{a}{c} \qquad \sin B = \frac{b}{c}$$

$$\cos A = \frac{b}{c} \qquad \cos B = \frac{a}{c}$$

$$\tan A = \frac{a}{b} \qquad \tan B = \frac{b}{a}$$

It might appear that each of these trigonometric ratios, for example sin A, depends upon the size of the right triangle that contains $\angle A$. However, this is not the case. In the following figure, consider the two right triangles ABC and $A'B'C'$ in which $\angle A = \angle A'$.

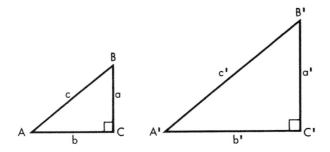

The lengths of the corresponding sides are different; however, $\angle A \cong \angle A'$. It follows that right triangle ABC is similar to right triangle $A'B'C'$ because they agree in two angles. Therefore, the lengths of the corresponding sides of these triangles are in proportion, giving $\dfrac{a}{c} = \dfrac{a'}{c'}$ or sin A = sin A'.

This proves that the number which is the value of sin A does not depend on the size of the right triangle which contains $\angle A$; it depends only on the measure of $\angle A$. The same reasoning is true for the two other trigonometric ratios.

Thus, with each acute angle we can associate one and only one number called the *sine*. Therefore, we have here an example of a function in which the first coordinate of every ordered pair is the measure of an acute angle and the second coordinate is the sine of that acute angle; that is, the set of ordered pairs $(A, \sin A)$ is a function. If the measure of $\angle A$ in degrees is represented by A, the domain of the function is the set of numbers between 0 and 90, $0 < A < 90$, and the range is the set of positive real numbers less than 1, $0 < \sin A < 1$.

$(\sin A = \dfrac{a}{c}$ must be less than 1 because in a right triangle the hypotenuse is always greater than either leg.) Similarly, the cosine and tangent ratios can be values of functions having the same domain $0 < A < 90$. Hence, these three ratios—sine, cosine, and tangent—can be used to define functions called **trigonometric functions.**

In future statements, the symbols $m \angle A = 50$ and $A = 50°$ will both mean "the measure of $\angle A$ is 50°."

~~~~~~~~~ *MODEL PROBLEMS* ~~~~~~~~~

1. In right triangle ABC, $a = 4$, $b = 3$, $c = 5$. Find the sine, cosine, and tangent of angle A and angle B.

Solution:

$$\sin A = \frac{a}{c} = \frac{4}{5} \qquad \sin B = \frac{b}{c} = \frac{3}{5}$$

$$\cos A = \frac{b}{c} = \frac{3}{5} \qquad \cos B = \frac{a}{c} = \frac{4}{5}$$

$$\tan A = \frac{a}{b} = \frac{4}{3} \qquad \tan B = \frac{b}{a} = \frac{3}{4}$$

2. In right triangle RST, $m \angle T = 90$, $r = 1$, and $s = 2$. Find $\sin R$, $\cos R$, and $\tan R$.

Solution: In order to find the required functions of $\angle R$, it is first necessary to find the length of the hypotenuse, t, by using the Pythagorean Theorem.

$$t^2 = r^2 + s^2$$
$$t^2 = 1 + 4$$
$$t^2 = 5$$
$$t = \sqrt{5}$$

$$\sin R = \frac{r}{t} = \frac{1}{\sqrt{5}} \cdot \frac{\sqrt{5}}{\sqrt{5}} = \frac{\sqrt{5}}{5}$$

$$\cos R = \frac{s}{t} = \frac{2}{\sqrt{5}} \cdot \frac{\sqrt{5}}{\sqrt{5}} = \frac{2\sqrt{5}}{5}$$

$$\tan R = \frac{r}{s} = \frac{1}{2}$$

Exercises

In 1–6, find the sine, cosine, and tangent of the acute angles in the right triangles:

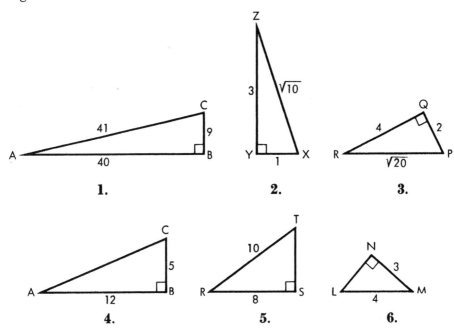

| 1. | 2. | 3. |

| 4. | 5. | 6. |

7. In right triangle RST, $m \angle T = 90$, $RS = 50$, and $ST = 30$. Find $\sin R$, $\cos R$, and $\tan R$.
8. The three sides of a right triangle are 8, 15, and 17. Find the sine, cosine, and tangent of the smaller acute angle.
9. In a right triangle, the hypotenuse is 4 and the shorter leg is 2. Find the sine, cosine, and tangent of the larger acute angle.
10. In right triangle RST, $m \angle T = 90$, $\tan S = \frac{5}{12}$, and $s = 10$. Find r.
11. In right triangle DEF, $m \angle F = 90$, $\cos D = \frac{15}{17}$, and $f = 68$. Find e.
12. In triangle ABC, $m \angle C = 90$. If $\cos A = \frac{5}{6}$ and $c = 30$, find b.
13. In triangle ABC, $m \angle C = 90$, $c = 51$, and $\sin B = \frac{8}{17}$. Find b.

2. Finding Trigonometric Ratios on a Calculator

In the past, mathematicians compiled tables of values of the sine, cosine, and tangent ratios for angle measures from 1° to 90° to be used as a reference. Today, trigonometric ratios can be found easily using the $\boxed{\text{SIN}}$, $\boxed{\text{COS}}$, and $\boxed{\text{TAN}}$ keys on a calculator. Values found are usually rounded to four decimal places unless otherwise specified.

To find tan 50° on some calculators, 50 is entered first and then the [TAN] key is pressed. On other calculators, 50 is entered after [TAN] is pressed. Try the following sequences of keys to determine which is correct for your calculator. Be sure that the calculator is in degree mode (DEG) before you start.

Method 1

   Enter: [TAN] 50 [ENTER]

Method 2

   Enter: 50 [TAN]

Display: [1.191753593]

So, tan 50° = 1.1918 rounded to four decimal places.

By trying different values on your calculator, you will find that, as the acute angle increases from 1° to 90°, the sine and the tangent of the angle increase; the cosine of the angle decreases. Note further that values of the sine and cosine functions of an acute angle are greater than 0 and less than 1.

Exercises

In 1–4, give the function value as a four-place decimal. Use a calculator.

1. sin 20°   2. tan 8°

3. sin 49°   4. cos 61°

In 5–7, state the change in the function values as angle A increases from 1° to 90°. Try several values for each function, using a calculator.

5. sin A 6. cos A 7. tan A

In 8–10, state the change in the function values as angle A decreases from 90° to 1°.

8. sin A 9. cos A 10. tan A

11. Which trigonometric function values of angle B increase as B increases from 1° to 90°?

12. Which trigonometric function values of angle D decrease as D increases from 1° to 90°?

13. As angle A changes from 1° to 90°, what number is sin A approaching?

14. Explain why sin A is less than 1 if A is an acute angle.

15. Explain why cos A is less than 1 if A is an acute angle.

16. Explain why tan A is less than 1 when A is less than 45°.

17. Explain why tan A is greater than 1 when A is greater than 45°.

18. Explain why sin A is less than tan A when A is an acute angle.

19. Which of the following numbers cannot be the value of cos A?

$$\tfrac{3}{5}, \tfrac{5}{3}, .9, \sqrt{2}, \tfrac{1}{2}\sqrt{2}$$

20. Which of the following numbers can be the value of sin A?

$$6, \tfrac{1}{6}, \tfrac{3}{4}, \tfrac{4}{3}, \sqrt{2}, .25, 2.5$$

21. If acute angle B is twice acute angle A, then sin B is always twice sin A. (Answer *true* or *false*.)

3. Finding the Measure of an Angle When a Function Value Is Given

If we know a function value of a positive acute angle, we can find the measure of the angle by using the inverse function keys on a calculator, $\boxed{\text{SIN}^{-1}}$, $\boxed{\text{COS}^{-1}}$, or $\boxed{\text{TAN}^{-1}}$. These keys are usually accessed by first pressing $\boxed{\text{2nd}}$ or $\boxed{\text{SHIFT}}$, and then following the sequences shown in the following two model problems. The calculator must be set in DEG mode to display the angle measure in degrees.

~~~~~~~ *MODEL PROBLEMS* ~~~~~~~

1. If sin A = .2182, find the degree measure of positive acute angle A, correct to the nearest degree.

Solution:

Enter: $\boxed{\text{2nd}}$ $\boxed{\text{SIN}^{-1}}$ .2182 $\boxed{\text{ENTER}}$

Display: $\boxed{\text{12.6033323}}$

Answer: $m\angle A = 13°$, correct to the nearest degree.

2. Given cos $A = \dfrac{\sqrt{3}}{3}$ (an exact function value), find A.

Solution:

Enter: [2nd] [COS⁻¹] [2nd] [√] 3 [)] [÷] 2 [ENTER]

Display: [30]

Note that in this case, the angle measure is exact.

Answer: $m\angle A = 30°$

KEEP IN MIND

1. We think of sin⁻¹ as "the angle whose sine is." Similar meanings exist for cos⁻¹ and tan⁻¹.

   Caution: Sin⁻¹ does *not* mean $\frac{1}{\sin}$ or the reciprocal of the sine function.

2. In most cases, we round approximate angle measures to the nearest degree unless otherwise instructed.

3. When we enter a real number greater than 1 as the value of sin A or cos A, the calculator displays an error message. For example, if we try to find A given sin $A = 2.5$, we get

   Enter: [2nd] [SIN⁻¹] 2.5 [ENTER]

   Display: [ERROR]

   There is no angle measure such that sin $A = 2.5$.

Exercises

In 1–12, use a calculator to find the degree measure of positive acute angle A to the nearest degree if:

1. sin $A = .4226$
2. cos $A = .9511$
3. tan $A = .7536$
4. sin $A = .7431$
5. cos $A = .3090$
6. tan $A = 2.9042$
7. sin $A = .6025$
8. cos $A = .0775$
9. tan $A = .4668$
10. sin $A = .9416$
11. cos $A = .1123$
12. tan $A = 8.2557$

4. Trigonometric Functions of 30°, 45°, and 60°

The values of the trigonometric functions of most angles are irrational numbers whose decimal approximations have been computed by mathematicians and can be found using calculators. However, the values of the trigonometric functions of 30°, 45°, and 60° can be determined without the use of calculators by applying geometric principles that should be familiar to us.

TRIGONOMETRIC FUNCTIONS OF 45°

In isosceles right triangle ABC, with $C = 90°$, let $a = 1$ unit and $b = 1$ unit. Then acute angles A and B have equal measures, each being 45°. Since $c^2 = a^2 + b^2$, $c^2 = 1 + 1 = 2$, or $c = \sqrt{2}$ units. The trigonometric functions of a 45° angle can now be expressed by applying their definitions in the 45°, 45°, 90° triangle.

$$\sin 45° = \frac{1}{\sqrt{2}} = \frac{1}{\sqrt{2}} \cdot \frac{\sqrt{2}}{\sqrt{2}} = \frac{\sqrt{2}}{2} = \frac{1}{2}\sqrt{2}$$

$$\cos 45° = \frac{1}{\sqrt{2}} = \frac{1}{\sqrt{2}} \cdot \frac{\sqrt{2}}{\sqrt{2}} = \frac{\sqrt{2}}{2} = \frac{1}{2}\sqrt{2}$$

$$\tan 45° = \frac{1}{1} = 1$$

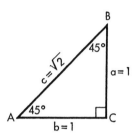

TRIGONOMETRIC FUNCTIONS OF 30° AND 60°

In right triangle ABC, $m\angle C = 90$, $m\angle A = 30$, and $m\angle B = 60$. Since in a 30°, 60°, 90° triangle, the side opposite the 30° angle is equal to one-half the hypotenuse, $a = \frac{1}{2}c$. If we let $c = 2$ units, then $a = 1$ unit. Since $b^2 = c^2 - a^2$, $b^2 = 4 - 1 = 3$, or $b = \sqrt{3}$ units. The trigonometric functions of a 30° angle and a 60° angle can now be expressed through the use of the 30°, 60°, 90° triangle.

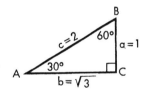

$$\sin 30° = \frac{1}{2}$$

$$\cos 30° = \frac{\sqrt{3}}{2} = \frac{1}{2}\sqrt{3}$$

$$\tan 30° = \frac{1}{\sqrt{3}} = \frac{1}{\sqrt{3}} \cdot \frac{\sqrt{3}}{\sqrt{3}} = \frac{\sqrt{3}}{3} = \frac{1}{3}\sqrt{3}$$

$$\sin 60° = \frac{\sqrt{3}}{2} = \frac{1}{2}\sqrt{3}$$

$$\cos 60° = \frac{1}{2}$$

$$\tan 60° = \frac{\sqrt{3}}{1} = \sqrt{3}$$

Summary of Values for Trigonometric Functions of 30°, 45°, and 60°

	30°	45°	60°
sin	$\frac{1}{2}$	$\frac{1}{2}\sqrt{2}$	$\frac{1}{2}\sqrt{3}$
cos	$\frac{1}{2}\sqrt{3}$	$\frac{1}{2}\sqrt{2}$	$\frac{1}{2}$
tan	$\frac{1}{3}\sqrt{3}$	1	$\sqrt{3}$

POWERS OF TRIGONOMETRIC FUNCTIONS

"The square of the sine of x" may be written as "$(\sin x)^2$" or more simply "$\sin^2 x$"; "the cube of the cosine of x" may be written "$\cos^3 x$"; and "the 5th power of the tangent of x" may be written "$\tan^5 x$". Thus, $\sin^2 30° = (\sin 30°)^2 = (\frac{1}{2})^2 = \frac{1}{4}$. Also, $\tan^5 45° = (\tan 45°)^5 = (1)^5 = 1$.

~~~~~~~~~~ *MODEL PROBLEMS* ~~~~~~~~~~

1. Show that the numerical value of $\sin^2 x + \cos^2 x$ is 1 when $x = 30°$.

Solution: When $x = 30°$, $\sin x = \frac{1}{2}$ and $\cos x = \frac{1}{2}\sqrt{3}$.

$$\sin^2 x + \cos^2 x = \sin^2 30° + \cos^2 30°$$

$$= (\tfrac{1}{2})^2 + (\tfrac{1}{2}\sqrt{3})^2 = \tfrac{1}{4} + \tfrac{3}{4} = 1 \quad Ans.$$

2. Find acute angle x when $2 \cos x - \sqrt{2} = 0$.

Solution: $2 \cos x - \sqrt{2} = 0$

$$2 \cos x = \sqrt{2} \qquad \text{Adding } \sqrt{2} \text{ to both sides.}$$
$$\cos x = \tfrac{1}{2}\sqrt{2} \qquad \text{Dividing both sides by 2.}$$

Hence, x must be the angle whose cosine is $\frac{1}{2}\sqrt{2}$. Therefore, $x = 45°$ *Ans.*

~~~~~~

Exercises

In 1–8, find the numerical value of the expression. (Leave answers containing radicals in radical form.)

1. $\cos 60° + \tan 45°$

2. $\tan 60° - \cos 60°$

3. $\sin 30° \cos 30°$

4. $3 \tan^3 45° - 2 \sin^2 45°$

5. $\dfrac{\sin 45°}{\tan 45°}$ 6. $\dfrac{\cos 45°}{\cos 60°}$ 7. $\dfrac{\sqrt{3} \tan 30°}{\tan 60°}$ 8. $\dfrac{10 \sin 60°}{\sin 30°}$

9. If $x = 45°$, find the value of $\sin^2 x + \cos^2 x$.
10. If $x = 30°$, show that $2 \sin x - 1 = 0$.
11. If $x = 30°$, show that $6 \sin^2 x + 7 \sin x = 5$.
12. If $x = 30°$, show that $2 \cos^2 x - 1 = 1 - 2 \sin^2 x$.

In 13–18, find the degree measure of acute angle x.

13. $\cos x = \frac{1}{2}$ **14.** $\sin x = \frac{1}{2}\sqrt{2}$ **15.** $\tan x = \sqrt{3}$
16. $\cos x = \frac{1}{2}\sqrt{3}$ **17.** $2 \sin x - 1 = 0$ **18.** $\tan x - 1 = 0$

5. Using Trigonometric Functions in Right Triangles

To solve a right triangle means to use the given sides and angles to find the remaining unknown sides and angles.

Procedure. To solve right triangles by trigonometry:
1. Make a careful diagram of the right triangle with the given parts.
2. Indicate the given measurements on the triangle.
3. Estimate the lines and angles to be found.
4. Find the remaining parts as follows:

If One Acute Angle and One Side Are Given:

a. Compute the remaining angle by finding the complement of the given angle.
b. Select the trigonometric function which relates the given angle and the ratio of the given side to the side to be found and write the equation expressing this relationship.
c. Find the second side by solving the equation and compare the answer to your previously found estimate.
d. Find the remaining side in the same way, using steps b and c.

If Two Sides Are Given:

a. Select a trigonometric ratio equal to a ratio of the two given sides and write the equation expressing this relationship.
b. Express the ratio of the two given sides as a four-place decimal.
c. Compute the angle by solving the equation.
d. Find the remaining acute angle by computing the complement of the first angle.
e. Find the third side by using the two given sides in the Pythagorean relationship, $a^2 + b^2 = c^2$.

ANGLE OF ELEVATION AND ANGLE OF DEPRESSION

If a person, using a telescope or some similar instrument, wishes to sight the top of the telephone pole above him (see the figure at the right), he must elevate (raise) the instrument from a horizontal position. $\overleftrightarrow{OT}$, the line joining the eye of the observer, O, to the top of the pole, T, is called the **line of sight**. The angle determined by the horizontal line ($\overleftrightarrow{OA}$) and the line of sight ($\overleftrightarrow{OT}$), $\angle AOT$, is called the **angle of elevation** of the top of the pole, T, from point O. (The horizontal line and the line of sight must be in the same vertical plane.)

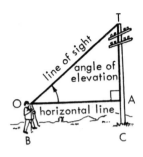

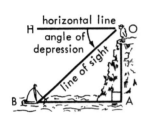

If a person, using a telescope or some similar instrument, wishes to sight the boat below him (see the figure at the left), he must depress (press down) the instrument from a horizontal position. $\overleftrightarrow{OB}$, the line joining the eye of the observer, O, and the boat, B, is called the **line of sight**. The angle determined by the horizontal line ($\overleftrightarrow{OH}$) and the line of sight ($\overleftrightarrow{OB}$), $\angle HOB$, is called the **angle of depression** of the boat, B, from point O. (The horizontal line and the line of sight must be in the same vertical plane.)

In the preceding figure if we measure the angle of elevation of O from B, $\angle OBA$, and also measure the angle of depression of B from O, $\angle HOB$, we discover that both angles contain the same number of degrees. We therefore say that the measure of the angle of elevation of O from B is equal to the measure of the angle of depression of B from O.

~~~~~~~~~~ *MODEL PROBLEMS* ~~~~~~~~~~

1. In right triangle ABC, $C = 90°$, $c = 25.0$, $A = 40°$; find B to the nearest degree; find a and b to the nearest tenth.

Given: $A = 40°$ *Find:* (1) B to the nearest degree
$\qquad\quad$ $C = 90°$ $\qquad\quad$ (2) a and b to the nearest
$\qquad\quad$ $c = 25.0$ $\qquad\qquad\quad$ tenth

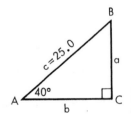

Solution: $B = 90° - A = 90° - 40° = 50°$

$$\sin A = \frac{a}{c} \qquad\qquad\qquad \cos A = \frac{b}{c}$$

$$\sin 40° = \frac{a}{25}$$

$$.6428 = \frac{a}{25}$$

$$a = 25 \,(.6428)$$
$$a = 16.07$$
Hence, $a = 16.1$ to the nearest tenth.

$$\cos 40° = \frac{b}{25}$$

$$b = 25 \cos 40°$$
$$b = 25 \,(.7660)$$
$$b = 19.15$$
Hence, $b = 19.2$ to the nearest tenth.

Answer: $B = 50°$, $a = 16.1$, $b = 19.2$

2. From the top of a lighthouse 180 feet above sea level, the angle of depression of a boat at sea measures 38°. Find, to the nearest foot, the distance from the boat to the foot of the lighthouse.

Given: $\angle LAB = 90°$
    $\angle HLB = 38°$
Distance $AL = 180$ ft.

Find: The distance BA
    to the nearest foot.

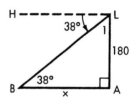

Solution: Since $\angle HLB$ is outside the triangle, find its congruent angle, $\angle LBA$, which is inside the triangle. Hence, $\angle LBA = 38°$.
    Let $x =$ the number of feet in distance BA.

Method 1

$$\tan B = \frac{LA}{BA}$$

$$\tan 38° = \frac{180}{x}$$

$$.7813 = \frac{180}{x}$$

$$.7813x = 180$$

$$x = \frac{180}{.7813} \approx 230.4$$

Hence, $x = 230$, to the nearest unit.

Method 2

$$\tan \angle 1 = \frac{BA}{LA}$$

$$\angle 1 = 90° - 38° = 52°$$

$$\tan 52° = \frac{x}{180}$$

$$x = 180 \tan 52°$$

$$x = 180 \,(1.2799)$$

$$x \approx 230.4$$

Hence, $x = 230$, to the nearest unit.

Answer: The distance is 230 feet.

Note. The result would have been the same if we had been given that the *angle of elevation* of the top of the lighthouse from point B measures 38°.

3. An approach to the overpass above a highway is 400 ft. long. The overpass is 30 ft. above the ground. Find, to the nearest degree, the angle at which the approach is inclined to the horizontal.

Given: $\angle BCA = 90°$, $BC = 30$ ft., $AB = 400$ ft.
Find: $\angle BAC$ or x to the nearest degree

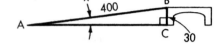

Solution: $\sin \angle BAC = \dfrac{BC}{AB}$

$\sin x = \dfrac{30}{400}$ Express $\dfrac{30}{400}$ as the decimal .0750.

$\sin x = .0750$ Use a calculator to find $\sin^{-1}$ of $(.0750)$.
$x \approx 4.301$

Answer: To the nearest degree, $x = 4°$.

~~~~~~

Exercises

In 1–8, find the lengths of the sides, to the nearest integer, and the measures of the angles, to the nearest degree. In the right triangle ABC, $\angle C = 90°$:

1. If $A = 35°$, $c = 20$, find a.   **2.** If $A = 40°$, $b = 36$, find a.
3. If $A = 42°$, $a = 35$, find c.   **4.** If $a = 20$, $c = 40$, find A.
5. If $a = 12$, $b = 5$, find A.   **6.** If $A = 28°$, $c = 30$, find B, a, b.
7. If $A = 39°$, $a = 27$, find B, b, c.   **8.** If $a = 15$, $b = 20$, find c, A, B.

9. In right triangle ABC, $\angle C = 90°$, $a = 6$, $c = 10$. Find $\angle A$, to the nearest degree.
10. In right triangle DEF, angle $E = 90°$, $f = 666$, $e = 999$. Find $\angle F$, to the nearest degree.

In 11–13, $\angle C$ is the right angle in right triangle ABC. Name a function which may be used:

11. to find a when A and c are given.
12. to find c when B and b are given.
13. to find A when a and b are given.

14. A captive balloon, fastened by a cable 1000 feet long, was blown by a wind so that the cable made an angle of 58° with the ground. Find, to the nearest foot, the height of the balloon.
15. A road is inclined at an angle of 10° with the horizontal. Find, to the nearest foot, the distance which must be driven on this road in order to be elevated 15 feet above the horizontal.

16. A plane takes off from a field and rises at an angle of 11° with the horizontal. Find, to the nearest foot, the height of the plane after it has traveled a horizontal distance of 1000 feet.

17. At a point 30 feet from the base of a tree, the angle of elevation of its top measures 53°. Find, to the nearest foot, the height of the tree.

18. A ladder 30 feet long leans against a building and makes an angle of 72° with the ground. Find, to the nearest foot, how high on the building the ladder reaches.

19. A lighthouse built at sea level is 170 feet high. From its top, the angle of depression of a buoy measures 25°. Find, to the nearest foot, the distance from the buoy to the foot of the lighthouse.

20. An artillery spotter in a plane at an altitude of 1000 feet observes the angle of depression of an enemy tank to measure 28°. How far, to the nearest foot, is the enemy tank from the point on the ground directly below the spotter?

21. A 20-foot ladder, AB, leans against a house that stands on level ground. The ladder, at A, makes an angle of 65° with the ground. How high, to the nearest foot, is B above the ground?

22. In triangle ABC, $C = 90°$, $c = 110$, $A = 42°$. Find b, to the nearest tenth.

23. In $\triangle ABC$, $C = 90°$, $c = 10$, $A = 22°$. Find b, correct to the nearest integer.

24. The lengths of two sides of a parallelogram are 6 inches and 10 inches and the angle between them measures 41°. What is the length of the altitude on the longer side? (Express your answer to the nearest inch.)

25. One leg of a trapezoid is 10 and makes an angle of 53° with the longer base. Find the altitude of the trapezoid. (Express your answer to the nearest integer.)

26. A plane takes off from a field and rises uniformly at an angle of 7° with the horizontal ground. Find, to the nearest 10 feet, the height of the plane after it has traveled over a horizontal distance of 1850 feet.

27. Find, to the nearest foot, the height of a church spire that casts a shadow of 60 feet when the angle of elevation of the sun measures 63°.

28. A lighthouse built at sea level is 160 feet high. From its top, the angle of depression of a buoy in the ocean measures 22°. Find, to the nearest 10 feet, the distance from the buoy to the foot of the lighthouse.

29. In right triangle ABC, if $a = 13$, $b = 20$, and $C = 90°$, find the measure of angle A, to the nearest degree.

30. Find, to the nearest degree, the measure of the angle of elevation of the sun when a tree 40 feet high casts a shadow 30 feet long.

31. A road rises 24 feet in a horizontal distance of 300 feet. Find, to the nearest degree, the measure of the angle that the road makes with the horizontal.

32. If a road rises 328 feet in a horizontal distance of 4000 feet, find, to the nearest degree, the measure of the angle that the road makes with the horizontal.

33. Find, to the nearest degree, the measure of the smaller acute angle of the right triangle whose legs are 20 inches and 25 inches.

34. If the vertex angle of an isosceles triangle measures 64° and each leg measures 10 inches, find, to the nearest tenth of an inch, the length of the altitude to the base.

35. In an isosceles triangle the vertex angle measures 50° and the length of the base is 30 inches. The length of the altitude drawn upon the base, to the nearest integer, is _____ inches.

36. The base of a rectangle measures 8 feet and the altitude measures 5 feet. Find, to the nearest degree, the measure of the angle that the diagonal makes with the base.

37. In a circle of radius 50 inches, a chord subtends an angle of 38° at the center. Find the distance of the chord from the center of the circle. (Express your answer to the nearest inch.)

38. Find, to the nearest hundredth of an inch, the length of a chord which subtends an angle of 144° at the center of a circle of radius 10 inches.

39. In a circle of radius 10 inches, a chord subtends a central angle of 48°. Find the length of the chord, to the nearest inch.

40. In an isosceles triangle, each of the congruent sides measures 220 and the base measures 275. Find, to the nearest degree, the measure of each of the congruent angles of the triangle.

6. Solving More Difficult Problems Involving Two Right Triangles

~~~~~~~~~~ *MODEL PROBLEM* ~~~~~~~~~~

An observer on the top of a cliff 2000 feet above sea level observes two ships due west of the foot of the cliff. The angles of depression of the ships measure 48° and 35°. Find, to the nearest foot, the distance between the ships.

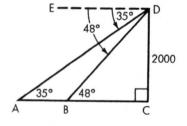

Fig. 1

Given: $\angle EDB = 48°$, $\angle EDA = 35°$, distance $DC = 2000$ ft.

Find: The distance between the ships, AB, to the nearest foot.

Solution: (It is useful to make separate diagrams of the two right triangles.)

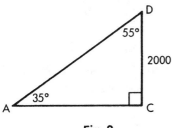

Fig. 2

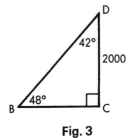

Fig. 3

In Fig. 2, since $\angle A = 35°$,
$\angle ADC = 55°$.
In right triangle CAD:

$$\tan 55° = \frac{AC}{2000}$$

Hence, $AC = 2000 \tan 55°$.

In Fig. 3, since $\angle B = 48°$,
$\angle BDC = 42°$.
In right triangle CBD:

$$\tan 42° = \frac{BC}{2000}$$

Hence, $BC = 2000 \tan 42°$.

In Fig. 1, $AB = AC - BC$.

$AB = 2000 \tan 55° - 2000 \tan 42° = 2000 (\tan 55° - \tan 42°)$
$AB = 2000 (1.4281 - .9004) = 2000 (.5277)$
$AB = 1055.4$

Hence, the distance $AB = 1055$ feet, to the nearest foot.

Answer: 1055 ft.

~~~~~~

Exercises

1. A woman on the top of a cliff 1000 feet above sea level observes two ships due west of the foot of the cliff. The angles of depression of the two ships measure 56° and 32°. Find, to the nearest foot, the distance between the ships.

2. $\overline{AB}$ is a flagpole which stands on top of a cliff BC. At point P, 310 feet from the foot of the cliff, the angle of elevation of B measures 21° and the angle of elevation of A measures 25°. Find, to the nearest foot, the length of the flagpole $\overline{AB}$.

3. A man observes the angle of elevation of the top of a vertical tower to measure 25°. He walks a distance of 250 feet toward the tower and in line with the foot of the tower and then observes the angle of elevation of the top of the tower to measure 50°. Find, to the nearest foot, the height of the tower.

4. A vertical tree is growing at the bank of a river. The angle of elevation of the top of the tree from a point directly across on the other bank measures 36°. From a second point 100 feet from the first and in line with the first point and the foot of the tree, the angle of elevation of the top of the tree measures 20°. Find, to the nearest foot, the width of the river.

~~~~~~~~ *CALCULATOR APPLICATIONS* ~~~~~~~~

Graphing calculators can graph trigonometric functions. Make sure that the calculator is in degree mode. For the problems below, set a ⃞WINDOW⃞ with Xmin = 0, Xmax = 90, Xscl = 5, Ymin = −1, Ymax = 2, Yscl = .1.

~~~~~~~~~~ *MODEL PROBLEMS* ~~~~~~~~~~

1. *a.* Graph $y = \sin x$ and $y = \cos x$, $0° \le x \le 90°$.
  b. Solve the equation $\sin x = \cos x$, $0° \le x \le 90°$.

Solution:

a. Enter: ⃞Y=⃞ ⃞SIN⃞ ⃞X,T,θ,*n*⃞ ⃞ENTER⃞ ⃞COS⃞ ⃞X,T,θ,*n*⃞ ⃞GRAPH⃞

   Display:

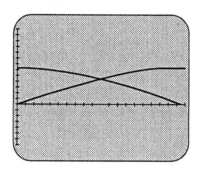

Note that the values of $y = \sin x$ increase from 0 to 1 on the interval $0° \le x \le 90°$, and the values of $y = \cos x$ decrease from 1 to 0.

b. Find the coordinates of the point where the two graphs intersect.

   Enter: ⃞2nd⃞ ⃞CALC⃞ 5

Move the cursor as close to the intersection point as possible. Then press ⃞ENTER⃞ ⃞ENTER⃞ ⃞ENTER⃞.

   Display:

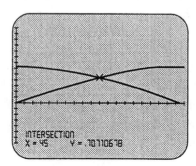

Answer: When $x = 45°$, $\sin x = \cos x \approx .707 \approx \dfrac{\sqrt{2}}{2}$.

2. Solve by graphing: $\sin x = .4$ if $0° \le x \le 90°$.

Solution: One method is to rewrite the equation as $\sin x - .4 = 0$. Then graph the function $y = \sin x - .4$ and look for the roots or zeros.

Enter: $\boxed{\text{Y=}}$ $\boxed{\text{SIN}}$ $\boxed{\text{X,T,θ,}n}$ $\boxed{)}$ $\boxed{-}$ .4 $\boxed{\text{GRAPH}}$

Display:

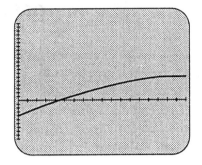

By looking at the graph, you can see there is one root in the interval. Use $\boxed{\text{2nd}}$ $\boxed{\text{CALC}}$, moving the cursor to enter a left bound and a right bound just on either side of the root; then press $\boxed{\text{ENTER}}$ for Guess? The calculator will display the zero as $x = 23.578178$.

Answer: To the nearest degree, $\sin x = .4$ for $x = 24°$. To check, use the $\boxed{\text{SIN}^{-1}}$ key to find the angle whose sine is .4.

CHAPTER XVII

Matrices

The English mathematician, James Joseph Sylvester, introduced the term *matrix* (plural, *matrices*) to the mathematical world in 1850. Arthur Cayley, also an English mathematician, began to develop the algebra of matrices in 1857. His work in this field became of tremendous value to nuclear physicists in quantum mechanics. Matrices are also used in modern industry, engineering, social sciences, and life sciences. Matrix operations are especially adaptable to computer performance.

1. Basic Properties of Matrices. Matrix Equality

THE DIMENSION OF A MATRIX

A *matrix* is any rectangular array of numbers, such as $\begin{bmatrix} 4 & 7 & -1 \\ 5 & 0 & 9 \end{bmatrix}$. Observe that this matrix has two (horizontal) rows and three (vertical) columns. Hence, we say that it is a 2×3 matrix, read, "a two-by-three matrix." We also say that it is a matrix of dimension 2×3. Note that, in the description of a matrix, the number of rows is stated first and the number of columns is stated second.

SPECIFYING A MATRIX

A capital letter such as A or B can be used to represent a matrix. For example,

$$A = \begin{bmatrix} 6 & -1 \\ 12 & 3 \end{bmatrix}, B = [2 \quad 15], C = \begin{bmatrix} 1 \\ 0 \end{bmatrix}, \quad \text{and} \quad D = \begin{bmatrix} 3 & 5 & 7 & 9 & 11 \\ 2 & 4 & 6 & 8 & 10 \\ -1 & 3 & 7 & 9 & 15 \end{bmatrix}$$

In order to indicate the dimensions of a matrix, subscripts can be used. Thus, for matrix A (above), we can write $A_{2 \times 2}$; for B, we can write $B_{1 \times 2}$; for C, we can write $C_{2 \times 1}$; and, for D, we can write $D_{3 \times 5}$.

In general, if a matrix E has m rows and n columns, we may write $E_{m \times n}$.

Every matrix is a member of a set of matrices whose members are matrices of the same dimension. For example, matrix A (above) is a member of the set $S_{2 \times 2}$; that is, $A \in S_{2 \times 2}$. Similarly, $B \in S_{1 \times 2}$; $C \in S_{2 \times 1}$; and $D \in S_{3 \times 5}$.

610

A matrix which has the same number of rows as columns is called a **square matrix.** For example, matrix A (above) is a square matrix of order 2 since it has 2 rows and 2 columns. In this chapter, we will deal almost exclusively with square matrices of order 2—matrices that are members of $S_{2 \times 2}$.

ZERO MATRIX

A **zero matrix** is a matrix, each of whose entries is zero. A zero matrix is denoted by $\mathbf{0}_{m \times n}$, or simply by $\mathbf{0}$. Hence, in the set $S_{2 \times 2}$, the zero matrix is $\begin{bmatrix} 0 & 0 \\ 0 & 0 \end{bmatrix}$.

Other examples of zero matrices are $[0 \quad 0 \quad 0]$, $\begin{bmatrix} 0 & 0 & 0 & 0 \\ 0 & 0 & 0 & 0 \end{bmatrix}$, and $\begin{bmatrix} 0 & 0 \\ 0 & 0 \\ 0 & 0 \end{bmatrix}$.

Note. The zero matrix should not be confused with the real number 0.

CORRESPONDING ENTRIES

If two matrices have the same dimensions, the entries in the same row and the same column are **corresponding entries**.

Thus, in matrices $A_{3 \times 2}$ and $B_{3 \times 2}$, the entry in the second row and first column of one of the two matrices corresponds to the entry in the second row and first column of the other.

NEGATIVE OF A MATRIX

The **negative of a matrix A,** denoted by $-A$, is a matrix of the same dimension, each of whose entries is the negative of the corresponding entry of A.

Thus, if $A = \begin{bmatrix} 0 & 1 \\ 2 & -3 \end{bmatrix}$, then $-A = \begin{bmatrix} 0 & -1 \\ -2 & 3 \end{bmatrix}$.

In general, if $A = \begin{bmatrix} a_1 & a_2 \\ a_3 & a_4 \end{bmatrix}$, then $-A = \begin{bmatrix} -a_1 & -a_2 \\ -a_3 & -a_4 \end{bmatrix}$.

TRANSPOSE OF A MATRIX

The **transpose of a matrix A,** denoted by A^T, is a matrix whose rows are the respective columns and whose columns are the respective rows of A. Read A^T as "A-transpose."

Thus, if $A = \begin{bmatrix} 2 & 4 \\ 5 & 7 \end{bmatrix}$, then $A^T = \begin{bmatrix} 2 & 5 \\ 4 & 7 \end{bmatrix}$.

In general, if $A = \begin{bmatrix} a_1 & a_2 \\ a_3 & a_4 \end{bmatrix}$, then $A^T = \begin{bmatrix} a_1 & a_3 \\ a_2 & a_4 \end{bmatrix}$.

EQUAL MATRICES

Two matrices are equal if and only if they are of the same dimension and their corresponding entries are equal.

Thus, if $A = \begin{bmatrix} x & 3 \\ 4 & y \end{bmatrix}$ and $B = \begin{bmatrix} 10 & z \\ w & -2 \end{bmatrix}$, then $A = B$ when $x = 10$, $y = -2$, $z = 3$, and $w = 4$.

In general, if $A = \begin{bmatrix} a_1 & a_2 \\ a_3 & a_4 \end{bmatrix}$ and $B = \begin{bmatrix} b_1 & b_2 \\ b_3 & b_4 \end{bmatrix}$, then $A = B$ if and only if $a_1 = b_1$, $a_2 = b_2$, $a_3 = b_3$, and $a_4 = b_4$.

~~~~~~~~~~ *MODEL PROBLEMS* ~~~~~~~~~~

1. If $C = \begin{bmatrix} 2x & \sqrt{y} \\ \dfrac{z}{2} & 2(w+2) \end{bmatrix}$ and $D = \begin{bmatrix} 1 & 2 \\ 3 & -4 \end{bmatrix}$, find replacements for the variables x, y, z, and w in order that $C = D$.

Solution: In order that $C = D$, the corresponding entries of C and D must be equal. Hence, $2x = 1$, $\sqrt{y} = 2$, $\dfrac{z}{2} = 3$, and $2(w+2) = -4$. Solving each equation, we obtain $x = \frac{1}{2}$, $y = 4$, $z = 6$, and $w = -4$.

Answer: $x = \frac{1}{2}$, $y = 4$, $z = 6$, $w = -4$

2. If $A = \begin{bmatrix} 5 & -6 \\ 7 & -8 \end{bmatrix}$ and $B = \begin{bmatrix} w & x \\ y & z \end{bmatrix}$, find replacements for the variables x, y, z, and w in order that $B^T = -A$.

Solution:

If $A = \begin{bmatrix} 5 & -6 \\ 7 & -8 \end{bmatrix}$, then $-A = \begin{bmatrix} -5 & 6 \\ -7 & 8 \end{bmatrix}$. To obtain $-A$, change the sign of each entry of A.

If $B = \begin{bmatrix} w & x \\ y & z \end{bmatrix}$, then $B^T = \begin{bmatrix} w & y \\ x & z \end{bmatrix}$. To obtain B^T, interchange the rows and the columns of B.

Hence, in order that $B^T = -A$, matrices $\begin{bmatrix} w & y \\ x & z \end{bmatrix}$ and $\begin{bmatrix} -5 & 6 \\ -7 & 8 \end{bmatrix}$ must be equal. Since matrices are equal if and only if their corresponding entries are equal, $w = -5$, $y = 6$, $x = -7$, and $z = 8$.

Answer: $x = -7$, $y = 6$, $z = 8$, $w = -5$

Exercises

In 1 and 2, use the matrix shown at the right. $\begin{bmatrix} 5 & -1 & 3 \\ 7 & 4 & -2 \\ 0 & 6 & 8 \end{bmatrix}$

1. Name the entry whose position in the matrix is:

   a. row 1, column 3   *b.* row 2, column 1

   c. row 3, column 1   *d.* row 1, column 2

2. Name the row and column of the indicated entry:

   a. 5   *b.* 6   *c.* -1 *d.* -2 *e.* 0

In 3–7, describe the dimensions of the matrix using the form $m \times n$ where m represents the number of rows and n represents the number of columns.

3. $\begin{bmatrix} 4 & 2 & -1 \end{bmatrix}$ **4.** $\begin{bmatrix} 0 & 5 \\ 3 & 2 \end{bmatrix}$

5. $\begin{bmatrix} 2 & -1 \\ 6 & 4 \\ -1 & 8 \end{bmatrix}$ **6.** $\begin{bmatrix} -6 \\ 4 \\ -8 \end{bmatrix}$ **7.** $\begin{bmatrix} 2 & -1 & 5 \\ 1 & 0 & 0 \\ 6 & -2 & -9 \end{bmatrix}$

In 8–12, state whether the two matrices are equal or not equal.

8. $\begin{bmatrix} 8 & 6 \\ 4 & 2 \end{bmatrix} \begin{bmatrix} 8 & 4 \\ 6 & 2 \end{bmatrix}$ **9.** $\begin{bmatrix} 4 & 2 & +5 \\ 8 & 8 & -2 \end{bmatrix} \begin{bmatrix} 5 & -1 & 7 \\ 7 & +1 & 6 \end{bmatrix}$

10. $\begin{bmatrix} 7 & 5 & 2 \end{bmatrix} \begin{bmatrix} 7 \\ 5 \\ 2 \end{bmatrix}$ **11.** $\begin{bmatrix} 1 & 3 & 5 \\ 7 & 9 & 11 \end{bmatrix} \begin{bmatrix} 5 & 3 & 1 \\ 11 & 9 & 7 \end{bmatrix}$

12. $\begin{bmatrix} 7 & \frac{1}{10} & \frac{1}{3} \\ 0 & -2 & \sqrt{-4} \\ 4 & 8 & -\sqrt{4} \end{bmatrix} \begin{bmatrix} \frac{1}{2}(20-6) & .10 & \frac{3}{9} \\ 5-\frac{1}{2}(10) & 5-7 & 2i \\ 2(1+1) & 2(2+2) & -2 \end{bmatrix}$

In 13–21, $A = \begin{bmatrix} 1 & 3 \\ 5 & 7 \end{bmatrix}$ and $B = \begin{bmatrix} -2 & 0 \\ 2 & -4 \end{bmatrix}$. Write the indicated matrix.

13. $-A$ **14.** $-B$ **15.** A^T **16.** B^T **17.** $-B^T$

18. $(-B)^T$ **19.** $-(-A)$ **20.** $-(-B)$ **21.** $(B^T)^T$

In 22–25, $A = \begin{bmatrix} q & r \\ s & t \end{bmatrix}$ and $B = \begin{bmatrix} -1 & 4 \\ 5 & -9 \end{bmatrix}$.

22. Find q if $A = -B$.   **23.** Find s if $A = B^T$.

24. Find r if $A = (-B)^T$.   **25.** Find r if $A = -B^T$.

In 26–31, $C = \begin{bmatrix} 2w & \sqrt{y} \\ \dfrac{z}{2} & \dfrac{1}{2}(x+2) \end{bmatrix}$ and $D = \begin{bmatrix} 0 & 1 \\ 2 & 3 \end{bmatrix}$.

26. Find w if $C = D$.   **27.** Find y if $C = D^T$.
28. Find z if $C = -D^T$.   **29.** Find x if $C = D$.
30. Find x if $C = -(-D)$.   **31.** Find x if $C = \mathbf{0}$, the zero matrix.

In 32–34, if $A = \begin{bmatrix} a_1 & a_2 \\ a_3 & a_4 \end{bmatrix}$, prove each given statement.

32. $-(-A) = A$ **33.** $(A^T)^T = A$ **34.** $-A^T = (-A)^T$

2. Matrix Addition and Matrix Subtraction

If A and B are matrices of the same dimension, then their sum, denoted by $A + B$, is a matrix of the same dimension, each of whose entries is the sum of the corresponding entries of A and B.

Thus, if $A = \begin{bmatrix} 2 & -4 \\ -3 & 5 \end{bmatrix}$ and $B = \begin{bmatrix} 0 & 4 \\ 3 & -2 \end{bmatrix}$, then

$$A + B = \begin{bmatrix} 2+0 & (-4)+4 \\ (-3)+3 & 5+(-2) \end{bmatrix} = \begin{bmatrix} 2 & 0 \\ 0 & 3 \end{bmatrix}.$$

In general, if $A = \begin{bmatrix} a_1 & a_2 \\ a_3 & a_4 \end{bmatrix}$ and $B = \begin{bmatrix} b_1 & b_2 \\ b_3 & b_4 \end{bmatrix}$, then

$$A + B = \begin{bmatrix} a_1 + b_1 & a_2 + b_2 \\ a_3 + b_3 & a_4 + b_4 \end{bmatrix}.$$

If A and B are matrices of the same dimension, then their difference, denoted by $A - B$, is a matrix of the same dimension, each of whose entries is the difference of the corresponding entries of A and B. (Each entry in B is subtracted from the corresponding entry in A.)

Thus, if $A = \begin{bmatrix} 8 & 7 \\ 6 & -2 \end{bmatrix}$ and $B = \begin{bmatrix} 4 & -3 \\ -2 & 0 \end{bmatrix}$, then

$$A - B = \begin{bmatrix} 8-4 & 7-(-3) \\ 6-(-2) & (-2)-0 \end{bmatrix} = \begin{bmatrix} 4 & 10 \\ 8 & -2 \end{bmatrix}.$$

In general, if $A = \begin{bmatrix} a_1 & a_2 \\ a_3 & a_4 \end{bmatrix}$ and $B = \begin{bmatrix} b_1 & b_2 \\ b_3 & b_4 \end{bmatrix}$, then

$$A - B = \begin{bmatrix} a_1 - b_1 & a_2 - b_2 \\ a_3 - b_3 & a_4 - b_4 \end{bmatrix}.$$

Under the definitions adopted here, matrices *cannot* be added or subtracted unless they are of the same dimension. For example, there is no matrix $A + B$ or matrix $A - B$ if $A = \begin{bmatrix} 3 & 5 & 1 \\ 7 & -1 & 9 \end{bmatrix}$ and $B = \begin{bmatrix} 3 & 7 \\ 5 & -1 \\ 1 & 9 \end{bmatrix}$.

ADDITIVE IDENTITY MATRIX

The *additive identity matrix* in the set $S_{m \times n}$ is the zero matrix, $\mathbf{0}_{m \times n}$, since the sum of $\mathbf{0}_{m \times n}$ and any matrix $A_{m \times n}$ is $A_{m \times n}$.

$$A_{m \times n} + \mathbf{0}_{m \times n} = \mathbf{0}_{m \times n} + A_{m \times n} = A_{m \times n}$$

For example, in the set $S_{2 \times 2}$, if $A = \begin{bmatrix} 1 & -2 \\ -3 & 4 \end{bmatrix}$:

$$\begin{bmatrix} 1 & -2 \\ -3 & 4 \end{bmatrix} + \begin{bmatrix} 0 & 0 \\ 0 & 0 \end{bmatrix} = \begin{bmatrix} 1+0 & -2+0 \\ -3+0 & 4+0 \end{bmatrix} = \begin{bmatrix} 1 & -2 \\ -3 & 4 \end{bmatrix}$$

In general, the zero matrix, $\begin{bmatrix} 0 & 0 \\ 0 & 0 \end{bmatrix}$, which is denoted by $\mathbf{0}$, is the unique additive identity matrix in the set $S_{2 \times 2}$.

ADDITIVE INVERSE MATRIX

The matrix $A_{m \times n}$ and its negative, $(-A)_{m \times n}$, are *additive inverses* of each other since their sum is the additive identity matrix, $\mathbf{0}_{m \times n}$.

$$A_{m \times n} + (-A)_{m \times n} = \mathbf{0}_{m \times n}$$

For example, in the set $S_{2 \times 2}$, if $A = \begin{bmatrix} 1 & -2 \\ -3 & 4 \end{bmatrix}$, then $-A = \begin{bmatrix} -1 & 2 \\ 3 & -4 \end{bmatrix}$.

Thus, A and $(-A)$ are additive inverses of each other, since

$$A + (-A) = \begin{bmatrix} 1+(-1) & (-2)+2 \\ (-3)+3 & 4+(-4) \end{bmatrix} = \begin{bmatrix} 0 & 0 \\ 0 & 0 \end{bmatrix} = \mathbf{0}.$$

In general, in the set $S_{2 \times 2}$, if $A = \begin{bmatrix} a_1 & a_2 \\ a_3 & a_4 \end{bmatrix}$, then its unique additive inverse is $-A = \begin{bmatrix} -a_1 & -a_2 \\ -a_3 & -a_4 \end{bmatrix}$.

It should be noted that if two matrices A and B are of the same dimension, then their difference, $A - B$, may be defined as the sum $A + (-B)$; that is, $A - B = A + (-B)$.

PROPERTIES INVOLVING THE ADDITION OF MATRICES OF THE SAME ORDER

In the matrix set $S_{2\times2}$, addition is closed, commutative, and associative; there is a unique additive identity element in the set; and for each member of the set there is a unique additive inverse.

These five properties are set forth as follows:

If A, B, and $C \in S_{2\times2}$, then:

1. $A + B \in S_{2\times2}$ (The closure property.)
2. $A + B = B + A$ (The commutative property with respect to addition.)
3. $(A + B) + C = A + (B + C)$ (The associative property with respect to addition.)
4. $A + 0 = 0 + A = A$ (The additive identity element is the zero matrix.)
5. $A + (-A) = (-A) + A = 0$ (The additive inverse element for each matrix A is $-A$.)

The five properties listed can be shown to apply in general to the addition of matrices of the same order. That is, the properties apply to matrices having the same number of rows and also the same number of columns, regardless of the number of rows and the number of columns.

〰〰〰〰〰 *MODEL PROBLEMS* 〰〰〰〰〰

In 1 and 2, $A = \begin{bmatrix} 3 & 7 \\ -2 & 0 \end{bmatrix}$, $B = \begin{bmatrix} -3 & 10 \\ 2 & -6 \end{bmatrix}$, and $C = \begin{bmatrix} 12 & 0 \\ -4 & -1 \end{bmatrix}$.

1. Find $A + B$.

Solution: $A + B = \begin{bmatrix} 3 + (-3) & 7 + 10 \\ (-2) + 2 & 0 + (-6) \end{bmatrix} = \begin{bmatrix} 0 & 17 \\ 0 & -6 \end{bmatrix}$

2. Find $C - B$.

Solution: $C - B = \begin{bmatrix} 12 - (-3) & 0 - 10 \\ -4 - 2 & -1 - (-6) \end{bmatrix} = \begin{bmatrix} 15 & -10 \\ -6 & 5 \end{bmatrix}$

3. Find the matrix X that satisfies the matrix equation

$$X + \begin{bmatrix} 2 & 5 \\ 1 & 0 \end{bmatrix} = \begin{bmatrix} 0 & -1 \\ 3 & -2 \end{bmatrix}.$$

Solution: (We will use the principle of substitution and properties of equality which also apply to matrices.) Add the additive inverse of $\begin{bmatrix} 2 & 5 \\ 1 & 0 \end{bmatrix}$ to each member to obtain:

$$X + \begin{bmatrix} 2 & 5 \\ 1 & 0 \end{bmatrix} + \begin{bmatrix} -2 & -5 \\ -1 & 0 \end{bmatrix} = \begin{bmatrix} 0 & -1 \\ 3 & -2 \end{bmatrix} + \begin{bmatrix} -2 & -5 \\ -1 & 0 \end{bmatrix}$$

Apply the associative property with respect to addition.

$$X + \left(\begin{bmatrix} 2 & 5 \\ 1 & 0 \end{bmatrix} + \begin{bmatrix} -2 & -5 \\ -1 & 0 \end{bmatrix} \right) = \begin{bmatrix} 0 + (-2) & (-1) + (-5) \\ 3 + (-1) & (-2) + 0 \end{bmatrix}$$

$$X + \begin{bmatrix} 0 & 0 \\ 0 & 0 \end{bmatrix} = \begin{bmatrix} -2 & -6 \\ 2 & -2 \end{bmatrix}$$

Answer: $X = \begin{bmatrix} -2 & -6 \\ 2 & -2 \end{bmatrix}$

4. If A and B are matrices of the set $S_{2 \times 2}$, prove that $A + B = B + A$.

Proof:

Let $A = \begin{bmatrix} a_1 & a_2 \\ a_3 & a_4 \end{bmatrix}$ and $B = \begin{bmatrix} b_1 & b_2 \\ b_3 & b_4 \end{bmatrix}$.

Then $A + B = \begin{bmatrix} a_1 + b_1 & a_2 + b_2 \\ a_3 + b_3 & a_4 + b_4 \end{bmatrix}$ and $B + A = \begin{bmatrix} b_1 + a_1 & b_2 + a_2 \\ b_3 + a_3 & b_4 + a_4 \end{bmatrix}$.

By the commutative property for the addition of real numbers,

$$a_1 + b_1 = b_1 + a_1 \qquad a_2 + b_2 = b_2 + a_2$$
$$a_3 + b_3 = b_3 + a_3 \qquad a_4 + b_4 = b_4 + a_4$$

Since each entry of $B + A$ is equal to the corresponding entry of $A + B$, then by the definition of the equality of matrices, $A + B = B + A$.

Exercises

In 1–6, if $A = \begin{bmatrix} 1 & 3 \\ 5 & 7 \end{bmatrix}$, $B = \begin{bmatrix} 0 & 2 \\ 4 & 6 \end{bmatrix}$, and $\mathbf{0} = \begin{bmatrix} 0 & 0 \\ 0 & 0 \end{bmatrix}$, write the indicated matrix.

1. $A + B$ **2.** $B + A$ **3.** $A - B$
4. $B - A$ **5.** $A + \mathbf{0}$ **6.** $\mathbf{0} - B$

In 7–12, if $A = \begin{bmatrix} -1 & 0 \\ 1 & 0 \end{bmatrix}$, $B = \begin{bmatrix} 1 & 0 \\ 0 & 1 \end{bmatrix}$, and $C = \begin{bmatrix} 0 & 1 \\ -1 & 0 \end{bmatrix}$, write the indicated matrix.

7. $(A + B) + C$ **8.** $A + (B + C)$ **9.** $(A + C) - B$
10. $(A - C) + B$ **11.** $(A + B) + (A - B)$ **12.** $(A + C) - (A - C)$

In 13–15, if $Q = \begin{bmatrix} 0 & 3 & 9 \\ 6 & -3 & 0 \end{bmatrix}$ and $R = \begin{bmatrix} -2 & 0 & 2 \\ -4 & 0 & 4 \end{bmatrix}$,

write the indicated matrix.

13. $Q + R$ **14.** $R - Q$ **15.** $Q^T + R^T$

In 16–18, if $A = \begin{bmatrix} 2 & 4 \\ 8 & 6 \end{bmatrix}$ and $B = \begin{bmatrix} 5 & 7 \\ 0 & 3 \end{bmatrix}$, find the matrix X that satisfies the

given matrix equation.

16. $X + A = B$ **17.** $X - A = B$

18. $B + X = A$ **19.** $X + B^T = -A$

In 20–27, if A, B, C, and $\mathbf{0}$ are matrices of the set $S_{2 \times 2}$, prove the given statement.

20. $A + B \in S_{2 \times 2}$ **21.** $(A + B) + C = A + (B + C)$

22. $A + \mathbf{0} = A$ **23.** $A + (-A) = \mathbf{0}$

24. $A^T + B^T = (A + B)^T$ **25.** $A^T + B = (A + B^T)^T$

26. $A - B^T = -(B^T - A)$ **27.** $(A - B) - C = A - (B + C)$

3. Multiplying a Real Number and a Matrix: Scalar Multiplication

The product of a real number and a matrix A is defined to be the matrix rA, each of whose entries is r times the corresponding entry of A. The number r is called a **scalar**.

Thus, if $A = \begin{bmatrix} 1 & -2 \\ 0 & 1\frac{1}{2} \end{bmatrix}$, then $10A = \begin{bmatrix} 10(1) & 10(-2) \\ 10(0) & 10(1\frac{1}{2}) \end{bmatrix} = \begin{bmatrix} 10 & -20 \\ 0 & 15 \end{bmatrix}$.

In general, if $A = \begin{bmatrix} a_1 & a_2 \\ a_3 & a_4 \end{bmatrix}$, then $rA = \begin{bmatrix} ra_1 & ra_2 \\ ra_3 & ra_4 \end{bmatrix}$.

Note that the product of a scalar and a matrix is a matrix of the same dimension as the original matrix.

〰〰〰〰〰〰 *MODEL PROBLEMS* 〰〰〰〰〰〰

1. If $A = \begin{bmatrix} 2 & -1 \\ 0 & 2\frac{1}{4} \end{bmatrix}$ and $B = \begin{bmatrix} -3 & -5 \\ 7 & 3\frac{1}{2} \end{bmatrix}$, find a 2×2 matrix equal to $6B - 12A$.

Solution: $6B - 12A = 6\begin{bmatrix} -3 & -5 \\ 7 & 3\frac{1}{2} \end{bmatrix} - 12\begin{bmatrix} 2 & -1 \\ 0 & 2\frac{1}{4} \end{bmatrix}$

$= \begin{bmatrix} -18 & -30 \\ 42 & 21 \end{bmatrix} - \begin{bmatrix} 24 & -12 \\ 0 & 27 \end{bmatrix}$

$$= \begin{bmatrix} -18-24 & -30-(-12) \\ 42-0 & 21-27 \end{bmatrix}$$

$$= \begin{bmatrix} -42 & -18 \\ 42 & -6 \end{bmatrix}$$

Answer: $\begin{bmatrix} -42 & -18 \\ 42 & -6 \end{bmatrix}$

2. If r is a real number and A and B are matrices of the set $S_{2\times 2}$, prove that $r(A+B)=rA+rB$.

Proof:

Let $A = \begin{bmatrix} a_1 & a_2 \\ a_3 & a_4 \end{bmatrix}$ and $B = \begin{bmatrix} b_1 & b_2 \\ b_3 & b_4 \end{bmatrix}$.

Since $A+B = \begin{bmatrix} a_1+b_1 & a_2+b_2 \\ a_3+b_3 & a_4+b_4 \end{bmatrix}$, then

$$r(A+B) = \begin{bmatrix} ra_1+rb_1 & ra_2+rb_2 \\ ra_3+rb_3 & ra_4+rb_4 \end{bmatrix}.$$

Since $rA = \begin{bmatrix} ra_1 & ra_2 \\ ra_3 & ra_4 \end{bmatrix}$ and $rB = \begin{bmatrix} rb_1 & rb_2 \\ rb_3 & rb_4 \end{bmatrix}$, then

$$rA+rB = \begin{bmatrix} ra_1+rb_1 & ra_2+rb_2 \\ ra_3+rb_3 & ra_4+rb_4 \end{bmatrix}.$$

Hence, $r(A+B)=rA+rB$.

Exercises

In 1–4, if $A = \begin{bmatrix} -1 & 0 \\ 0 & 1 \end{bmatrix}$ and $B = \begin{bmatrix} 2 & -1 \\ 1 & 0 \end{bmatrix}$, find a matrix equal to the given expression.

1. $3A$ **2.** $-2B$ **3.** $4A^T$ **4.** $(-5B)^T$

In 5–8, if $C = \begin{bmatrix} 1 & 0 \\ 0 & 1 \end{bmatrix}$ and $D = \begin{bmatrix} 0 & 1 \\ -1 & 0 \end{bmatrix}$, find a matrix equal to the given expression.

5. $C+2D$ **6.** $2C-D$ **7.** $3C+2D$ **8.** $1\frac{1}{2}D-\frac{1}{2}C$

In 9–11, if $A = \begin{bmatrix} -2 & 1 \\ 4 & 7 \end{bmatrix}$ and $B = \begin{bmatrix} 0 & 1 \\ 1 & 0 \end{bmatrix}$, find the matrix X that satisfies the given matrix equation.

9. $X+2B=A$ **10.** $X-A=3B$ **11.** $B-X=2A$

In 12–14, if $A = \begin{bmatrix} 5 & 7 \\ 2 & 4 \end{bmatrix}$, $B = \begin{bmatrix} 0 & 10 \\ \frac{1}{2} & -1 \end{bmatrix}$, $c = 4$, and $d = 10$, show that the given statement is true.

12. $(c + d)A = cA + dA$ **13.** $c(dB) = cd(B)$

14. $d(B - A) = dB - dA$

In 15–19, if A, B, and C are matrices of the set $S_{2 \times 2}$, $\mathbf{0}$ is the zero matrix of the same set, and c and d are real numbers, prove that the given statement is true.

15. $c\mathbf{0} = \mathbf{0}$ **16.** $1A = A$

17. $-1C = -C$ **18.** $(c + d)A = cA + dA$

19. $c(dB) = cd(B)$

4. Multiplying Matrices

In the last section, we found the product of a real number and a matrix. Now, we will learn how to find the product of two matrices. In the following illustration, the product of A and B is found where A and B are matrices of the set $S_{2 \times 2}$. The product of A and B is denoted by $A \times B$, $A \cdot B$, or AB.

If $A = \begin{bmatrix} 2 & 3 \\ 4 & 5 \end{bmatrix}$ and $B = \begin{bmatrix} 10 & 11 \\ 12 & 20 \end{bmatrix}$, then $AB = \begin{bmatrix} 2 & 3 \\ 4 & 5 \end{bmatrix} \cdot \begin{bmatrix} 10 & 11 \\ 12 & 20 \end{bmatrix}$.

Observe in the following product matrix AB that each entry is the sum of two products. Each product is obtained by multiplying the entries of a row of A by the entries of a column of B. Thus, in multiplying matrix A by matrix B, the entries of the first row of A and the first column of B determine the entry in the first row and first column of AB. Similar statements can be made about the remaining entries of AB.

$$AB = \begin{bmatrix} 2 \cdot 10 + 3 \cdot 12 & 2 \cdot 11 + 3 \cdot 20 \\ 4 \cdot 10 + 5 \cdot 12 & 4 \cdot 11 + 5 \cdot 20 \end{bmatrix} = \begin{bmatrix} 20 + 36 & 22 + 60 \\ 40 + 60 & 44 + 100 \end{bmatrix} = \begin{bmatrix} 56 & 82 \\ 100 & 144 \end{bmatrix}$$

Thus, $AB = \begin{bmatrix} 56 & 82 \\ 100 & 144 \end{bmatrix}$.

In general, the product of two matrices A and B in $S_{2 \times 2}$ is the following defined matrix AB:

$$\text{If } A = \begin{bmatrix} a_1 & a_2 \\ a_3 & a_4 \end{bmatrix} \text{ and } B = \begin{bmatrix} b_1 & b_2 \\ b_3 & b_4 \end{bmatrix},$$

$$\text{then } AB = \begin{bmatrix} a_1 b_1 + a_2 b_3 & a_1 b_2 + a_2 b_4 \\ a_3 b_1 + a_4 b_3 & a_3 b_2 + a_4 b_4 \end{bmatrix}.$$

SQUARING A MATRIX

The square of a matrix A is the matrix $A \cdot A$, which may be denoted by A^2. If $A = \begin{bmatrix} a_1 & a_2 \\ a_3 & a_4 \end{bmatrix}$, then, using the definition for the product of two matrices,

$$A^2 \text{ or } A \cdot A = \begin{bmatrix} a_1 & a_2 \\ a_3 & a_4 \end{bmatrix} \cdot \begin{bmatrix} a_1 & a_2 \\ a_3 & a_4 \end{bmatrix} = \begin{bmatrix} a_1 a_1 + a_2 a_3 & a_1 a_2 + a_2 a_4 \\ a_3 a_1 + a_4 a_3 & a_3 a_2 + a_4 a_4 \end{bmatrix}.$$

Note that the entries of A^2, the square of matrix A, are *not* the squares of the entries of matrix A.

~~~~~~~~~ *MODEL PROBLEM* ~~~~~~~~~

If $A = \begin{bmatrix} 1 & 3 \\ 5 & 7 \end{bmatrix}$ and $B = \begin{bmatrix} 2 & 4 \\ 6 & 8 \end{bmatrix}$, find AB.

How To Proceed	*Solution*
1. Obtain the entry in the 1st row-1st column of AB by adding the products of the entries in row 1 of A and column 1 of B.	1. $1 \cdot 2 + 3 \cdot 6$ $= 2 + 18 = 20$ $AB = \begin{bmatrix} 20 & \end{bmatrix}$
2. Obtain the entry in the 1st row-2nd column of AB by adding the products of the entries in row 1 of A and column 2 of B.	2. $1 \cdot 4 + 3 \cdot 8$ $= 4 + 24 = 28$ $AB = \begin{bmatrix} 20 & 28 \end{bmatrix}$
3. Obtain the entry in the 2nd row-1st column of AB by adding the products of the entries in row 2 of A and column 1 of B.	3. $5 \cdot 2 + 7 \cdot 6$ $= 10 + 42 = 52$ $AB = \begin{bmatrix} 20 & 28 \\ 52 & \end{bmatrix}$
4. Obtain the entry in the 2nd row-2nd column of AB by adding the products of the entries in row 2 of A and column 2 of B.	4. $5 \cdot 4 + 7 \cdot 8$ $= 20 + 56 = 76$ $AB = \begin{bmatrix} 20 & 28 \\ 52 & 76 \end{bmatrix}$

Answer: $\begin{bmatrix} 20 & 28 \\ 52 & 76 \end{bmatrix}$

MULTIPLYING MATRICES OF DIFFERENT DIMENSIONS

The definition given for the product of two matrices in $S_{2 \times 2}$ can be extended to the multiplication of matrices having different dimensions, provided that the *number of columns forming the matrix that is the first factor is equal to the number of rows forming the matrix that is the second factor*. Thus, in the following illustration, the product AB can be obtained since matrix A has two columns and matrix B has two rows.

$$\text{If } A = \begin{bmatrix} 1 & 2 \\ 3 & 4 \\ 5 & 6 \end{bmatrix} \text{ and } B = \begin{bmatrix} 10 & 20 & 30 \\ 40 & 50 & 60 \end{bmatrix},$$

$$\text{then } AB = \begin{bmatrix} 1 \cdot 10 + 2 \cdot 40 & 1 \cdot 20 + 2 \cdot 50 & 1 \cdot 30 + 2 \cdot 60 \\ 3 \cdot 10 + 4 \cdot 40 & 3 \cdot 20 + 4 \cdot 50 & 3 \cdot 30 + 4 \cdot 60 \\ 5 \cdot 10 + 6 \cdot 40 & 5 \cdot 20 + 6 \cdot 50 & 5 \cdot 30 + 6 \cdot 60 \end{bmatrix}.$$

$$\text{Hence, } AB = \begin{bmatrix} 10 + 80 & 20 + 100 & 30 + 120 \\ 30 + 160 & 60 + 200 & 90 + 240 \\ 50 + 240 & 100 + 300 & 150 + 360 \end{bmatrix} = \begin{bmatrix} 90 & 120 & 150 \\ 190 & 260 & 330 \\ 290 & 400 & 510 \end{bmatrix}.$$

MULTIPLICATIVE IDENTITY MATRIX IN SQUARE MATRICES

The multiplicative identity matrix in the set $S_{2 \times 2}$ is $I = \begin{bmatrix} 1 & 0 \\ 0 & 1 \end{bmatrix}$ since, for any matrix A in the set $S_{2 \times 2}$,

$$AI = IA = A$$

We can verify that I is the multiplicative identity element in $S_{2 \times 2}$ as follows:

If $A = \begin{bmatrix} a_1 & a_2 \\ a_3 & a_4 \end{bmatrix}$, then

$$IA = \begin{bmatrix} 1 & 0 \\ 0 & 1 \end{bmatrix} \cdot \begin{bmatrix} a_1 & a_2 \\ a_3 & a_4 \end{bmatrix} = \begin{bmatrix} 1 \cdot a_1 + 0 \cdot a_3 & 1 \cdot a_2 + 0 \cdot a_4 \\ 0 \cdot a_1 + 1 \cdot a_3 & 0 \cdot a_2 + 1 \cdot a_4 \end{bmatrix} = \begin{bmatrix} a_1 & a_2 \\ a_3 & a_4 \end{bmatrix} = A$$

It is left to the student to show that $AI = A$. Hence, matrix multiplication by the multiplicative identity I is commutative.

In the case of a square matrix in $S_{3 \times 3}$, the multiplicative identity element is $\begin{bmatrix} 1 & 0 & 0 \\ 0 & 1 & 0 \\ 0 & 0 & 1 \end{bmatrix}$. Study the **principal diagonal**, the diagonal that runs from the upper left corner to the lower right corner. Note that each entry on the principal diagonal is 1, whereas every other entry is 0. In general, in the case of a square matrix in $S_{n \times n}$, the multiplicative identity element is a square matrix of the same order having 1 as each entry on the principal diagonal and 0 as every other entry.

In general, if M is a square matrix and I is the identity matrix of the same order, then

$$MI = IM = M$$

DIFFERENCES BETWEEN THE MULTIPLICATION OF MATRICES AND THE MULTIPLICATION OF REAL OR COMPLEX NUMBERS

Unlike the multiplication of real or complex numbers, matrix multiplication is not in general commutative. Also, it is possible for two nonzero matrices to have the zero matrix for a product.

1. *Multiplication of Matrices Is Not Necessarily Commutative.*

If $A = \begin{bmatrix} 2 & 3 \\ 4 & 5 \end{bmatrix}$ and $B = \begin{bmatrix} 4 & 0 \\ 0 & 1 \end{bmatrix}$,

then $AB = \begin{bmatrix} 2 & 3 \\ 4 & 5 \end{bmatrix} \cdot \begin{bmatrix} 4 & 0 \\ 0 & 1 \end{bmatrix} = \begin{bmatrix} 2 \cdot 4 + 3 \cdot 0 & 2 \cdot 0 + 3 \cdot 1 \\ 4 \cdot 4 + 5 \cdot 0 & 4 \cdot 0 + 5 \cdot 1 \end{bmatrix} = \begin{bmatrix} 8 & 3 \\ 16 & 5 \end{bmatrix}$.

Also, $BA = \begin{bmatrix} 4 & 0 \\ 0 & 1 \end{bmatrix} \cdot \begin{bmatrix} 2 & 3 \\ 4 & 5 \end{bmatrix} = \begin{bmatrix} 4 \cdot 2 + 0 \cdot 4 & 4 \cdot 3 + 0 \cdot 5 \\ 0 \cdot 2 + 1 \cdot 4 & 0 \cdot 3 + 1 \cdot 5 \end{bmatrix} = \begin{bmatrix} 8 & 12 \\ 4 & 5 \end{bmatrix}$.

From the above, it follows that $AB \neq BA$. Hence, the multiplication of matrices is not necessarily commutative.

2. *The Product of Two Nonzero Matrices May Be the Zero Matrix.*

If $A = \begin{bmatrix} -2 & 1 \\ -4 & 2 \end{bmatrix}$ and $B = \begin{bmatrix} 2 & 3 \\ 4 & 6 \end{bmatrix}$,

then $AB = \begin{bmatrix} -2 & 1 \\ -4 & 2 \end{bmatrix} \cdot \begin{bmatrix} 2 & 3 \\ 4 & 6 \end{bmatrix} = \begin{bmatrix} -4 + 4 & -6 + 6 \\ -8 + 8 & -12 + 12 \end{bmatrix} = \begin{bmatrix} 0 & 0 \\ 0 & 0 \end{bmatrix}$.

From the above, it follows that $AB = \mathbf{0}$. Hence, the product of two nonzero matrices may be the zero matrix, $\mathbf{0}$.

PROPERTIES INVOLVING THE MULTIPLICATION OF 2 × 2 MATRICES

Although the commutative law of multiplication does not apply generally in the set of 2 × 2 matrices, the following is a list of properties that do hold in $S_{2 \times 2}$.

If A, B, and $C \in S_{2 \times 2}$, and r is a real number (scalar), then:

1. $AB \in S_{2 \times 2}$ (The closure property.)
2. $A(B + C) = AB + AC$ (The left distributive property.)
3. $(A + B)C = AC + BC$ (The right distributive property.)
4. $(AB)C = A(BC)$ (The associative property of multiplication.)
5. $r(AB) = (rA)B = A(rB)$ (The associative property of multiplication involving a real number, r.)
6. $AI = IA = A$ (The multiplicative identity element is the matrix I.)
7. $A0 = 0A = 0$ (Multiplication by the zero matrix results in the zero matrix.)

~~~~~ MODEL PROBLEMS ~~~~~

In 1 and 2, $A = \begin{bmatrix} 2 & 3 \\ 4 & 0 \end{bmatrix}$ and $B = \begin{bmatrix} -2 & 0 \\ 0 & -4 \end{bmatrix}$.

1. Find $(A + B)^2$.

Solution: Since $A + B = \begin{bmatrix} 2 + (-2) & 3 + 0 \\ 4 + 0 & 0 + (-4) \end{bmatrix} = \begin{bmatrix} 0 & 3 \\ 4 & -4 \end{bmatrix}$,

then $(A + B)^2 = \begin{bmatrix} 0 & 3 \\ 4 & -4 \end{bmatrix} \cdot \begin{bmatrix} 0 & 3 \\ 4 & -4 \end{bmatrix}$

$= \begin{bmatrix} 0 \cdot 0 + 3 \cdot 4 & 0 \cdot 3 + 3(-4) \\ 4 \cdot 0 + (-4)4 & 4 \cdot 3 + (-4)(-4) \end{bmatrix}$

$= \begin{bmatrix} 12 & -12 \\ -16 & 28 \end{bmatrix}$

Answer: $\begin{bmatrix} 12 & -12 \\ -16 & 28 \end{bmatrix}$

2. Find $A^2 + B^2$.

Solution: Since $A = \begin{bmatrix} 2 & 3 \\ 4 & 0 \end{bmatrix}$,

then $A^2 = \begin{bmatrix} 2 & 3 \\ 4 & 0 \end{bmatrix} \cdot \begin{bmatrix} 2 & 3 \\ 4 & 0 \end{bmatrix} = \begin{bmatrix} 2 \cdot 2 + 3 \cdot 4 & 2 \cdot 3 + 3 \cdot 0 \\ 4 \cdot 2 + 0 \cdot 4 & 4 \cdot 3 + 0 \cdot 0 \end{bmatrix} = \begin{bmatrix} 16 & 6 \\ 8 & 12 \end{bmatrix}$.

Since $B = \begin{bmatrix} -2 & 0 \\ 0 & -4 \end{bmatrix}$, then $B^2 = \begin{bmatrix} -2 & 0 \\ 0 & -4 \end{bmatrix} \cdot \begin{bmatrix} -2 & 0 \\ 0 & -4 \end{bmatrix}$

$= \begin{bmatrix} (-2)(-2) + 0 \cdot 0 & (-2)0 + 0(-4) \\ 0(-2) + (-4)0 & 0 \cdot 0 + (-4)(-4) \end{bmatrix} = \begin{bmatrix} 4 & 0 \\ 0 & 16 \end{bmatrix}$.

Hence, $A^2 + B^2 = \begin{bmatrix} 16 & 6 \\ 8 & 12 \end{bmatrix} + \begin{bmatrix} 4 & 0 \\ 0 & 16 \end{bmatrix} = \begin{bmatrix} 16+4 & 6+0 \\ 8+0 & 12+16 \end{bmatrix}$

$$= \begin{bmatrix} 20 & 6 \\ 8 & 28 \end{bmatrix}$$

Answer: $\begin{bmatrix} 20 & 6 \\ 8 & 28 \end{bmatrix}$

3. If $A = \begin{bmatrix} a_1 & a_2 \\ a_3 & a_4 \end{bmatrix}$ and $B = \begin{bmatrix} b_1 & b_2 \\ b_3 & b_4 \end{bmatrix}$, prove that $(AB)^T = B^T A^T$.

Proof:

Since $AB = \begin{bmatrix} a_1 b_1 + a_2 b_3 & a_1 b_2 + a_2 b_4 \\ a_3 b_1 + a_4 b_3 & a_3 b_2 + a_4 b_4 \end{bmatrix}$,

then $(AB)^T = \begin{bmatrix} a_1 b_1 + a_2 b_3 & a_3 b_1 + a_4 b_3 \\ a_1 b_2 + a_2 b_4 & a_3 b_2 + a_4 b_4 \end{bmatrix}$.

Since $B^T = \begin{bmatrix} b_1 & b_3 \\ b_2 & b_4 \end{bmatrix}$ and $A^T = \begin{bmatrix} a_1 & a_3 \\ a_2 & a_4 \end{bmatrix}$,

then $B^T A^T = \begin{bmatrix} b_1 a_1 + b_3 a_2 & b_1 a_3 + b_3 a_4 \\ b_2 a_1 + b_4 a_2 & b_2 a_3 + b_4 a_4 \end{bmatrix}$.

Since each entry of A and B is a real number, and the commutative property of multiplication applies to real numbers, each entry of $(AB)^T$ equals the corresponding entry of $B^T A^T$. Hence, $(AB)^T = B^T A^T$.

Exercises

In 1–8, find the product.

1. $\begin{bmatrix} 2 & 1 \\ 7 & 4 \end{bmatrix} \cdot \begin{bmatrix} 1 & 0 \\ 0 & 1 \end{bmatrix}$

2. $\begin{bmatrix} 2 & -1 \\ -7 & 4 \end{bmatrix} \cdot \begin{bmatrix} 0 & 0 \\ 0 & 0 \end{bmatrix}$

3. $\begin{bmatrix} 2 & 1 \\ 7 & 4 \end{bmatrix} \cdot \begin{bmatrix} 4 & -1 \\ -7 & 2 \end{bmatrix}$

4. $\begin{bmatrix} 2 & -1 \\ 7 & 4 \end{bmatrix} \cdot \begin{bmatrix} 2 & -1 \\ 7 & 4 \end{bmatrix}$

5. $\begin{bmatrix} 1 & 3 \\ 2 & 0 \end{bmatrix} \cdot \begin{bmatrix} 1 & -3 \\ 0 & 2 \end{bmatrix}$

6. $\begin{bmatrix} 1 & -3 \\ 0 & 2 \end{bmatrix} \cdot \begin{bmatrix} 1 & 3 \\ 2 & 0 \end{bmatrix}$

7. $\begin{bmatrix} 2 & 3 \\ -4 & -6 \end{bmatrix} \cdot \begin{bmatrix} 3 & 9 \\ -2 & -6 \end{bmatrix}$

8. $\begin{bmatrix} 1 & 2 & 3 \\ 4 & 5 & 4 \\ 3 & 2 & 1 \end{bmatrix} \cdot \begin{bmatrix} 1 & 0 & 0 \\ 0 & 1 & 0 \\ 0 & 0 & 1 \end{bmatrix}$

In 9–12, let $A = \begin{bmatrix} 0 & 2 \\ 4 & -2 \end{bmatrix}$ and $I = \begin{bmatrix} 1 & 0 \\ 0 & 1 \end{bmatrix}$.

9. Find AI.   **10.** Find IA.   **11.** Find $1\frac{1}{2}AI$.   **12.** Find $-2AI$.

In 13–17, let $B = \begin{bmatrix} 1 & 0 \\ 2 & -1 \end{bmatrix}$.

13. Find B^2.   **14.** Find $(-B)^2$.   **15.** Find $-B^2$.

16. Find matrix X if $BX = B$.   **17.** Find matrix X if $XB = 2B$.

In 18–22, let $A = \begin{bmatrix} 0 & 1 \\ -1 & 2 \end{bmatrix}$ and $B = \begin{bmatrix} 1 & 2 \\ 3 & 0 \end{bmatrix}$.

18. Find AB.   **19.** Find BA.   **20.** Find $A^2 + B^2$.

21. Find $A^2 B^2$.   **22.** Find $(A + B)^2$.

23. If $A = \begin{bmatrix} \frac{1}{2} & -\frac{1}{2}\sqrt{3} \\ \frac{1}{2}\sqrt{3} & \frac{1}{2} \end{bmatrix}$, show that $A \cdot A^T = I$.

24. Let $A = \begin{bmatrix} 3 & 2 \\ 1 & 0 \end{bmatrix}$ and $B = \begin{bmatrix} 0 & 1 \\ 2 & -4 \end{bmatrix}$. Show that:

  a. $(A + B)(A + B) \neq A^2 + 2AB + B^2$
  b. $(A + B)(A - B) \neq A^2 - B^2$

25. Let $A = \begin{bmatrix} 1 & 2 \\ 0 & 3 \end{bmatrix}$, $B = \begin{bmatrix} 2 & 3 \\ -1 & 5 \end{bmatrix}$, and $C = \begin{bmatrix} -1 & 0 \\ -2 & 3 \end{bmatrix}$.

  a. Find $A(BC)$.
  b. Find $(AB)C$.
  c. What property for the multiplication of matrices seems to hold true?
  d. Find $A(B + C)$.
  e. Find $AB + AC$.
  f. What property for the multiplication of matrices seems to hold true?

In 26–31, let $A = \begin{bmatrix} 1 \\ 2 \\ 3 \end{bmatrix}$, $B = \begin{bmatrix} 1 & 4 \\ 2 & 5 \\ 3 & 0 \end{bmatrix}$, $C = \begin{bmatrix} 1 & 2 & 3 \\ 4 & 3 & 0 \end{bmatrix}$, and $D = \begin{bmatrix} 0 & 1 & 2 \\ 1 & 0 & 2 \\ 2 & 1 & 0 \end{bmatrix}$.

26. Find BC.   **27.** Find CB.   **28.** Find CD.

29. Find DC.   **30.** Find DB.   **31.** Find DA.

In 32–35, if A, B, and C are matrices of the set $S_{2 \times 2}$ and r is a real number, prove that the given statement is true.

32. $A(B + C) = AB + AC$ **33.** $(A + B)C = AC + BC$

34. $(AB)C = A(BC)$ **35.** $r(AB) = (rA)B$

5. Using Matrices To Represent Complex Numbers

It is startling but nevertheless true that the matrix $\begin{bmatrix} a & b \\ -b & a \end{bmatrix}$ can be used to represent the complex number $a + bi$. Thus, the complex number $5 + 3i$ can be.

represented by the matrix $\begin{bmatrix} 5 & 3 \\ -3 & 5 \end{bmatrix}$ and the complex number $1 - 2i$ can be

represented by the matrix $\begin{bmatrix} 1 & -2 \\ 2 & 1 \end{bmatrix}$. If this is done, matrices representing complex numbers can then be used to perform operations upon complex numbers instead of using the complex numbers themselves.

For example, $(5 + 3i) + (1 - 2i) = 6 + i$. The sum $6 + i$ may be found by using matrices instead:

$$\begin{bmatrix} 5 & 3 \\ -3 & 5 \end{bmatrix} + \begin{bmatrix} 1 & -2 \\ 2 & 1 \end{bmatrix} = \begin{bmatrix} 5+1 & 3+(-2) \\ (-3)+2 & 5+1 \end{bmatrix}$$
$$= \begin{bmatrix} 6 & 1 \\ -1 & 6 \end{bmatrix}$$

The matrix $\begin{bmatrix} 6 & 1 \\ -1 & 6 \end{bmatrix}$ represents $6 + i$.

Also, $(5 + 3i)(1 - 2i) = 11 - 7i$. The product $11 - 7i$ may be found by using matrices instead:

$$\begin{bmatrix} 5 & 3 \\ -3 & 5 \end{bmatrix} \cdot \begin{bmatrix} 1 & -2 \\ 2 & 1 \end{bmatrix} = \begin{bmatrix} 5 \cdot 1 + 3 \cdot 2 & 5(-2) + 3 \cdot 1 \\ (-3) \cdot 1 + 5 \cdot 2 & (-3)(-2) + 5 \cdot 1 \end{bmatrix}$$
$$= \begin{bmatrix} 11 & -7 \\ 7 & 11 \end{bmatrix}$$

The matrix $\begin{bmatrix} 11 & -7 \\ 7 & 11 \end{bmatrix}$ represents $11 - 7i$.

Consider the equality of two complex numbers, $a + bi$ and $c + di$. We know that if $a + bi = c + di$, then $a = c$ and also $b = d$. Similarly, using matrices that represent $a + bi$ and $c + di$, if

$$\begin{bmatrix} a & b \\ -b & a \end{bmatrix} = \begin{bmatrix} c & d \\ -d & c \end{bmatrix},$$

then $a = c$ and also $b = d$.

Since the matrix $\begin{bmatrix} a & b \\ -b & a \end{bmatrix}$ represents the complex number $a + bi$, it follows that:

(1) the multiplicative identity matrix $\begin{bmatrix} 1 & 0 \\ 0 & 1 \end{bmatrix}$, denoted by I, represents $1 + 0i$, or 1.

(2) the matrix $\begin{bmatrix} 0 & 1 \\ -1 & 0 \end{bmatrix}$, denoted by J, represents $0 + 1i$, or the imaginary unit, i.

(3) the matrix $\begin{bmatrix} a & b \\ -b & a \end{bmatrix}$, which represents $a + bi$, can be shown to be equal to $aI + bJ$ in the following manner:

$$aI + bJ = a\begin{bmatrix} 1 & 0 \\ 0 & 1 \end{bmatrix} + b\begin{bmatrix} 0 & 1 \\ -1 & 0 \end{bmatrix}$$

$$= \begin{bmatrix} a & 0 \\ 0 & a \end{bmatrix} + \begin{bmatrix} 0 & b \\ -b & 0 \end{bmatrix}$$

$$= \begin{bmatrix} a & b \\ -b & a \end{bmatrix}.$$

MODEL PROBLEMS

1. Using matrices, find $(2 + 3i)^2$.

Solution:

Let $\begin{bmatrix} 2 & 3 \\ -3 & 2 \end{bmatrix}$ represent $2 + 3i$.

Then, $\begin{bmatrix} 2 & 3 \\ -3 & 2 \end{bmatrix}^2 = \begin{bmatrix} 2 & 3 \\ -3 & 2 \end{bmatrix} \cdot \begin{bmatrix} 2 & 3 \\ -3 & 2 \end{bmatrix} = \begin{bmatrix} 4-9 & 6+6 \\ -6-6 & -9+4 \end{bmatrix}$

$$= \begin{bmatrix} -5 & 12 \\ -12 & -5 \end{bmatrix}.$$

The resulting matrix corresponds to $-5 + 12i$.

Answer: $-5 + 12i$

2. Show, using matrices, that the sum of $a + bi$ and its conjugate is $2a$.

Solution:

The conjugate of $a + bi$ is $a - bi$. Using matrices,

$$\begin{bmatrix} a & b \\ -b & a \end{bmatrix} + \begin{bmatrix} a & -b \\ b & a \end{bmatrix} = \begin{bmatrix} a+a & b+(-b) \\ -b+b & a+a \end{bmatrix} = \begin{bmatrix} 2a & 0 \\ 0 & 2a \end{bmatrix}$$

The resulting matrix corresponds to $2a + 0i$, or $2a$.

Answer: $2a$

Exercises

In 1–12, state the matrix that corresponds to the complex number.

1. $3 + 7i$ **2.** $-6 + 4i$ **3.** $2 - 5i$ **4.** $-8 - i$

5. $7i$ **6.** $-i$ **7.** -1 **8.** $\sqrt{3}$

9. $\frac{1}{2} + i$ **10.** $5 + \sqrt{2}i$ **11.** $\sqrt{5} - 2i$ **12.** $\frac{1}{2}\sqrt{13} + \sqrt{17}i$

In 13–20, state the complex number that corresponds to the matrix.

13. $\begin{bmatrix} 3 & 2 \\ -2 & 3 \end{bmatrix}$ **14.** $\begin{bmatrix} 7. & -2 \\ 2 & 7 \end{bmatrix}$ **15.** $\begin{bmatrix} -1 & 5 \\ -5 & -1 \end{bmatrix}$

16. $\begin{bmatrix} -4 & -1 \\ 1 & -4 \end{bmatrix}$ **17.** $\begin{bmatrix} 0 & -3 \\ 3 & 0 \end{bmatrix}$ **18.** $\begin{bmatrix} \sqrt{3} & 0 \\ 0 & \sqrt{3} \end{bmatrix}$

19. $\begin{bmatrix} 0 & -\sqrt{5} \\ \sqrt{5} & 0 \end{bmatrix}$ **20.** $\begin{bmatrix} 3 & -\sqrt{2} \\ \sqrt{2} & 3 \end{bmatrix}$

In 21–34, use matrices to perform the indicated operation on the complex numbers. Express the answer both in matrix form and in the form $a + bi$.

21. $(3 + 4i) + (2 + 5i)$ **22.** $(2 + i) + (-4 + 5i)$

23. $(2 + i) + (2 - i)$ **24.** $(-6 + 2i) + (6 - 2i)$

25. $(1 - 3i) - 4i$ **26.** $10 - (8 + \sqrt{2}i)$

27. $(2 + 3i)(5 + 6i)$ **28.** $(-2 + 8i)(3 - 7i)$

29. $(1 + i)(1 - i)$ **30.** $(-2 + i)(2 - i)$ **31.** $(1 - i)^2$

32. i^2 **33.** $(1 + i)^3$ **34.** i^3

35. Using matrices, show that $(a + bi) - (a - bi) = 2bi$.

36. Using matrices, show that the product of $a + bi$ and its conjugate is $a^2 + b^2$.

37. The matrix $J = \begin{bmatrix} 0 & 1 \\ -1 & 0 \end{bmatrix}$ corresponds to the imaginary unit i. Using matrices, show that $i^3 = -i$.

6. Multiplicative Inverse of a Matrix in $S_{2 \times 2}$

Recall that every nonzero real number r has a unique multiplicative inverse $\frac{1}{r}$; that is, $r \cdot \frac{1}{r} = 1$ or $r \cdot r^{-1} = 1$.

Similarly, every 2×2 matrix has a unique multiplicative inverse with an exception to be noted later. If A^{-1} is the multiplicative inverse matrix of A, then, by definition, it must be true that $AA^{-1} = I$, the multiplicative identity matrix. It must also be true that $A^{-1}A = I$. Since matrices are not generally commutative, both matrix equations must be satisfied. Thus, since the product of $\begin{bmatrix} 3 & 5 \\ 1 & 2 \end{bmatrix}$ and $\begin{bmatrix} 2 & -5 \\ -1 & 3 \end{bmatrix}$ is $\begin{bmatrix} 1 & 0 \\ 0 & 1 \end{bmatrix}$, and since the product of $\begin{bmatrix} 2 & -5 \\ -1 & 3 \end{bmatrix}$ and $\begin{bmatrix} 3 & 5 \\ 1 & 2 \end{bmatrix}$ is also $\begin{bmatrix} 1 & 0 \\ 0 & 1 \end{bmatrix}$, then each of the matrix factors is the multiplicative inverse of the other.

Now, let us discover how to find the multiplicative inverse of a given 2×2 matrix. Let $A = \begin{bmatrix} a & b \\ c & d \end{bmatrix}$ be the representation of the general 2×2 matrix. Also, let $A^{-1} = \begin{bmatrix} w & x \\ y & z \end{bmatrix}$ be the representation of the multiplicative inverse of matrix A.

By the definition of the multiplicative inverse of a matrix, $AA^{-1} = I$, where $I = \begin{bmatrix} 1 & 0 \\ 0 & 1 \end{bmatrix}$.

$$\text{Hence,} \quad \begin{bmatrix} a & b \\ c & d \end{bmatrix} \cdot \begin{bmatrix} w & x \\ y & z \end{bmatrix} = \begin{bmatrix} 1 & 0 \\ 0 & 1 \end{bmatrix}.$$

$$\text{Then,} \quad \begin{bmatrix} aw + by & ax + bz \\ cw + dy & cx + dz \end{bmatrix} = \begin{bmatrix} 1 & 0 \\ 0 & 1 \end{bmatrix}.$$

Since corresponding entries of equal matrices are equal:

1. $aw + by = 1$

2. $cw + dy = 0$

3. $ax + bz = 0$

4. $cx + dz = 1$

Solve equations 1 and 2 for w.

$adw + bdy = d$ (In eq. 1, M_d.)

$bcw + bdy = 0$ (In eq. 2, M_b.)

$\overline{adw - bcw = d}$

$(ad - bc)w = d$

$$w = \frac{d}{ad - bc} \quad (ad - bc \neq 0)$$

Solve equations 3 and 4 for x.

$adx + bdz = 0$ (In eq. 3, M_d.)

$bcx + bdz = b$ (In eq. 4, M_b.)

$\overline{adx - bcx = -b}$

$(ad - bc)x = -b$

$$x = \frac{-b}{ad - bc} \quad (ad - bc \neq 0)$$

Solve equations 1 and 2 for y.

$acw + ady = 0$ (In eq. 2, M_a.)

$acw + bcy = c$ (In eq. 1, M_c.)

$\overline{ady - bcy = -c}$

$(ad - bc)y = -c$

$$y = \frac{-c}{ad - bc} \quad (ad - bc \neq 0)$$

Solve equations 3 and 4 for z.

$acx + adz = a$ (In eq. 4, M_a.)

$acx + bcz = 0$ (In eq. 3, M_c.)

$\overline{adz - bcz = a}$

$(ad - bc)z = a$

$$z = \frac{a}{ad - bc} \quad (ad - bc \neq 0)$$

In general, if $A = \begin{bmatrix} a & b \\ c & d \end{bmatrix}$, its multiplicative inverse is

$$A^{-1} = \begin{bmatrix} \dfrac{d}{ad - bc} & \dfrac{-b}{ad - bc} \\[2ex] \dfrac{-c}{ad - bc} & \dfrac{a}{ad - bc} \end{bmatrix}, \quad ad - bc \neq 0.$$

Since $\dfrac{1}{ad-bc}$ is a common factor of each entry,

$$A^{-1}=\frac{1}{ad-bc}\begin{bmatrix} d & -b \\ -c & a \end{bmatrix}, \qquad ad-bc\neq 0.$$

Using matrix multiplication, we can also show that $A^{-1}A=I.$

Note that we can obtain $\begin{bmatrix} d & -b \\ -c & a \end{bmatrix}$, the matrix factor of the multiplicative inverse of matrix $A=\begin{bmatrix} a & b \\ c & d \end{bmatrix}$, by interchanging in matrix A the entries a and d and by replacing b and c by their negatives.

~~~~~~~ **MODEL PROBLEMS** ~~~~~~~

In 1 and 2, find the multiplicative inverse of the given matrices.

1. $\begin{bmatrix} 2 & 4 \\ 3 & 7 \end{bmatrix}$ **2.** $\begin{bmatrix} 2 & -3 \\ -1 & 0 \end{bmatrix}$

How To Proceed	*Solution*	*Solution*
To find the multiplicative inverse of $A=\begin{bmatrix} a & b \\ c & d \end{bmatrix}$:		
1. Interchange a and d:	1. $\begin{bmatrix} 7 & \\ & 2 \end{bmatrix}$	1. $\begin{bmatrix} 0 & \\ & 2 \end{bmatrix}$
2. Replace b and c with their negatives:	2. $\begin{bmatrix} 7 & -4 \\ -3 & 2 \end{bmatrix}$	2. $\begin{bmatrix} 0 & 3 \\ 1 & 2 \end{bmatrix}$
3. Multiply the resulting matrix by $\dfrac{1}{ad-bc}$:	3. $(ad-bc$ $=14-12=2)$ $\dfrac{1}{2}\begin{bmatrix} 7 & -4 \\ -3 & 2 \end{bmatrix}$ $=\begin{bmatrix} 3\frac{1}{2} & -2 \\ -1\frac{1}{2} & 1 \end{bmatrix}$ *Ans.*	3. $(ad-bc$ $=0-3=-3)$ $-\dfrac{1}{3}\begin{bmatrix} 0 & 3 \\ 1 & 2 \end{bmatrix}$ $=\begin{bmatrix} 0 & -1 \\ -\frac{1}{3} & -\frac{2}{3} \end{bmatrix}$ *Ans.*

Exercises

In 1–5, state the multiplicative inverse of the matrix.

1. $\begin{bmatrix} 8 & 2 \\ 7 & 2 \end{bmatrix}$ **2.** $\begin{bmatrix} 5 & -4 \\ -2 & -3 \end{bmatrix}$ **3.** $\begin{bmatrix} 0 & -1 \\ 1 & 0 \end{bmatrix}$

4. $\begin{bmatrix} 1 & 1 \\ 1 & 0 \end{bmatrix}$ **5.** $\begin{bmatrix} 1 & -1 \\ -1 & 0 \end{bmatrix}$

In 6–8, show that there is no multiplicative inverse of the matrix.

6. $\begin{bmatrix} 1 & -1 \\ -1 & 1 \end{bmatrix}$ **7.** $\begin{bmatrix} 4 & 2 \\ 6 & 3 \end{bmatrix}$ **8.** $\begin{bmatrix} 10 & 2 \\ -5 & -1 \end{bmatrix}$

In 9–11, $A = \begin{bmatrix} 3 & 4 \\ -1 & 2 \end{bmatrix}$.

9. Find A^{-1} **10.** Find $(-A)^{-1}$ **11.** Find $(A^T)^{-1}$

12. Find matrix X if $\begin{bmatrix} 3 & 4 \\ 5 & 7 \end{bmatrix} X = \begin{bmatrix} 1 & 2 \\ 3 & 4 \end{bmatrix}$. (*Hint:* Multiply each member by the multiplicative inverse of $\begin{bmatrix} 3 & 4 \\ 5 & 7 \end{bmatrix}$.)

13. Using matrices, show that the multiplicative inverse of the complex number $a + bi$ is

$$\frac{a}{a^2 + b^2} - \frac{b}{a^2 + b^2} i$$

7. Using Matrices to Solve Systems of Equations

Inverse matrices can be used to solve systems of equations. Consider the following system of equations:

$$3x + 7y = 18$$
$$2x + 5y = 13$$

A system of equations such as the one above can be written in matrix form as follows:

Let $A = \begin{bmatrix} 3 & 7 \\ 2 & 5 \end{bmatrix}$ This is the *coefficient matrix*.

$X = \begin{bmatrix} x \\ y \end{bmatrix}$ This is the *variable matrix*.

$B = \begin{bmatrix} 18 \\ 13 \end{bmatrix}$ This is the *constant matrix*.

Since the matrix product AX gives

$$\begin{bmatrix} 3 & 7 \\ 2 & 5 \end{bmatrix} \begin{bmatrix} x \\ y \end{bmatrix} = \begin{bmatrix} 3x + 7y \\ 2x + 5y \end{bmatrix}$$

the system of equations can be represented by the **matrix equation**

$$\begin{bmatrix} 3 & 7 \\ 2 & 5 \end{bmatrix} \cdot \begin{bmatrix} x \\ y \end{bmatrix} = \begin{bmatrix} 18 \\ 13 \end{bmatrix}$$

or more simply by

$$AX = B$$

Then

$AX = B$	
$A^{-1}AX = A^{-1}B$	Multiplying both sides by A^{-1}
$IX = A^{-1}B$	$A^{-1}A = I$ Inverse property
$X = A^{-1}B$	$IX = X$ Identity property

Therefore, the matrix X, which is the solution matrix, can be found by multiplying A^{-1} times B with A^{-1} on the left (remember that matrix multiplication is *not* commutative.) This method can be used only if the inverse of the coefficient matrix exists. In fact, a system $AX = B$ has a unique solution if and only if A^{-1} exists. The steps of the method are illustrated in the following problems.

~~~~~~~~~~~~~~~~ MODEL PROBLEMS ~~~~~~~~~~~~~

1. Use a matrix equation to solve the system of equations.

$$3x + 7y = 18$$
$$2x + 5y = 13$$

Solution:

1. Write the matrix equation $AX = B$. $\begin{bmatrix} 3 & 7 \\ 2 & 5 \end{bmatrix} \cdot \begin{bmatrix} x \\ y \end{bmatrix} = \begin{bmatrix} 18 \\ 13 \end{bmatrix}$

2. Find the inverse of the coefficient matrix A.

   • Interchange a and d. $\begin{bmatrix} 5 & \\ & 3 \end{bmatrix}$

   • Replace b and c with their negatives. $\begin{bmatrix} 5 & -7 \\ -2 & 3 \end{bmatrix}$

   • Multiply the resulting matrix by $\dfrac{1}{ad - bc}$. Here, $ad - bc = 15 - 14 = 1$ $\dfrac{1}{1}\begin{bmatrix} 5 & -7 \\ -2 & 3 \end{bmatrix} = \begin{bmatrix} 5 & -7 \\ -2 & 3 \end{bmatrix} = A^{-1}$

3. Multiply the constant matrix by the inverse matrix.

$$X = \begin{bmatrix} 5 & -7 \\ -2 & 3 \end{bmatrix} \cdot \begin{bmatrix} 18 \\ 13 \end{bmatrix} = \begin{bmatrix} -1 \\ 3 \end{bmatrix} = \begin{bmatrix} x \\ y \end{bmatrix}$$

Answer: $x = -1$ and $y = 3$

2. Solve for x and y.

$$2x + 2y = 4$$
$$4x + 5y = 6$$

Solution:

1. Write the matrix equation $AX = B$.

$$\begin{bmatrix} 2 & 2 \\ 4 & 5 \end{bmatrix} \cdot \begin{bmatrix} x \\ y \end{bmatrix} = \begin{bmatrix} 4 \\ 6 \end{bmatrix}$$

2. Find the inverse of matrix A.
   Here, $ad - bc = 10 - 8 = 2$.

$$\frac{1}{2}\begin{bmatrix} 5 & -2 \\ -4 & 2 \end{bmatrix} = \begin{bmatrix} \frac{5}{2} & -1 \\ -2 & 1 \end{bmatrix} = A^{-1}$$

3. Multiply matrix B by A^{-1}.

$$\begin{bmatrix} \frac{5}{2} & -1 \\ -2 & 1 \end{bmatrix} \cdot \begin{bmatrix} 4 \\ 6 \end{bmatrix} = \begin{bmatrix} 4 \\ -2 \end{bmatrix} = \begin{bmatrix} x \\ y \end{bmatrix}$$

Answer: $x = 4$ and $y = -2$

The same method can be used to solve a system of three equations in three unknowns, but requires computation of inverses of 3×3 matrices.

~~~~~~

Exercises

In 1–3, write a matrix equation to represent each system.

1. $7x - 5y = 12$
$\quad\;\; 2x - 3y = 6$

2. $2x = 4 + 5y$
$\quad\;\; x + 2y = 3$

3. $x + 2 = y$
$\quad\;\; y + 1 = -2x$

In 4–9, use a matrix equation to solve the system of equations. If the system does not have a unique solution, state this. Check the solutions.

4. $2x + 3y = 19$
$\quad\;\; 4x - y = 3$

5. $4x - 2y = 4$
$\quad\;\; x - 2y = -2$

6. $3x - 2y = 6$
$\quad\;\; 4y = 8$

7. $3x - y = 10$
$\quad\;\; 6x - 2y = 5$

8. $2x = 2 + y$
$\quad\;\; 3x + 4y = 6$

9. $5x = 12 - 3y$
$\quad\;\; 2y = 17 - 3x$

10. The inverse of the matrix $\begin{bmatrix} 1 & 1 & -1 \\ 1 & 2 & 3 \\ 2 & -1 & -13 \end{bmatrix}$ is $\begin{bmatrix} -23 & 14 & 5 \\ 19 & -11 & -4 \\ -5 & 3 & 1 \end{bmatrix}$.

Use this inverse to solve the system

$$x + y - z = 0$$
$$x + 2y + 3z = -5$$
$$2x - y - 13z = 17$$

8. Determinants of the Second Order

Every 2×2 matrix A with real number entries has associated with it a unique real number called the **determinant of A**, symbolized by det A, or $|A|$. For example, associated with the matrix $\begin{bmatrix} a & b \\ c & d \end{bmatrix}$ is a number which is defined as follows:

$$\det \begin{bmatrix} a & b \\ c & d \end{bmatrix} = \begin{vmatrix} a & b \\ c & d \end{vmatrix} = ad + (-bc) = ad - bc$$

The value $ad - bc$ is found as follows: Find the product of the entries along the *principal diagonal*, which runs from the upper left to the lower right. This product is ad. Then, find the *negative* of the product of the entries along the other diagonal. The negative of this product is $-bc$. Finally, add these two values, getting $ad + (-bc) = ad - bc$.

Thus, the number represented by $\begin{vmatrix} 4 & 1 \\ 3 & 2 \end{vmatrix}$ is 5, since $4 \cdot 2 - 1 \cdot 3 = 8 - 3 = 5$.

Also, the number represented by $\begin{vmatrix} 6 & -2 \\ -7 & 3 \end{vmatrix}$ is 4, since $6 \cdot 3 - (-2)(-7) = 18 - 14 = 4$.

A determinant can be written in the same form as a square matrix. Note that the brackets of the matrix are replaced by vertical bars. The order of the matrix is also called the **order** of the determinant. The entries (**elements**) in the rows and columns of the matrix are called the entries (elements) in the rows and columns of the determinant.

USING 2 × 2 DETERMINANTS TO SOLVE A SYSTEM OF TWO LINEAR EQUATIONS IN TWO VARIABLES (CRAMER'S RULE)

Let us solve the following general system of linear equations in standard form:

$$a_1 x + b_1 y = c_1 \quad \text{(eq. 1)}$$
$$a_2 x + b_2 y = c_2 \quad \text{(eq. 2)}$$

$a_1 b_2 x + b_1 b_2 y = b_2 c_1$ (In 1, M_{b_2}.)
$a_2 b_1 x + b_1 b_2 y = b_1 c_2$ (In 2, M_{b_1}.)
$\overline{a_1 b_2 x - a_2 b_1 x = b_2 c_1 - b_1 c_2}$
$(a_1 b_2 - a_2 b_1)x = b_2 c_1 - b_1 c_2$
$$x = \frac{b_2 c_1 - b_1 c_2}{a_1 b_2 - a_2 b_1}$$

$a_1 a_2 x + a_1 b_2 y = a_1 c_2$ (In 2, M_{a_1}.)
$a_1 a_2 x + a_2 b_1 y = a_2 c_1$ (In 1, M_{a_2}.)
$\overline{a_1 b_2 y - a_2 b_1 y = a_1 c_2 - a_2 c_1}$
$(a_1 b_2 - a_2 b_1)y = a_1 c_2 - a_2 c_1$
$$y = \frac{a_1 c_2 - a_2 c_1}{a_1 b_2 - a_2 b_1}$$

In each case, $a_1 b_2 - a_2 b_1 \neq 0$.

If determinants are used to express x and y, we have:

$$x = \frac{\begin{vmatrix} c_1 & b_1 \\ c_2 & b_2 \end{vmatrix}}{\begin{vmatrix} a_1 & b_1 \\ a_2 & b_2 \end{vmatrix}} \qquad y = \frac{\begin{vmatrix} a_1 & c_1 \\ a_2 & c_2 \end{vmatrix}}{\begin{vmatrix} a_1 & b_1 \\ a_2 & b_2 \end{vmatrix}}$$

Observe that the determinant $\begin{vmatrix} a_1 & b_1 \\ a_2 & b_2 \end{vmatrix}$ is the denominator of both the fraction that represents x and the fraction that represents y. The entries in this determinant are, respectively, the coefficients of the x-terms and y-terms when the equations of the system are written in standard form. To obtain the determinant that is the numerator of the fraction that represents x, replace a_1 and a_2, the x-coefficients in the denominator-determinant, by the corresponding constant terms c_1 and c_2. Similarly, to obtain the determinant that is the numerator of the fraction which represents y, replace b_1 and b_2, the y-coefficients in the denominator-determinant, by the corresponding constant terms c_1 and c_2.

We can simplify the writing of the expressions for x and y by letting N_x represent the numerator-determinant of x, letting N_y represent the numerator-determinant of y, and letting D represent the denominator-determinant of both x and y, as follows:

$$x = \frac{N_x}{D} \qquad y = \frac{N_y}{D} \qquad (D \neq 0)$$

This method is called Cramer's Rule.

Thus we can solve the system $\begin{aligned} 5x + 2y &= 9 \\ 3x - 5y &= -7 \end{aligned}$ in the following manner:

$$D = \begin{vmatrix} 5 & 2 \\ 3 & -5 \end{vmatrix} \qquad N_x = \begin{vmatrix} 9 & 2 \\ -7 & -5 \end{vmatrix} \qquad N_y = \begin{vmatrix} 5 & 9 \\ 3 & -7 \end{vmatrix}$$

Since $D = -25 - 6 = -31$, $N_x = -45 + 14 = -31$, and $N_y = -35 - 27 = -62$, then:

$$x = \frac{N_x}{D} = \frac{-31}{-31} = 1 \qquad y = \frac{N_y}{D} = \frac{-62}{-31} = 2$$

The solution of the system is $x = 1$, $y = 2$. The check is left to the student.

WHEN THE DENOMINATOR-DETERMINANT IS ZERO

If $D = 0$, then $\begin{vmatrix} a_1 & b_1 \\ a_2 & b_2 \end{vmatrix} = 0$, or $a_1 b_2 - a_2 b_1 = 0$. When the denominator $D = 0$, x and y are undefined. In this case, $a_1 b_2 = a_2 b_1$ and $\dfrac{a_1}{b_1} = \dfrac{a_2}{b_2}$. Since the slope of the line, $a_1x + b_1y = c$, is $-\dfrac{a_1}{b_1}$; and since the slope of the line, a_2x

$+ b_2y = c_2$, is $-\dfrac{a_2}{b_2}$; it follows that, when $D = 0$, the slopes of the lines are

equal. Because the slopes are equal, the lines are parallel or coincide. If the lines are parallel, there is no common solution and the equations are inconsistent. In case the lines coincide, there is an infinite number of solutions and the equations are dependent. Dependent equations result when $\dfrac{a_1}{a_2} = \dfrac{b_1}{b_2} = \dfrac{c_1}{c_2}$.

Thus, in the system $\begin{array}{l} 5x + 2y = 9 \\ 10x + 4y = 6 \end{array}$, $D = \begin{vmatrix} 5 & 2 \\ 10 & 4 \end{vmatrix} = 5(4) - (2)10 = 0$. Here, there is no common solution.

Also, in the system $\begin{array}{l} 5x + 2y = 9 \\ 10x + 4y = 18 \end{array}$, $D = 0$. However, since $\dfrac{5}{10} = \dfrac{2}{4} = \dfrac{9}{18}$, there is an infinite number of solutions in this case. When there is an infinite number of solutions, either linear equation can be transformed into the other.

USING DETERMINANTS TO WRITE THE MULTIPLICATIVE INVERSE OF A MATRIX

Recall that the multiplicative inverse of the matrix $\begin{bmatrix} a & b \\ c & d \end{bmatrix}$ is

$$\frac{1}{ad - bc}\begin{bmatrix} d & -b \\ -c & a \end{bmatrix}.$$

Since $ad - bc$ is equal to the determinant $\begin{vmatrix} a & b \\ c & d \end{vmatrix}$, the multiplicative inverse of

the matrix $\begin{bmatrix} a & b \\ c & d \end{bmatrix}$ may be written as $\dfrac{1}{\begin{vmatrix} a & b \\ c & d \end{vmatrix}}\begin{bmatrix} d & -b \\ -c & a \end{bmatrix}$, or $\dfrac{\begin{bmatrix} d & -b \\ -c & a \end{bmatrix}}{\begin{vmatrix} a & b \\ c & d \end{vmatrix}}$.

Thus, the multiplicative inverse of $\begin{bmatrix} -1 & 0 \\ 3 & -2 \end{bmatrix}$ is $\dfrac{\begin{bmatrix} -2 & 0 \\ -3 & -1 \end{bmatrix}}{\begin{vmatrix} -1 & 0 \\ 3 & -2 \end{vmatrix}}$.

∿∿∿ MODEL PROBLEMS ∿∿∿

1. Find x if $\begin{vmatrix} x & 2 \\ x+2 & 3 \end{vmatrix} = 12$.

How To Proceed	*Solution*

1. Multiply the entries of the principal diagonal:

1. $\begin{vmatrix} x & \\ & 3 \end{vmatrix} = 3x$

2. Multiply the entries of the other diagonal:

2. $\begin{vmatrix} & 2 \\ x+2 & \end{vmatrix} = 2x + 4$

3. Replace the determinant by the difference of the products obtained in (1) and (2); then solve:

3. $3x - (2x + 4) = 12$
$3x - 2x - 4 = 12$
$x = 16$

4. Check:

4. Let $x = 16$:

$$\begin{vmatrix} 16 & 2 \\ 18 & 3 \end{vmatrix} \overset{?}{=} 12$$

$$16(3) - 18(2) \overset{?}{=} 12$$
$$48 - 36 \overset{?}{=} 12$$
$$12 = 12 \checkmark$$

Answer: $x = 16$

2. Use determinants to find the solution set of the system

$$3x + 2y - 4 = 0$$
$$2x = 3y + 7$$

How To Proceed	*Solution*

1. Transform the equations into standard form:

1. $\begin{cases} 3x + 2y = 4 \\ 2x - 3y = 7 \end{cases}$

2. Evaluate the determinants:

$$D = \begin{vmatrix} a_1 & b_1 \\ a_2 & b_2 \end{vmatrix}$$

$$N_x = \begin{vmatrix} c_1 & b_1 \\ c_2 & b_2 \end{vmatrix}$$

$$N_y = \begin{vmatrix} a_1 & c_1 \\ a_2 & c_2 \end{vmatrix}$$

2. $D = \begin{vmatrix} 3 & 2 \\ 2 & -3 \end{vmatrix} = -9 - 4 = -13$

$N_x = \begin{vmatrix} 4 & 2 \\ 7 & -3 \end{vmatrix} = -12 - 14 = -26$

$N_y = \begin{vmatrix} 3 & 4 \\ 2 & 7 \end{vmatrix} = 21 - 8 = 13$

3. Find the values of the variables:

$$x = \frac{N_x}{D}, \qquad y = \frac{N_y}{D}$$

3. $x = \dfrac{N_x}{D} = \dfrac{-26}{-13} = 2$

$y = \dfrac{N_y}{D} = \dfrac{13}{-13} = -1$

4. Check:

| 4. Let $x = 2$ and $y = -1$: |

$$3x + 2y - 4 = 0 \qquad\qquad 2x = 3y + 7$$
$$6 + (-2) - 4 \overset{?}{=} 0 \qquad\qquad 4 \overset{?}{=} (-3) + 7$$
$$0 = 0 \checkmark \qquad\qquad 4 = 4 \checkmark$$

Answer: $\{(2, -1)\}$

Exercises

In 1–4, if A is the given matrix, evaluate the determinant of A.

1. $\begin{bmatrix} 4 & 2 \\ 3 & 7 \end{bmatrix}$
  2. $\begin{bmatrix} -1 & 5 \\ 3 & 2 \end{bmatrix}$
  3. $\begin{bmatrix} 4 & 0 \\ 6 & -2 \end{bmatrix}$
  4. $\begin{bmatrix} 8 & \frac{2}{3} \\ 6 & -\frac{1}{2} \end{bmatrix}$

In 5–8, evaluate the determinant.

5. $\begin{vmatrix} 0 & 1 \\ 1 & 3 \end{vmatrix}$
  6. $\begin{vmatrix} 1 & -1 \\ 0 & -2 \end{vmatrix}$
  7. $\begin{vmatrix} -2 & 0 \\ 0 & 3 \end{vmatrix}$
  8. $\begin{vmatrix} a & -b \\ b & a \end{vmatrix}$

In 9–14, solve for x.

9. $2x + \begin{vmatrix} 6 & 1 \\ 2 & 3 \end{vmatrix} = 36$
  10. $x + \begin{vmatrix} 1 & -1 \\ 0 & -2 \end{vmatrix} = 2 \begin{vmatrix} -2 & 0 \\ 0 & 3 \end{vmatrix}$

11. $2x - 3 \begin{vmatrix} 0 & 1 \\ 1 & 3 \end{vmatrix} = 5 \begin{vmatrix} 1 & -1 \\ 0 & -2 \end{vmatrix}$
  12. $\begin{vmatrix} x & 1 \\ x+1 & 2 \end{vmatrix} = 9$

13. $\begin{vmatrix} x & x+2 \\ 5 & 6 \end{vmatrix} = -13$
  14. $\begin{vmatrix} x & 8 \\ 9 & 2x \end{vmatrix} = 72$

In 15–20, solve the system of equations by using determinants.

15. $x + 3y = 11$
$2x - y = 8$

16. $3x - 2y = 13$
$2x - 7y = 20$

17. $4x + 4y = 7$
$x = 1 + 2y$

18. $\dfrac{x}{4} + 3y = -4$
$2x = 13 - \dfrac{3y}{2}$

19. $2(x + 1) = 3(y - 2)$
$x - 1 = y$

20. $\dfrac{x-2}{y+2} = 1$
$x - 1 = 2y$

In 21–23, write the multiplicative inverse of the matrix (*a*) using only a matrix and (*b*) using a matrix and a determinant.

21. $\begin{bmatrix} 5 & 3 \\ 1 & 2 \end{bmatrix}$
  22. $\begin{bmatrix} 4 & 0 \\ -1 & 2 \end{bmatrix}$
  23. $\begin{bmatrix} 0 & -2 \\ -5 & 0 \end{bmatrix}$

9. Determinants of the Third Order

Associated with each 3×3 matrix with real number entries is a unique real number called the determinant of the matrix. For example, associated with the

third-order matrix $\begin{bmatrix} a_1 & b_1 & c_1 \\ a_2 & b_2 & c_2 \\ a_3 & b_3 & c_3 \end{bmatrix}$ is the third-order determinant $\begin{vmatrix} a_1 & b_1 & c_1 \\ a_2 & b_2 & c_2 \\ a_3 & b_3 & c_3 \end{vmatrix}$,

which is defined as follows:

$$\begin{vmatrix} a_1 & b_1 & c_1 \\ a_2 & b_2 & c_2 \\ a_3 & b_3 & c_3 \end{vmatrix} = a_1b_2c_3 + a_2b_3c_1 + a_3b_1c_2 - a_1b_3c_2 - a_2b_1c_3 - a_3b_2c_1$$

You need not memorize the six products that are involved in the expansion of this third-order determinant. Instead, you can use the following procedure to obtain the expansion:

1. Copy the columns of the determinant, and repeat the first two columns after the third column. Draw the diagonals as shown at the right.

   1.
      $(-) \ (-) \ (-) \ \ (+) \ (+) \ (+)$

2. Find the products of the entries along each diagonal that descends from left to right.

   2. $a_1b_2c_3$
      $b_1c_2a_3$
      $c_1a_2b_3$

3. Find the products of the entries along each diagonal that descends from right to left.

   3. $c_1b_2a_3$
      $a_1c_2b_3$
      $b_1a_2c_3$

4. Add the products obtained in step 2 and the *negatives* of the products obtained in step 3.

   4. $a_1b_2c_3 + b_1c_2a_3 + c_1a_2b_3$
      $- c_1b_2a_3 - a_1c_2b_3 - b_1a_2c_3$

Note that the result obtained in step 4 is equal to the expression previously given as the definition of the determinant:

$a_1b_2c_3 + b_1c_2a_3 + c_1a_2b_3 - c_1b_2a_3 - a_1c_2b_3 - b_1a_2c_3$ is equal to

$a_1b_2c_3 + a_2b_3c_1 + a_3b_1c_2 - a_1b_3c_2 - a_2b_1c_3 - a_3b_2c_1.$

Thus, to evaluate $\begin{vmatrix} 1 & -2 & 0 \\ 4 & 5 & 1 \\ 2 & 3 & 4 \end{vmatrix}$, repeat the first two columns of the determinant after the third column and evaluate as follows:

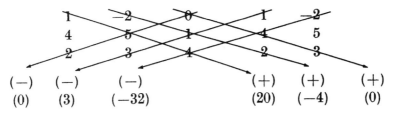

(−) (−)   (−)   (+)   (+)   (+)
(0) (3) (−32)   (20) (−4)   (0)

The value of the determinant is $20 + (-4) + 0 - 0 - 3 - (-32) = 45$.

USING 3 × 3 DETERMINANTS TO SOLVE A SYSTEM OF THREE FIRST-DEGREE EQUATIONS (CRAMER'S RULE)

A method similar to the one used to solve a general system of two first-degree equations in two variables (page 636) can be used to solve a general system of three first-degree equations in three variables. If these solutions are expressed in terms of determinants, we find that the values of each variable can be expressed as a fraction whose numerator is a third-order determinant, and whose denominator is also a third-order determinant. The determinant in each numerator and in each denominator can be written following the same rules that were used in the case of solving two first-degree equations in two variables (page 636).

Hence, the solution of the system
$$a_1 x + b_1 y + c_1 z = d_1$$
$$a_2 x + b_2 y + c_2 z = d_2$$
$$a_3 x + b_3 y + c_3 z = d_3$$
is represented by:

$$x = \frac{N_x}{D} \qquad y = \frac{N_y}{D} \qquad z = \frac{N_z}{D}$$

where N_x, N_y, N_z, and D ($D \neq 0$) are the following determinants:

$$D = \begin{vmatrix} a_1 & b_1 & c_1 \\ a_2 & b_2 & c_2 \\ a_3 & b_3 & c_3 \end{vmatrix}$$, the determinant containing the coefficients of x, y, and z.

$$N_x = \begin{vmatrix} d_1 & b_1 & c_1 \\ d_2 & b_2 & c_2 \\ d_3 & b_3 & c_3 \end{vmatrix}$$, obtained by replacing, in D, the coefficients of x with the constant terms.

$$N_y = \begin{vmatrix} a_1 & d_1 & c_1 \\ a_2 & d_2 & c_2 \\ a_3 & d_3 & c_3 \end{vmatrix}$$, obtained by replacing, in D, the coefficients of y with the constant terms.

$$N_z = \begin{vmatrix} a_1 & b_1 & d_1 \\ a_2 & b_2 & d_2 \\ a_3 & b_3 & d_3 \end{vmatrix}$$, obtained by replacing, in D, the coefficients of z with the constant terms.

Thus, we can solve the system
$$\begin{aligned} x + 2y - 3z &= 4 \\ 2x - y + 2z &= 6 \\ 3x + 4y - 5z &= 12 \end{aligned}$$
in the following manner:

$$D = \begin{vmatrix} 1 & 2 & -3 \\ 2 & -1 & 2 \\ 3 & 4 & -5 \end{vmatrix} = 5 + 12 + (-24) - 9 - 8 - (-20) = -4$$

$$N_x = \begin{vmatrix} 4 & 2 & -3 \\ 6 & -1 & 2 \\ 12 & 4 & -5 \end{vmatrix} = 20 + 48 + (-72) - 36 - 32 - (-60) = -12$$

$$N_y = \begin{vmatrix} 1 & 4 & -3 \\ 2 & 6 & 2 \\ 3 & 12 & -5 \end{vmatrix} = (-30) + 24 + (-72) - (-54) - 24 - (-40) = -8$$

$$N_z = \begin{vmatrix} 1 & 2 & 4 \\ 2 & -1 & 6 \\ 3 & 4 & 12 \end{vmatrix} = (-12) + 36 + 32 - (-12) - 24 - 48 = -4$$

Note. In evaluating each determinant, first repeat the first two columns after the third column.

Hence, $x = \dfrac{N_x}{D} = \dfrac{-12}{-4} = 3$; $y = \dfrac{N_y}{D} = \dfrac{-8}{-4} = 2$; $z = \dfrac{N_z}{D} = \dfrac{-4}{-4} = 1$. The solution of the system is $x = 3$, $y = 2$, $z = 1$. The check is left to the student.

As in the case of the system of two first-degree equations, if in a system of three first-degree equations the determinant D equals 0, there is no unique solution. If $D = 0$, there are no solutions when the equations are inconsistent; there are an infinite number of solutions when the equations are dependent.

~~~~~~~~~ **MODEL PROBLEMS** ~~~~~~~~~

1. Find x if $\begin{vmatrix} x & 1 & 2 \\ x+1 & 2 & -1 \\ 0 & 3 & -2 \end{vmatrix} = 29$.

How To Proceed	*Solution*
1. Copy the columns of the determinant and repeat the first two columns after the third column. Then, find the product of the entries along each diagonal that descends from left to right.	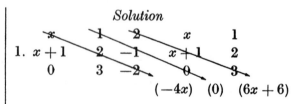

2. Find the product of the entries along each diagonal that descends from right to left.

3. Add the products obtained in step 1 and the negatives of the products obtained in step 2.

4. Solve the resulting equation.

5. Check.

Answer: $x = 3$

2. $\begin{array}{ccc} x & 1 & 2 \\ x+1 & 2 & -1 \\ 0 & 3 & -2 \end{array}$ $\begin{array}{ccc} x & 1 \\ x+1 & 2 \\ 0 & 3 \end{array}$

$(0) \quad (-3x) \quad (-2x-2)$

3. $(-4x) + 0 + (6x+6)$
$-0 - (-3x) - (-2x-2) = 7x + 8$

4. $7x + 8 = 29$
$$7x = 21, \text{ and } x = 3$$

5. Let $x = 3$. Then:

$$\begin{vmatrix} 3 & 1 & 2 \\ 4 & 2 & -1 \\ 0 & 3 & -2 \end{vmatrix} \stackrel{?}{=} 29$$

$$-12 + 0 + 24$$
$$-0 - (-9) - (-8) \stackrel{?}{=} 29$$
$$29 = 29 \checkmark$$

2. Using determinants, find the solution $\begin{aligned} 4x - 3y + z &= 14 \\ 5x + y &= 9 \\ x + 3z &= 11 \end{aligned}$

How To Proceed	*Solution*

1. Evaluate the following determinants:

1. Evaluating:

$$D = \begin{vmatrix} a_1 & b_1 & c_1 \\ a_2 & b_2 & c_2 \\ a_3 & b_3 & c_3 \end{vmatrix}$$

$$D = \begin{vmatrix} 4 & -3 & 1 \\ 5 & 1 & 0 \\ 1 & 0 & 3 \end{vmatrix} = \begin{array}{l} 12 + 0 + 0 - 1 \\ -0 - (-45) \\ = 56 \end{array}$$

$$N_x = \begin{vmatrix} d_1 & b_1 & c_1 \\ d_2 & b_2 & c_2 \\ d_3 & b_3 & c_3 \end{vmatrix}$$

$$N_x = \begin{vmatrix} 14 & -3 & 1 \\ 9 & 1 & 0 \\ 11 & 0 & 3 \end{vmatrix} = \begin{array}{l} 42 + 0 + 0 - 11 \\ -(-81) - 0 \\ = 112 \end{array}$$

$$N_y = \begin{vmatrix} a_1 & d_1 & c_1 \\ a_2 & d_2 & c_2 \\ a_3 & d_3 & c_3 \end{vmatrix}$$

$$N_y = \begin{vmatrix} 4 & 14 & 1 \\ 5 & 9 & 0 \\ 1 & 11 & 3 \end{vmatrix} = \begin{array}{l} 108 + 0 + 55 - 9 \\ -210 - 0 = -56 \end{array}$$

$$N_z = \begin{vmatrix} a_1 & b_1 & d_1 \\ a_2 & b_2 & d_2 \\ a_3 & b_3 & d_3 \end{vmatrix}$$

$$N_z = \begin{vmatrix} 4 & -3 & 14 \\ 5 & 1 & 9 \\ 1 & 0 & 11 \end{vmatrix} = \begin{array}{l} 44 + (-27) + 0 \\ -14 - (-165) \\ -0 = 168 \end{array}$$

2. Find the values of the variables as follows:

$$x = \frac{N_x}{D} \quad y = \frac{N_y}{D} \quad z = \frac{N_z}{D}$$

2. Substituting:

$$x = \frac{N_x}{D} = \frac{112}{56} = 2$$

$$y = \frac{N_y}{D} = \frac{-56}{56} = -1$$

$$z = \frac{N_z}{D} = \frac{168}{56} = 3$$

3. Check.

3. Let $x = 2$, $y = -1$, and $z = 3$.

$$\begin{array}{ll} 4x - 3y + z = 14 & 5x + y = 9 \\ 8 + 3 + 3 \stackrel{?}{=} 14 & 10 - 1 \stackrel{?}{=} 9 \\ 14 = 14\checkmark & 9 = 9\checkmark \end{array}$$

$$\begin{array}{l} x + 3z = 11 \\ 2 + 9 \stackrel{?}{=} 11 \\ 11 = 11\checkmark \end{array}$$

Answer: $\{(2, -1, 3)\}$

~~~~~~

Exercises

In 1–3, evaluate the determinant.

1. $\begin{vmatrix} 1 & -1 & 0 \\ 0 & 1 & 2 \\ 2 & 0 & -1 \end{vmatrix}$
   2. $\begin{vmatrix} 1 & 3 & 5 \\ -3 & 1 & -5 \\ 0 & 0 & 3 \end{vmatrix}$
   3. $\begin{vmatrix} 2 & -4 & 0 \\ 1 & 0 & 3 \\ 0 & 4 & -2 \end{vmatrix}$

4. Solve for x: $\quad x + \begin{vmatrix} 2 & -1 & 0 \\ 0 & 1 & 2 \\ 1 & 0 & -1 \end{vmatrix} = \begin{vmatrix} 1 & -4 & 0 \\ 2 & 0 & 3 \\ 0 & 4 & -2 \end{vmatrix}$.

5. Solve for x: $\quad \dfrac{x}{3} - 3 \begin{vmatrix} 2 & -4 & 0 \\ 1 & 0 & -2 \\ 0 & 4 & 2 \end{vmatrix} = \begin{vmatrix} 1 & 3 & -5 \\ -3 & 1 & 5 \\ 0 & 0 & 3 \end{vmatrix}$.

In 6–8, find the solution set of the given equation.

6. $\begin{vmatrix} 3 & x & 5 \\ -1 & x+1 & 0 \\ 0 & 2x & 1 \end{vmatrix} = 1$
   7. $\begin{vmatrix} x & x+2 & -x \\ 1 & 0 & 5 \\ -2 & 1 & -3 \end{vmatrix} = -18$

8. $\begin{vmatrix} 4 & -2 & -3 \\ 3 & 0 & 1 \\ 2x & x+2 & 0 \end{vmatrix} = 0$

In 9–11, using determinants, find the solution set of the system.

9.
$$x + 2y - z = 6$$
$$2x - 4y + 3z = -9$$
$$4x - 2y - z = 1$$

10.
$$2x - 2y + 3z = 6$$
$$3x + y - 2z = -13$$
$$-x + 3y + 4z = 11$$

11.
$$x + y - z = 10$$
$$2x + 2y - z = 16$$
$$-3x + y + z = -8$$

In 12–14, after transforming the system of equations into standard form, use determinants to find the solution set of the system.

12.
$$x + y = z$$
$$2x + z = 5$$
$$2y - 3x = 1$$

13.
$$x + 2z = 3 - y$$
$$z = y - 3x + 1$$
$$8 = 2x - 4z + 3y$$

14.
$$\frac{x}{3} + \frac{y}{2} = z + 7$$
$$\frac{x}{4} + \frac{z}{2} = \frac{3y}{2} - 6$$
$$\frac{x}{6} = \frac{y}{4} + \frac{z}{3} + 1$$

CALCULATOR APPLICATIONS

With a graphing calculator, you can add or multiply matrices, evaluate determinants, and find inverse matrices. To enter a matrix into the calculator, use the $\boxed{\text{MATRX}}$ key to display the matrix menu, then choose the Edit feature. Enter the name of the matrix, its dimensions, then its elements.

MODEL PROBLEM

a. Enter each matrix in a graphing calculator.

$$A = \begin{bmatrix} 4 & 2 \\ 3 & 5 \end{bmatrix} \qquad B = \begin{bmatrix} 5 & 3 \\ 1 & 2 \end{bmatrix}$$

b. Find $A + 2B$, AB, B^2, the determinant of B, and the inverse of B.

Solution: The $\boxed{\text{ENTER}}$ key fills the rows from left to right.

   a. Enter: $\boxed{\text{MATRX}}$ $\boxed{\blacktriangleright}$ $\boxed{\blacktriangleright}$ $\boxed{\text{ENTER}}$ 2 $\boxed{\text{ENTER}}$ 2 $\boxed{\text{ENTER}}$

                4 $\boxed{\text{ENTER}}$ 2 $\boxed{\text{ENTER}}$ 3 $\boxed{\text{ENTER}}$ 5 $\boxed{\text{ENTER}}$

     Display:

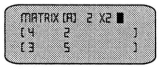

                $\boxed{\text{MATRX}}$ $\boxed{\blacktriangleright}$ $\boxed{\blacktriangleright}$ 2 2 $\boxed{\text{ENTER}}$ 2 $\boxed{\text{ENTER}}$

                5 $\boxed{\text{ENTER}}$ 3 $\boxed{\text{ENTER}}$ 1 $\boxed{\text{ENTER}}$ 2 $\boxed{\text{ENTER}}$

     Display:

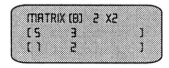

   b. Find $A + 2B$.

     Enter: $\boxed{\text{MATRX}}$ 1 $\boxed{+}$ 2 $\boxed{\text{MATRX}}$ 2 $\boxed{\text{ENTER}}$

     Display:

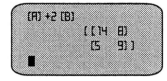

    Find AB.

Enter: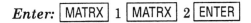

Display:

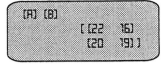

Find the determinant of B (det B).

Enter:

Display:

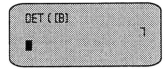

Find the inverse of B (B^{-1}).

Enter: MATRX 2 X^{-1} ENTER

The display shows entries that are repeating decimals, so convert to fraction form.

MATH 1 ENTER

Display:

```
[B]⁻¹
[[.2857142857    …
  [-.1428571429  …
ANS▶FRAC
     [[2/7  -3/7]
      [-1/7  5/7 ]]
```

CHAPTER XVIII

Polynomial Functions and Polynomial Equations of a Higher Degree than Second Degree

1. The General Form of a Polynomial Function and a Polynomial Equation

THE POLYNOMIAL AND THE POLYNOMIAL FUNCTION

Previously, we have studied expressions such as $3x + 5$ and $4x^2 - 3x + 8$. These expressions are called *polynomials in x*. The expression $3x + 5$ is called a *first-degree polynomial*; the expression $4x^2 - 3x + 8$ is called a *second-degree polynomial*. In this chapter, we shall consider polynomials of degree higher than the second, such as $x^3 + 2x^2 - 3x + 10$ and $2x^5 + x^2 - 12$.

The general form of a polynomial of degree n may be represented as follows:

$$a_0 x^n + a_1 x^{n-1} + a_2 x^{n-2} + \cdots + a_{n-2} x^2 + a_{n-1} x + a_n \quad (a_0 \neq 0)$$

In this form:

1. n and all the exponents of x are non-negative integers. If $n = 2$, the polynomial is a second-degree, or quadratic, polynomial.

    If $n = 1$, the polynomial is a first-degree, or linear, polynomial.

    If $n = 0$, the polynomial is a constant, such as 25.

2. The coefficients, $a_0, a_1, \ldots, a_{n-1}, a_n$, are members of the same number system. If no mention is made of the nature of the coefficients, we will assume that they are members of the set of complex numbers. Any or all of the coefficients except a_0 may be zero. The coefficient a_0 is called the *leading coefficient*. The coefficient a_n is called the *constant term*.

3. The degree of the polynomial is n. If $n = 0$, the polynomial consists of the number a_0, and its degree is 0. The polynomial 0 does not have a degree.

Thus, in the polynomial $x^3 + 2x^2 - 3x + 8$, the coefficients are 1, 2, -3, and 8; the constant term is 8; the degree of the polynomial is 3. We say that this polynomial is a third-degree, or cubic, polynomial.

Note that, in the general form of a polynomial, the powers of the variable are written in descending order. If any power is missing, the coefficient of that power is 0. Also, the letter used to represent the variable in a polynomial is arbitrary. For example, the letter might be g or t, etc. Thus, in the fourth-degree, or quartic, polynomial $\frac{1}{4}s^4 - \sqrt{2}$, the terms in s^3, s^2, and s are missing. Hence, the coefficients are $\frac{1}{4}$, 0, 0, 0, and $-\sqrt{2}$.

If the variable x of a polynomial is replaced by a number, we obtain a number which is called the **value** of the polynomial. For example, if $f(x)$ represents the polynomial $x^3 + 2x^2 - 3x + 8$, we can find the value of the polynomial when $x = 10$ by substituting 10 for x in the following manner:

$$f(x) = x^3 + 2x^2 - 3x + 8$$
$$f(10) = (10)^3 + 2(10)^2 - 3(10) + 8$$
$$= 1000 + 200 - 30 + 8 = 1178$$

In general, if $f(x)$ represents a polynomial in x, and b is a number, then $f(b)$ represents the value of $f(x)$ at $x = b$. Hence, a polynomial can be used to define a function called a **polynomial function**.

THE POLYNOMIAL EQUATION

When a polynomial in general form is equated to zero, the resulting equation is called the general form of a **polynomial equation**.

The general form of a polynomial equation of degree n may be represented as follows:

$$a_0 x^n + a_1 x^{n-1} + a_2 x^{n-2} + \cdots + a_{n-1}x + a_n = 0 \quad (a_0 \neq 0)$$

For example, $2x^3 + 6x^2 - 3x + 4 = 0$ is a polynomial equation in general form.

〜〜〜〜〜 *MODEL PROBLEMS* 〜〜〜〜〜

1. If $f(y) = y^3 - y - 5$, find (*a*) $f(2)$ and (*b*) $f(-2)$.

Solution:

(*a*) Substituting 2 for y, $f(2) = 2^3 - 2 - 5 = 8 - 2 - 5 = 1$

Answer: $f(2) = 1$

(*b*) Substituting -2 for y, $f(-2) = (-2)^3 - (-2) - 5 = -8 + 2 - 5 = -11$

Answer: $f(-2) = -11$

2. If $g(x) = x^4 + x^2 - k$, and $i = \sqrt{-1}$, find the value of k for which:

    a. $g(3) = 2$ *b.* $g(\sqrt{2}) = 5$ *c.* $g(i) = -4$

Solution:

    a. Substitute 3 for x and let $g(3) = 2$.

$$x^4 + x^2 - k = g(x)$$
$$3^4 + 3^2 - k = g(3)$$
$$81 + 9 - k = 2 \qquad \text{Hence, } k = 88$$

Answer: 88

    b. Substitute $\sqrt{2}$ for x and let $g(\sqrt{2}) = 5$.

$$x^4 + x^2 - k = g(x)$$
$$(\sqrt{2})^4 + (\sqrt{2})^2 - k = g(\sqrt{2})$$
$$4 + 2 - k = 5 \qquad \text{Hence, } k = 1$$

Answer: 1

    c. Substitute i for x and let $g(i) = -4$.

$$x^4 + x^2 - k = g(x)$$
$$i^4 + i^2 - k = g(i)$$
$$1 - 1 - k = -4 \qquad \text{Hence, } k = 4$$

Answer: 4

3. If $h(t) = 2t^3 + at + b$, find the value of a and the value of b for which $h(1) = 5$ and $h(-1) = 3$.

Solution:

Substitute 1 for t and let $h(1) = 5$.
$$2(1)^3 + a(1) + b = 5$$
$$2 + a + b = 5 \qquad \text{Hence, } a + b = 3$$

Substitute -1 for t and let $h(-1) = 3$.
$$2(-1)^3 + a(-1) + b = 3$$
$$-2 - a + b = 3 \qquad \text{Hence, } -a + b = 5$$

Add: $a + b = 3$
$$\underline{-a + b = 5}$$
$$2b = 8$$
$$b = 4$$

Since $a + b = 3$ and $b = 4$, then $a + 4 = 3$, and $a = -1$.

Answer: $a = -1$, $b = 4$

~~~~~~

Exercises

In 1–3, find f(1), f(-1) and f(0) for the given polynomial.

1. $f(x) = x^3 + x^2 + x + 10$ **2.** $f(y) = y^4 + y^2 + 10$

3. $f(z) = z^5 + z^3 + z$

In 4–6, find g(10) and g(-10) for the given polynomial.

4. $g(x) = 2x^3 + 3x^2 + 4x + 5$ **5.** $g(x) = 2x^4 + 3x^2 + 4$

6. $g(x) = 2x^5 + 3x^3 + 4x$

In 7–10, if $f(s) = s^3 + 3s^2 - 5s + 7$, find the indicated function value.

7. f(0)   **8.** f(-1)   **9.** f($\frac{1}{2}$)   **10.** f($-\frac{1}{2}$)

In 11–13, if $g(x) = x^4 + x^2 - 7$, find the indicated function value.

11. g($\sqrt{2}$)   **12.** g($-\sqrt{3}$)   **13.** g($\frac{1}{2}\sqrt{2}$)

In 14–16, if $h(x) = x^5 + x^4 + x^3 + x^2 + x + 1$, and $i = \sqrt{-1}$, find the indicated function value.

14. h(i)   **15.** h($-i$)   **16.** h($2i$)

17. If $g(x) = x^3 + 2x^2 + 3x - k$, find the value of k for which:

    a. g(1) = 10   *b.* g(-1) = -10 *c.* g(10) = 1200

    d. g(-10) = 1200

18. If $g(y) = y^3 + ry - s$, find the value of r and the value of s for which:

    a. g(2) = 10 and g(0) = 5

    b. g(1) = 10 and g(-1) = -5

    c. g(3) = 30 and g(-3) = 23

2. Synthetic Division

In forthcoming units of this chapter, we shall often find it necessary to divide a polynomial $p(x)$ by a linear factor of the form $x - a$. In such cases, most of the steps in conventional long division can be eliminated and the operation can be reduced to a convenient short form called **synthetic division**.

In order to show how the process of synthetic division is developed, we shall first divide $2x^3 - 3x^2 + 4x + 5$ by $x - 2$, using the conventional method. Then, we shall indicate the many steps that can be eliminated. Note that the division is continued until the remainder does not contain a term in x.

1. Here is the conventional method of long division (the divisor is of the form $x - a$):

CONVENTIONAL DIVISION

$$
\begin{array}{r}
2x^2 + x\ +6 \\
x-2\ {\overline{\smash{\big)}\,2x^3 - 3x^2 +4x +\ 5}} \\
\underline{2x^3 - 4x^2} \\
+\ x^2 +4x \\
\underline{+\ x^2 -2x} \\
+6x +\ 5 \\
\underline{+6x -12} \\
+17
\end{array}
$$

2. Here, nonessential terms in step 1, such as the repeated terms and x in the divisor, are eliminated:

ELIMINATION OF NON-ESSENTIAL TERMS

$$
\begin{array}{r}
2x^2 + x\ +6 \\
-2\ {\overline{\smash{\big)}\,2x^3 - 3x^2 +4x +\ 5}} \\
\underline{-4x^2} \\
+\ x^2 \\
-2x \\
\underline{+6x} \\
-12 \\
\underline{} \\
+17
\end{array}
$$

3. Here, the arrangement in step 2 is "compressed" after powers of x, such as x^3, x^2, and x, are eliminated and their coefficients retained:

4. By replacing "-2" in the divisor (in step 3) by its additive inverse, "$+2$," and replacing the numbers in row 2 (step 3) by their additive inverses, the numbers that are now in row 2 can be added instead of subtracted:

COMPRESSION AND USE OF COEFFICIENTS

$$
\begin{array}{r}
2 +1 +6 \\
-2\ {\overline{\smash{\big)}\,2 -3 +4 +\ 5}} \\
\underline{-4 -2 -12}\quad \leftarrow \text{row 2} \\
+1 +6 +17
\end{array}
$$

CHANGING SUBTRACTION TO ADDITION

$$
\begin{array}{r}
2 +1 +6 \\
+2\ {\overline{\smash{\big)}\,2 -3 +4 +\ 5}} \\
\underline{+4 +2 +12}\quad \leftarrow \text{row 2 can} \\
+1 +6 +17 \qquad \text{be added}
\end{array}
$$

5. In this final step, we can copy the 2, which is the first number in the dividend in step 4, as the first number in the last line. Now, the first three numbers in the last line are the same as the numbers in the quotient. Therefore, we can eliminate the coefficients of the quotient. Note that the last number in the last line is the remainder.

ELIMINATION OF THE TOP ROW

$$
\begin{array}{r}
+2\ {\overline{\smash{\big)}\,2 -3 +4 +\ 5}} \\
\underline{+4 +2 +12} \\
2 +1 +6 +17
\end{array}
$$

$\underbrace{}$ $\underbrace{}$
coefficients   remain-
of the   der
quotient

We are now ready to understand the process of synthetic division. We will apply it to the preceding division, stating the quotient and the remainder.

To divide $2x^3 - 3x^2 + 4x + 5$ by $x - 2$, using synthetic division:

How To Proceed	*Solution*

1. Arrange the terms in the dividend and divisor in descending order of exponents.

    1. $x - 2\overline{)2x^3 - 3x^2 + 4x + 5}$

2. In row 1, retain the coefficients of the terms of the dividend, each with its proper sign, and write 0 as the coefficient of any missing term. In the divisor, omit x and replace " -2 " by " $+2$," which we will call the **synthetic divisor.**

    2. $+2\overline{)2 - 3 + 4 + 5}$ $\leftarrow$ row 1

3. In row 3, write the first coefficient of the dividend, 2. From this point, we *multiply and add repeatedly*, as follows:

    3.
$$+2\overline{)\begin{array}{l} 2 - 3 + 4 + \;\;5 \leftarrow \text{row 1} \\ +4 + 2 + 12 \leftarrow \text{row 2} \\ \hline 2 + 1 + 6 + 17 \leftarrow \text{row 3} \end{array}}$$

               $\underbrace{}_{\substack{\text{coefficients} \\ \text{of the} \\ \text{quotient}}}$ $\underbrace{}_{\text{remainder}}$

    (1) Multiply 2 in row 3 by the synthetic divisor, 2, and add the result to -3 to obtain $+1$. Write $+1$ in row 3.

    (2) Multiply $+1$ in row 3 by the synthetic divisor, 2, and add the result to $+4$ to obtain $+6$. Write $+6$ in row 3.

    (3) Multiply $+6$ in row 3 by the synthetic divisor, 2, and add the result to $+5$ to obtain $+17$. Write $+17$ in row 3.

4. The last number obtained, $+17$, is the remainder, while the other numbers in row 3 are the coefficients of the quotient. Note that the degree of the quotient is always *one less* than the degree of the dividend.

    4. The quotient is $2x^2 + x + 6$. The remainder is 17. The answer may be written as
$$2x^2 + x + 6 + \frac{17}{x - 2}.$$

Thus, $(2x^3 - 3x^2 + 4x + 5) \div (x - 2) = 2x^2 + x + 6 + \dfrac{17}{x - 2}$.

GENERALIZED MULTIPLICATION AND ADDITION PROCEDURE IN SYNTHETIC DIVISION

After the first coefficient of the dividend has been written in row 3, each number obtained in row 3 is multiplied by the synthetic divisor. The product, which is written in the next column of row 2 is then added to the number above it, and the sum is then entered in row 3. The process of multiplying and then adding is repeated as long as there are numbers in row 1.

~~~~~~~~ *MODEL PROBLEMS* ~~~~~~~~

1. Using synthetic division, divide $3x^2 - 7x - 40$ by $x - 5$, stating the quotient and the remainder.

Solution: $\underline{+5 \vert 3 -\ 7 - 40}$

$$ +15 + 40$$
$$\overline{ 3 +\ 8 +\ 0}$$

Answer: The quotient is $3x + 8$, the remainder is 0.

Note. The zero remainder shows that $x - 5$ is a factor of $3x^2 - 7x - 40$. The quotient $3x + 8$ is also a factor of $3x^2 - 7x - 40$ because, by the definition of division, $(x - 5)(3x + 8) = 3x^2 - 7x - 40$. (Suggested check: Show that $(x - 5)(3x + 8) = 3x^2 - 7x - 40$.)

2. Using synthetic division, divide $3x^3 - x - 7x^2 - 23$ by $x - 3$.

Solution: First, rearrange the terms of the dividend in descending order, obtaining $3x^3 - 7x^2 - x - 23$.

$$\underline{+3 \vert 3 - 7 - 1 - 23}$$
$$ +9 + 6 + 15$$
$$\overline{ 3 + 2 + 5 -\ 8}$$

Answer: The quotient is $3x^2 + 2x + 5$, the remainder is -8. The answer may also be written as $3x^2 + 2x + 5 + \dfrac{-8}{x - 3}$.

3. Using synthetic division, divide $2x^4 + 125x + 15$ by $x + 4$.

Solution: First, supply the missing terms to obtain the complete dividend: $2x^4 + 0x^3 + 0x^2 + 125x + 15$. Note that the divisor, $x + 4$, is the same as $x - (-4)$.

$$\underline{-4 \vert 2 + 0 +\ 0 + 125 + 15}$$
$$ -8 + 32 - 128 + 12$$
$$\overline{ 2 - 8 + 32 -\ \ 3 + 27}$$

Answer: The quotient is $2x^3 - 8x^2 + 32x - 3$; the remainder is 27. The answer may also be written as $2x^3 - 8x^2 + 32x - 3 + \dfrac{27}{x + 4}$.

4. Using synthetic division, find the value of k so that, when $x^3 + 12x + k$ is divided by $x - 2$, the remainder is 0.

Solution: First, supply the missing term to obtain the complete dividend: $x^3 + 0x^2 + 12x + k$.

$$
\begin{array}{r}
+2\,|\,1 + 0 + 12 + k \\
\underline{+2 +\ \ 4 + 32} \\
1 + 2 + 16 + (k + 32)
\end{array}
$$

Since the remainder, which is represented by $k + 32$, must be zero, then $k + 32 = 0$. Hence, $k = -32$.

Answer: $k = -32$

5. Using synthetic division show that, when the general quadratic polynomial $f(x) = a_0 x^2 + a_1 x + a_2$ is divided by $x - d$, the remainder obtained is equal to f(d).

Solution:

$$
\begin{array}{r}
\underline{d\,|\,a_0 + a_1 \qquad\qquad + a_2} \\
\underline{+ a_0 d \qquad\quad + (a_0 d^2 + a_1 d)} \\
a_0 + (a_0 d + a_1) + (a_0 d^2 + a_1 d + a_2)
\end{array}
$$

The remainder obtained by synthetic division is $a_0 d^2 + a_1 d + a_2$. Exactly the same expression is obtained when we find f(d) by substituting d for x in $a_0 x^2 + a_1 x + a_2$.

~~~~~~

Exercises

In 1–8, find the quotient and the remainder, using synthetic division.

1. $(x^3 - 3x^2 + 5x - 4) \div (x - 1)$ 2. $(x^3 - 6x^2 + 8x - 3) \div (x + 1)$
3. $(2x^3 - 3x^2 + 8x - 7) \div (x - 2)$ 4. $(x^3 - 6x^2 + 7) \div (x - 3)$
5. $(x^3 - 6x + 9) \div (x + 3)$ 6. $(2x^3 + 5x - 4) \div (x - 4)$
7. $(x^4 + 3x^2 - 4) \div (x + 1)$ 8. $(x^3 - 8) \div (x - 2)$

In 9–13, use synthetic division to divide $4x^3 + 2x^2 - 6x + 5$ by the given divisor.

9. $x - 1$ 10. $x + 1$ 11. $x - 2$ 12. $x + 2$ 13. $x - \frac{1}{2}$

In 14–19, using synthetic division, find the value of k needed to obtain a remainder equal to 0.

14. $(x^3 - x^2 + x + k) \div (x - 1)$ 15. $(x^3 + x - k) \div (x + 1)$
16. $(x^4 + x^2 + 2k) \div (x - 2)$ 17. $(x^3 - k) \div (x - 1)$
18. $(x^3 - 3x^2 + kx - 5) \div (x + 2)$ 19. $(3x^3 - 11x^2 + 2kx - 12) \div (x - 3)$

20. Using synthetic division, show that the remainder obtained when $f(x) = ax^5 - b$ is divided by $x - d$ is equal to f(d).

3. The Remainder Theorem

In Model Problems 1, 2, and 3 of the previous section, synthetic division was used. The remainders that were obtained are shown in the following table:

Division Problem	Synthetic Divisor, d	Remain- der	$f(x)$ when $x = d$
$(3x^2 - 7x - 40) \div (x - 5)$	$+5$	0	$f(5) = 0$
$(3x^3 - x - 7x^2 - 23) \div (x - 3)$	$+3$	-8	$f(3) = -8$
$(2x^4 + 125x + 15) \div (x + 4)$	-4	27	$f(-4) = 27$

Note in the table that, in each case, the remainder obtained by division is exactly the same as the result obtained when the value of the synthetic divisor, d, is substituted for x in f(x). These examples illustrate the following theorem:

REMAINDER THEOREM

If d is a constant and if a polynomial f(x) is divided by a binomial of the form $(x - d)$ until the remainder contains no term in x, then the remainder equals f(d).

Proof:

If f(x) is divided by $(x - d)$ until there is no term in x in the remainder, R, and the quotient is represented by q(x), the following identity applies:

$$f(x) = [q(x)](x - d) + R$$

Since an identity is true for all values of the variable, we may substitute d for x.

$$f(d) = [q(d)](d - d) + R$$
$$f(d) = [q(d)]0 + R$$
$$f(d) = 0 + R$$

Hence, f(d) = R

Thus, if $x^{65} + 5$ is divided by $(x - 1)$, then the remainder R = f(1). Hence, R = f(1) = $(1)^{65} + 5 = 1 + 5 = 6$.

The remainder is obtainable by conventional or synthetic division, but with much lengthier computation.

For some polynomial functions, however, synthetic division is a convenient method of obtaining a function value. When synthetic division is used for this purpose, the process is known as *synthetic substitution.*

~~~~~~~~~~ **MODEL PROBLEMS** ~~~~~~~~~~

1. Using the Remainder Theorem, find the remainder when $x^3 - 2x^2 + 3x - 4$ is divided by:

   a. $x - 2$ *b.* $x + 1$ *c.* $x - k$

  Solution: Let $f(x) = x^3 - 2x^2 + 3x - 4$. When $f(x) \div (x - d)$, then the remainder $R = f(d)$.

  a. Since the divisor is $(x - 2)$, then the remainder $R = f(2)$.

$$R = f(2) = 2^3 - 2(2^2) + 3(2) - 4$$
$$= 8 - 8 + 6 - 4 = 2$$

  Answer: 2

  b. Since the divisor is $(x + 1)$, or $[x - (-1)]$, the remainder $R = f(-1)$.

$$R = f(-1) = (-1)^3 - 2(-1)^2 + 3(-1) - 4$$
$$= -1 - 2 - 3 - 4 = -10$$

  Answer: -10

  c. Since the divisor is $(x - k)$, then the remainder $R = f(k)$.

$$R = f(k) = k^3 - 2k^2 + 3k - 4$$

  Answer: $k^3 - 2k^2 + 3k - 4$

2. Find the value of k for which the remainder equals 0 when $3x^3 - 2x^2 + k$ is divided by $(x - 10)$.

  Solution: Let $f(x) = 3x^3 - 2x^2 + k$. When $f(x) \div (x - 10)$, then the remainder $R = f(10)$.

$$R = f(10) = 3(10)^3 - 2(10)^2 + k$$
$$= 3000 - 200 + k = 2800 + k$$

  Since $R = 0$, then $2800 + k = 0$. Hence, $k = -2800$.

  Answer: -2800

3. Show that the remainder is 0 when $x^3 - y^3$ is divided by $(x - y)$, $x \neq y$.

  Solution: Let $f(x) = x^3 - y^3$. When $f(x)$ is divided by $(x - y)$, the remainder $R = f(y)$. $R = f(y) = y^3 - y^3 = 0$. The resulting remainder is 0.

4. Show that the remainder is 0 when $x^4 - y^4$ is divided by $(x + y)$, $x \neq (-y)$.

  Solution: Let $f(x) = x^4 - y^4$. When $f(x)$ is divided by $(x + y)$, or $[x - (-y)]$, the remainder $R = f(-y)$. $R = f(-y) = (-y)^4 - y^4 = y^4 - y^4 = 0$. The resulting remainder is 0.

5. If $f(x) = x^4 - 2x^3 + 7x^2 - 4x + 5$, find $f(6)$ by synthetic substitution. Check by direct substitution.

Solution: When $f(x)$ is divided by $x - 6$, the remainder is $f(6)$.

$$
\begin{array}{r|rrrrr}
6 & 1 & -2 & 7 & -4 & 5 \\
  &   & 6 & 24 & 186 & 1092 \\
\hline
  & 1 & 4 & 31 & 182 & 1097
\end{array}
$$

The remainder is 1097, so $f(6) = 1097$.

By direct substitution:

$$
\begin{aligned}
f(6) &= 6^4 - 2(6)^3 + 7(6)^2 - 4(6) + 5 \\
     &= 1296 - 432 + 252 - 24 + 5 \\
     &= 1097
\end{aligned}
$$

Answer: So, $f(6) = 1097$, the same value found by synthetic substitution.

Exercises

In 1–3, use the Remainder Theorem to find the remainder when $x^3 + 2x^2 - 3x - 8$ is divided by the given binomial.

1. $x - 2$ **2.** $x + 1$ **3.** $x - k$

In 4–7, use the Remainder Theorem to find the remainder when the first polynomial is divided by the second polynomial.

4. $3x^4 - 6x^3 + 8$; $x - 2$ **5.** $2x^5 - 8x^2 + 7x + 4$; $x + 2$

6. $2x^4 + 3x^3 + 6x^2 + 3x - 10$; $x - 5$

7. $5x^4 + 10x^3 - 7x^2 + 8x - 4$; $x + 3$

In 8–10, find the remainder when the given polynomial is divided by $(x + 1)$.

8. $x^{10} + 1$ **9.** $x^9 + 9x$ **10.** $x^{99} - 99$

In 11–13, find the value of k needed to obtain a remainder of 0 when:

11. $(x^2 + k)$ is divided by $(x - 1)$.

12. $(2x^3 + k)$ is divided by $(x + 1)$.

13. $(6x^4 + x^3 + x^2 + x + k)$ is divided by $(x + 2)$.

In 14–17, show that the remainder is 0 for the indicated division.

14. $(x^3 + y^3) \div (x + y)$, $x \neq (-y)$

15. $(x^6 - y^6) \div (x - y)$, $x \neq y$

16. $(x^n - y^n) \div (x - y)$, when n is a positive integer, and $x \neq y$

17. $(x^n + y^n) \div (x + y)$, when n is a positive odd integer, and $x \neq -y$

In 18–20, use synthetic substitution to find f(3) and f(–2) for each function. Check by direct substitution.

18. $x^4 + 3x^3 - 10x^2 - 7x + 9$

19. $5x^3 - 6x + 11$

20. $2x^4 - 4x^3 + 8x^2 - 21$

4. The Factor Theorem

Now, we will study an important corollary of the Remainder Theorem.

If in $f(x) = x^2 - 7x + 12$, we substitute 3 for x, we get $f(3) = (3)^2 - 7(3) + 12$, or $f(3) = 0$. According to the Remainder Theorem, $f(3)$ is the remainder when $f(x)$ is divided by $(x - 3)$. Since $f(3) = 0$, then when $x^2 - 7x + 12$ is divided by $(x - 3)$, the remainder is 0. Therefore, $(x - 3)$ is a factor of $x^2 - 7x + 12$.

This conclusion is expressed in a general form by the **Factor Theorem,** a corollary of the Remainder Theorem.

THE FACTOR THEOREM

If f(x) is a polynomial and *r* is a number such that f(*r*) = 0, then (*x* − *r*) is a factor of f(x).

Proof: According to the Remainder Theorem, $f(r)$ equals the remainder when $f(x)$ is divided by $(x - r)$. If $f(r) = 0$, then the remainder when $f(x)$ is divided by $(x - r)$ equals 0. Hence, $(x - r)$ is a factor of $f(x)$.

We now go one step further in our study. If $x - 3$ is a factor of $x^2 - 7x + 12$, then when $x^2 - 7x + 12$ is divided by $x - 3$, the remainder $f(3) = 0$. Since $f(3) = 0$, 3 must be a root of $x^2 - 7x + 12 = 0$, applying the definition of the root of an equation. This conclusion may be stated in general form as follows:

THEOREM

If (*x* − *r*) is a factor of the polynomial f(*x*), then *r* is a root of f(*x*) = 0.

Proof: Since $(x - r)$ is a factor of $f(x)$ when $f(x)$ is divided by $(x - r)$, the remainder is 0. Since, by the Remainder Theorem, this remainder equals $f(r)$, we have $f(r) = 0$. Since $f(r) = 0$, r is a root of $f(x) = 0$, applying the definition of a root of an equation.

Thus, if $(x - 4)$ is a factor of $x^2 - 7x + 12$, then 4 is a root of $x^2 - 7x + 12 = 0$.

ZERO OF A POLYNOMIAL

A *zero of a polynomial* $f(x)$ is a root of the equation $f(x) = 0$.

For example, in the case of the equation $x^2 - 7x + 12 = 0$, represented as $f(x) = 0$, $f(4) = (4)^2 - 7(4) + 12 = 0$. Since $f(4) = 0$, 4 is a root of $x^2 - 7x + 12 = 0$, and 4 is a zero of the polynomial $x^2 - 7x + 12$.

~~~~~~~~~~ **MODEL PROBLEMS** ~~~~~~~~~~

1. Show, without using division, that $x + 1$ is a factor of $x^3 - 7x + 6$.

   Solution: According to the Factor Theorem, if $f(-1) = 0$, then $x - (-1)$ or $x + 1$ is a factor of $f(x)$.

   Let $f(x) = x^3 - 7x + 6$.

   $$f(-1) = (-1)^3 - 7(-1) + 6 = -1 + 7 - 6 = 0$$

   Hence, $x + 1$ is a factor of $x^3 - 7x + 6$.

2. Find a value of k for which $x - 3$ will be a factor of $2x^3 - 11x^2 + 12x + k$.

   Solution: According to the Factor Theorem, if $f(3) = 0$, then $x - 3$ is a factor of $f(x)$.

   Let $f(x) = 2x^3 - 11x^2 + 12x + k$.
   $f(3) = 2(3)^3 - 11(3)^2 + 12(3) + k = 54 - 99 + 36 + k = k - 9$
   If $f(3) = 0$, then $k - 9 = 0$. Hence, $k = 9$.

   Answer: 9

3. Find a value of k such that -10 will be a root of $x^4 - 10x^2 + x + k = 0$.

   Solution: According to the definition of a root of an equation, given the equation $f(x) = 0$ and $f(-10) = 0$, then -10 is a root of $f(x) = 0$.

   Let $f(x) = x^4 - 10x^2 + x + k$.
   $f(-10) = (-10)^4 - 10(-10)^2 + (-10) + k$
   $= 10{,}000 - 1000 - 10 + k = 8990 + k$
   If $f(-10) = 0$, then $8990 + k = 0$. Hence, $k = -8990$.

   Answer: -8990

4. Show, without using division, that -2 is a zero of $x^3 + 2x^2 + 2x + 4$.

   Solution: If $f(-2) = 0$, then -2 is a zero of $f(x)$.

   Let $f(x) = x^3 + 2x^2 + 2x + 4$.
   $f(-2) = (-2)^3 + 2(-2)^2 + 2(-2) + 4 = -8 + 8 - 4 + 4 = 0$
   Hence, -2 is a zero of $x^3 + 2x^2 + 2x + 4$.

~~~~~~

Exercises

In 1 and 2 show, without using division, that the first polynomial is a factor of the second.

1. $x - 1$; $x^3 + 5x^2 + 10x - 16$ **2.** $x + 1$; $6x^3 - 8x^2 - 8x + 6$

3. Show that 2 is a zero of $x^3 + 2x^2 - 5x - 6$ and that $x - 2$ is a factor of this polynomial.

4. Show that -3 is a zero of $x^4 + 3x^3 - 6x - 18$ and that $x + 3$ is a factor of this polynomial.

In 5–8 show, without using division, that the number in parentheses is a root of the given equation.

5. $x^3 + 2x^2 - 5x - 6 = 0$; (-3) **6.** $x^3 - x + 6 = 0$; (-2)

7. $6x^3 - 19x^2 + x + 6 = 0$; (3) **8.** $x^4 - 3x^3 - x^2 - 11x - 4 = 0$; (4)

In 9–12, determine the value of k so that the given binomial will be a factor of $x^3 - 8x^2 + 10x + k$.

9. $x - 1$ **10.** $x + 1$ **11.** $x - 10$ **12.** $x + 10$

In 13–16, determine the value of k so that the first polynomial will be a factor of the second.

13. $x - 2$; $x^3 + 5x + k$ **14.** $x + 2$; $x^3 + x^2 - k$

15. $x - 10$; $2x^3 - 12x + k$ **16.** $x + 10$; $10x^4 - 100x^2 + k$

In 17–20, determine the value of k so that the number in parentheses will be the root of the given equation.

17. $x^3 + x^2 - x + k = 0$; (2) **18.** $2x^3 - 5x - k = 0$; (6)

19. $x^4 + 2x^3 - 20x + k = 0$; (-3) **20.** $x^5 + x^3 - 6x^2 + 2x + k = 0$; (-1)

In 21–23, determine the value of k so that the number in parentheses will be a zero of the given polynomial.

21. $2x^3 - 2x + k$; (-1) **22.** $x^5 + 2x^3 - k$; (1)

23. $x^4 - kx + 5$; (-2)

5. Factoring Higher Degree Polynomials Using the Factor Theorem

It can be proved that if an integer r is a zero of a polynomial whose coefficients are integers, then r must be a factor of the constant term of the polynomial. For example, if an integer r is a zero of the polynomial $x^3 + 3x^2 - 6x - 8$, then the integer r must be a factor of the constant term, -8. Hence, the only integers that can be zeros of $x^3 + 3x^2 - 6x - 8$ are ± 1, ± 2, ± 4, and ± 8. This theorem, together with the Factor Theorem, is useful in factoring a polynomial $f(x)$.

～～～～～ *MODEL PROBLEM* ～～～～～

Factor $x^3 + 3x^2 - 6x - 8$ completely.

How To Proceed	*Solution*
1. Factor the constant term to discover the possible zeros of the polynomial f(x).	1. The factors of the constant term, -8, are: $\pm 1, \pm 2, \pm 4, \pm 8$
2. Test the factors of the constant term to find a number r which is a zero of f(x).	2. $f(x) = x^3 + 3x^2 - 6x - 8$ $f(1) = (1)^3 + 3(1)^2 - 6(1) - 8$ $\quad = -10$ Hence, 1 is not a zero of f(x). $f(-1) = (-1)^3 + 3(-1)^2 -$ $\qquad\qquad 6(-1) - 8 = 0$ Hence, -1 is a zero of f(x).
3. If r is a zero of f(x), then $x - r$ is a factor of f(x).	3. Since -1 is a zero of f(x), then $x - (-1)$, or $x + 1$, is a factor of f(x).
4. Divide f(x) by $x - r$ to discover the quotient which is the other factor of f(x). Use synthetic division.	4. $\underline{-1\lfloor 1 + 3 - 6 - 8}$ $\quad\quad\underline{-1 - 2 + 8}$ $\quad\quad 1 + 2 - 8 + 0$ The quotient is $x^2 + 2x - 8$.
5. Write f(x) as the product of ($x - r$) and the quotient q(x).	5. $x^3 + 3x^2 - 6x - 8$ $\quad = (x + 1)(x^2 + 2x - 8)$
6. Factor q(x) and write f(x) as the product of ($x - r$) and the factors of q(x).	6. The factors of $x^2 + 2x - 8$ are ($x + 4$) and ($x - 2$). Hence, $x^3 + 3x^2 - 6x - 8 =$ $(x + 1)(x + 4)(x - 2)$

Answer: $(x + 1)(x + 4)(x - 2)$

Exercises

In 1–15, factor the polynomial completely.

1. $x^3 + x^2 - 5x - 2$
2. $x^3 + 2x^2 + 5x + 4$
3. $y^3 - 5y^2 + 4y + 6$
4. $x^3 + 2x^2 - x - 2$
5. $x^3 + 3x^2 - 4x - 12$
6. $a^3 + a^2 - 2a - 8$
7. $t^3 - 7t - 6$
8. $y^3 - 12y - 16$
9. $x^3 - 1$
10. $x^3 - 27$
11. $y^3 + 8$
12. $c^3 + 64$
13. $2x^3 - 7x^2 + 7x - 2$
14. $3y^3 + 4y^2 - 13y + 6$
15. $2z^3 - 2z^2 - z - 6$

6. Solving Polynomial Equations of Higher Degree

A. SOLVING POLYNOMIAL EQUATIONS WHICH HAVE KNOWN RATIONAL ROOTS

Suppose we are asked to find the solution set of $x^3 - 6x^2 + 3x + 10 = 0$ and we are told that one of the roots is 5. We find the remaining roots as follows:

If 5 is a root of $x^3 - 6x^2 + 3x + 10 = 0$, then 5 is a zero of $x^3 - 6x^2 + 3x + 10$, and $x - 5$ is a factor of $x^3 - 6x^2 + 3x + 10$. Using synthetic division, we can determine the other factor, which is the quotient obtained when $x^3 - 6x^2 + 3x + 10$ is divided by $x - 5$.

$$\underline{5 | 1 - 6 + 3 + 10}$$
$$\underline{+5 - 5 - 10}$$
$$1 - 1 - 2 + \ \ 0$$

The other factor is the quotient, $x^2 - x - 2$.

The original equation $x^3 - 6x^2 + 3x + 10 = 0$ can now be written as

$$(x - 5)(x^2 - x - 2) = 0$$

Hence, $x - 5 = 0$, or $x^2 - x - 2 = 0$. From $x - 5 = 0$, we obtain the root $x = 5$, which we already know. Therefore, we can disregard this equation at this time.

The remaining roots of the original equation can be found from the other equation, $x^2 - x - 2 = 0$, which is called the ***depressed equation.***

$$x^2 - x - 2 = 0$$
$$(x - 2)\,(x + 1) = 0$$
$$x - 2 = 0 \ | \ x + 1 = 0$$
$$x = 2 \ | \qquad x = -1$$

The solution set of $x^3 - 6x^2 + 3x + 10 = 0$ is $\{5, 2, -1\}$.

Note in the above example that the original equation is a third-degree equation, while the resulting depressed equation is a second-degree equation. Had the original equation been an equation of degree 4, the resulting depressed equation would have been of degree 3. In general, if the original equation is of degree n, then the depressed equation is of degree $n - 1$. Furthermore, if r is a root of $f(x) = 0$ and $[q(x)](x - r) = f(x)$, then the remaining roots of $f(x)$ can be found by solving the depressed equation, $q(x) = 0$.

~~~~~~~~~ *MODEL PROBLEMS* ~~~~~~~~~

1. Find the solution set of $x^3 - 5x^2 - 6 = 22x$ if -3 is one of the roots.

Solution: First, transform the equation into the form $f(x) = 0$. In this case,

$x^3 - 5x^2 - 22x - 6 = 0$. Since -3 is a root of the equation, then $x - (-3)$, or $x + 3$, is a factor of its left member. Use -3 as the synthetic divisor to find the other factor of the left member.

$$\underline{-3} | 1 - 5 - 22 + 6$$
$$\quad\quad -3 + 24 - 6$$
$$\overline{\quad 1 - 8 + \ 2 + 0}$$

The quotient is $x^2 - 8x + 2$.

Hence, the quotient $x^2 - 8x + 2$ is the other factor of the left member. The depressed equation is $x^2 - 8x + 2 = 0$; its roots are the two remaining roots of the original equation. Applying the quadratic formula:

$$x = \frac{-b \pm \sqrt{b^2 - 4ac}}{2a} \quad\quad a = 1 \quad b = -8 \quad c = 2$$

$$x = \frac{8 \pm \sqrt{64 - 8}}{2} = \frac{8 \pm \sqrt{56}}{2} = \frac{8 \pm 2\sqrt{14}}{2} = 4 \pm \sqrt{14}$$

Answer: The solution set is $\{-3, \ 4 + \sqrt{14}, \ 4 - \sqrt{14}\}$.

2. Find the solution set of $12x^3 - 77x + 28x + 12 = 0$ if $-\frac{1}{4}$ is one of the roots.
    Solution: Since $-\frac{1}{4}$ is a root of the equation, then $x - (-\frac{1}{4})$, or $x + \frac{1}{4}$, is a factor of its left member. Use $-\frac{1}{4}$ as the synthetic divisor to find the other factor of the left member.

$$\underline{-\tfrac{1}{4}} | 12 - 77 + 28 + 12$$
$$\quad\quad\quad -3 + 20 - 12$$
$$\overline{\quad 12 - 80 + 48 + \ 0}$$

The quotient is $12x^2 - 80x + 48$.

Hence, the quotient $12x^2 - 80x + 48$ is the other factor of the left member. The depressed equation is $12x^2 - 80x + 48 = 0$. Its roots are the two remaining roots of the original equation. Dividing each side by 4 and then factoring:

$$3x^2 - 20x + 12 = 0$$
$$(x - 6)(3x - 2) = 0$$
$$x - 6 = 0 \quad | \quad 3x - 2 = 0$$
$$x = 6 \quad\quad | \quad\quad x = \tfrac{2}{3}$$

Answer: The solution set is $\{-\frac{1}{4}, \ 6, \ \frac{2}{3}\}$.

3. Find the solution set of $x^3 + 2x^2 + 2x + 4 = 0$ if -2 is one of the roots.
    Solution: Since -2 is a root of the equation, then $x - (-2)$, or $x + 2$, is a factor of its left member. Use -2 as the synthetic divisor to find the other factor of the left member.

$$-2\,\underline{|\,1 + 2 + 2 + 4}$$
$$\underline{-2 + 0 - 4}$$
$$1 + 0 + 2 + 0$$

The quotient is $x^2 + 2$.

Hence, the quotient $x^2 + 2$ is the other factor of the left member. The depressed equation is $x^2 + 2 = 0$; its roots are the two remaining roots of the original equation. Thus,

$$x^2 + 2 = 0$$
$$x^2 = -2$$
$$x = \pm\sqrt{-2}, \text{ or } x = \pm i\sqrt{2}$$

Answer: The solution set is $\{-2,\, i\sqrt{2},\, -i\sqrt{2}\}$.

Exercises

In 1–6, find the solution set if the number in parentheses is a root of the given equation.

1. $x^3 + 3x^2 - 4x - 12 = 0$; (-2) **2.** $x^3 - 2x^2 - 5x + 6 = 0$; (3)
3. $x^3 - 7x = 6$; (-1) **4.** $y^3 - 5y^2 + 20 = 4y$; (5)
5. $z^3 - z^2 = 5z + 3$; (-1) **6.** $2r^3 - 3r^2 + 1 = 0$; $(-\frac{1}{2})$

In 7–12, find the solution set if the number in parentheses is a root of the given equation.

7. $x^3 - x^2 - x - 15 = 0$; (3) **8.** $y^3 + 2y^2 - 7y = 2$; (2)
9. $z^3 + 3z^2 - 2z = 8$; (-2) **10.** $x^3 - 10x^2 - 2x + 20 = 0$; (10)
11. $x^3 - 4x^2 + 4x = 3$; (3) **12.** $x^3 - 10x^2 + x - 10 = 0$; (10)

In 13–16, find the solution set if the numbers in parentheses are roots of the given equation.

13. $x^4 + 12x^2 - 6x^3 = 10x - 3$; $(1, 3)$
14. $y^4 - 2y^3 = 4x^2 - 11x + 6$; $(1, 2)$
15. $2z^4 + 5z^3 + 3z^2 = 2 - z$; $(\frac{1}{2}, -2)$
16. $9x^4 - 6x^3 + 10x^2 - 6x + 1 = 0$; $(\frac{1}{3}, \frac{1}{3})$

B. SOLVING POLYNOMIAL EQUATIONS WHICH HAVE A RATIONAL ROOT THAT IS NOT KNOWN IN ADVANCE

The following theorem, which is proved in more advanced texts, can be used in solving a polynomial equation which has a rational number as a root.

THEOREM

If a polynomial equation that has integral coefficients is represented as $a_0x^n + a_1x^{n-1} + a_2x^{n-2} + \cdots + a_{n-1}x + a_n = 0$, $a \neq 0$, and if this equation has a root $\frac{p}{q}$ with p and q being integers which have no common factor other than 1 and with q not equal to 0, then: (1) p is a factor of the constant term, a_n; and (2) q is a factor of the leading coefficient, a_0.

For example, if the equation $2x^3 - 3x^2 - 8x - 3 = 0$ has a rational fraction $\frac{p}{q}$ as a root, p must be a factor of the constant term, -3, and q must be a factor of the leading coefficient, 2. Hence, the only possible rational numbers that can be roots of this equation are $\pm\frac{1}{2}$, $\pm\frac{3}{2}$, $\pm\frac{1}{1}$, or $\pm\frac{3}{1}$. Note that $+\frac{1}{1}$, $-\frac{1}{1}$, $+\frac{3}{1}$, and $-\frac{3}{1}$ may be represented as the integers $+1$, -1, $+3$, and -3, which are the factors of the constant term, -3.

The general principle set forth above has an important corollary which can be applied to polynomial equations such as $x^3 + 2x^2 - 11x - 12 = 0$, in which the leading coefficient is 1. Where the coefficient of the highest power of the variable x is 1, the denominator of any rational fraction which is a root of the equation must be 1. Hence, the rational roots of such an equation can only be integers which are factors of the constant term.

For example, the only possible rational roots of $x^3 + 2x^2 - 11x - 12 = 0$ are the integers which are factors of the constant term, -12; namely, the integers $\pm1, \pm2, \pm3, \pm4, \pm6$, and ±12.

In general terms, we have the following corollary:

COROLLARY

If a polynomial equation that has integral coefficients is represented as $x^n + a_1x^{n-1} + a_2x^{n-2} + \cdots + a_{n-1}x + a_n = 0$, and if this equation has a rational root, then this root is an integer which is a factor of the constant term, a_n.

~~~~~ *MODEL PROBLEMS* ~~~~~

1. Find the solution set of $6x^3 + 5x^2 = 3x + 2$.

How To Proceed	*Solution*
1. Transform the given equation into the general form of a polynomial equation, f(x) = 0.	1. Transform into : $6x^3 + 5x^2 - 3x - 2 = 0$

2. Factor the constant term to find the possible numerators of the rational roots.

2. The factors of -2 are ±1 and ±2.

3. Factor the leading coefficient to find the possible denominators of the rational roots.

3. The factors of 6 are ±1, ±2, ±3, and ±6.

4. Form the possible rational roots, expressed in lowest terms.

4. The possible rational roots are ±1, ±2, $\pm\frac{1}{2}$, $\pm\frac{1}{3}$, $\pm\frac{1}{6}$, $\pm\frac{2}{3}$.

5. Find one of the rational roots by testing the possibilities.

5. Using synthetic division:

$$\frac{2}{3} \underline{| 6 + 5 - 3 - 2}$$
$$\phantom{\frac{2}{3}|} \underline{+4 + 6 + 2}$$
$$\phantom{\frac{2}{3}|} 6 + 9 + 3 + 0$$

Since the remainder is 0, then $\frac{2}{3}$ is a root of $6x^3 + 5x^2 - 3x - 2 = 0$.

6. Write the depressed equation.

6. The depressed equation is
$$6x^2 + 9x + 3 = 0.$$

7. Solve the depressed equation.

7.
$$6x^2 + 9x + 3 = 0$$
$$D_3: \ 2x^2 + 3x + 1 = 0$$
$$(2x + 1)(x + 1) = 0$$
$$2x + 1 = 0 \quad | \quad x + 1 = 0$$
$$x = -\tfrac{1}{2} \quad | \quad x = -1$$

Answer: The solution set is $\{\frac{2}{3}, -\frac{1}{2}, -1\}$.

2. Find the solution set of $x^3 - 3x^2 = 2x - 6$.

How To Proceed	*Solution*

1. Transform the given equation into the general form of a polynomial equation: $f(x) = 0$.

1. Transform into:
$$x^3 - 3x^2 - 2x + 6 = 0.$$

2. Since the coefficients are all integers and the leading coefficient is 1, factor the constant term to find the possible integral roots.

2. A possible integral root must be one of the following factors of $+6$:
$$\pm1, \ \pm2, \ \pm3, \ \pm6$$

3. Find one of the rational roots by testing the possibilities.

3. Using synthetic division:

$$3 \underline{| 1 - 3 - 2 + 6}$$
$$ \underline{+3 + 0 - 6}$$
$$ 1 + 0 - 2 + 0$$

Since the remainder is 0, then 3 is a root of $x^3 - 3x^2 - 2x + 6 = 0$.

4. Write the depressed equation.

4. The depressed equation is
$$x^2 - 2 = 0.$$

5. Solve the depressed equation. | 5. $x^2 - 2 = 0$

$$x^2 = 2$$
$$x = \pm\sqrt{2}$$

Answer: The solution set is $\{3, \sqrt{2}, -\sqrt{2}\}$.

~~~~~~

Exercises

In 1–8, list the rational numbers which might possibly be roots of the given equation. Find by testing which, if any, of these numbers actually are roots of the equation.

1. $x^3 - x^2 + 2x - 2 = 0$ **2.** $x^3 + 3x^2 + x - 15 = 0$
3. $2y^3 + 3y^2 - 3y - 2 = 0$ **4.** $6w^3 + 5w^2 - 17w - 6 = 0$
5. $x^3 + 9x^2 = 8 - 10x$ **6.** $y^3 + 7y = 12 + 2y^2$
7. $4x^3 + 3 = 4x^2 + 5x$ **8.** $3z^3 - 17z = 4z^2 - 6$

9. Select the number which cannot possibly be a root of the equation $4x^3 + px^2 + qx - 6 = 0$, where p and q are integers.
   (a) $\frac{1}{2}$ (b) $\frac{1}{4}$ (c) $\frac{3}{2}$ (d) 4

10. Select the number which cannot possibly be a root of the equation $6x^3 + px^2 + qx + 4 = 0$, where p and q are integers.
   (a) $\frac{1}{3}$ (b) $\frac{1}{2}$ (c) $\frac{2}{3}$ (d) $\frac{3}{2}$

11. Select the number which cannot possibly be a root of the equation $x^3 + px^2 + qx + 8 = 0$, where p and q are integers.
   (a) 1   (b) -2 (c) $\frac{1}{2}$ (d) 8

In 12–29, find the solution set of the equation.

12. $x^3 - 2x^2 - x + 2 = 0$ **13.** $x^3 - 7x - 6 = 0$
14. $2x^3 - 11x^2 + 8x + 7 = 0$ **15.** $5x^3 - 8x^2 - 2x + 3 = 0$
16. $y^3 + 2y^2 - 5y - 6 = 0$ **17.** $x^3 - 3x^2 - 7x + 21 = 0$
18. $x^3 - 3x^2 + 5x - 6 = 0$ **19.** $x^3 - 3x^2 - 13x + 15 = 0$
20. $x^3 - 2x^2 + 26x = 0$ **21.** $3x^3 - 10x^2 + 18 = 51x$
22. $6x^3 + 1 = 7x^2$ **23.** $5x^3 - 8x^2 = 2x - 3$
24. $x^3 + 8x = 10 - 7x^2$ **25.** $y^3 - 18 = 2y^2 - 9y$
26. $8x + 7 = 11x^2 - 2x^3$ **27.** $x^4 - 3x^3 - x + 3 = 0$
28. $2x^4 - 5x^3 + 4x^2 - 5x + 2 = 0$ **29.** $3x^4 + 14x^3 + 2x^2 - 13x - 6 = 0$

30. The length of a rectangular box is 5 inches more than its width, and its height is 1 inch more than its width. If the capacity of the box is 96 cubic inches, find its dimensions.

31. The height of a rectangular bin is 1 foot more than its width, and the length is 2 feet more than the width. If the capacity of the bin is 210 cubic feet, find the dimensions of the bin.

32. The length of a rectangular bin exceeds its width by 2 feet, and its height exceeds twice its width by 1 foot. If the capacity of the bin is 216 cubic feet, find the dimensions of the bin.

33. The length of a rectangular box is 3 inches more than twice the width, and the depth is 1 inch less than the width. The capacity of the box is 132 cubic inches. Find its dimensions.

34. A rectangular box with a square base has a height which is 2 inches greater than each edge of the base. If the volume of the box is 45 cubic inches, find the dimensions of the box.

35. A square piece of tin measuring 8 inches on each side is made into an open-top box by cutting squares out of the corners and folding up the edges. The volume of the box that is formed is 32 cubic inches. Find the length of a side of each square that is cut out.

⁓⁓⁓⁓⁓ CALCULATOR APPLICATIONS ⁓⁓⁓⁓⁓

In Chapter XI, the zeros of quadratic functions were found by graphing. Since the graphing calculator makes it easy to display the graphs of higher-order polynomial functions, the real roots or zeros can be found by identifying the x-intercepts of the graph. Rational roots can be found exactly, while irrational real roots can be closely approximated.

⁓⁓⁓⁓⁓ MODEL PROBLEMS ⁓⁓⁓⁓⁓

1. Find the roots of $x^3 - 6x^2 + 3x + 10 = 0$ by graphing.

   Solution: To see all the features of this graph, use a WINDOW with Xmin = –10, Xmax = 10, Xscl = 1, Ymin = –16, Ymax = 16, Yscl = 1.

   Enter: Y=   X, T, θ, n ∧ 3 – 6 X, T, θ, n x^2 +

         3 X, T, θ, n + 10 GRAPH

   Display:

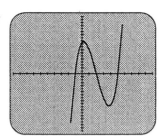

   The graph has three x-intercepts. Enter 2nd   CALC and select the zero option to determine each value. You should find that $x = -1$, 2, and 5 (working from left to right) which agrees with the algebraic solution on page 663.

2. Find the roots of $\frac{1}{3}x^3 - x^2 - 3x + 2 = 0$ by graphing.

   Solution: Rewriting the equation with integral coefficients as $x^3 - 3x^2 - 9x + 6$ and applying the Rational Root Theorem will show that there are no rational zeros. Graph the original equation using a standard WINDOW .

   Enter: Y=   (X, T, θ, n ∧ 3 ÷ 3)   –   X, T, θ, n x^2 –

         3 X, T, θ, n + 2 GRAPH

The graph shows there are three x-intercepts, so there are three real roots. Use the zero option on the CALC menu as before or TRACE to find an approximate value for each irrational root.

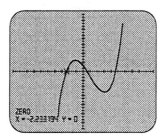

The approximate zeros are $x = -2.234$, $x = .577$, and $x = 4.656$.

Index